Remote Sensing of Climate

AF723133

Remote Sensing of Climate

Edited by

Timothy Dube
Department of Earth Science, Institute for Water Studies, University of the Western Cape, Cape Town, South Africa

Munyaradzi Davis Shekede
Department of Geography Geospatial Sciences and Earth Observation, University of Zimbabwe, Harare, Zimbabwe

Cletah Shoko
Division of Geography, School of Geography, Archaeology and Environmental Studies, University of Witwatersrand, Johannesburg, South Africa

Terence Darlington Mushore
Department of Space Science and Applied Physics, Faculty of Science, University of Zimbabwe, Harare, Zimbabwe

ELSEVIER

Elsevier
Radarweg 29, PO Box 211, 1000 AE Amsterdam, Netherlands
125 London Wall, London EC2Y 5AS, United Kingdom
50 Hampshire Street, 5th Floor, Cambridge, MA 02139, United States

Copyright © 2024 Elsevier Inc. All rights are reserved, including those for text and data mining, AI training, and similar technologies.

Publisher's note: Elsevier takes a neutral position with respect to territorial disputes or jurisdictional claims in its published content, including in maps and institutional affiliations.

No part of this publication may be reproduced or transmitted in any form or by any means, electronic or mechanical, including photocopying, recording, or any information storage and retrieval system, without permission in writing from the publisher. Details on how to seek permission, further information about the Publisher's permissions policies and our arrangements with organizations such as the Copyright Clearance Center and the Copyright Licensing Agency, can be found at our website: www.elsevier.com/permissions.

This book and the individual contributions contained in it are protected under copyright by the Publisher (other than as may be noted herein).

Notices
Knowledge and best practice in this field are constantly changing. As new research and experience broaden our understanding, changes in research methods, professional practices, or medical treatment may become necessary.

Practitioners and researchers must always rely on their own experience and knowledge in evaluating and using any information, methods, compounds, or experiments described herein. In using such information or methods they should be mindful of their own safety and the safety of others, including parties for whom they have a professional responsibility.

To the fullest extent of the law, neither the Publisher nor the authors, contributors, or editors, assume any liability for any injury and/or damage to persons or property as a matter of products liability, negligence or otherwise, or from any use or operation of any methods, products, instructions, or ideas contained in the material herein.

ISBN: 978-0-443-21731-9

For information on all Elsevier publications
visit our website at https://www.elsevier.com/books-and-journals

Publisher: Candice Janco
Acquisitions Editor: Peter Llewellyn
Editorial Project Manager: Rupinder Heron
Production Project Manager: Nadhiya Sekar
Cover Designer: Vicky Pearson

Typeset by STRAIVE, India

Transferred to Digital Printing in 2024

Working together to grow libraries in developing countries
www.elsevier.com • www.bookaid.org

Contents

Contributors

Numbers in parenthesis indicate the pages on which the authors' contributions begin.

Iqra Atif (29,183), Division of Geography, School of Geography, Archaeology and Environmental Studies, University of Witwatersrand, Johannesburg, South Africa

Trisha D. Bhaga (159), Department of Earth Science, Institute for Water Studies, University of the Western Cape, Cape Town, South Africa

Abel Chemura (51), Faculty of Geo-Information Science and Earth Observation (ITC), University of Twente, Enschede, The Netherlands

Chantel Chiloane (133), Department of Earth Science, Institute for Water Studies, University of the Western Cape, Cape Town, South Africa

Mwazvita Sachikonye Dalu (183), School of Biology and Environmental Sciences, University of Mpumalanga, Nelspruit, South Africa

Tatenda Dalu (133,313), Aquatic Systems Research Group, School of Biology and Environmental Sciences, University of Mpumalanga, Nelspruit; South African Institute for Aquatic Biodiversity, Makhanda, South Africa

Farai Dondofema (133,313), Department of Geography and Environmental Sciences, University of Venda, Thohoyandou, South Africa

Timothy Dube (1,17,133,159,183,279,313,349,377,407), Department of Earth Science, Institute for Water Studies, University of the Western Cape, Cape Town, South Africa

Charity Gororo (261,339), Department of Space Science and Applied Physics, Faculty of Science, University of Zimbabwe, Harare, Zimbabwe

Onalenna Gwate (107), Department of Geography and Geo-Information Sciences, Lupane State University, Lupane, Zimbabwe

Isaiah Gwitira (51), Department of Geography Geospatial Sciences and Earth Observation, University of Zimbabwe, Harare, Zimbabwe

Siyamthanda Gxokwe (17), Department of Earth Science, Institute for Water Studies, University of the Western Cape, Cape Town, South Africa

Mahlatse Kganyago (85,241), Department of Geography, Environmental Management and Energy Studies, University of Johannesburg, Johannesburg, South Africa

Samuel Kusangaya (205), Department of Geography Geospatial Sciences and Earth Observation, University of Zimbabwe, Harare, Zimbabwe

Elsie N. Leshaba (133), Aquatic Systems Research Group, School of Biology and Environmental Sciences, University of Mpumalanga, Nelspruit, South Africa

Muhammad Ahsan Mahboob (29), Sibanye-Stillwater Digital Mining Laboratory (DigiMine), Wits Mining Institute (WMI), Faculty of Engineering and the Built Environment, University of the Witwatersrand, Johannesburg, South Africa

Tapiwa Michelle Masawi (261,339), Department of Space Science and Applied Physics, Faculty of Science, University of Zimbabwe, Harare, Zimbabwe

Katlego Mashiane (85), Department of Geography, University of the Free State, Phuthaditjhaba, South Africa

Emmanuel Mashonjowa (261,339), Department of Space Science and Applied Physics, Faculty of Science, University of Zimbabwe, Harare, Zimbabwe

Fannie M. Masina (133), Aquatic Systems Research Group, School of Biology and Environmental Sciences, University of Mpumalanga, Nelspruit, South Africa

Mark Matsa (183), Department of Geography, Environmental Sustainability and Resilience Building, Midlands State University, Gweru, Zimbabwe

Teddious Mhizha (251,261,339), Department of Space Science and Applied Physics, Faculty of Science, University of Zimbabwe, Harare, Zimbabwe

Mthunzi Mndela (85), Department of Livestock and Pasture, University of Fort Hare, Alice, South Africa

Kudzai S. Mpakairi (313), Department of Earth Science, Institute for Water Studies, University of the Western Cape, Cape Town, South Africa; School of Wildlife Conservation, African Leadership University, Kigali, Rwanda

Chipo P. Mungenge (313), Department of Zoology and Entomology, Rhodes University, Makhanda, South Africa

Linton F. Munyai (133), Aquatic Systems Research Group, School of Biology and Environmental Sciences, University of Mpumalanga, Nelspruit, South Africa

Oshneck Mupepi (349), Department of Geography, Environmental Sustainability and Resilience Building, Midlands State University, Gweru, Zimbabwe

Godfrey Muroyiwa (251), Department of Space Science and Applied Physics, Faculty of Science, University of Zimbabwe, Harare, Zimbabwe

Tatenda Musasa (279,377), Department of Earth Science, Institute for Water Studies, University of the Western Cape, Cape Town, South Africa

Terence Darlington Mushore (251,261,279,339), Department of Space Science and Applied Physics, Faculty of Science, University of Zimbabwe, Harare, Zimbabwe; Discipline of Geography, School of Agricultural, Earth and Environmental Sciences, University of KwaZulu-Natal, Pietermaritzburg, South Africa

Onisimo Mutanga (261,339), Discipline of Geography, School of Agricultural, Earth and Environmental Sciences, University of KwaZulu-Natal, Pietermaritzburg, South Africa

Glen T. Nwaila (29), Sibanye-Stillwater Digital Mining Laboratory (DigiMine), Wits Mining Institute (WMI), Faculty of Engineering and the Built Environment, University of the Witwatersrand, Johannesburg, South Africa

John Odindi (261,339), Discipline of Geography, School of Agricultural, Earth and Environmental Sciences, University of KwaZulu-Natal, Pietermaritzburg, South Africa

Munyaradzi Davis Shekede (1,17,51,107,205,279,407), Department of Geography Geospatial Sciences and Earth Observation, University of Zimbabwe; Department of Geospatial Sciences and Earth Observation, Zimbabwe National Geospatial and Space Agency, Harare, Zimbabwe

Cletah Shoko (17,29,183,279,407), Division of Geography, School of Geography, Archaeology and Environmental Studies, University of Witwatersrand, Johannesburg, South Africa

Kgabo Humphrey Thamaga (85,241), Department of GIS and Remote Sensing, University of Fort Hare, Alice, South Africa

Gilberta Luba Thwala (261,339), Mining and Mineral Processing Engineering, Faculty of Engineering, Manicaland State University of Applied Sciences, Mutare, Zimbabwe

About the editors

Timothy Dube is a Professor of Water Resources and Environmental Management at the University of the Western Cape, South Africa, with a notable adjunct Professorship at the United Nations FLORES University, Germany. He has received esteemed awards and nominations, including recognition from the Egyptian Academy of Scientific Research and Technology and consecutive University of the Western Cape—Best Researcher Awards. His extensive research, spanning over 250 publications, focuses on utilizing big data for environmental and water resources assessment. Additionally, he actively contributes to international committees like the African Association of Remote Sensing of the Environment and the Global Monitoring for Environment and Security and Africa. His involvement in projects such as the Southern African Regional Technical Committee for the Green Hydrogen Project showcases his commitment to impactful research in global environmental change and sustainability.

Munyaradzi Davis Shekede is a Spatial Ecologist and a Senior Lecturer at the University of Zimbabwe with experience in Geographic Information Science and Earth observation applications in natural resources management. Dr. Shekede's research interests are in the applications of Earth observation and Geoinformation science to understanding Earth system dynamics, focusing on climate change, socioecological systems, and water resources.

Cletah Shoko is a Senior Lecturer in Geospatial Sciences in the Department of Geography at the University of the Witwatersrand, South Africa. She focuses on Geoinformation science and Earth observation applications in environmental assessment and monitoring, as well as their response to environmental change. Her research interests also include the development and testing of geospatial-based models to derive essential information that can be used in biodiversity conservation, climate change mitigation and adaptation, and disaster analysis.

Terence Darlington Mushore is an experienced Research scientist in remote sensing, disaster management, environmental management, and climate science. Recently, he was involved in a national project that revised agroecological regions of Zimbabwe using geospatial techniques. His research interests include efforts to understand and minimize the impacts of adverse climate on socioeconomic activities. Currently, his research focuses on observation, monitoring, and modeling the interaction between land surface and the thermal environment in urban areas.

About the editors

Chapter 1

Remote sensing of climate variability: An introduction

Munyaradzi Davis Shekede[a,b] and Timothy Dube[c]
[a]*Department of Geography Geospatial Sciences and Earth Observation, University of Zimbabwe, Harare, Zimbabwe,* [b]*Department of Geospatial Sciences and Earth Observation, Zimbabwe National Geospatial and Space Agency, Harare, Zimbabwe,* [c]*Department of Earth Science, Institute for Water Studies, University of the Western Cape, Cape Town, South Africa*

Chapter outline

1 Introduction

The profound changes in the biophysical and socioeconomic systems emanating from climate variability and change require the development of methods and approaches for an understanding of their impacts as well as developing mitigation and adaptation options in sectors affected therein. Over the years, climate variability and change have been characterized by extreme events, namely drought and floods, with adverse impacts on socio-economic development and the environment (Winsemius et al., 2016; Shekede et al., 2016; Sharara et al., 2022). For instance, 11 million human deaths have been attributed to droughts in the past while two billion people have been directly affected by drought-induced food shortages (FAO, 2016). The African continent has the highest drought-induced fatalities, with the continent having experienced 800,000 deaths since 1900, with about 262 million directly affected (Masih et al., 2014; Sharara et al., 2022). On the other hand, floods are responsible for losses worth tens of billions of United States dollars and thousands of fatalities per year (Kundzewicz et al., 2019). Africa exhibits the most extreme climatic contrasts characterized by the expansive Sahara Desert, which is regarded

Remote Sensing of Climate. https://doi.org/10.1016/B978-0-443-21731-9.00006-5
Copyright © 2024 Elsevier Inc. All rights are reserved, including those for text and data mining, AI training, and similar technologies.

as the most extreme desert in the world, and Mount Cameroon, one of the rainiest places in the world. As extreme events such as floods and droughts induced by climate change and variability increase in frequency and intensity through alterations in precipitation and evapotranspiration patterns (Trenberth et al., 2014), it will be important to understand how components of the earth system such as ecosphere, hydrosphere, and systems dependent thereon respond to climate variability and change. This chapter provides definitions and explanations of key concepts related to climate variability and change. This introductory chapter provides the context within which remote sensing of climate variability and change are discussed in the book.

2 Drought

Drought is a complex natural phenomenon whose dimensions are still poorly understood despite several years of research (Frischen et al., 2020). The complexity of drought emanates from the plethora of definitions that have emerged over the years, resulting in operational definitions being applied. However, common among the definitions of drought is an element of precipitation deficiency over a specified period, such as a month, resulting in an imbalance between water supply and societal or environmental demands (Wilhite and Buchanan-Smith, 2005). It is thus a period of marked precipitation deficiency relative to the local long-term average that impacts the earth's system functions (Gao et al., 2019; Slette et al., 2023). Thus, four main categories have been used to characterize drought and understand its occurrence in space and over time and these are meteorological, hydrological, agricultural, and socio-economic drought (Wilhite and Glantz, 1985). It is, therefore, apparent that the operational definition is dependent on the affected biophysical or socio-economic system (Winkler et al., 2017). Within each of the drought categories, several dimensions have been used to determine whether and to what extent water supply fails to meet demand, and these include duration, geographical extent, intensity, and severity (Wilhite and Glantz, 1985; Rossi et al., 1992). For operational purposes, drought indices are used to measure and track these four dimensions. Over the years, remotely sensed derived indices have been used to characterize drought and even formed the basis for specific drought management measures (Kogan, 1997; Wei et al., 2021; Jalayer et al., 2023; Zarei et al., 2023). This is because of its ability to provide observations over wider regions in a consistent and repetitive manner (Patel et al., 2023), thereby helping in understanding their spatial and temporal evolution, which is key to identifying areas at risk and disaster risk reduction strategies.

3 Drivers of climate variability and climate change

Climate denotes the long-term average weather patterns observed in a particular region over 30 years (IPCC, 2012). Elements that constitute climate include precipitation, temperature, wind patterns, and other atmospheric properties. Several statistical properties are used to characterize the climate of a given region, and these include means, variability, and extremes. On the other hand, climate variability refers to the variations in climate parameters such as precipitation, temperature, and wind patterns over a range of time scales from monthly to decadal scales or even much longer (WMO). Climate variability is driven by either internal (within the climate system (e.g., oceanic and atmospheric interactions) or external factors (e.g., solar radiation and volcanic eruptions). The following sections provide an overview of the natural and anthropogenic drivers of climate variability, focusing on (1) El Niño and La Niña, (2) solar activity, (3) volcanic eruptions, and (4) anthropogenic activity.

3.1 El Niño and La Nina

El Niño and La Nina are two natural climate patterns that develop in the Pacific Ocean with a profound influence on ocean and atmospheric temperature, which in turn have a bearing on global weather patterns (Marjani et al., 2019). Occurring every 2–7 years in the central and eastern tropical Pacific Ocean, El Niño is associated with the warming of the sea surface temperature and weakening of the trade winds, thereby influencing global atmospheric circulation. It can therefore lead to global changes in rainfall patterns, temperature, and atmospheric conditions, including droughts, floods, and shifts in storm tracks. Recently, studies have shown a climate change-driven intensification of El Niño events expressed through high atmospheric and energy resulting in extreme weather events such as drought and floods (Wu et al., 2023; Trascasa-Castro et al., 2023). In contrast, La Niña is characterized by below-average cold ocean temperatures in the tropical Pacific, leading to the strengthening of the trade winds, causing an intensification of the normal atmospheric circulation patterns (Sharmila et al., 2023). The phenomenon is associated with increased rainfall in some regions, while other areas may experience droughts, colder temperatures, and altered weather patterns. However, specific impacts of El Niño and La Niña globally vary depending on the location of the region and time of the year (Geng et al., 2023). The observation that El Niño and La Niña events are being influenced by climate change has important implications for global weather patterns in the coming years.

3.2 Solar activity

Studies based on long-term temperature data have shown that the Sun's energy output is not constant but varies with time, thereby affecting the Earth's incident electromagnetic radiation and climate (Pasachoff et al., 2023; Hossain, 2022). Specifically, studies have shown that variations in the Sun's output of energy, particularly in terms of electromagnetic radiation, solar winds, and the occurrence of sunspots, have a major influence on the Earth's climate system. For instance, sunspots, which are indicative of intense magnetic activity, have been shown to follow an 11-year cycle (Jayalekshmi et al., 2022). During this solar cycle, the number of sunspots varies such that when solar activity is high (more sunspots), the Sun emits more energy, and the opposite is true during periods of low solar activity. Thus, high solar activity is associated with more energy and warming of the Earth, while periods of low solar activity are characterized by less energy and cooler temperatures (Hazra and Risner, 2023). It is important to note that the nexus between solar activity and climate variability is complex and requires considerable scientific research. However, scientists concur that solar activity's influence on the Earth's climate system is relatively minimal compared with anthropogenic-driven global warming.

3.3 Volcanic eruptions

Volcanic eruptions result from the expulsion of ash and gases from molten rock when a volcano erupts. When large, these volcanic eruptions eject significant amounts of ash and gases into the atmosphere, with the potential to affect the Earth's radiation balance by blocking sunlight and cooling the Earth's surface (Hegerl et al., 2003; Aubry et al., 2021). For instance, sulfate aerosols from volcanic eruptions are known to be resident in the stratosphere for up to 3 years, during which they scatter sunlight, thereby leading to a net negative radiative forcing at the top of the atmosphere and cooling at the surface (Robock, 2000). This is especially the case for large-magnitude volcanic eruptions where ejections reach 10Tg of sulfur dioxide (SO_2) (Sigl et al., 2015) with significant impact on climate as witnessed after the Mount Tambora (1815) and Mount Pinatubo (1991) eruptions (Anderson et al., 2017). In contrast, significant amounts of carbon dioxide are released into the atmosphere during volcanic eruptions, leading to global warming. Thus, overall impact of volcanic eruptions on climate is dependent on several factors, including the magnitude of the eruption, the composition of the erupted material, and the altitude of the volcano (Aubry et al., 2022). Studies have shown that the cooling effect of volcanic eruptions can persist for months to years, depending on the size and duration of the eruption (McGregor et al., 2020). At the same time, volcanic eruptions release enormous amounts of energy, which results in intense heat, explosive eruptions, and the creation of lava flows (Moitra and Sonder, 2022). While volcanic eruptions can influence short-term climate variability, their impact is temporary compared to long-term climate change driven by human activities, such as greenhouse gas emissions.

3.4 Anthropogenic activities

The IPCC identifies human activities such as burning of fossil fuels and deforestation as key drivers responsible for the increase in the concentration of greenhouse gases in the atmosphere. In fact, Stocker et al. (2013) has demonstrated unequivocally that human activities have been the major driver of climate change since the mid-twentieth century. These GHGs trap heat and cause global temperatures to rise, leading to global warming. Chou and Lan (2012) have shown that global warming increases the water-holding capacity of the atmosphere, leading to an intensification of the global water cycle. Global warming has profound impacts such as sea-level rise, disruption of ecosystems as well as an increase in the frequency and severity of climate extremes, such as heat waves, hurricanes, droughts, and extreme weather events across the globe (Lehmann et al., 2015; Fu et al., 2023). These extreme events have a propensity to occur simultaneously in different regions of the world (Zhou et al., 2023), further triggering considerable economic and ecological damage and, in the worst cases, increased mortality (Lee et al., 2021). Masson-Delmotte et al. (2021) has demonstrated, with high confidence, an increase in the frequency and intensity of agriculture and ecological droughts in some regions, as well as medium confidence in the increase in meteorological and hydrological droughts. While natural processes like volcanic eruptions and variations in solar radiation can influence short-term climate variability, the long-term trends in climate are primarily driven by human activities (Ballew et al., 2019). Thus, mitigating climate change requires concerted global efforts to reduce GHGs and transitioning to cleaner energy sources. Some of the global efforts include the promotion of renewable energy, energy efficiency, sustainable agriculture, and reforestation to help stabilize the climate. However, the development of effective strategies and policies for adapting to and mitigating climate change impacts requires continuous scientific research, monitoring, and international cooperation.

4 Evolution of climate science

Although knowledge of the climate system and its drivers has improved over the past few decades, understanding the intricate interactions between solar activity, climate variability, and climate change is still ongoing research (Mishra and Dubey, 2023). Climate science has evolved over the years to an interdisciplinary field aiming at understanding the Earth's climate system as a coupled system consisting of components, processes, and interactions. The scope of climate science include the study of interactions of atmosphere, oceans, land surfaces, ice sheets, and biological systems and their influence on Earth's climate over a range of temporal scales (Giorgi, 2019). Several focal areas of climate science have emerged over years, with a particular focus on understanding the causes, mechanisms, and impacts of climate change. Specifically, these studies have examined the role of greenhouse gases, aerosols, natural climate variability, and human activities, such as burning fossil fuels and deforestation, in driving changes to the climate system (Grassl, 1999; Nica et al., 2019; Bozorg, 2022). Climate scientists have also developed computer-based climate models and projections that have significantly improved not only our understanding of the

Earth's climate system but also assisted in simulating future climate change as driven by natural and anthropogenic perturbations (Tebaldi and Knutti, 2010; Kashinath et al., 2021). Further climate projections have also helped in the simulation of potential future climate scenarios, including changes in temperature, precipitation patterns, sea-level rise, and extreme weather events (Donat et al., 2020). An important field of climate science that has shaped climate actions globally relates to modeling climate change impacts and vulnerability in a range of sectors such as agriculture, ecosystems, health, and water, and this has profoundly influenced IPCC reporting approaches (Ishtiaque et al., 2022). Through such studies, scientists have also developed mitigation and adaptation strategies aimed at minimizing risks and enhancing resilience to climate-related shocks. Over the past few decades, great interest has been generated in understanding carbon-related climate forcing including feedbacks. Under this theme, the focus has been on unraveling the complex interactions between the carbon cycle and climate system by examining GHG (particularly carbon dioxide) sources and sinks the sources. These studies aim at increasing our knowledge of feedback mechanisms that can either amplify or mitigate climate change (Coppola et al., 2022; Reichle, 2023). A significant number of studies have adopted a paleoclimatic approach to understanding the climate system through interrogating a rich archive of climate records reposited in ice cores, tree rings, sediments, and other natural archives to gain insights into long-term climate patterns and the natural variability of Earth's climate system (Erb et al., 2022; Smerdon et al., 2023). Using paleoclimatic research, scientists have generated useful insights into past climate, which have helped in understanding and validating current climate observations and model simulations (Wallace and Dee, 2022; Jackson, 2023). Finally, climate scientists have significantly contributed to the generation of knowledge that has shaped the development of climate policies but have been key in crafting mitigation strategies such as the Kyoto Protocol, the Paris Agreement, and related interventions (Bhandari, 2022). These studies have provided policymakers with the scientific basis and options for reducing greenhouse gas emissions, transitioning to renewable energy sources, and implementing sustainable practices to combat climate change (Gills and Morgan, 2022; Tan et al., 2022). It can thus be observed that science is ever-evolving, incorporating new methodologies, technologies, and datasets, thereby providing a dynamic but robust basis for understanding and addressing shocks related to climate change for the benefit of present and future generations.

5 Evidence of climate change and variability

While our knowledge of climate variability and change has improved over the years, the evidence related to these two phenomena has been equivocal. However, lately, there seems to be substantial consensus among scientists. Masson-Delmotte et al. (2021), which synthesizes the current understanding of climate-related issues, asserts that multiple and independent global temperature records demonstrate a 0.8°C increase in temperatures since 1880. The increase in global temperature has been primarily attributable to GHGs. For example, Anthropogenic greenhouse gas emissions caused a significant 1.07°C warming from 1850–1900 to 2010–2019, significantly impacting the Earth's climate system. Anthropogenic GHG emissions are responsible for approximately 1.07°C of warming from 1850–1900 to 2010–2019, and this had a profound effect on the Earth's climate system (Adak et al., 2023). The increasing consensus on climate change and its drivers has been driven by improved observational datasets to assess historical warming as well as progress in scientific understanding of the response of the climate system to human-caused greenhouse gas emissions (Lee et al., 2023). However, questions have been raised regarding whether the length of climate records is long enough to allow drawing of long-term trends. The evidence is regarded as limited due to sparse or even absence of observation networks and research in several regions of the world, such as the Tibetan Plateau, as well as arid and semiarid areas (Chen et al., 2022).

The melting of Arctic sea ice and glaciers at unprecedented rates, especially in the last three decades, has also been cited as indicative of climate change (Vaughan et al., 2013). For instance, using satellite images, Slater et al., 2021 report the Earth's cryosphere lost approximately 28 trillion tons of ice between 1994 and 2017. Similarly, satellite data from NASA's Gravity Recovery and Climate Experiment show that between 1993 and 2019, annual ice losses from Greenland and Antarctica were ~279 billion and 148 billion tons, respectively (Velicogna et al., 2020). This has subsequently resulted in sea-level rise with significant reductions in planetary albedo (Thackeray and Hall, 2019) and significant effects on freshwater resources (Immerzeel et al., 2020).

Another line of evidence of climate change makes reference to the warming of the ocean emanating from increasing anthropogenic greenhouse gas concentrations. Ocean warming arises from the absorption of significant amounts of carbon dioxide in the atmosphere, which often leads to ocean warming and acidification. According to Cheng et al. (2022), human-driven ocean warming has been observed since the 1950s, with record values recorded during 2012–2021. Similarly, Masson-Delmotte et al. (2021) has identified global ocean warming as the key indicator of anthropogenically driven climate change. In fact, oceans are responsible for 91% of the energy increase in the climate system between 1971 and 2018 (Silvy et al., 2022). However, there has been a differential change in ocean heat content over the past 50 years, characterized by increase, decrease, and no change in some regions (Fox-Kemper et al., 2021). Due to the lag response of the oceans to GHG forcing, especially at the subsurface level, it is anticipated that there will be accelerated warming due to current perturbations in the future.

The increase in extreme weather events such as heat waves, droughts, wildfires, floods, hurricanes, and other storms has been attributed to climate change (Salas and Ebi, 2021; Clarke et al., 2022). Most of the relationships between extreme weather events and climate change have been generated through extreme event attribution studies. Extreme event attribution studies refer to methods that have been developed to assess and quantify the role of climate change in an individual event (Van Oldenborgh et al., 2021). Over the past two decades, studies have demonstrated the role of climate change in extreme events (Clarke et al., 2022; Carbon Brief, 2021). Masson-Delmotte et al. (2021) report that human-induced climate change has affected several weather and climate extremes in every region across the globe. Specifically, the following was observed:

- hot extremes have become more frequent and more intense across most land regions since the 1950s,
- cold extremes have become less frequent and less severe,
- marine heatwaves have doubled in frequency since the 1980s,
- major tropical cyclone occurrence has increased over the last four decades, with peak intensity shifting northward,
- the frequency and intensity of heavy precipitation events have increased since the 1950s over most land areas.

All these indicators have been used as evidence of climate change.

6 Remote sensing of climate change and variability

The past decade has witnessed the deployment of Earth observation satellites in space with a view of collecting and analyzing information about the Earth's surface, atmosphere, cryosphere, oceans, and carbon cycle using sensors (Zhao et al., 2021). The sensors are mounted on various platforms that are either space-borne or air-borne, and they are further complemented with in situ measurements (ESA, 2020). The information is acquired and transmitted in near real time to ground stations for processing and archiving, thereby providing long-time series data that is handy for climate applications. These satellites range from public to commercial. For instance, the Union of Concerned Scientists (UCS) reports that in 1957, when Sputnik was launched, there were two satellites that were dedicated to earth observation, but this number increased to 15,946 at the start of 2023. In 2013 alone, 1354 satellites had been launched by the first half of 2023 (UCS, 2023); these satellites were dedicated to optical imaging (480), Meteorology (179), Radar imaging (113), earth science (75), and hyperspectral/multispectral imaging (43) among other applications. Table 1 summarizes the satellite missions that have been deployed to monitor and study the climate system.

The proliferation of these satellites has provided scientists with an opportunity for not only understanding the climate system but its dynamics at various spatial and temporal scale including planetary scale. For instance, remote sensing capability in monitoring the components of the Earth's system such as earth's surface, the ocean, and the atmosphere at a range of spatiotemporal scales has opened avenues for climate system observation, thereby enabling

TABLE 1 Satellite programs dedicated to observing the climate system.

Satellite mission	Description	Typical variables measured
NASA Satellites (Terra and Aqua)	These provide global observations of the atmosphere, oceans, land, and cryosphere.	These satellites observe a range of variables including atmospheric temperature, water vapor, and sea surface temperature.
NOAA's Geostationary Operational Environmental Satellite (GOES) series	GOES provides atmospheric, hydrologic, oceanic, climatic, solar, and space data for detection and observation of environmental phenomena such as severe storms.	Atmospheric temperature, water vapor, rainfall, particulate matter, sea surface temperature, lightning, fire.
European Space Agency's (ESA) Climate Change Initiative (CCI)	The program uses multiple satellites to provide data on the Earth's climate system, including temperature, sea level, hydrology, and atmospheric composition.	Essential climate variables including precipitation, pressure, surface radiation, temperature, surface wind speed, lightning, water vapor, near-surface carbon dioxide, and methane.
Japan Aerospace Exploration Agency's Global Change Observation Mission (GCOM)	GCOM Mission provides global, long-term observation of Earth's environment including monitoring global water circulation and climate change and examining the health of Earth from space.	The mission has several satellites that monitor ocean color, vegetation, snow and ice sea surface temperature, precipitation, carbon cycle, and radiation budget, such as clouds and aerosols.
Sentinel satellites (Sentinel 1, 2, 3)	The satellite series are part of the European Union's Copernicus Program and provide data on land use, sea surface temperature, and air quality.	Sea ice and oil spills, sea surface temperature, ocean color, and sea level, as well as land surface temperature and vegetation indices.
China's Fengyun series of meteorological satellites, Tiangong-2 and Tiangong-1 space labs	Missions have focused on monitoring and studying of weather conditions, atmospheric phenomena, and the Earth's	Monitoring severe weather events such as typhoons and droughts, as well as studying the

Continued

TABLE 1 Satellite programs dedicated to observing the climate system—cont'd

Satellite mission	Description	Typical variables measured
	environment. The Tiangong 1 and 2 were prototype space labs that served as an important milestone in China's plan to establish a modular space station.	impact of climate change on the planet.
INSAT-3D, INSAT-3DR, and SCATSAT-1 (oceanographic mission)	SCATSAT-1 provides information on wind speed and direction over the ocean surface.	Variables monitored include temperature, humidity, winds, precipitation, and cloud properties.

the understanding of climate processes as well as coupling or interconnections among these components. In fact, Global Climate Observing System defines and identifies a set of physical, chemical, or biological variables or a group of linked variables that critically contribute to the characterization of Earth' s climate, known as the Essential climate variables (ECVs) (Bojinski et al., 2014). Thus, 54 ECVs are relevant for understanding the climate system and evolution, scalability, feasibility, and cost-effectiveness (Bojinski et al., 2014). These ECVs focus on the following components: (1) atmosphere including the surface, upper air, and atmospheric composition; (2) the land covering the hydrosphere, cryosphere, biosphere, and anthroposphere; and (3) the ocean targeting the physical, biogeochemical, and ecosystems. Some of the climate-related data or information collected by satellites include temperature, precipitation, groundwater, vegetation patterns, soil carbon, sea level, glaciers, and anthropogenic and atmospheric gases. This is especially important considering the ability of remote sensing systems to acquire analysis-ready data, over extended periods, in inaccessible, poorly monitored, and dangerous regions at scales never imagined before such as planetary scale (ESA, 2020). The incorporation of remote sensing data into models provides critical insights into potential future climate patterns and trends. Information on future climate trends not only provides a better understanding of how climate patterns are changing over time but also helps in understanding their impacts on the environment. This is critical for informing mitigation and adaptation strategies across different sectors and regions of the globe.

Overall, remote sensing has emerged as an inevitable element of climate science by providing valuable data and insights into various aspects of the Earth's climate system. As previously mentioned, remote sensing has been the preferred approach for continuous and systematic monitoring of climate variables such as temperature, precipitation, sea surface temperature, sea ice extent, vegetation dynamics, and land cover changes. At the same time, remote sensing is instrumental in the mapping and estimation of greenhouse gas emissions including carbon dioxide (CO_2) and methane (CH_4) concentration in the atmosphere. With such data, scientists have characterized the sources and sinks of GHGs and tracked their changes over time. This has helped in understanding GHG impacts on the climate system as well as monitoring mitigation efforts such as reforestation and restoration of degraded lands. A critical analytical capability of remote sensing in climate studies lies in its ability to characterize climate patterns and variability. This is achieved by analyzing long-term series remote sensing datasets, thereby yielding information on climate trends, change, and extremes in climate variables, such as temperature and precipitation. Thus, a number of breakthroughs in climate science, especially in understanding climate change and variability including their drivers have been greatly shaped by the invaluable datasets acquired by remote sensors such as NASA's AVHRR. Other functionalities enhanced by remote sensing include assessment of climate change impacts, climate model validation, and the development of early warning systems for hazards such as floods and droughts in near real time. These remote sensing capabilities have thus improved our understanding of large-scale patterns and trends in the climate system as well as providing scientific evidence required for climate change mitigation and adaptation.

7 Conclusion

This chapter has provided an overview of key concepts that are used in climate studies. In summary, remote sensing has revolutionized our ability to study the Earth's climate system and its changes. By providing detailed and accurate data on various environmental parameters, remote sensing allows researchers to make informed decisions about how best to manage and protect the planet's resources.

References

Adak, S., Mandal, N., Mukhopadhyay, A., Maity, P.P., Sen, S., 2023. Current state and prediction of future global climate change and variability in terms of CO_2 levels and temperature. In: Enhancing Resilience of Dryland Agriculture Under Changing Climate: Interdisciplinary and Convergence Approaches. Springer Nature Singapore, Singapore, pp. 15–43.

Anderson, W., Seager, R., Baethgen, W., Cane, M., 2017. Life cycles of agriculturally relevant ENSO teleconnections in North and South America. Int. J. Climatol. 37, 3297–3318.

Aubry, T.J., Staunton-Sykes, J., Marshall, L.R., et al., 2021. Climate change modulates the stratospheric volcanic sulfate aerosol lifecycle and radiative forcing from tropical eruptions. Nat. Commun. 12, 4708.

Aubry, T.J., Farquharson, J.I., Rowell, C.R., Watt, S.F., Pinel, V., Beckett, F., Fasullo, J., Hopcroft, P.O., Pyle, D.M., Schmidt, A., Sykes, J.S., 2022. Impact of climate change on volcanic processes: current understanding and future challenges. Bull. Volcanol. 84 (6), 58.

Ballew, M.T., Leiserowitz, A., Roser-Renouf, C., Rosenthal, S.A., Kotcher, J.E., Marlon, J.R., Lyon, E., Goldberg, M.H., Maibach, E.W., 2019. Climate change in the American mind: data, tools, and trends. Environment 61 (3), 4–18.

Bhandari, M.P., 2022. Getting the Climate Science Facts Right: The Role of the IPCC. CRC Press.

Bojinski, S., Verstraete, M., Peterson, T.C., Richter, C., Simmons, A., Zemp, M., 2014. The concept of essential climate variables in support of climate research, applications, and policy. Bull. Am. Meteorol. Soc. 95, 1431–1443.

Bozorg, O., 2022. Climate change drivers. In: Climate Change in Sustainable Water Resources Management. Springer, p. 59.

Carbon Brief, 2021. Mapped: How Climate Change Affects Extreme Weather Around the World. Carbon Brief, p. 21.

Chen, R., Duan, K., Shang, W., Shi, P., Meng, Y., Zhang, Z., 2022. Increase in seasonal precipitation over the Tibetan Plateau in the 21st century projected using CMIP6 models. Atmos. Res. 277, 106306.

Cheng, L., von Schuckmann, K., Abraham, J.P., Trenberth, K.E., Mann, M.E., Zanna, L., England, M.H., Zika, J.D., Fasullo, J.T., Yu, Y., Pan, Y., 2022. Past and future ocean warming. Nat. Rev. Earth Environ. 3 (11), 776–794.

Chou, C., Lan, C.W., 2012. Changes in the annual range of precipitation under global warming. J. Clim. 25 (1), 222–235.

Clarke, B., Otto, F., Stuart-Smith, R., Harrington, L., 2022. Extreme weather impacts of climate change: an attribution perspective. Environ. Res.: Clim. 1 (1), 012001.

Coppola, A.I., Wagner, S., Lennartz, S.T., Seidel, M., Ward, N.D., Dittmar, T., Santín, C., Jones, M. W., 2022. The black carbon cycle and its role in the Earth system. Nat. Rev. Earth Environ. 3 (8), 516–532.

Donat, M.G., Sillmann, J., Fischer, E.M., 2020. Changes in climate extremes in observations and climate model simulations. From the past to the future. In: Climate Extremes and Their Implications for Impact and Risk Assessment. Elsevier, pp. 31–57.

Erb, M.P., McKay, N.P., Steiger, N., Dee, S., Hancock, C., Ivanovic, R.F., Gregoire, L.J., Valdes, P., 2022. Reconstructing Holocene temperatures in time and space using paleoclimate data assimilation. Clim. Past 18 (12), 2599–2629.

ESA, 2020. Compendium of EO Contributions to the SDG Targets and Indicators. Available online: https://eo4society.esa.int/wp-content/uploads/2021/01/EO_Compendium-for-SDGs.pdf. (Accessed 12 October 2023).

FAO, 2016. The State of Food and Agriculture 2016: Climate Change, Agriculture, and Food Security. FAO, Rome, Italy.

Fox-Kemper, B., Hewitt, H.T., Xiao, C., Aðalgeirsdóttir, G., Drijfhout, S.S., Edwards, T.L., Golledge, N.R., Hemer, M., Kopp, R.E., Krinner, G., Mix, A., 2021. Ocean, cryosphere, and sea level change. In: Climate Change 2021: The Physical Science Basis. Contribution of Working Group I to the Sixth Assessment Report of the Intergovernmental Panel on Climate Change. Cambridge University Press, pp. 1211–1361.

Frischen, J., Meza, I., Rupp, D., Wietler, K., Hagenlocher, M., 2020. Drought risk to agricultural systems in Zimbabwe: a spatial analysis of hazard, exposure, and vulnerability. Sustainability 12 (3), 752.

Fu, S., Zhang, H., Zhong, Q., Chen, Q., Liu, A., Yang, J., Pang, J., 2023. Spatiotemporal variations of precipitation concentration influenced by large-scale climatic factors and potential links to flood-drought events across China 1958–2019. Atmos. Res. 282, 106507.

Gao, J., Zhang, L., Tang, Z., Wu, S., 2019. A synthesis of ecosystem aboveground productivity and its process variables under simulated drought stress. J. Ecol. 107 (6), 2519–2531.

Geng, T., Jia, F., Cai, W., Wu, L., Gan, B., Jing, Z., Li, S., McPhaden, M.J., 2023. Increased occurrences of consecutive La Niña events under global warming. Nature 619 (7971), 774–781.

Gills, B., Morgan, J., 2022. Global climate emergency: after COP24, climate science, urgency, and the threat to humanity. In: Economics and Climate Emergency. Routledge, pp. 253–270.

Giorgi, F., 2019. Thirty years of regional climate modeling: where are we and where are we going next? J. Geophys. Res. Atmos. 124 (11), 5696–5723.

Grassl, H., 1999. Global climate change. Interdiscip. Sci. Rev. 24 (3), 185–194.

Hazra, S., Risner, J., 2023. Solar activity variation and its effects on ionospheric ion densities. Am. J. Appl. Sci. 5 (08), 13–20.

Hegerl, G.C., Crowley, T.J., Baum, S.K., Kim, K.-.Y., Hyde, W.T., 2003. Detection of volcanic, solar and greenhouse gas signals in paleo-reconstructions of Northern Hemispheric temperature. Geophys. Res. Lett. 30 (1242), 5.

Hossain, E., 2022. The sun and the earth. In: The Sun, Energy, and Climate Change. Springer Nature Switzerland, Cham, pp. 1–67.

Immerzeel, W.W., Lutz, A.F., Andrade, M., Bahl, A., Biemans, H., Bolch, T., Hyde, S., Brumby, S., Davies, B.J., Elmore, A.C., Emmer, A., Feng, M., Fernández, A., Haritashya, U., Kargel, J.S., Koppes, M., Kraaijenbrink, P.D.A., Kulkarni, A.V., Mayewski, P.A., Nepal, S., Pacheco, P., Painter, T.H., Pellicciotti, F., Rajaram, H., Rupper, S., Sinisalo, A., Shrestha, A.B., Viviroli, D., Wada, Y., Xiao, C., Yao, T., Baillie, J.E.M., 2020. Importance and vulnerability of the world's water towers. Nature 577, 364–369. https://doi.org/10.1038/s41586-019-1822-y.

IPCC, 2012. Glossary of terms. In: Field, C.B., Barros, V., Stocker, T.F., Qin, D., Dokken, D.J., Ebi, K.L., Midgley, P.M. (Eds.), Managing the Risks of Extreme Events and Disasters to Advance Climate Change Adaptation. A Special Report of Working Groups I and II of the Intergovernmental Panel on Climate Change (IPCC), Cambridge University Press, Cambridge, UK, and New York, NY, USA, pp. 555–564.

Ishtiaque, A., Estoque, R.C., Eakin, H., Parajuli, J., Rabby, Y.W., 2022. IPCC's current conceptualization of 'vulnerability' needs more clarification for climate change vulnerability assessments. J. Environ. Manage. 303, 114246.

Jackson, K.S., 2023. Predicting Paleoclimate and Checking Prediction Performance. North Carolina State University.

Jalayer, S., Sharifi, A., Abbasi-Moghadam, D., Tariq, A., Qin, S., 2023. Assessment of spatiotemporal characteristic of droughts using in situ and remote sensing-based drought indices. IEEE J. Sel. Top. Appl. Earth Obs. Remote Sens. 16, 1483–1502.

Jayalekshmi, G.L., Pant, T.K., Prince, P.R., 2022. Sunspot-cycle evolution of major periodicities of solar activity. Sol. Phys. 297 (7), 85.

Kashinath, K., Mustafa, M., Albert, A., Wu, J.L., Jiang, C., Esmaeilzadeh, S., Azizzadenesheli, K., Wang, R., Chattopadhyay, A., Singh, A., Manepalli, A., 2021. Physics-informed machine learning: case studies for weather and climate modelling. Philos. T. R. Soc. A 379 (2194), 20200093.

Kogan, F.N., 1997. Global drought watch from space. Bull. Am. Meteorol. Soc. 78 (4), 621–636.

Kundzewicz, Z.W., Szwed, M., Pińskwar, I., 2019. Climate variability and floods—a global review. Water 11 (7), 1399.

Lee, H., Calvin, K., Dasgupta, D., Krinner, G., Mukherji, A., Thorne, P., Trisos, C., Romero, J., Aldunce, P., Barret, K., Blanco, G., 2023. In: Core Writing Team, Lee, H., Romero, J. (Eds.), IPCC, 2023: Climate Change 2023: Synthesis Report, Summary for Policymakers. Contribution of Working Groups I, II and III to the Sixth Assessment Report of the Intergovernmental Panel on Climate Change. IPCC, Geneva, Switzerland.

Lee, O., Seo, J., Won, J., Choi, J., Kim, S., 2021. Future extreme heat wave events using Bayesian heat wave intensity-persistence day-frequency model and their uncertainty. Atmos. Res. 255, 105541.

Lehmann, J., Coumou, D., Frieler, K., 2015. Increased record-breaking precipitation events under global warming. Clim. Change 132, 501–515.

Marjani, S., Alizadeh-Choobari, O., Irannejad, P., 2019. Frequency of extreme El Niño and La Niña events under global warming. Clim. Dynam. 53, 5799–5813.

Masih, I., Maskey, S., Mussá, F.E.F., Trambauer, P., 2014. A review of droughts on the African continent: a geospatial and long-term perspective. Hydrol. Earth Syst. Sci. 18 (9), 3635–3649.

Masson-Delmotte, V.P., Zhai, P., Pirani, A., Connors, S.L., Péan, C., Berger, S., Caud, N., Chen, Y., Goldfarb, L., Gomis, M.I., Huang, M., Leitzell, K., Lonnoy, E., Matthews, J.B.R., Maycock, T.K., Waterfield, T., Yelekçi, O., Yu, R., Zhou, B., 2021. IPCC, 2021: Summary for policymakers. Climate Change 2021: The Physical Science Basis. Contribution of Working Group I to the Sixth Assessment Report of the Intergovernmental Panel on Climate Change. Cambridge University Press, Cambridge.

McGregor, S., Khodri, M., Maher, N., Ohba, M., Pausata, F.S., Stevenson, S., 2020. The effect of strong volcanic eruptions on ENSO. In: El Niño Southern Oscillation in a Changing Climate. Wiley, pp. 267–287.

Mishra, R.K., Dubey, S.C., 2023. Solar activity cause and effect of climate variability and their various impacts. Br. J. Multidiscip. Adv. Stud. 4 (2), 21–38.

Moitra, P., Sonder, I., 2022. Vapor bubbles and velocity control on the cooling rates of lava and pyroclasts during submarine eruptions. J. Geophys. Res.: Solid Earth 127 (8). e2022JB024665.

Nica, A., Popescu, A., Ibanescu, D.C., 2019. Human influence on the climate system. Curr. Trends Nat. Sci. 8 (15), 209–215.

Pasachoff, J.M., Seaton, D.B., Reardon, K.P., 2023. Sunspots and their cycle. Phys. Today 76 (2), 54–55.

Patel, N.R., Pokhariyal, S., Singh, R.P., 2023. Advancements in remote sensing based crop yield modelling in India. J. Agrometeorol. 25 (3), 343–351.

Reichle, D.E., 2023. The Global Carbon Cycle and Climate Change: Scaling Ecological Energetics From Organism to the Biosphere. Elsevier.

Robock, A., 2000. Volcanic eruptions and climate. Rev. Geophys. 38, 191–219.

Rossi, G., Benedini, M., Tsakiris, G., Giakoumakis, S., 1992. On regional drought estimation and analysis. Water Resour. Manage. 6, 249–277.

Salas, R.N., Ebi, K.L., 2021. The health benefits of urgent upstream action on climate change. Ann. Intern. Med. 174, 1612–1613.

Sharara, A., Shekede, M.D., Gwitira, I., Masocha, M., Dube, T., 2022. Fine-scale multi-temporal and spatial analysis of agricultural drought in agro-ecological regions of Zimbabwe. Geomat. Nat. Haz. Risk 13 (1), 1342–1365.

Sharmila, S., Hendon, H., Alves, O., Weisheimer, A., Balmaseda, M., 2023. Contrasting El Niño–La Niña predictability and prediction skill in 2-year reforecasts of the twentieth century. J. Climate 36 (5), 1269–1285.

Shekede, M.D., Murwira, A., Masocha, M., Zengeya, F.M., 2016. Decadal changes in mean annual rainfall drive long-term changes in bush-encroached southern African savannas. Austral Ecol. 41 (6), 690–700.

Sigl, M., Winstrup, M., McConnell, J.R., Welten, K.C., Plunkett, G., Ludlow, F., Büntgen, U., Caffee, M., Chellman, N., Dahl-Jensen, D., Fischer, H., 2015. Timing and climate forcing of volcanic eruptions for the past 2,500 years. Nature 523 (7562), 543–549.

Silvy, Y., Sallée, J.B., Guilyardi, E., Mignot, J., Rousset, C., 2022. What causes anthropogenic ocean warming to emerge from internal variability in a coupled model? J. Climate 35 (22), 7435–7454.

Slater, T., Lawrence, I.R., Otosaka, I.N., Shepherd, A., Gourmelen, N., Jakob, L., Tepes, P., Gilbert, L., Nienow, P., 2021. Review article: Earth's ice imbalance. Cryosphere 15, 233–246. https://doi.org/10.5194/tc-15-233-2021.

Slette, I.J., Hoover, D.L., Smith, M.D., Knapp, A.K., 2023. Repeated extreme droughts decrease root production, but not the potential for post-drought recovery of root production, in a mesic grassland. Oikos 2023 (1), e08899.

Smerdon, J.E., Cook, E.R., Steiger, N.J., 2023. The historical development of large-scale paleoclimate field reconstructions over the Common Era. Rev. Geophys. 61 (4). e2022RG000782.

Stocker, T.F., Qin, D., Plattner, G.-K., Tignor, M., Allen, S.K., Boschung, J., Nauels, A., Xia, Y., Bex, V., Midgley, P.M. (Eds.), 2013. Climate Change 2013: The Physical Science Basis. Contribution of Working Group I to the Fifth Assessment Report of the Intergovernmental Panel on Climate Change. Cambridge University Press, Cambridge.

Tan, X.C., Wang, Y., Gu, B.H., Kong, L.S., Zeng, A., 2022. Research on the national climate governance system toward carbon neutrality—a critical literature review. Fundam. Res. 2 (3), 384–391.

Tebaldi, C., Knutti, R., 2010. Climate models and their projections of future changes. In: Climate Change and Food Security: Adapting Agriculture to a Warmer World. Springer, pp. 31–56.

Thackeray, C.W., Hall, A., 2019. An emergent constraint on future Arctic sea-ice albedo feedback. Nat. Clim. Change 9, 972–978. https://doi.org/10.1038/s41558-019-0619-1.

Trascasa-Castro, P., Maycock, A.C., Ruprich-Robert, Y., Turco, M., Staten, P.W., 2023. Atlantic multidecadal variability modulates the climate impacts of El Niño–Southern Oscillation in Australia. Environ. Res. Lett. 18 (8), 084029.

Trenberth, K.E., Dai, A., van der Schrier, G., Jones, P.D., Barichivich, J., et al., 2014. Global warming and changes in drought. Nat. Clim. Change 4, 17–22.

UCS (Union of Concerned Scientists), 2023. UCS Satellite Database. https://www.ucsusa.org/resources/satellite-database. (Accessed 2 December 2023).

Van Oldenborgh, G.J., van Der Wiel, K., Kew, S., Philip, S., Otto, F., Vautard, R., King, A., Lott, F., Arrighi, J., Singh, R., van Aalst, M., 2021. Pathways and pitfalls in extreme event attribution. Clim. Change 166 (1–2), 13.

Vaughan, D.G., Comiso, J.C., Allison, I., Carrasco, J., Kaser, G., Kwok, R., Mote, P., Murray, T., Paul, F., Ren, J., Rignot, E., Solomina, O., Steffen, K., Zhang, T., 2013. Observations: cryosphere. In: Stocker, T.F., Qin, D., Plattner, G.-K., Tignor, M., Allen, S.K., Boschung, J., Midgley, P.M. (Eds.), Climate Change 2013: The Physical Science Basis. Contribution of Working Group I to the Fifth Assessment Report of the Intergovernmental Panel on Climate Change. Cambridge University Press, Cambridge, United Kingdom and New York, NY, USA, pp. 317–382.

Velicogna, I., Mohajerani, Y., Geruo, A., Landerer, F., Mouginot, J., Noel, B., Rignot, E., Sutterly, T., van den Broeke, M., Wessem, M., Wiese, D., 2020. Continuity of ice sheet mass loss in Greenland and Antarctica from the GRACE and GRACE Follow-On missions. Geophys. Res. Lett. 47 (8). https://doi.org/10.1029/2020GL087291. e2020GL087291.

Wallace, E.J., Dee, S.G., 2022. Tropical cyclone frequency: turning paleoclimate into projections. Environ. Res. Clim. 1 (2), 023002.

Wei, W., Zhang, J., Zhou, L., Xie, B., Zhou, J., Li, C., 2021. Comparative evaluation of drought indices for monitoring drought based on remote sensing data. Environ. Sci. Pollut. Res. 28, 20408–20425.

Wilhite, D.A., Buchanan-Smith, M., 2005. Drought as hazard: understanding the natural and social context. In: Drought and Water Crises: Science, Technology, and Management Issues. vol. 3. CRC Press, p. 29.

Wilhite, D.A., Glantz, M.H., 1985. Understanding: the drought phenomenon: the role of definitions. Water Int. 10 (3), 111–120.

Winkler, K., Gessner, U., Hochschild, V., 2017. Identifying droughts affecting agriculture in Africa based on remote sensing time series between 2000–2016: rainfall anomalies and vegetation condition in the context of ENSO. Remote Sens. (Basel) 9 (8), 831.

Winsemius, H.C., Aerts, J.C., Van Beek, L.P., Bierkens, M.F., Bouwman, A., Jongman, B., Kwadijk, J.C., Ligtvoet, W., Lucas, P.L., Van Vuuren, D.P., Ward, P.J., 2016. Global drivers of future river flood risk. Nat. Clim. Change 6 (4), 381–385.

Wu, H., Su, X., Singh, V.P., Zhang, T., 2023. Compound climate extremes over the globe during 1951–2021: changes in risk and driving factors. J. Hydrol., 130387.

Zarei, A.R., Mokarram, M., Mahmoudi, M.R., 2023. Comparison of the capability of the meteorological and remote sensing drought indices. Water Resour. Manage. 37 (2), 769–796.

Zhao, T., Cosh, M.H., Roy, A., Mu, X., Qiu, Y., Shi, J., 2021. Remote sensing experiments for earth system science. Int. J. Digit. Earth 14 (10), 1237–1242.

Zhou, S., Yu, B., Zhang, Y., 2023. Global concurrent climate extremes exacerbated by anthropogenic climate change. Sci. Adv. 9 (10), eabo1638.

Chapter 2

Remote sensing technological advancements and applications in climate variability analysis studies

Siyamthanda Gxokwe[a], Timothy Dube[a], Cletah Shoko[b], and Munyaradzi Davis Shekede[c,d]

[a]*Department of Earth Science, Institute for Water Studies, University of the Western Cape, Cape Town, South Africa,* [b]*Division of Geography, School of Geography, Archaeology and Environmental Studies, University of Witwatersrand, Johannesburg, South Africa,* [c]*Department of Geography Geospatial Sciences and Earth Observation, University of Zimbabwe, Harare, Zimbabwe,* [d]*Department of Geospatial Sciences and Earth Observation, Zimbabwe National Geospatial and Space Agency, Harare, Zimbabwe*

Chapter outline

1 Introduction

Over the years, issues relating to climate variability have attracted the attention of scientists, researchers, decision-makers, and the public. Although this is the case, there are still differences in views among and within countries regarding the nature and severity of problems associated with climate variability and change and whether there is anything that needs to be done. Such uncertainties are caused by a limited understanding of long-term climatic trends in various parts of the world. Understanding such changes over longer time periods is critical in establishing adaptation strategies to the negative impacts associated with climate variability (Crane et al., 2011). Moreover, understanding such trends enables us to put measures in place to prevent anthropogenic driving factors to climate variability trends, as these changes tend to threaten food sustainability, water availability, and human livelihoods, among others (Thornton et al., 2014).

Remote Sensing of Climate. https://doi.org/10.1016/B978-0-443-21731-9.00007-7
Copyright © 2024 Elsevier Inc. All rights are reserved, including those for text and data mining, AI training, and similar technologies.

Most climate variability and change studies addressing the potential impacts and potential adaptation strategies are largely based on modeling technologies and approaches (Reyes-García et al., 2016). Although modeling approaches are useful in analyzing the potential future outcomes and evaluating options for potential adaptation, they do not adequately represent and integrate adaptive human agency. Moreover, these models are based on field data that is scanty, incomplete, or missing in certain parts of the world due to the lack of extensive field observational network (Govender et al., 2022). This, therefore, results in uncertainties in the representativeness and predictions of these models, as the modeling process requires quality data with distinct spatial and temporal resolutions for setup, calibration, and validation to improve its confidence. Often, challenges in acquiring adequate data for the modeling task at hand and the spatial distribution of the data may render certain analyses impossible, for instance, in accounting for small geographic areas with high topographical variability.

The availability of remote sensing data, approaches, and tools offers unprecedented opportunities to address issues associated with field data limitations for analyzing climate variability trends over a larger spatial extent. However, such data and approaches have been contested by nonremote sensing communities because of the unavailability of cloud-free images over certain parts of the year, thus limiting continuous monitoring of climatic changes and variability using remote sensing data (Bhaga et al., 2020). A study by Gxokwe et al. (2022), highlighted the problem of obtaining all year-round images for assessing LULC for a wetland over a 20-year period; their study had to rely on only dry season images due to a lack of cloud-free images for wet season in their study area. These dry season images used did not completely enhance some wetland features, thus introducing misclassifications in their model outputs. Another study by Gokool et al. (2019) focusing on the assessment of evapotranspiration (ET) dynamics reported that lack of good quality images was among the limitations of the use of remote sensing data in ET dynamics. Another challenge, with remote sensing data applications in monitoring climate variability indicators, is the spatial resolution of the remotely sensed data, which tends to influence the accuracy of monitoring of such indicators, particularly with the use of freely accessible remote sensing products.

Advancements in remote sensing data analytics tools such as the introduction of cloud computing platforms like Google Earth Engine platforms, offer opportunities to address limitations associated with using remote sensing data in understanding the trends in climate variability indicators. With advanced data integration tools, filtering, and processing approaches in these platforms, it is possible to perform remote sensing data integrations from different sources, thus combining the strengths of the combined data, which can enhance the monitoring of the climate variable indicators. Moreover, platforms like Google Earth Engine have 40 years and more of petabyte remote sensing data in addition to the global climate data, therefore making it possible to monitor climate variability trends over time and over a large spatial extent.

This chapter provides an overview of remote sensing applications in climate variability analysis studies with a specific focus on the use of advanced technological tools and data applications in understanding climate variability at various scales. The first section of this chapter covers, in a broader sense, an overview of climate variability from a global perspective. The second section gives an overview of the available remote sensing data used to analyze climate variability in various climatic zones, and the last section provides an overview of remote sensing technological advancements and their associated use in climate variability studies.

2 Overview of climate variability analysis and their associated impacts

Climate refers to the average weather conditions for a certain area for a prolonged period of time. These conditions tend to vary over time and space, and such variations are driven to a great extent by the interactions between the oceans, atmosphere, and land surface (Wang, 2019). Variations in climate resulting from these interactions may have profound impacts on the Earth and atmosphere; as such, it is an important component of the economy. For example, in various parts of Africa, it has been reported that temperatures have been increasing while rainfall trends have been declining due to associated impacts of climate variability, thus causing significant negative impacts on the environment, livelihoods, food, and energy (Baninla et al., 2022; Muluneh, 2021). The reported consequences in Africa include, among others, the loss of US$1.4 billion annually on food crops (Akinsanola et al., 2021), aggregate annual production losses of 8.9%, which translates to translating to 2.3 million MT of wheat lost (Sileshi and Gebeyehu, 2021; Berck et al., 2018). Furthermore, studies estimate that by the end of 2023, over $1.4 trillion of Africa's GDP will be vulnerable to climate change, which will translate to 48% of the entire continent's GDP (Sileshi and Gebeyehu, 2021). This, therefore, necessitates for science-based mitigation and adaptation strategies to minimize the negative impacts of climate variability both on the economy, environment, and human livelihoods.

The basis for sound and evidence-based mitigation and adaptation strategies to negative impacts associated with climate variability is long-term and large-scale analysis of key climate variability indicators (Table 1). The monitoring of these will aid in the identification of the extent to which climate has changed and the associated impacts of those changes on the environment, human

TABLE 1 Key climate variability indicators.

Land indicators	Atmospheric and hydrological indicators	Ocean indicators
Land temperatures	Rainfall regimes	Sea levels
Soil moisture changes	Greenhouse gases change	Sea pH
Impacts on ecosystems	Floods	Ocean heat content
Forest systems changes	River flows changes	Marine heatwaves

Modified from Reyes-García, V., Fernández-Llamazares, Á., Guèze, M., Garcés, A., Mallo, M., Vila-Gómez, M., Vilaseca, M., 2016. Local indicators of climate change: the potential contribution of local knowledge to climate research. Wiley Interdiscip. Rev. Clim. Change 7, 109–124. https://doi.org/10.1002/wcc.374.

livelihoods, and economy, among others. Moreover, understanding such trends enables the prediction of future climate variability scenarios and their associated impacts. Such information is critical when developing mitigation measures and adaptation strategies to cope with negative impacts associated with climate variability (Seyisi et al., 2023).

Globally, significant research strides have been made in analyzing the trends in climate variability indicators on longer time period and large spatial extent. This research mostly uses field-based models which analyze trends in climate indicators over a large spatial extent, their drivers as well as future scenarios. For example, Adamowski and Sun (2010) developed an Artificial Neural Network (ANN)-based model to predict flows in nonperennial rivers-dominated watersheds. The findings underscored the relevance of the ANN-based model in forecasting flows in the nonperennial rivers dominated, although the model was limited by the unavailability of some climatic variables. Another study by Mahmood et al. (2019) analyzed the climatic variable (temperature and precipitation) trends of the Lake Chad Basin in Africa between 1951 and 2015 using a synergy of field climate data, linear regression model, polynomial model, Sen's slope, and Mann-Kendal test. Although the findings showed that the studied basin is highly sensitive to perturbed natural climate, causing water resources of the basin to be highly vulnerable to climate change, the study was limited by field observed station data. Both studies indicate that field-based models are limited by the availability of field-based data, thus indicating that such models may not perform well in data-scarce regions such as sub-Saharan Africa, where some basins are not equipped with climate data measuring stations.

3 Remote sensing data applications and associated challenges for climate variability analysis

The availability of remote sensing data, tools, and approaches offers unprecedented opportunities to monitor climate variability trends with their ability to provide multiscale information on the atmosphere, Earth's surface, and subsurface from sensors without contacting the target directly. Remote sensing approaches have been recognized as indispensable observation methods, particularly due to their capability of providing timely and dynamic spatiotemporal continuous information globally (Zhao et al., 2023). Such approaches detect much information on variables related to climate change, including clouds, aerosols, GHGs, land use and land cover changes, glacier ablation, water and energy flux, and sea level change, with input data from various satellite sensors such as the Landsat series, Sentinel, Moderate Resolution Imaging Spectroradiometer (MODIS), among others, which are orbiting the Earth providing information about the changes on Earth and associated climatic conditions. The data provided by these satellites vary in terms of their sensing characteristics (Table 2), which tend to influence the level of accuracy when monitoring climate variability indicators and their associated impacts.

Various studies have used data from these sensors in studying various climate variabilities indicators such as rainfall patterns, land surface temperatures, soil moisture changes, forest changes, flooding susceptibility, and droughts, among others, in various parts of the world (Avand et al., 2021; Camberlin et al., 2019; Halder et al., 2021; Jalilvand et al., 2019; Negassa et al., 2020; Shahzaman et al., 2021). Although these studies demonstrate the feasibility of using remote sensing data to understand trends in climate variability indicators, there are some challenges with these studies. Such challenges include the coarseness in spatial resolution of some of the remotely sensed data, which tends to introduce inaccuracies in analysis of such trends. For example, Avand et al. (2021) studied soil moisture changes and their relationships with agricultural irrigation in the west Azerbaijan province of Iran. In the study, Avand et al. (2021) reported that coarseness in the spatial resolution (50 km) of the remotely sensed data used resulted in difficulty in capturing the irrigation water of small irrigated croplands. Unreliable rainfall and evapotranspiration data from some of the remote sensing products also lead to the over/underestimation of irrigation water-related soil moisture. The use of very high-resolution remote sensing data offers solutions to challenges associated with the sensing characteristics of the commonly used data; however, the use of such data has cost implications and is therefore unfeasible for large-scale and long-term monitoring of climate variability trends. Another issue with the use of remote sensing data is lack of seamless cloud-free images for certain times of the year, which makes it difficult to continuously monitor climate variability trends over time (Gxokwe et al., 2023). Fortunately, advancements in remote sensing data analytics tools and free accessibility of higher-resolution remotely sensed data offer opportunities to overcome challenges associated with the use of higher-resolution images.

TABLE 2 Commonly used remote sensing products in climate variability studies (Gxokwe et al., 2020).

Resolution	Landsat MSS	Landsat TM	Landsat ETM+	Landsat OLI	Sentinel-1	Sentinel-2	MODIS
Spectral bands (μm)					C-band (3.75–75 cm)		
Band 1		0.45–0.52	0.45–0.515	0.43–0.45		0.443	0.62–0.67
Band 2		0.52–0.62	0.525–0.605	0.45–0.51		0.490	0.841–0.876
Band 3		0.63–0.69	0.63–0.69	0.53–0.59		0.560	0.459–0.479
Band 4	0.5–0.6	0.76–0.90	0.775–0.90	0.64–0.67		0.665	0.545–0.565
Band 5	0.6–0.7	1.55–1.75	1.55–1.75	0.85–0.88		0.705	1.23–1.25
Band 6	0.7–0.8	10.4–12.5	10.4–12.5	1.57–1.65		0.740	1.628–1.652
Band 7	0.8–1.1	2.08–2.35	2.08–2.35	2.11–2.29		0.783	2.105–2.155
Band 8			0.52–0.9	0.50–0.68		0.842	
Band 8A						0.865	
Band 9				1.36–1.38		0.945	0.438–0.448
Band 10				10.6–11.19		1.375	
Band 11				11.5–12.5		1.610	
Band 12						2.190	
Band 19							0.915–0.965

Band 31							10.78–11.28
Band 32							11.77–12.27
Temporal	180 days	16 days	16 days	16 days	12 days	5 days	1–2 days
Spatial (pixel-sizes)	80 m	30 and 120 m for Band 6	30, 15 m B8, and 60 m B6	30 m B1–7 and 9 15 m B8 100 m B10–11	5 m × 5 m	60 m B1,9,10 10 m B2,3,4,8 20 m B5,6,7,11,12	250 m B1–2, 500 m B8–36 1000 m B8–36
Period	1972–1992	1982–Present	2003–Present	2013–Present	2014–Present for 1A 2016–Present For 1B	2015–Present for 2A and 2017 for 2B	2000–Present for Terra 2002–Present for Aqua

4 Technological advancements in remote sensing tools and application in climate variability studies

The advancement in data analytic tools such as the introductions of cloud computing platforms such as Google Earth Engine and the availability of free access to higher-resolution remotely sensed data from these platforms, offer unprecedented opportunities to address issues associated with challenges associated with the use of remote sensing data in remote sensing of climate variability trends. Cloud computing platforms like Google Earth Engine host 40+ years of petabytes of remotely sensed data ranging from course to higher-resolution images, thus addressing the issues of limited data. In addition, the platforms are equipped with advanced data filtering, integrations, and processing, which assist in selection and consolidation of various remotely sensed data (Adeli et al., 2020; Tamiminia et al., 2020). Therefore, combining the strengths of various datasets enhances climate variability monitoring from remote sensing. Over the past decade, there has been an increase in the use of these cloud computing approaches in analyzing the climate variability trends in various parts of the world. The bibliometric review by Velastegui-Montoya et al. (2023) reported a dramatic increase in the number of studies that have been published between the period of 2017 and 2022, with a total of 2764 papers globally. A proportion of 30.2% and 16% of those papers focused on Earth and plenary themes as well as Environmental science, where climate variability analysis studies are captured in these themes. Another metaanalysis review by Tamiminia et al. (2020) also showed a dramatic increase in the number of studies that utilized cloud computing platforms in various environmental assessment studies. From the review, it is noted that between the periods of 2013 and 2019, a total of 230 papers using the Google Earth Engine Cloud computing platform were recorded. Further, from the published papers, 30 were in the forest change analysis, 14 in soil studies, 62 in water resources analysis, and 28 in disaster analysis. All the studies cited in these reviews reported high overall accuracies in their analysis (>70%), thus underscoring the relevance of these technologies coupled with freely accessible higher-resolution data in monitoring environmental changes including climate variability indicators. Although studies indicated the success of such platforms in analyzing trends in climate variability, there are some limitations. These limitations include, among others, storage limits for noncommercial license holders, programming limitations for nonprogrammers as some of these platforms, like Google Earth Engine, use coding language, and the black box approach used by the in-built algorithms in these platforms.

5 Conclusions

This chapter aimed to provide an overview of remote sensing applications in climate variability analysis studies with a specific focus on the use of advanced technological tools and data applications in understanding climate variability indicators. The first section provided an overview of climate variability indicators, which can be modeled using remote sensing data and approaches. The second part of this section covered the approaches commonly used to study climate variability, where climate models with their associated challenges were discussed. One of the challenges identified were the fact that these models rely on extensive field data, which is scanty to nonexistent in certain parts of the world, which in turn result in nonrepresentative models that tend to under/ overestimate the climate variability indicators being modeled. The second section introduced the use of remote sensing data and tools as one of the feasible ways to analyze trends in climate variability indicators in the absence of extensive field climate models. The section highlighted some advantages of using remote sensing approaches, including the fact that climate variability indicators can be monitored at no cost and over large spatial extents. In this section, the chapter also highlighted some challenges associated with the remote sensing of climate variability indicators, and these included coarseness in spatial resolution of some of the remotely sensed data, which tends to introduce inaccuracies in the analysis of such trends as well as the unavailability of seamless and cloud-free images which therefore impact the continuous monitoring of the climate variability trends. The last section covered advancements in remote sensing data analytic tools as solutions to some of the limitations of remote sensing of climate variability analysis where cloud computing approaches are introduced with their advantages. Moreover, their limitations are also discussed.

References

Adamowski, J., Sun, K., 2010. Development of a coupled wavelet transform and neural network method for flow forecasting of non-perennial rivers in semi-arid watersheds. J. Hydrol. 390, 85–91. https://doi.org/10.1016/j.jhydrol.2010.06.033.

Adeli, S., Salehi, B., Mahdianpari, M., Quackenbush, L.J., Brisco, B., Tamiminia, H., Shaw, S., 2020. Wetland monitoring using SAR data: a meta-analysis and comprehensive review. Remote Sens. 12. https://doi.org/10.3390/rs12142190.

Akinsanola, A.A., Ongoma, V., Kooperman, G.J., 2021. Evaluation of CMIP6 models in simulating the statistics of extreme precipitation over Eastern Africa. Atmos. Res. 254, 105509. https://doi.org/10.1016/j.atmosres.2021.105509.

Avand, M., Moradi, H., lasboyee, M.R., 2021. Using machine learning models, remote sensing, and GIS to investigate the effects of changing climates and land uses on flood probability. J. Hydrol. 595, 125663. https://doi.org/10.1016/j.jhydrol.2020.125663.

Baninla, Y., Sharifi, A., Allam, Z., Tume, S.J.P., Gangtar, N.N., George, N., 2022. An overview of climate change adaptation and mitigation research in Africa. Front. Clim. 4, 1–24. https://doi.org/10.3389/fclim.2022.976427.

Berck, P., Berck, C.S., Jacobson, T.N., 2018. A Changing Climate in a Changing Land, Agricultural Adaptation to Climate Change in Africa. Routledge, London, UK.

Bhaga, T.D., Dube, T., Shekede, M.D., Shoko, C., 2020. Impacts of climate variability and drought on surface water resources in sub-Saharan Africa using remote sensing: a review. Remote Sens. 12, 1–34. https://doi.org/10.3390/rs12244184.

Camberlin, P., Barraud, G., Bigot, S., Dewitte, O., Makanzu Imwangana, F., Maki Mateso, J.C., Martiny, N., Monsieurs, E., Moron, V., Pellarin, T., Philippon, N., Sahani, M., Samba, G.,

2019. Evaluation of remotely sensed rainfall products over Central Africa. Q. J. R. Meteorol. Soc. 145, 2115–2138. https://doi.org/10.1002/qj.3547.

Crane, T.A., Roncoli, C., Hoogenboom, G., 2011. Adaptation to climate change and climate variability: the importance of understanding agriculture as performance. NJAS—Wageningen J. Life Sci. 57, 179–185. https://doi.org/10.1016/j.njas.2010.11.002.

Gokool, S., Riddell, E.S., Jarmain, C., Chetty, K.T., Feig, G., Thenga, H., 2019. Evaluating the accuracy of satellite-derived evapotranspiration estimates acquired during conditions of water stress. Int. J. Remote Sens. 41 (2), 704–742. https://doi.org/10.1080/01431161.2019.1646940.

Govender, T., Dube, T., Shoko, C., 2022. Remote sensing of land use-land cover change and climate variability on hydrological processes in sub-Saharan Africa: key scientific strides and challenges. Geocarto Int. 37, 10925–10949. https://doi.org/10.1080/10106049.2022.2043451.

Gxokwe, S., Dube, T., Mazvimavi, D., 2020. Multispectral remote sensing of wetlands in semi-arid and arid areas: a review on applications, challenges and possible future research directions. Remote Sens. 12, 1–19. https://doi.org/10.3390/rs12244190.

Gxokwe, S., Dube, T., Mazvimavi, D., Grenfell, M., 2022. Using cloud computing techniques to monitor long-term variations in ecohydrological dynamics of small seasonally-flooded wetlands in semi-arid South Africa. J. Hydrol. 612, 128080. https://doi.org/10.1016/j.jhydrol.2022.128080.

Gxokwe, S., Dube, T., Mazvimavi, D., 2023. Available satellite data for monitoring small and seasonally flooded wetlands in semi-arid. Ecohydrology, 1–16. https://doi.org/10.1002/eco.2605.

Halder, B., Bandyopadhyay, J., Banik, P., 2021. Evaluation of the climate change impact on urban heat island based on land surface temperature and geospatial indicators. Int. J. Environ. Res. 15, 819–835. https://doi.org/10.1007/s41742-021-00356-8.

Jalilvand, E., Tajrishy, M., Ghazi Zadeh Hashemi, S.A., Brocca, L., 2019. Quantification of irrigation water using remote sensing of soil moisture in a semi-arid region. Remote Sens. Environ. 231, 111226. https://doi.org/10.1016/j.rse.2019.111226.

Mahmood, R., Jia, S., Zhu, W., 2019. Analysis of climate variability, trends, and prediction in the most active parts of the Lake Chad basin, Africa. Sci. Rep. 9, 1–18. https://doi.org/10.1038/s41598-019-42811-9.

Muluneh, M.G., 2021. Impact of climate change on biodiversity and food security: a global perspective—a review article. Agric. Food Secur. 10, 1–25. https://doi.org/10.1186/s40066-021-00318-5.

Negassa, M.D., Mallie, D.T., Gemeda, D.O., 2020. Forest cover change detection using Geographic Information Systems and remote sensing techniques: a spatio-temporal study on Komto Protected forest priority area, East Wollega Zone, Ethiopia. Environ. Syst. Res. 9, 1–14. https://doi.org/10.1186/s40068-020-0163-z.

Reyes-García, V., Fernández-Llamazares, Á., Guèze, M., Garcés, A., Mallo, M., Vila-Gómez, M., Vilaseca, M., 2016. Local indicators of climate change: the potential contribution of local knowledge to climate research. Wiley Interdiscip. Rev. Clim. Change 7, 109–124. https://doi.org/10.1002/wcc.374.

Seyisi, E., Mantlana, B., Ndhleve, S., 2023. Indicators for monitoring and evaluating climate change adaptation efforts in South Africa. Jamba J. Disaster Risk Stud. 15, 1–9. https://doi.org/10.4102/JAMBA.V15I1.1426.

Shahzaman, M., Zhu, W., Ullah, I., Mustafa, F., Bilal, M., Ishfaq, S., Nisar, S., Arshad, M., Iqbal, R., Aslam, R.W., 2021. Comparison of multi-year reanalysis, models, and satellite remote sensing products for agricultural drought monitoring over South Asian countries. Remote Sens. 13. https://doi.org/10.3390/rs13163294.

Sileshi, G.W., Gebeyehu, S., 2021. Emerging infectious diseases threatening food security and economies in Africa. Glob. Food Sec. 28, 100479. https://doi.org/10.1016/j.gfs.2020.100479.

Tamiminia, H., Salehi, B., Mahdianpari, M., Quackenbush, L., Adeli, S., Brisco, B., 2020. Google Earth Engine for geo-big data applications: a meta-analysis and systematic review. ISPRS J. Photogramm. Remote Sens. 164, 152–170. https://doi.org/10.1016/j.isprsjprs.2020.04.001.

Thornton, P.K., Ericksen, P.J., Herrero, M., Challinor, A.J., 2014. Climate variability and vulnerability to climate change: a review. Glob. Chang. Biol. 20, 3313–3328. https://doi.org/10.1111/gcb.12581.

Velastegui-Montoya, A., Montalván-Burbano, N., Carrión-Mero, P., Rivera-Torres, H., Sadeck, L., Adami, M., 2023. Google Earth Engine: a global analysis and future trends. Remote Sens. 15. https://doi.org/10.3390/rs15143675.

Wang, C., 2019. Three-ocean interactions and climate variability: a review and perspective. Clim. Dyn. 53, 5119–5136. https://doi.org/10.1007/s00382-019-04930-x.

Zhao, S., Liu, M., Tao, M., Zhou, W., Lu, X., Xiong, Y., Li, F., Wang, Q., 2023. The role of satellite remote sensing in mitigating and adapting to global climate change. Sci. Total Environ. 904, 166820. https://doi.org/10.1016/j.scitotenv.2023.166820.

Chapter 3

Big data and analytical algorithms for climate variability

Iqra Atif[a], Muhammad Ahsan Mahboob[b], Cletah Shoko[a], and Glen T. Nwaila[b]

[a]*Division of Geography, School of Geography, Archaeology and Environmental Studies, University of Witwatersrand, Johannesburg, South Africa,* [b]*Sibanye-Stillwater Digital Mining Laboratory (DigiMine), Wits Mining Institute (WMI), Faculty of Engineering and the Built Environment, University of the Witwatersrand, Johannesburg, South Africa*

Chapter outline

1 Introduction

In the current era of digital transformation, the combination of big data analytics and analytical algorithms has become a crucial factor in understanding and dealing with the complexities of climate variability. This chapter explores the complex relationship between advanced data analytics and the various aspects of climate change, providing a thorough summary of how modern technology is transforming our approach to one of the most critical issues of our era.

Climate variability describes the changes in the Earth's climate system that occur over time durations ranging from months to decades. It is a complex and ever-changing phenomenon. The changes influenced by both natural processes

Remote Sensing of Climate. https://doi.org/10.1016/B978-0-443-21731-9.00005-3
Copyright © 2024 Elsevier Inc. All rights are reserved, including those for text and data mining, AI training, and similar technologies.

and human activities, have significant impacts on ecosystems, agriculture, water resources, and human societies. Understanding climate variability is a challenging task due to the complicated interactions of atmospheric, oceanic, terrestrial, and cryospheric factors (Wang et al., 2022). However, the emergence of big data analytics has introduced new possibilities in the field of climatology, providing innovative knowledge of these challenging processes.

Big data in climate science applies to the vast quantities of data produced through satellite observations, ground-based sensors, ocean beacons, and climate models. This data is characterized not only by its volume but also by its variety, velocity, and veracity (Alia Johnson, 2021). Conventional techniques for processing data are not sufficient to handle this enormous amount of information, which requires advanced analytical algorithms that can extract significant patterns and insights. These algorithms, which include machine learning models and multilayered statistical methodologies, empower researchers to identify patterns, generate forecasts, and understand the nonlinear and uncertain characteristics of the climate system. The application of analytical algorithms in climate science involves more than just data processing; it involves an extensive interaction of mathematics, statistics, and computer science. These algorithms were designed specifically to handle the natural variations in climate data over space and time, enabling the extraction of meaningful patterns from random fluctuations. They offer the tools needed to measure uncertainties, which are a fundamental component of climate projections, therefore providing a more reliable interpretation of climatic variability. Moreover, the incorporation of big data analytics into climate studies is transforming the methodology for climate modeling. Advanced geospatial models with high resolution, powered by extensive observational data, are providing an unparalleled understanding of climate systems. These models, which are equipped with advanced algorithms, could simulate complex climatic interactions. They offer insights into potential future climate scenarios based on different levels of greenhouse gas emissions and numerous other human and environmental variables.

2 Trends in big data analytics for climate change

Several modern developments related to big data analytics are influencing the understanding and adaptation toward climate change, focusing on scientific methods and practical implementations of these advancements. However, the most significant and most relevant ones are discussed below.

An emerging trend in climate research is the increasing adoption of machine learning and artificial intelligence (AI). These technologies provide powerful tools for analyzing complex datasets, allowing researchers to detect patterns and connections that were previously undetected. Machine learning algorithms, especially those applying neural networks, have demonstrated exceptional abilities in analyzing and understanding extensive climate data, hence enabling improved climate modeling and prediction reliability (Ardabili et al., 2020; Bochenek and Ustrnul, 2022; Chattopadhyay et al., 2020). These models serve a crucial role in interpreting the complexities of climate systems, offering valuable insights into potential climatic scenarios depending on many environmental and human-related factors.

Another prominent trend involves the application of powerful sensor technology and remote sensing approaches. The rapid growth of Earth observation satellites, ground-based sensors, and ocean vessels has resulted in a significant rise in the amount of climate data (Simoes et al., 2021). The collection of this data, which includes a range of factors like temperature, precipitation, ocean salinity, and atmospheric gases, is essential for the purpose of monitoring and determining the Earth's changing climate.

Big data analytics has enabled the development of high-resolution climate models, which is a significant trend in the field. These models are becoming more advanced, utilizing extensive datasets that simulate complex relationships within the climate system. Highly detailed models are essential for accurately predicting the unique impacts of climate change in specific regions, which is necessary for creating focused plans to adapt and mitigate its consequences (Sebestyén et al., 2021). They provide a more detailed explanation of climate phenomena, such as severe weather events, the increase in sea levels, and fluctuations in precipitation patterns. The importance of data integration and interoperability has grown in climate research. By incorporating a variety of data sources such as satellite observations, in situ measurements, and socioeconomic data, a more thorough understanding of the effects of climate change can be obtained. Big data analytics allows for the integration and combination of diverse datasets, making it easier to conduct interdisciplinary research that is crucial for addressing the complex difficulties caused by climate change.

Geographical Information System (GIS) is a powerful technology that integrates spatial data and analytical models, enabling researchers and scientists to visually represent, analyze, and interpret data to better understand spatial patterns and connections, particularly regarding climatic variability (Atif et al., 2019; Gintamo et al., 2021). Interactive GIS software programs and maps can communicate complex climate data to a larger audience, hence improving public understanding of climate change issues. Such applications have the capability to transform abstract climate data into concrete visual representations that are clearer and more effective for both the general public and decision-makers.

Moreover, there is an increasing focus on the availability of climate data and ensuring its accessibility to a wider range of stakeholders. The accessibility and analysis of climate data are being facilitated by open data efforts and cloud-based platforms, benefiting scientists, policymakers, and the public. The democratization of this process is essential for promoting collaborative research, well-informed policymaking, and raising public awareness about climate change challenges.

3 Types of big data analytics

Several analytical algorithms have been developed over time. These algorithms, ranging from simple linear regressions to complex deep learning networks, serve as the tools that analyze large and comprehensive datasets. In the field of climate variability, these algorithms play an important part in effectively managing and evaluating a wide range of diverse data sources. Their primary objective is to identify regular trends, forecast future climate conditions, and contribute to understanding the complex interrelationships within the global climate system. The following are the general types of analytical algorithms:

3.1 Descriptive analytics

Descriptive analytics serves as the foundational component of business intelligence and data analysis, providing preliminary insight into the vast datasets that organizations and researchers come across (Sharma et al., 2022). The start of descriptive analytics can be dated back to past civilizations, where data was systematically collected and recorded for agricultural, commercial, and tax purposes, and the 20th century had a notable change in the application of descriptive analytics due to the emergence of the digital era and the introduction of computers (Moreira et al., 2018). This allowed for the processing of larger datasets and the development of more comprehensive reports. The primary purpose of descriptive analytics is to summarize and interpret historical data to understand changes over time. It provides an in-depth evaluation of an organization's activities or a specific issue, such as climate trends, user habits, or financial results. Within the domain of big data, descriptive analytics functions as the primary stage of analysis, filtering through unprocessed data to construct an objective foundation of past events.

The methods used in descriptive analytics have progressed from basic spreadsheets and charting techniques to more advanced applications and platforms. Modern technologies include advanced Business Intelligence (BI) software, data visualization tools, and dashboard systems that consolidate large amounts of data and present them in a readily understandable format for decision-makers.

3.2 Diagnostic analytics

Diagnostic analytics is an important development from basic descriptive analytics, as it involves a more thorough analysis of data that reveals the underlying causes of previous events and behaviors. The start of diagnostic analytics can be linked to several scientific and business disciplines, where the need to understand the underlying causes of events became significant (Gudivada, 2017; Sharma et al., 2022). Due to the development of advanced statistical techniques and the rise of big data, diagnostic analytics has become a more precise and powerful technique. This allows analysts to reveal patterns and connections that were previously hidden. It is similar to scientific work in data analysis, when the emphasis shifts from determining "what" happened to understanding "why" it happened. This type of analytics frequently includes hypothesis testing, root cause analysis, and drill-down techniques. Diagnostic analytics is also important in the field of big data. The application of complex tools such as data mining software, OLAP (Online Analytical Processing) tools (Bernier et al., 2009), and advanced visualization platforms are used. These technologies enable analysts to efficiently analyze extensive datasets, detect irregularities, and investigate relationships between data points.

The applications of diagnostic analytics are particularly significant when considering climatic variability. Scientists apply it to understand the underlying factors contributing to unusual weather patterns or climate anomalies (Abdullah et al., 2020). Researchers can analyze past climate data to identify the causes behind the increased frequency of droughts in specific areas or the rising in the intensity of tropical storms. Through the identification of these trends and their underlying causes, scientists and policymakers may develop more effective policies for mitigating the impacts of climate change.

3.3 Prescriptive analytics

Prescriptive analytics is the center of data analytics, when the focus shifts from understanding "what" and "why" to deciding "what should be done next." The origins of this analytics approach may be traced back to decision science and operations research, fields that have traditionally aimed to maximize results by considering different limitations and demands. The rise of vast data and powerful computing has led to the increased use of prescriptive analytics, enabling advanced and impactful decision-making (Lepenioti et al., 2020). Prescriptive analytics aims to provide recommendations for actions that can result in desired outcomes or reduce risks. It involves utilizing mathematical and computational sciences to provide decision recommendations. This is accomplished by methodologies such as optimization, simulation, and scenario analysis, which can help in assessing possible actions and their potential outcomes. With the advent of big data, the technologies used for prescriptive analytics become increasingly advanced and diverse. These elements include optimization techniques, machine learning models, and AI-powered decision systems (Digalakis, 2023). These technologies have the ability to process complex datasets and provide informative guidance rapidly.

Prescriptive analytics has significant implications in the context of climatic variability. For example, it can be utilized in environmental management to determine the most effective techniques for mitigating carbon emissions or to optimize the allocation of resources in disaster management. Prescriptive analytics utilizes climate data analysis to recommend the most efficient strategies for adapting to or reducing the consequences of climate change.

3.4 Cognitive analytics

Cognitive analytics, frequently linked to AI and cognitive computing, signifies an important step in the progression of data analytics. The origins of the technology may be traced back to the early stages of AI research, with the objective of reproducing human cognitive processes in a machine (Gudivada et al., 2016). The emergence of big data and improved computer tools has greatly expanded the field, enabling advanced and accurate analysis. The primary goal of cognitive analytics is to overcome the limitations of conventional data analysis methods, enabling the development of deeper and more predictive insights. It has several applications in climate change studies, mainly in analyzing the vast and complex datasets that are characteristic of environmental studies. This includes information obtained from satellites, sensors, and climate models, which often lack structure and are extensive in size. Cognitive analytics uses advanced methods such as neural networks, natural language processing, and sentiment analysis to extract significant patterns from the overwhelming amount of data (Hurwitz et al., 2015). Deep learning, a type of machine learning, is particularly known for its capability to handle extensive and complex datasets. Neural networks, namely convolutional neural networks (CNNs) and recurrent neural networks (RNNs), adept in analyzing data with spatial and temporal characteristics, respectively. Interpreting data from satellite imagery and time-series data from climatic measurements is of greatest significance. These methods not only analyze numerical data but also assess textual information, such as reports and research papers, to offer an integrated understanding of climate events.

3.5 Behavioral analytics

Behavioral analytics is a specialized analytical algorithm that focuses on analysis of human behavior patterns, preferences, and actions in relation to climate change. This field has been growing in significance in the past two decades, along with the rise of big data technologies. The foundation of this discipline of research can be traced back to the necessity of understanding the reciprocal relationship between human activities and decisions, and the effects of climate change (Motiwalla et al., 2019). The focus is to identify patterns in human behavior pertaining to the adaptation and mitigation of climate change. The aim is to address issues regarding the perception of climate risks by individuals and groups, their ability to adapt to changing climate conditions, and the factors that impact their decisions regarding environmental conservation and reducing carbon footprint (Gifford et al., 2011). Behavioral analytics utilizes techniques such as data mining, statistical analysis, and predictive modeling. These techniques utilize the power of vast amounts of information collected from diverse sources such as surveys, social media, transactional data, and sensor data. Advanced statistical techniques such as regression analysis, cluster analysis, and time-series analysis play a vital role in detecting trends and relationships in human behavior concerning climate change.

Behavioral analytics are essential in understanding and predicting how the general public will react to climate policies and activities, which is a significant use of this discipline in climate variability. Analyzing social media data can provide insights into the general perspective of the people toward the acceptance of renewable energy or opposition to climate policy (Giest and Mukherjee, 2018). This knowledge is important for leaders to develop more efficient and widely accepted environmental legislation and campaigns. Another significant application falls under the field of energy consumption behavior. This analytical algorithm can assist in the identification of energy use patterns, which are crucial for devising methods to enhance energy efficiency and mitigate carbon emissions. Smart meter data, when paired with demographic information, can offer valuable insights into the energy consumption behaviors of households (Cruz et al., 2021). This, in turn, allows for customized measures to be implemented to minimize energy waste.

3.6 Quantitative analytics

Quantitative analytics, an interdisciplinary field that combines mathematics, statistics, and computer technology, has experienced significant advancements throughout its history. The origins of this can be traced back to the early 20th century, coinciding with advancements in statistical theory and computing technology. Originally, it was used in domains such as economics and engineering, but its usage has subsequently broadened considerably (Albright and Winston, 2020). This analytical algorithm plays a significant part in the field of big data and climate variability. It enhances the understanding of extensive and complex datasets, enabling more knowledgeable and precise forecasts regarding climate patterns and anomalies (Popoola and Nuamah, 2018). The emergence of big data technology has fundamentally transformed this domain, providing innovative tools and approaches for managing vast quantities of climate-related data.

Machine learning is a key technique for quantitative analytics, particularly when processing and analyzing extensive datasets. Machine learning algorithms could detect hidden patterns and trends in data, which can offer significant insights into climatic phenomena (Park et al., 2019; Sidhu et al., 2023). Statistical modeling is a prominent method applied to predict future climate situations by analyzing past data. This involves the development of models capable of simulating various climate circumstances, assisting in understanding possible future shifts. The summary of all the analytical algorithms is given in Table 1.

TABLE 1 Comprehensive analysis of big data analytics frameworks.

Type of big data analytics	Purpose	Methods/techniques	Tools	Examples	Outcome
Descriptive analytics	Understand what is happening now based on incoming data	Data aggregation, data mining, and data visualization	BI tools, dashboards, reporting software	Sales and marketing trend reports, financial statements	Insight into the current state of things
Diagnostic analytics	Diagnose reasons for past outcomes	Drill-down, data discovery, root cause analysis, and correlations	Query tools, OLAP, visualization software	Troubleshooting problems, error analysis	Understanding of past performance and behaviors
Predictive analytics	Predict future outcomes based on historical data	Statistical analysis, forecasting, machine learning, regression models	Data mining tools, statistical software	Credit scoring, demand forecasting	Forecast of future trends and behaviors
Prescriptive analytics	Recommend actions to affect desired future outcomes	Optimization, simulation, scenario analysis, decision trees	Simulation software, optimization software	Route optimization, inventory management	Actionable insights and strategic guidance for future
Cognitive analytics	To simulate human thought processes in a computerized model	Machine learning, natural language processing, data mining	AI platforms, cognitive computing tools	Chatbots, personal assistants, risk assessment systems	Enhanced decision-making, automated reasoning and learning

Behavioral analytics	To analyze data about how people behave and predict how they will behave in the future	Statistical analysis, predictive modeling, pattern recognition	Behavioral targeting tools, customer analytics software	Customer segmentation, product recommendations, user experience research	Improved customer engagement, tailored marketing strategies
Quantitative analytics	To use mathematical and statistical modeling, measurement, and research to understand behavior	Mathematical models, statistical analysis, hypothesis testing	Mathematical tools, statistical software	Financial models, econometrics, operational research	Informed decision-making, and optimized operational processes

4 Types of analytical algorithms

4.1 Regression algorithms

Regression algorithms, within the realm of big data analytics and their application in analyzing climate variability, are statistical techniques employed for modeling and examining the correlation between a dependent variable (typically representing a specific climate characteristic) and one or more independent variables (predictors) (Bonaccorso, 2017). These algorithms play an important part in analyzing large and complex datasets that are typical of climate studies. Linear regression is the basic form of regression analysis that considers a linear relationship between the dependent and independent variables. Due to its direct application and ease of interpretation, it is an essential part of climate research. Linear regression models are highly successful in revealing the relationship between certain factors, such as carbon dioxide emissions, and their influence on global temperatures over a period (Skytt et al., 2020). They serve as a basis for predicting and understanding the steady shifts in climate patterns.

Logistic regression can be applied when the outcome is categorical, such as predicting the chance of extreme weather events, instead of linear relationships. The use of probability-based methods is essential in climate research, particularly when predicting binary outcomes such as the existence or lack of a given event, based on a set of variables. Logistic regression models assist in evaluating the risks and probabilities associated with these severe events, which are increasingly occurring with more intensity and frequency as a result of climate change (Semenza et al., 2011). By contrast, polynomial regression extends the capabilities of regression models by incorporating nonlinear relationships, which are frequently observed in climate changes. This technique is essential for representing the complex relationship among different climatic elements, as these interactions frequently deviate from an ordinary linear trend. Polynomial regression models are extremely important for gaining a more precise and comprehensive knowledge of the complex nature of climate variability, as they can effectively capture and represent irregularities in the data.

4.2 Classification algorithms

Classification algorithms are a particular set of techniques within the field of machine learning that focus on the task of assigning data into specified classes or categories. These algorithms serve as a key part in pattern recognition and are used to determine the category to which a new observation belongs. This is done by analyzing a training set of data that includes observations with defined category members (Black et al., 2023). The origins of classification algorithms can be dated back to the early stages of machine learning and statistics. Early techniques primarily depended on statistical techniques, such as linear discriminant analysis, which was first published by R.A. Fisher in 1936. With the emergence of computers and the digital era, these algorithms have seen significant advancements, integrating more complex techniques such as decision trees, neural networks, and support vector machines (SVMs). Classification algorithms play an important role in big data analytics, particularly when it comes to exploring climate variability. They are commonly used when the output variable (response variable) falls into categories (Suthaharan, 2016). For example, within the context of climate variability, these algorithms could be used to identify meteorological patterns, classify climate zones, or predict weather-related events such as hurricanes or droughts. They possess a high level of skill in managing structured data, such as numerical or categorical variables, which are frequently available in climate-related research. Nevertheless, when more advanced methods emerge, these algorithms are progressively utilized for unstructured data, such as text and images, particularly in the context of big data analysis. For efficient classification in big data analytics, it is necessary to have accurately labeled and complete data. The data needs to go through preprocessing and transformation to be in a format suitable for algorithmic analysis, ensuring that the classification algorithms can successfully learn from and accurately predict based on the provided data (Atif et al., 2021). There are several types of classification algorithms, each with unique characteristics and suitability for different kinds of data; the following are the most used ones:

Decision Trees: These are straightforward, tree-like models of decisions and their possible consequences. They are very helpful for managing categorical data and have a relatively simple interpretation.

Random Forest: A collection of decision trees specifically created to enhance the accuracy of classification. Random forests provide robustness against overfitting and provide efficiency in managing huge amounts of data with higher dimensionality.

Support Vector Machines: These are effective in high-dimensional spaces and are particularly suitable for datasets where the number of dimensions exceeds the number of samples.

K-Nearest Neighbors: This algorithm classifies data based on similarity measures, such as distance functions. It is best used for datasets where the data points naturally group together.

Neural Networks: Particularly relevant in the era of big data, these algorithms are inspired by the neural structures of the brain and are capable of learning complex patterns through layers of processing units.

4.3 Clustering algorithms

Clustering algorithms are a category of methods employed in data analysis to group a collection of objects in an order where objects within the same group, or cluster, indicate greater similarity to each other compared to those in different groups. These algorithms serve an essential part in the domain of big data analytics, facilitating the arrangement and understanding of extensive information

(Ahuja et al., 2020). The origins of clustering algorithms may be traced back to the mid-20th century, with significant progress in statistics and computer science playing a key part in their evolution. Clustering algorithms are mostly used to detect fundamental structures in data, enabling the identification of patterns or groups without any prior knowledge of these categories. Unsupervised learning is differentiated from classification algorithms by its lack of dependence on preexisting classes for grouping (Ikotun et al., 2023).

Clustering algorithms perform a crucial role in climate variability studies conducted using big data analytics. These tools are used for analyzing complex and extensive datasets, such as temperature measurements, precipitation trends, or atmospheric information, in order to identify climate patterns, classify climate regions, or identify abnormal weather events (Hidayati et al., 2020). The type of data suitable for clustering in climate studies includes both structured and unstructured data. Structured data may consist of numerical values like temperature readings, while unstructured data can include satellite images or textual climate reports. These algorithms help in segmenting the data into meaningful clusters, thereby facilitating a better understanding of climate trends and variability.

There are several types of clustering algorithms, each suited to different kinds of data and applications:

K-Means Clustering: This algorithm splits the data into K distinct and non-overlapping subsets or clusters, with no internal structure within each cluster. It is most appropriate for large data sets that include unique, distinguishable clusters.

Hierarchical Clustering: This creates a tree of clusters. It is particularly useful for understanding nested relationships in data and provides a visual representation of these groupings.

DBSCAN (Density-Based Spatial Clustering of Applications with Noise): This algorithm can identify clusters with high density and is effective at detecting clusters with arbitrary shapes. It is particularly useful for spatial data analysis.

Mean Shift Clustering: Based on centroid shifting, this algorithm is used to discover clusters in a smooth density of samples. It is beneficial in situations where the number of clusters is not known explicitly. Table 2 summaries the types of algorithms along with other significant characteristics.

5 Applications of big data analytics in climate change

5.1 Predictive modeling for extreme weather events

The key component of predictive modeling is the comprehensive assessment of various and extensive data sources, encompassing satellite observations, atmospheric data, oceanic patterns, and historical weather records. Through the integration of diverse datasets, models have the ability to identify small climatic

TABLE 2 Summary of algorithm types and key characteristics.

Type of algorithm	Purpose	Methods/ techniques	Tools	Examples	Outcome
Regression algorithms	Forecasting climate trends and predicting variables	Linear Regression, Logistic Regression	R, Python, SAS	Predicting temperature changes	Accurate forecasts of climate parameters
Classification algorithms	Categorizing climate data into specific classes	Decision Trees, SVM, Naive Bayes	Python (Scikit-Learn), MATLAB	Classifying types of extreme weather events	Improved understanding of weather patterns
Clustering algorithms	Grouping similar climate data points	K-Means, Hierarchical, DBSCAN	Python (Scikit-Learn), R	Identifying climate zones	Better regional climate analysis
Association rule learning	Finding relationships in climate data	Apriori, Eclat	R, Python, Weka	Discovering patterns in climate change	Insights into how different factors are connected
Dimensionality reduction	Simplifying climate data analysis	PCA, LDA	R, Python, MATLAB	Reducing climate data dimensions	More efficient data analysis
Ensemble algorithms	Improving predictive accuracy	Random Forest, GBM, AdaBoost	Python (Scikit-Learn), R	Predicting the likelihood of droughts	Enhanced prediction accuracy and reliability
Neural networks	Complex climate pattern recognition	CNN, RNN, Autoencoders	TensorFlow, PyTorch, Keras	Analyzing satellite images	Advanced pattern recognition in climate data
Time series analysis	Analyzing chronological climate data	ARIMA, LSTM	R, Python	Forecasting rainfall patterns	Accurate time-based predictions

Continued

TABLE 2 Summary of algorithm types and key characteristics—cont'd

Type of algorithm	Purpose	Methods/techniques	Tools	Examples	Outcome
Optimization algorithms	Optimizing climate models and simulations	Genetic Algorithms, Simulated Annealing	MATLAB, Python	Optimizing resource allocation for mitigation	Efficient use of resources in climate response
Reinforcement learning	Learning from climate simulations	Q-learning, SARSA	Python, TensorFlow	Adaptive climate strategy development	Development of robust climate strategies

shifts and trends that happen before major weather events (Bag et al., 2022). The resulting analysis empowers meteorologists and scientists to predict events such as cyclones, heatwaves, and storms with enhanced accuracy. Advanced analytical algorithms, such as those based on machine learning, play an important role in improving these predictive models. They provide an integration of real-time data, continually improving and updating predictions. The adoption of this dynamic strategy is essential in a world where climate patterns are becoming more unpredictable and uncertain because of global warming. Furthermore, predictive modeling includes more than just weather predictions. It serves an essential part in reducing the impact of disasters by providing valuable information for the development of emergency planning and response tactics. Accurate forecasts of flood patterns allow for prompt evacuations and deployment of resources, resulting in a substantial reduction in potential loss and damage (Keum et al., 2020; Sun et al., 2020). Moreover, the utilization of drought forecasts facilitates the effective management of water resources and the strategic planning of agricultural activities, thereby providing valuable assistance to communities residing in areas highly susceptible to climate variations (Nandgude et al., 2023). In addition, predictive modeling is utilized in climate research to provide valuable insights into long-term climatic patterns and their possible implications. These insights are crucial for formulating sustainable policies and implementing practices, particularly for sectors that are highly susceptible to weather changes, such as agriculture, energy, and transportation.

5.2 Energy efficiency and intelligence

The application of big data analytics serves a crucial role in enhancing energy efficiency and promoting the development of sustainable energy methods. The increasing number of smart meters and IoT devices in homes, businesses, and industries has led to the collection of vast quantities of energy consumption data (Tomazzoli et al., 2023). Examining this data enables the detection of regularities and discrepancies in energy usage, enabling the formulation of strategies to improve energy efficiency. For example, by using predictive analytics, utility firms may accurately predict periods of high demand, allowing them to effectively manage the load and alleviate stress on the grid. Big data analytics play a crucial role in optimizing the efficiency of renewable energy sources like solar and wind power in the field of green energy (Marinakis et al., 2023). Through analysis of weather data, topographical information, and historical energy production data, algorithms can enhance the location and functioning of renewable energy projects. This optimization ensures the most efficient utilization of renewable energy supplies, hence leading to a decrease in greenhouse gas emissions. Moreover, big data analytics promotes the migration of distributed energy resources into the energy grid (Deepa et al., 2022). The challenges of controlling the energy flow from varied sources such as microgrids and home-based renewable energy systems are growing due to their increasing popularity. Big data analytics offer the means to handle this complexity, maintaining the stability and efficiency of energy supply and supporting the shift toward a more distributed and environmentally friendly energy system. Energy intelligence, enabled by the analysis of enormous amounts of data, extends beyond optimization and includes the capability to predictively maintain energy infrastructure. Through the analysis of data obtained from sensors and other monitoring systems, it is possible to proactively identify and resolve potential problems in power plants and distribution networks. Predictive maintenance reduces downtime and increases the lifespan of energy infrastructure, hence enhancing energy efficiency.

5.3 Smart farming, agriculture, and forestry

Smart farming is a growing field that combines agriculture and technology, using big data analytics to improve farming methods. Predictive models can be developed by analyzing extensive datasets that include soil parameters, weather patterns, crop health, and pest infestations (Alfred et al., 2021). These models assist farmers in making well-informed decisions, including crop selection, fertilization, and irrigation management. For example, the use of these analytics in precision agriculture allows for the customization of farming techniques to meet the requirements of specific crop areas, resulting in the optimization of crop production while minimizing the use of resources. Big data analytics are also essential in the field of forestry for the purpose of monitoring and controlling the health and biodiversity of forests (Singh et al., 2022). The integration of satellite imagery and drone data, along with ground-based sensor information, offers a full understanding of forest ecosystems. Analytical algorithms are used to analyze this data, enabling the monitoring of changes over time. These algorithms can identify regions that are prone to deforestation, pest infestation, or fire outbreaks. Proactive monitoring is essential for the preservation of forest ecosystems, which play a critical role as carbon sinks in the quest against climate change (Zhang and Wei, 2021). Moreover, big data analytics allows the adaptation of agriculture and forestry activities to the changing patterns of climate variability. Considering the impact of climate change and weather patterns, including the rising frequency of extreme events such as droughts and floods, it is essential to depend on data-driven insights to build agricultural and forestry systems that can sustain these challenges. Models that forecast the impact of climate change on particular crops can provide valuable insights for optimizing crop rotation plans, scheduling irrigation, and advancing the cultivation of drought-tolerant crop types (Gul et al., 2020). In addition, big data analytics improves by measuring the carbon footprints of agricultural and forestry techniques. These technologies help measure the amount of carbon that can be stored in the soil through various land management systems. This information is useful for developing practices that not only guarantee food security but also help reduce the impact of climate change.

5.4 Sustainable urban planning and infrastructure

Urban environments, including complex systems, are particularly vulnerable to the effects of climate variability. The impacts of climate change, such as heat islands, flooding, and pollution, are aggravated. Therefore, creative methods are required in urban design and infrastructure development. Big data analytics provides a means to tackle these difficulties by employing well-informed and evidence-driven measures. A major application of big data in this particular situation is for the strategic design and enhancement of urban infrastructure. Urban planners can enhance the efficiency and sustainability of cities by evaluating extensive datasets from many sources, such as traffic patterns, energy usage, population trends, and environmental data (Al Batayneh et al., 2021). For instance, data analytics can provide insights for the development of public transport systems that effectively reduce traffic congestion and decrease greenhouse gas emissions. Big data analytics allows cities to plan for and tackle climate-related risks in the field of infrastructure resilience (Rojo, 2020). Flood risk models, which are based on historical data and predictive analytics, can provide guidelines for the development of flood mitigation strategies and emergency response techniques. Moreover, the utilization of data-driven analysis on urban heat islands can effectively guide the application of cooling measures, such as the introduction of green spaces and the use of reflective building materials. Moreover, big data also improves the sustainability of urban water management systems. Cities can enhance water conservation and wastewater management in water-scarce regions affected by climate change by conducting an in-depth assessment of data related to water usage, precipitation patterns, and infrastructure problems.

5.5 Natural disaster and disease assessment

Big data analytics in natural disaster assessment involves the comprehensive collection and analysis of vast datasets derived from various sources like satellite imagery, sensor networks, and climate models (Hajjaji et al., 2021; Sarker et al., 2020). The integration of multiple information enables accurate forecasting of severe weather phenomena, such as hurricanes, floods, and wildfires. Through the utilization of predictive models, governments and organizations have the ability to strategically allocate resources, minimize response times, and improve the effectiveness of disaster preparedness measures. Furthermore, these analytics enable the identification and mapping of places that are susceptible to disasters, thereby assisting in the development of strong infrastructure and measures to enhance community resilience. Big data analytics also plays a key part in disease assessment by helping researchers recognize the connection between climate change and the transmission of infectious diseases (Galetsi et al., 2020). Climate variability directly affects the spread of vector-borne diseases such as malaria and dengue by altering temperature, precipitation, and humidity. Predictive models can predict epidemics by analyzing environmental data, health records, and population dynamics, so providing guidance for public health measures and resource allocation. This strategy not only assists in reducing the transmission of diseases but also in developing long-term health policies and strategies adapted to the changing climate conditions. In addition, big data analytics allows the real-time integration of social media, news reporting, and mobile data for disease surveillance. Data science in this context improves the ability to promptly address disease outbreaks, particularly in areas that are geographically isolated and lack sufficient resources (Alamo et al., 2020; Atif et al., 2020). By providing information on social mobility and public opinion, it facilitates the development of improved communication tactics and community involvement in disease prevention and control.

6 Conclusion

This chapter provides an in-depth overview of the crucial role that big data analytics and analytical algorithms serve in understanding and addressing climate variability. The integration of extensive data derived from multiple sources, including as satellite observations, sensors, and climate models, has fundamentally transformed our methodology in the field of climate science. Advanced analytical tools, which include machine learning models and advanced statistical techniques, play an essential part in understanding complicated, nonlinear, and unpredictable elements of climate systems. These techniques have improved our capability to predict weather patterns, understand climatic relationships, and simulate future climate scenarios. The emergence of machine learning and AI in climate research represents a significant trend, offering enhanced capabilities in climate modeling and prediction. The integration of powerful sensor technology and remote sensing methods has also escalated the amount and quality of climate data available, further advancing our understanding of the Earth's changing climate. Moreover, the development of high-resolution climate models utilizing extensive datasets has been crucial in providing detailed insights into specific regional impacts of climate change. This detailed modeling is essential for formulating targeted strategies for climate adaptation and mitigation. Another important aspect is the availability and openness of climate data, ensuring its accessibility to a wider range of people, which is significant for collaborative research, informed policymaking, and increasing public awareness of climate issues. Technologies like GIS have made it possible to visually represent and analyze complex and huge amounts of climate data, enhancing its understanding and impact.

References

Abdullah, M.F., Amin, M.Z.M., Zainol, Z., Ideris, M.M., 2020. Big Data Analytics as Game Changer in Dealing Impact of Climate Change in Malaysia: Present and Future Research. IoTBDS.

Ahuja, R., Chug, A., Gupta, S., Ahuja, P., Kohli, S., 2020. Classification and clustering algorithms of machine learning with their applications. In: Nature-Inspired Computation in Data Mining and Machine Learning. Springer, pp. 225–248.

Alamo, T., Reina, D.G., Millán, P., 2020. Data-driven methods to monitor, model, forecast and control Covid-19 pandemic: leveraging data science, epidemiology and control theory. arXiv preprint arXiv:2006.01731.

Al Batayneh, R.M., Taleb, N., Said, R.A., Alshurideh, M.T., Ghazal, T.M., Alzoubi, H.M., 2021. IT governance framework and smart services integration for future development of Dubai infrastructure utilizing AI and big data, its reflection on the citizens standard of living. In: The International Conference on Artificial Intelligence and Computer Vision.

Albright, S.C., Winston, W.L., 2020. Business Analytics: Data Analysis and Decision Making. Cengage Learning, Inc.

Alfred, R., Obit, J.H., Chin, C.P.-Y., Haviluddin, H., Lim, Y., 2021. Towards paddy rice smart farming: a review on big data, machine learning, and rice production tasks. IEEE Access 9, 50358–50380.

Alia Johnson, A.N., 2021. Climate science and AI: transforming environmental big data analysis. Int. J. New Media Stud. 8 (1), 14–19.

Ardabili, S., Mosavi, A., Dehghani, M., Várkonyi-Kóczy, A.R., 2020. Deep learning and machine learning in hydrological processes climate change and earth systems a systematic review. In: Engineering for Sustainable Future: Selected Papers of the 18th International Conference on Global Research and Education Inter-Academia—2019, p. 18.

Atif, I., Iqbal, J., Su, L.-j., 2019. Modeling hydrological response to climate change in a data-scarce glacierized high mountain Astore Basin using a fully distributed TOPKAPI model. Climate 7 (11), 127.

Atif, I., Cawood, F.T., Mahboob, M.A., 2020. The role of digital technologies that could be applied for prescreening in the mining industry during the COVID-19 pandemic. Trans. Indian Natl. Acad. Eng. 5, 663–674.

Atif, I., Cawood, F.T., Mahboob, M.A., 2021. Development of interactive dashboards and intelligent data analytics for visual decision-making in the underground mining environment: The Sterkfontein Cave Case Study. In: 2021 International Bhurban Conference on Applied Sciences and Technologies (IBCAST).

Bag, S., Rahman, M.S., Srivastava, G., Chan, H.-L., Bryde, D.J., 2022. The role of big data and predictive analytics in developing a resilient supply chain network in the South African mining industry against extreme weather events. Int. J. Prod. Econ. 251, 108541.

Bernier, E., Gosselin, P., Badard, T., Bédard, Y., 2009. Easier surveillance of climate-related health vulnerabilities through a web-based spatial OLAP application. Int. J. Health Geogr. 8 (1), 1–18.

Black, J.E., Kueper, J.K., Williamson, T.S., 2023. An introduction to machine learning for classification and prediction. Fam. Pract. 40 (1), 200–204.

Bochenek, B., Ustrnul, Z., 2022. Machine learning in weather prediction and climate analyses—applications and perspectives. Atmosphere 13 (2), 180.

Bonaccorso, G., 2017. Machine Learning Algorithms. Packt Publishing Ltd.

Chattopadhyay, A., Hassanzadeh, P., Pasha, S., 2020. Predicting clustered weather patterns: a test case for applications of convolutional neural networks to spatio-temporal climate data. Sci. Rep. 10 (1), 1317.

Cruz, C., Palomar, E., Bravo, I., Aleixandre, M., 2021. Behavioural patterns in aggregated demand response developments for communities targeting renewables. Sustain. Cities Soc. 72, 103001.

Deepa, N., Pham, Q.-V., Nguyen, D.C., Bhattacharya, S., Prabadevi, B., Gadekallu, T.R., Maddikunta, P.K.R., Fang, F., Pathirana, P.N., 2022. A survey on blockchain for big data: approaches, opportunities, and future directions. Future Gener. Comput. Syst. 131, 209–226.

Digalakis, V., 2023. Analytics Under Variability, Volume, and Velocity With Applications to Sustainability and Healthcare. Massachusetts Institute of Technology.

Galetsi, P., Katsaliaki, K., Kumar, S., 2020. Big data analytics in health sector: theoretical framework, techniques and prospects. Int. J. Inf. Manag. 50, 206–216.

Giest, S., Mukherjee, I., 2018. Behavioral instruments in renewable energy and the role of big data: a policy perspective. Energy Policy 123, 360–366.

Gifford, R., Kormos, C., McIntyre, A., 2011. Behavioral dimensions of climate change: drivers, responses, barriers, and interventions. Wiley Interdiscip. Rev. Clim. Change 2 (6), 801–827.

Gintamo, T.T., Mengistu, H., Kanyerere, T., 2021. GIS-based modelling of climate variability impacts on groundwater quality: Cape Flats aquifer, Cape Town, South Africa. Groundw. Sustain. Dev. 15, 100663.

Gudivada, V.N., 2017. Data analytics: fundamentals. In: Data Analytics for Intelligent Transportation Systems. Elsevier, pp. 31–67.

Gudivada, V.N., Irfan, M.T., Fathi, E., Rao, D.L., 2016. Cognitive analytics: going beyond big data analytics and machine learning. In: Handbook of Statistics. vol. 35. Elsevier, pp. 169–205.

Gul, F., Ahmed, I., Ashfaq, M., Jan, D., Fahad, S., Li, X., Wang, D., Fahad, M., Fayyaz, M., Shah, S. A., 2020. Use of crop growth model to simulate the impact of climate change on yield of various wheat cultivars under different agro-environmental conditions in Khyber Pakhtunkhwa, Pakistan. Arab. J. Geosci. 13, 1–14.

Hajjaji, Y., Boulila, W., Farah, I.R., Romdhani, I., Hussain, A., 2021. Big data and IoT-based applications in smart environments: a systematic review. Comput. Sci. Rev. 39, 100318.

Hidayati, I.C., Nalaratih, N., Shabrina, A., Wahyuni, I.N., Latifah, A.L., 2020. Correlation of climate variability and burned area in Borneo using Clustering Methods. For. Soc. 4 (2), 280–293.

Hurwitz, J., Kaufman, M., Bowles, A., Nugent, A., Kobielus, J.G., Kowolenko, M.D., 2015. Cognitive Computing and Big Data Analytics. vol. 288 Wiley Online Library.

Ikotun, A.M., Ezugwu, A.E., Abualigah, L., Abuhaija, B., Heming, J., 2023. K-means clustering algorithms: a comprehensive review, variants analysis, and advances in the era of big data. Inf. Sci. 622, 178–210.

Keum, H.J., Han, K.Y., Kim, H.I., 2020. Real-time flood disaster prediction system by applying machine learning technique. KSCE J. Civ. Eng. 24 (9), 2835–2848.

Lepenioti, K., Bousdekis, A., Apostolou, D., Mentzas, G., 2020. Prescriptive analytics: literature review and research challenges. Int. J. Inf. Manag. 50, 57–70.

Marinakis, V., Doukas, H., Psarras, J., 2023. Energy management 4.0. In: Handbook of Research on Artificial Intelligence, Innovation and Entrepreneurship. Edward Elgar Publishing, pp. 121–134.

Moreira, J., Carvalho, A., Horvath, T., 2018. A General Introduction to Data Analytics. John Wiley & Sons.

Motiwalla, L., Deokar, A.V., Sarnikar, S., Dimoka, A., 2019. Leveraging data analytics for behavioral research. Inf. Syst. Front. 21, 735–742.

Nandgude, N., Singh, T., Nandgude, S., Tiwari, M., 2023. Drought prediction: a comprehensive review of different drought prediction models and adopted technologies. Sustainability 15 (15), 11684.

Park, J.H., Yoo, S., Nadiga, B., 2019. Machine learning climate variability. In: Proceedings of the 33rd Conference on Neural Information Processing Systems (NeurIPS), Vancouver, BC, Canada.

Popoola, O., Nuamah, N., 2018. New trends in modelling climate change in the era of big data. Ann. Comput. Sci. Ser. 16 (2), 69–74.

Rojo, J., 2020. Urban Climate Resilience and the Promise of Big Data Solutions: Assessing Big Data Application Into Madrid's Urban Climate Change Resilience Scenario.

Sarker, M.N.I., Peng, Y., Yiran, C., Shouse, R.C., 2020. Disaster resilience through big data: way to environmental sustainability. Int. J. Disaster Risk Reduct. 51, 101769.

Sebestyén, V., Czvetkó, T., Abonyi, J., 2021. The applicability of big data in climate change research: the importance of system of systems thinking. Front. Environ. Sci. 9, 70.

Semenza, J.C., Ploubidis, G.B., George, L.A., 2011. Climate change and climate variability: personal motivation for adaptation and mitigation. Environ. Health 10, 1–12.

Sharma, A.K., Sharma, D.M., Purohit, N., Rout, S.K., Sharma, S.A., 2022. Analytics techniques: descriptive analytics, predictive analytics, and prescriptive analytics. In: Decision Intelligence Analytics and the Implementation of Strategic Business Management. Springer, pp. 1–14.

Sidhu, B.S., Mehrabi, Z., Ramankutty, N., Kandlikar, M., 2023. How can machine learning help in understanding the impact of climate change on crop yields? Environ. Res. Lett. 18 (2), 024008.

Simoes, R., Camara, G., Queiroz, G., Souza, F., Andrade, P.R., Santos, L., Carvalho, A., Ferreira, K., 2021. Satellite image time series analysis for big earth observation data. Remote Sens. 13 (13), 2428.

Singh, R., Gehlot, A., Akram, S.V., Thakur, A.K., Buddhi, D., Das, P.K., 2022. Forest 4.0: digitalization of forest using the internet of things (IoT). J. King. Saud. Univ. Comput. Inf. Sci. 34 (8), 5587–5601.

Skytt, T., Nielsen, S.N., Jonsson, B.-G., 2020. Global warming potential and absolute global temperature change potential from carbon dioxide and methane fluxes as indicators of regional sustainability—a case study of Jämtland, Sweden. Ecol. Indic. 110, 105831.

Sun, W., Bocchini, P., Davison, B.D., 2020. Applications of artificial intelligence for disaster management. Nat. Hazards 103 (3), 2631–2689.

Suthaharan, S., 2016. Machine learning models and algorithms for big data classification. Integr. Ser. Inf. Syst. 36, 1–12.

Tomazzoli, C., Scannapieco, S., Cristani, M., 2023. Internet of things and artificial intelligence enable energy efficiency. J. Ambient Intell. Humaniz. Comput. 14 (5), 4933–4954.

Wang, H., Dai, Y., Yang, S., Li, T., Luo, J., Sun, B., Duan, M., Ma, J., Yin, Z., Huang, Y., 2022. Predicting climate anomalies: a real challenge. Atmos. Ocean. Sci. Lett. 15 (1), 100115.

Zhang, M., Wei, X., 2021. Deforestation, forestation, and water supply. Science 371 (6533), 990–991.

Chapter 4

Climate variability and agriculture

Munyaradzi Davis Shekede[a,b], Isaiah Gwitira[a], and Abel Chemura[c]
[a]*Department of Geography Geospatial Sciences and Earth Observation, University of Zimbabwe, Harare, Zimbabwe,* [b]*Department of Geospatial Sciences and Earth Observation, Zimbabwe National Geospatial and Space Agency, Harare, Zimbabwe,* [c]*Faculty of Geo-Information Science and Earth Observation (ITC), University of Twente, Enschede, The Netherlands*

Chapter outline

1 Introduction

Although several factors such as soils, technology, genetics, as well as field management practices such as fertilizer applications and crop hybrid selection, climate is regarded as the most important determinant of global-level spatial and temporal variability in agricultural production (Ahmad et al., 2020a,b). Specifically, temperature and precipitation are key contributing factors to agriculture production and productivity (Quaye et al., 2018). In this regard, climate conditions explain 20%–49% of the variance of global aggregated yield

Remote Sensing of Climate. https://doi.org/10.1016/B978-0-443-21731-9.00010-7
Copyright © 2024 Elsevier Inc. All rights are reserved, including those for text and data mining, AI training, and similar technologies.

anomalies, with temperature-related predictors being more profound than precipitation-related predictors (Vogel, 2022). The agricultural sector is regarded as the most sensitive to climate variability and change as well as climate extremes due to its overdependence on weather variables, its interconnectedness with other sectors that can be disrupted by climate, heavy reliance on land, water, and other natural resources affected by climate, occupies large areas of land meaning that it is exposed across a large area, and many crops and livestock are in the field for an extended period of time. Zhu et al. (2022) note that the sector, particularly crop yield is sensitive to in-season and interannual climate variability.

Malhi et al. (2021) reports that virtually all cultivated areas across the globe have been affected by climate change, with only 3.5% regarded as "safe" from climate change-related environmental limitations. The sensitivity of the sector is high in developing countries due to heavy dependency on rainfed agriculture and the disproportionate effects of climate change in these countries (IPCC, 2014). However, the impact of climate variability and climate change are interwoven and are dependent on crop or livestock systems, management practices, and location (Ferrara and Lindberg, 2023). Understanding the impacts of climate change and variability on the agricultural systems across the globe is imperative as they influence the attainment of sustainable development goals such as zero hunger, clean water, and maintaining life on land and in water (Ortiz et al., 2021). Moreover, humanity is faced with the formidable challenge of increasing agricultural food production to ensure food security for a ballooning global population expected to reach 10 billion people by 2050 under a changing climate (Delgado et al., 2019). Thus, there is a need to understand the extent to which climate change and variability affect agriculture as a first toward the development of mitigation and adaptation measures.

Considerable research has been undertaken to explore the impacts of climate variability, including changes in temperature and precipitation, on crop production systems across the globe. So far, studies have shown, with high confidence, the likelihood of more negative than positive impacts on crop production (IPCC, 2014). For instance, Ray et al. (2019) used linear regression analysis alongside long-term weather and crop data to assess the potential impact of observed climate change on the yields of the ten global most important crops, namely barley, cassava, maize, oil palm, rapeseed, rice, sorghum, soybean, sugarcane, and wheat. The study obtained mixed results showing that climate change impacts on crop yields have been predominantly negative in Europe, Southern Africa, and Australia but generally positive in Latin America. In another global study, Vogel et al. (2019) found general negative impacts of increasing unusually warm days and unusually cold daily minimum temperatures on crop yields. Similarly, Lobell et al. (2011) demonstrated a 7% reduction in global wheat yields with a 1°C increase in temperature during the growing season. In a related study, Asseng et al. (2015) noted the negative effects of climate change-driven temperature extremes on wheat production, with estimates of up to 6% reduction in wheat yield associated with each 1°C rise in temperature. Schlenker and Roberts (2009) also observed that a 1°C increase in temperature induced a 10% reduction in maize yields in the United States.

According to Boyer (1982) climate change has been responsible for ~70% decline in global yield since 1982 due to climate change. Pruski and Nearing (2002) reported that extreme weather events increase soil erosion potential in agricultural systems and estimate the doubling of soil erosion rates for every 1% increase in total rainfall. Results of metaanalyses project an 8% decline in mean yield across the whole of Africa and 11% in West Africa in the absence of adaptation (Sultan et al., 2019). There is overwhelming evidence showing major reductions in global staple crop yield under climate change without significant adaptation initiatives. Fig. 1 illustrates some of the positive and negative impacts of climate change and variability on the agricultural sector.

Several studies in China quantified climate-yield relationships using statistical models and have reported mixed results. For instance, warming climate and rising temperatures have been associated with maize yield reduction in some parts of China (Tao et al., 2008; Chen and Pang, 2020) while associated with some positive effects in selected regions such as the northeastern region of China (Liu et al., 2013; Zhang and Huang, 2013). Similarly, extremely high levels of precipitation have a significant negative effect on maize yield through increasing insects and disease (Raza et al., 2019).

The impacts of climate change and variability are not limited to the cropping system but extend to the animal production system in many regions of the

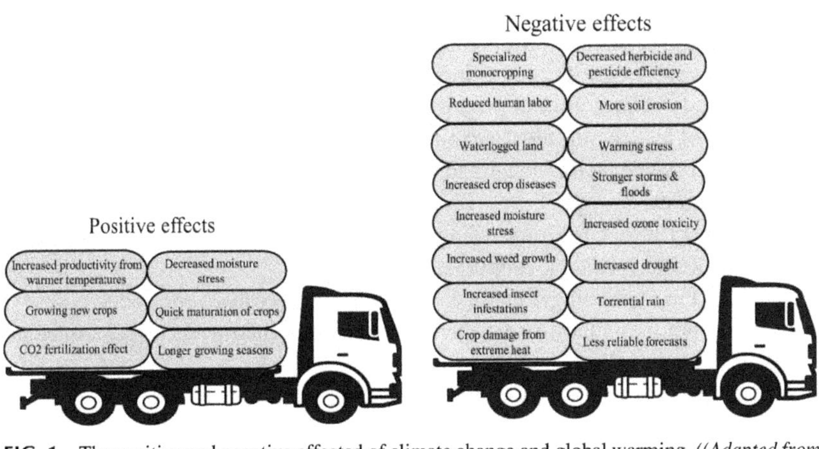

FIG. 1 The positive and negative effected of climate change and global warming. *((Adapted from Raza, A., Razzaq, A., Mehmood, S.S., Zou, X., Zhang, X., Lv, Y., Xu, J., 2019. Impact of climate change on crops adaptation and strategies to tackle its outcome: a review. Plants 8(2), 34.)*

world. Studies have shown that climate change and variability affect livestock production directly by reducing rangeland productivity, bush encroachment (displacement of herbaceous by often unpalatable woody species), and impacting forage distribution in space and time (Habte et al., 2022). Moreover, climate change impacts livestock production by creating an environment conducive to livestock disease proliferation and even via thermal stress due to heat shock (Bett et al., 2017). Heat stress reduces livestock productivity, fertility, and susceptibility to disease and even increases mortality (Godde et al., 2021; Thornton et al., 2021).

Rangeland's productivity is influenced by climate change and variability through the contraction of growing periods of herbaceous plants and reducing pasture yields (Parry, 2007). Climate change and variability have been associated with increasing frequency and intensity of droughts, changes in water availability, and increasing patterns of temperature and rainfall variability, which are all cumulatively impacting livestock production and livelihood dependent thereon (Ulrichs et al., 2019). However, whether and to what extent livestock is vulnerable is a function of the adaptive capacity of livestock species exposed to climate extremes. For instance, Wako et al. (2017) report disproportionate cattle deaths from climate change-induced droughts relative to livestock such as goats and sheep. The effects of the livestock sector and rangeland resources will likely further worsen in the future with projections of temperature increases, uneven rainfall patterns, and heat waves (IPCC, 2014).

Godde et al., 2021 provide a detailed overview of the potential differential effects of climate change on the livestock food supply chain. The study highlighted that climate change is likely to:

- Adversely affects productivity, nutritional quality, and availability of feed resources.
- Increased pressure on water resources in regions already water-stressed and reduced water quality due to increased temperature and flooding.
- Negatively impact animal production, welfare, and life expectancy through decreased feed availability and quality, heat stress, diseases, and mortality at a global scale with localized beneficial impacts, especially in cold environments.
- Negatively affect the processing, storage, transportation, and retailing of livestock and livestock products.
- Reduce the availability, quality, and safety of livestock products.
- Reduce labor availability and productivity through climate change-induced heat stress and disease outbreaks.
- Increase price volatility along the livestock supply chain.

Combined, these results suggest that the effects of climate variability and change are complex, idiosyncratic, and scale-dependent, thus warrant multi-scale analysis.

2 Monitoring climate variability and its impacts on agriculture

2.1 Monitoring climate variables

Climate change and variability have affected virtually every component of the agricultural system, and with projections of a warmer and more variable future climate characterized by increasing frequency and intensity of extreme events, the capability of the agricultural sector to provide food, feed, and fiber to an ever-growing population will likely be compromised. This is particularly of great concern given that shifting weather patterns are already impacting food systems across the globe (Vermeulen et al., 2012). It is therefore critical to accurately monitor and evaluate climate change impacts on the agriculture sector at scales ranging from the field to the global scale. Traditionally, field-based methods have been used to acquire information on indicators of climate change. However, climate change related impacts on the agriculture system require data at large spatial and temporal scales thereby making field-based methods inappropriate.

In this regard, remote sensing and associated geospatial technologies provide invaluable tools for tracking changes in the agriculture sector induced by climate change and variability. Thus, scientists can use a suite of remote sensing platforms such as satellite remote sensing, airborne remote sensing, unmanned aerial vehicles, and ground-based remote sensing to collect multispectral, hyperspectral, synthetic aperture radar, and even lidar data to monitor changes in climate patterns and evaluate their impacts on agriculture. Using a combination of satellite imagery and remotely sensed data, scientists can measure various climate parameters such as precipitation, temperature, humidity, and radiation, yielding information that is critical for monitoring vegetation health, crop productivity, and water availability associated with climate variability. Understanding the influence of climate-related impacts on the agriculture sector is of great importance in developing and implementing adaptation and mitigation measures for agricultural systems.

One of the key tasks that scientists are interested in understanding climate change impacts and variability on the agricultural system is to determine relevant variables to monitor. Hatfield et al. (2020) provide a conceptual framework for identifying indicators of climate change in agricultural systems (Fig. 2).

Based on the conceptual framework, climate variables such as carbon dioxide, temperature, solar radiation, and precipitation have a direct impact on dynamics in agricultural systems including grassland, cropping systems, livestock production, and pests. Regarding precipitation, Hatfield et al. (2020) note that the water cycle is a key component of agricultural systems whose variability, as mediated by temperature, has a direct bearing on the grassland or cropping system (Hatfield et al., 2011; Izaurralde et al., 2011). It is further observed that there are linkages and feedback among the components representing the direct effects of climate on production and pests. In addition, it highlights

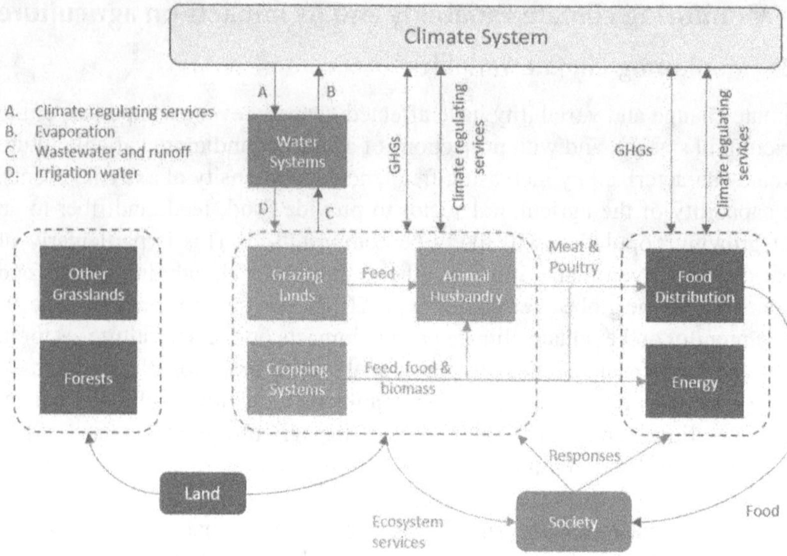

FIG. 2 Conceptual diagram of potential indicators of climate impacts on agricultural systems (Hatfield et al., 2020).

the indirect impacts of human activities on the environment and how they affect the services provided by ecosystems. It further demonstrates the ways in which society's needs and desires for energy and food production can result in various ecological responses. Thus, Hatfield et al. (2020) identified the following as key indicators of climate change in agricultural systems:

- heat stress,
- soil erosion related to changes in precipitation,
- soil carbon changes in response to increasing carbon dioxide and soil management practices,
- economic response to climate change in agricultural production and crop progress and productivity.

2.2 Approaches for monitoring droughts

Of the different effects of climate variability on agriculture, droughts seem to be the most profound and widely studied. Drought is regarded as one of the most complex global natural hazards with impacts across several sectors include agriculture, water, and the environment (Dikshit et al., 2022; Sharara et al., 2022; Wilhite, 2000). In recent years, the frequency and magnitude of droughts have been amplified by climate change and variability. Despite years of research, the dimensions of drought are still poorly understood (Frischen et al., 2020). This is partly attributed to hundreds of drought definitions that have emerged over the years, leading to the adoption of operational definitions.

A review of the definitions undisputedly shows that all droughts emanate from some form of precipitation deficiency, which, in turn, leads to an imbalance between water supply and societal or environmental demands (Wilhite and Buchanan-Smith, 2005). For instance, agricultural drought is a condition of insufficient soil water content in the root zone to such an extent that the percentage is lower than the minimum requirement for a plant to grow (Liu et al., 2016). Thus, four main drought categories have been used to understand drought, that is, meteorological, hydrological, agricultural, and socioeconomic drought (Wilhite and Buchanan-Smith, 2005). These classifications are based on their causality as follows: (i) meteorological drought (precipitation loss), (ii) agricultural drought (soil moisture loss), (iii) hydrological drought (surface runoff, total water, and groundwater loss), and (iv) socioeconomic drought (water supply and demand loss) (Fig. 3).

The four drought types are sequential, with agricultural drought occurring after meteorological and hydrological droughts (Liu et al., 2016). Wilhite (1993) defines meteorological drought as a temporal condition during which the amount of rainfall received within a specific period is less than the long-term average for a specific area. However, the severity of the meteorological drought is quantified by comparing the ratio between the amount of precipitation received versus the long-term normal for a specific period and area (Siswanto et al., 2022). In the event of an anomaly, that is, the amount of precipitation received for a given period is less than the long-term average, a hydrological drought then ensues, but its magnitude is a function of the extent of the precipitation anomaly. Thus, a hydrological drought is characterized by

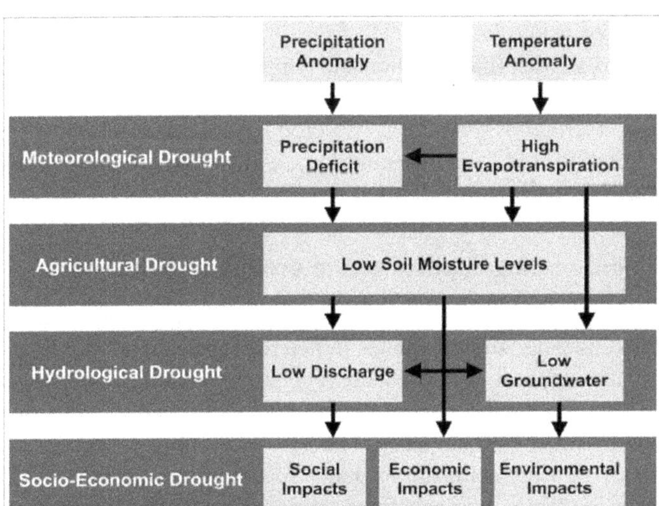

FIG. 3 The mechanism and classification of four drought types as commonly presented in literature (Van Loon and Van Lanen, 2012).

a substantial decrease in surface and groundwater supply and is expressed through a significant lowering of water levels in river, reservoirs, and groundwater (Dai, 2011).

On the other hand, agricultural drought is a culmination of temporal precipitation shortage that reduces soil moisture levels and is usually accompanied by temperature increases that cause an increase in evapotranspiration levels above the water supply (Hazaymeh and Hassan, 2016). These two factors result in a deficit of plant-available water with the potential to compromise plant growth and, ultimately, crop yields (Sánchez et al., 2018). Combined, the three drought types culminate in socioeconomic drought, whose impacts are felt through decline in social and economic activities as well as disturbance of natural processes occurring in the environment. Therefore, these droughts are interlinked, though agricultural drought is regarded as one of the most potent due to its implications on the economy, food security, and social stability across the globe (Sánchez et al., 2018).

Agricultural drought has traditionally been monitored using in situ methods, which are reliant on data collected from ground-based climatic, agricultural, and hydrologic stations. Typical data collected from these ground stations include precipitation, temperature, relative humidity, and soil water content which is in turn used to give quantitative and qualitative information on drought (Hazaymeh and Hassan, 2016). While these datasets were initially used to determine meteorological drought they have been applied in characterizing agricultural drought. Overall, drought indices derived from hydro-meteorological ground stations are deemed accurate at locations where data are collected but suffer from uneven and sparse distribution, rendering these ineffective and uncertain away from the measured points (Maes, 2012). Some of the drought indices derived using data collected from ground stations include (i) palmer drought severity index (precipitation and temperature); (ii) crop moisture index (soil moisture, precipitation, and temperature); (iii) crop water stress index (actual and potential evaporation); (iv) crop specific drought index (temperature, precipitation, and evapotranspiration); and (iv) standardized precipitation index (precipitation regimes) (Fig. 4).

2.3 Remote sensing applications in drought monitoring

Advances in satellite observation technology have provided an effective approach to characterizing dynamics in heterogeneous agricultural landscapes at large spatiotemporal scales. Since vegetation (including species found in rangeland and croplands) is a key determinant of exchange of energy, water, and carbon between the land surface and the atmosphere (Chen et al., 2020), most remote sensing applications have focused on monitoring vegetation dynamics as a viable approach for detecting climate change and variability impacts on the agricultural system. In fact, the increase in satellite missions has led to the proliferation of remote sensing datasets that have supported

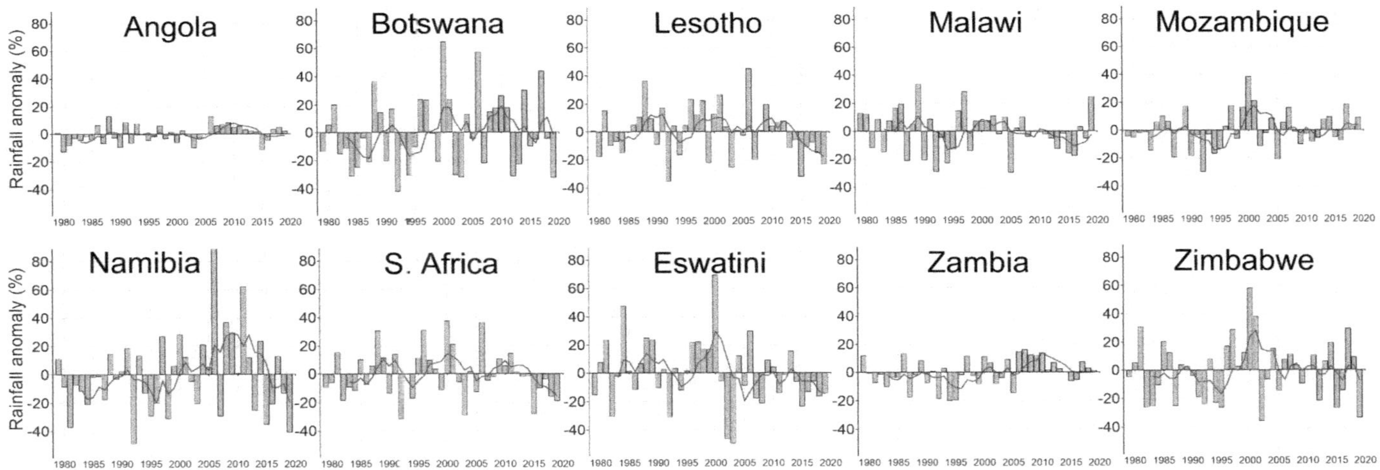

FIG. 4 Rainfall anomalies across southern African countries showing annual climate variability. ((Source: *CHIRPS data*.)

the development of several indices and models for evaluating climate variability in agricultural landscapes across the globe. The basis for using spectral signatures for agricultural drought monitoring lies in the unique reflectance of soil and crops across the electromagnetic spectrum, particularly in the red, near-infrared, shortwave infrared, and thermal spectral bands (Hazaymeh and Hassan, 2016).

Drought induces changes in biophysical and chemical properties including soil moisture, organic matter, vegetation biomass, chlorophyll, and canopy and soil temperature thereby changing the spectral and thermal characteristics of soil and vegetation. Sánchez et al. (2018) observe that remote sensing-based indices are critically important for effectively monitoring climate variability proxies such as drought at large spatial extents due to their ability to integrate soil moisture and or vegetation information into agricultural drought indicators overcoming difficulties associated with direct field observations.

Over the years, scientists have developed remotely sensed methods for detecting and monitoring the duration, extent, magnitude, and severity of droughts. Hazaymeh and Hassan (2016) provide a comprehensive overview of remote sensing applications in agriculture and divide these into (i) optical remote sensing methods, (ii) thermal remote sensing methods, (iii) microwave remote sensing methods, (iv) combined remote sensing methods, and (v) synergy between in situ and remote sensing-based methods (Table 1). Optical remote sensing approaches exploit the visible (red), near-infrared, and shortwave infrared portions of the electromagnetic spectrum. In these spectral regions, vegetation and soil differentially reflect incident radiation depending on the greenness of the vegetation and the moisture status of the soil (Qin et al., 2021). Hazaymeh and Hassan (2016) divided optical remote sensing agricultural indices into three categories based on the aspect being measured, that is, (a) soil drought monitoring indices, (b) vegetation drought monitoring indices, and (c) soil and vegetation drought indices. Evidence from previous shows that indices for monitoring soil drought are more effective on bare soil surfaces than in vegetated areas. Overall, optical remote sensing approaches do not integrate temperature anomalies in calculating agricultural drought indices.

Microwave remote sensing detects changes in the dielectric constants between water, soil, and vegetation to determine soil and water content (Wang and Qu, 2009). Passive microwave remote sensing approaches estimate surface water content by measuring the intensity of microwave emission from soil and vegetation (Moran, 2004). Typical models/indices developed using passive microwave remote sensing include (i) bare soil emission models, (ii) vegetative areas emission models, and (iii) soil and vegetation models (Meesters et al., 2005). Unlike passive microwave remote sensing, active microwave sensors, such as Synthetic Aperture Radar sensors, propagate microwave energy and receive backscattered pulses in different frequencies such as C-band, L-band, and X-band. Using the backscattered pulses, scientists measure the backscattering coefficient to derive the water content of soils and

TABLE 1 **Commonly used optical, thermal, microwave, and integrated remote sensing approaches for assessing agricultural drought.**

Remote sensing approach	Type	Index	Advantages	Disadvantages
Optical remote sensing	Soil drought index	Perpendicular drought index (PDI)	Simple and effective for calculating drought conditions	Unable to provide high accuracy over variable land cover types
	Vegetation drought index	Normalized difference vegetation index (NDVI)	Provides a measure of vegetation health or greenness conditions	Sensitive to dark and wet soil conditions; has lag response to soil moisture
	Vegetation drought index	Moisture stress index (MSI)	More sensitive at canopy level rather than leaf level Applicable for densely vegetated areas	
		Simple ratio water index (SRWI)		
		Normalized difference water index (NDWI)	Effective in monitoring vegetation water content	Uncertainties increased considerably in the presence of soil and sparsely vegetated or bare surfaces
		Normalized difference infrared index (NDII)		
		Land surface water index (LSWI)		
		Vegetation condition index (VCI)	Provides vegetation greenness conditions	Requires data over longer time period

Continued

TABLE 1 Commonly used optical, thermal, microwave, and integrated remote sensing approaches for assessing agricultural drought—cont'd

Remote sensing approach	Type	Index	Advantages	Disadvantages
	Soil and vegetation drought index	Modified perpendicular drought index (MPDI)	Applicable over variable topography, soil types, and ecosystems	Assumption of fixed soil line but highly dependent on the soil type, level of fertilization, and soil moisture
	Soil and vegetation drought index	Shortwave infrared water stress index (SIWSI)	Effective in monitoring vegetation water content	Uncertainties increased considerably in the presence of soil and sparsely vegetated or bare surfaces
		Normalized multiband drought index (NMDI)	Appropriate for estimating both vegetation and soil water content	Requires further investigation over moderately dense vegetation
		Visible and shortwave drought index (VSDI)	Applicable for estimating both vegetation and soil water content	Performs unwell if temperature is more dominant over the precipitation
Thermal remote sensing		Apparent thermal inertia (ATI)	Suitable for bare land areas	Not applicable over vegetated regions
		Temperature condition index (TCI)	Easy to get the required input data	Requires clear-sky conditions at the time of imaging
		Normalized difference temperature index (NDTI)	Able to accurately reflect the spatial-temporal variations of soil moisture	Requires other input variables (e.g., solar radiation, wind speed, and leaf area index) that are not complicated to acquire

Microwave remote sensing		TRMM precipitation condition index (PCI)	Works well at a regional scale under most weather conditions	Low spatial resolution and inability to acquire images at higher latitudes
		Soil moisture condition index (SMCI)		
		Microwave polarization difference index (MPDI)	Able to provide soil moisture and vegetation optical depth	Low spatial resolution and requires further processing for daytime applications
		Normalized backscatter moisture index (NBMI)	Provides high spatial resolution in comparison to PCI, SMCI, and MPDI	Poor temporal resolution and unable to penetrate vegetation canopy
Combined remote sensing-based methods	Combined optical based indices	Normalized difference drought index (NDDI)	Combines both vegetation greenness and wetness conditions	Not applicable for short-term drought monitoring
		Normalized moisture index (NMI)		
	Ts-VIs mathematical approach	Vegetation health index (VHI)	Provides more comprehensive drought monitoring capabilities	Requires appropriate data fusion for VCI and TCI
		Temperature-vegetation index (TVX)	Depicts drought conditions at regional scale	TVX/VSWI slopes may vary in space
		Vegetation water supply index (VWSI)		

Continued

TABLE 1 Commonly used optical, thermal, microwave, and integrated remote sensing approaches for assessing agricultural drought—cont'd

Remote sensing approach	Type	Index	Advantages	Disadvantages
	Ts-VIs scatter plot approach	Vegetation temperature condition index (VTCI)	Works better at regional scale	Unable to calculate: (i) over small study area and (ii) variable topography
		Temperature-vegetation dryness index (TVDI)		
		Water deficit index (WDI)		
		Temperature-vegetation wetness index (TVWI)	Eliminates the issue of variable topography	Unable to calculate over small study area
	Combined microwave and/or optical/thermal approach	Microwave integrated drought index (MIDI)	Applicable over semi-arid regions in particular	Nonsensitive to water content over different levels of vegetation covers
		Scaled drought condition index (SDCI)	Applicable over both arid and humid regions	The coefficients are ecosystem-specific

Adapted from Hazaymeh, K., Hassan, Q.K., 2016. Remote sensing of agricultural drought monitoring: a state of art review. AIMS Environ. Sci. 3(4), 604–630.

vegetation. This is done by exploiting the contrast of the dielectric constants between the three main landcover types, that is, bare soil, vegetation, and water (Moran, 2004). Some of the approaches developed using active microwave remote sensing include theoretical, empirical, and semiempirical approaches (Dubois et al., 1995).

Methodologies for measuring agricultural drought have applied indices that incorporate surface temperature retrieved from remote sensing systems. These applications are based on the principle of thermal inertia, that is, a physical property of materials describing their impedance to temperature change. However, remote sensing applications for agriculture make use of apparent thermal inertia (ATI), which approximates thermal inertia calculated as one minus albedo divided by diurnal temperature range and radiant temperatures (Badarinath and Chand, 2007). Research has shown that temperature-based agriculture drought monitoring indices perform better in sparsely vegetated or bare areas than in densely vegetated areas. Examples of indices that incorporate temperature include the temperature condition index (TCI) and the normalized difference temperature index (NDTI).

Although drought has been commonly assessed using single data source drought indices, the integration of various drought indices from different data sources is regarded as superior to the use of one (Hazaymeh and Hassan, 2016; Li et al., 2024; Wei et al., 2024). Studies integrating meteorological, hydrological, and agricultural drought indices are becoming more common just as those that are exploiting different remote sensing datasets such as optical, thermal, microwave, and in situ data (Du et al., 2013). Thus, the integration of various remote sensing-based indices overcomes the weaknesses associated with each of the indices developed for monitoring agricultural drought conditions.

The growing need to understand the impacts of climate change and variability on cropping systems has made satellite missions a fundamental data source for agriculture monitoring. This has seen the proliferation of both open-source and commercial satellite missions in the form of multispectral, hyperspectral, and radar satellite missions. Within remote sensing, the trend has been toward the launching of very high-resolution nanosatellite constellations, which has increased the availability of remotely sensed images. Moreover, due to the highly dynamic nature of agriculture in space and time, missions are also focusing on developing satellites with high temporal resolution (at least every 8 days) and fine-to-moderate (FTM $<$ 100 m) spatial resolution (Whitcraft et al., 2015). Moreover, the increasing availability of cloud-computing platforms, such as Amazon and Google Earth Engine, have increased capabilities of satellite data processing and information extraction techniques including integration with meteorological, soil, and agriculture-related data (Wu et al., 2023). Table 2 provides an overview of satellites that have been applied in monitoring cropping systems across the globe. The review shows the increasing use of a diverse suite of remote sensing images ranging from optical to microwave.

TABLE 2 Satellite data sources and satellite indices for monitoring agricultural drought across different regions of the world.

Satellite data source	Index	Reference study	Key result	Region
Moderate Resolution Imaging Spectroradiometer (MODIS) NDVI (MODIS13Q1V6)	NDVI, VCI	Moussa Kourouma et al., 2021	Monthly, seasonal, and annual normalized difference vegetation index (NDVI) and vegetation condition index (VCI) were correlated with crop yield to determine the drought impacts on crop productivity.	Ethiopia
Landsat L8 Operational Land Imager/Thermal Infrared Sensor (TIRS)	NDVI, land surface temperature (LST)	Ahmad et al., 2020a,b	The study developed satellite remote sensing-based estimates of maize acreage and statistical-empirical model for prediction of maize yields before assessing the influence of temperature on interannual variability in maize yields across a decade.	Pakistan
GIMMS and MODIS deviation of NDVI	NDVI deviation, VCI	Bhavani et al., 2017	Long-term NDVI, in situ climate and soil moisture data to understand climate variability impact on the agriculture growth/stress, length of the growing period, and projected agriculture NDVI under a future climate.	India

In situ data	SPEI, SPI	Tirivarombo et al., 2018	Long-term rainfall data was used to characterize drought using SPI and SPEI in the Zambezi River Basin in Africa to establish the relationship between agricultural drought and food security. SPI picked moderate to severe droughts while SPEI detected extreme droughts.	Zambezi Basin, southern Africa
MODIS and Tropical Rainfall Measuring Mission (TRMM) satellite	Scaled drought condition index (SDCI) temperature (LST) data and the normalized difference vegetation index (NDVI)	Rhee et al., 2010	The scaled drought condition index (SDCI) which combines satellite-based LST, NDVI, and precipitation data, performed better than existing indices such as NDVI and vegetation health index (VHI) across a rainfall gradient.	USA
(MODIS) Soil Moisture and Ocean Salinity (SMOS) mission	Soil moisture agricultural drought index (SMADI) land surface temperature (LST)	Sánchez et al., 2018	The study integrates remotely sensed datasets of LST and NDVI with the surface soil moisture from the soil moisture and ocean salinity (SMOS) mission at various spatial scales and observed a close match between agricultural droughts detected using SMADI and in situ measurements.	Globe
Satellite Pour l'Observation de la Terre (SPOT) Vegetation (VGT) and Project for On-Board Autonomy-Vegetation (PROBA-V)	Vegetation condition index (VCI)	Sharara et al., 2022	Satellite-derived vegetation condition index (VCI) was used to characterize agriculture and validated, using maize yield and meteorological station-derived standardized precipitation index (SPI).	Zimbabwe

Continued

TABLE 2 Satellite data sources and satellite indices for monitoring agricultural drought across different regions of the world—cont'd

Satellite data source	Index	Reference study	Key result	Region
Multiple sensors	Optical and thermal indices including their integration	Qin et al., 2021	The study reviews four categories of agricultural drought monitoring approaches, i.e., optical, thermal, optical and thermal, and multisource and identifies mechanisms through which drought is detected by each approach.	Global
CHIRPS precipitation Global Inventory Monitoring and Modelling Studies (GIMMS) NDVI; and terrestrial water storage (TWS) from MERRA-2 and GRACE	VCI and terrestrial water storage (TWS)	Agutu et al., 2017	The study showed CHIRPS, GPCC, FLDAS, and VCI were suitable for agricultural drought monitoring/ characterization in east Africa.	East Africa
	Standardized precipitation index (SPI), standardized soil moisture index (SSI), and standardized solar-induced chlorophyll fluorescence index (SSIFI)	Kumar and Chu, 2024	The study integrates precipitation, soil moisture, and solar-induced chlorophyll fluorescence datasets from 2001 to 2020 to characterize drought.	India
Daily MODIS surface reflectance and thermal data	Standard precipitation index (SPI), vegetation health index (VHI), soil drought condition index (SMCI)	Chang et al., 2024	Standard precipitation index (SPI) effectively detected meteorological drought, while the vegetation health index (VHI) coupled with temperature data was optimal for vegetation drought monitoring in the study region.	Kyrgyzstan

3 Climate change adaptation strategies for agricultural systems

Agriculture is likely to be impacted negatively by climate change more than any other economic sector since the sector is climate sensitive (Brown et al., 2012). Agricultural production in most parts of the developing world is heavily reliant on rainfall, yet according to the IPCC, the variability of rainfall and temperature is increasing because of climate change (Team et al., 2014). The negative effects of climate change on agricultural systems are likely to include, among other effects (i) a rapid decline in agricultural productivity resulting in severe food insecurity, (ii) an increase in production costs due to reduced irrigation system efficiency and power generation, and (iii) shift in the geographic distribution of suitable habitat for major crops (Amin et al., 2010). In this regard, there is need for countries to develop adaptation strategies for the sector to deal with the negative impacts of climate change, enhancing the resilience of farmers and reducing their vulnerability.

IPCC (2014) defines adaptation as an adjustment of natural and human systems to actual or expected climatic stimuli or their effects, resulting in the intensification of harmful effects (Adger et al., 2009). In order to reduce the negative effects of climate change, there is a need to adopt adaptation strategies that include a comprehensive program that addresses most issues related to adaptation (Jones and Boyd, 2011). In most developing countries, the agricultural sector can develop adaptation strategies that range from crop and livestock variation, community-based actions, water harvesting and storage, expanding irrigation, water-conservation techniques, and planting of drought-resistant crop varieties (Mugambiwa, 2018). Other strategies include changing sowing dates, increasing crop densities, and optimizing fertilization levels when planting crops. These measures are important in an attempt to make sure agriculture remains the backbone of most developing countries, despite current and anticipated adverse impacts of climate change (Silva, 2018).

Observations from studies in developing countries such as Zimbabwe have suggested that there is a need to integrate indigenous knowledge systems in adaptation planning. This is because the advent of climate change has rendered less applicable to the large body of knowledge within local communities on how they can cope with climate variability. In countries such as Zimbabwe, a key adaptation measure is the implementation of climate-smart agriculture accompanied by the strengthening of agricultural value chains and markets. The government of Zimbabwe is promoting and rolling out gender-sensitive climate-smart agriculture technologies and practices that include land and water resources conservation, sustainable mechanization, agro-ecology, water-efficient irrigation, renewable energy, and energy efficiency, climate adapted crop types/varieties and livestock types/breeds, crop/livestock diversification, agro-forestry, integrated pest and disease management, postharvest technologies, improved livestock management, fodder production and livestock feeding strategies,

Silvi-pastoral systems. These actions are expected to increase adaptive capacity by providing the technology and tools necessary to increase the efficiency of agricultural production. In addition, adaptive capacity would also be increased through the provision of tools to anticipate future changes in climate and adjust production accordingly. This will have the net effect of reducing sensitivity to climate change through expanding the use of climate-resilient breeds of livestock and crop varieties. In most countries, climate change adaptation strategies succeed if they are linked with coordinated efforts to alleviate poverty and enhance food security and water availability. In addition, the strategies need to be linked to combating land degradation and soil erosion and reducing the loss of biological diversity, which will improve the food production chain within the framework of sustainable development.

3.1 Types of adaptation strategies in agriculture

There are different types of adaptation strategies that can be employed in agriculture, and these depend on the timing, objectives, and motivation of implementation. In this case, adaptation strategies can be reactive or anticipatory, private or public, planned or autonomous, as outlined by Chanza (2015). Reactive adaptation is a strategy adopted by farmers after the initial impacts of climate change have occurred, while anticipatory adaptation takes place before impacts become apparent. Private or public adaptation strategies are based on whether adaptation is motivated by private (individual households and companies) or public (government). Planned and autonomous is when adaptation is a result of deliberate policy taking into consideration changing circumstances or an anticipation that things might change so that some action is required to maintain the desired state. This type of adaptation involves changes that take place in agricultural systems in response to changing climate, although policy might not have changed. Fig. 5 provides a summary of the types of adaptation strategies that can be employed in the agricultural system.

3.2 Opportunities and barriers for adapting to climate change in agriculture

The advent of climate change has provided both opportunities and barriers when farmers attempt to adapt to its negative effects. According to Biesbroek et al. (2011), barriers are impediments that individuals face when attempting to adopt climate change adaptation strategies. These barriers, if not properly addressed, will render adaptation strategies ineffective (Kruse et al., 2013). Climate change strategies in agriculture are often hampered by various barriers; hence, the selection of appropriate strategies should be based on barriers-opportunities analyses. Several factors hinder the ability of farmers to adapt to climate change, and these range from social, technological, economic, institutional, informational, and natural barriers (Jones and Boyd, 2011).

		Anticipatory	Reactive
Natural System			• Changes in length of growing season • Changes in ecosystem composition • Wetland migration
Human System	Private	• Purchace of insurance • Construction of houses on stilts • Redesign of oil rigs	• Changes in farm practices • Changes in insurance premiums • Purchase of air-conditioning
	Public	• Early-warning systems • New building codes, design standards • Incentives for relocation	• Compensatory payments, subsidies • Enforcement of building codes • Beach nourishment

FIG. 5 Classification of adaptation strategies depending on whether they are anticipatory or reactive and whether they are public or private. *(Adapted from Field, C.B., Barros, V.R., Dokken, D.J., Mach, K.J., Mastrandrea, M.D., Bilir, T.E., et al. (Eds.), Climate Change 2014: Impacts, Adaptation, and Vulnerability. Part A: Global and Sectoral Aspects. Contribution of Working Group II to the Fifth Assessment Report of the Intergovernmental Panel on Climate Change. Cambridge University Press, Cambridge, UK and New York, NY, p. 1132.)*

A systematic typology of barriers to climate change is provided by the, and these include lack of knowledge, awareness, and technology; the physical environment; biological tolerances; economic factors; financial factors; human resources; social and cultural factors; and governance and institutional processes (Klein et al., 2014). The barriers to climate change adaptation in agriculture include (a) land tenure systems resulting in constraints on land and its availability, (b) limited access to climate change information and agricultural extension service, (c) high costs of farm inputs and equipment, (d) high cost of irrigation facilities, (e) poor government responsiveness to climate change risk management, (f) credit constraints, (g) labor constraints, and (h) low income from sale of agriculture products (Kang et al., 2009). Based on the outlined constraints, farmers need to be supported for them to be able to effectively adapt to the negative impacts of climate change.

One of the major barriers to adapting to climate change in most developing countries is high levels of poverty among the population (Gifford et al., 2011). Most farmers do not have adequate economic resources which can anchor their ability to adapt to climate change. This is exacerbated by the fact that in most developing countries, social protection systems are weak, and farmers have limited access to credit or loan facilities which could improve their financial reserves. In addition to high levels of poverty, there is limited access to climate change adaptation information. In most developing countries, climate change information is scant, and where it is available, it is marred by mixed perceptions as rural farmers grapple to understand the concept and its dynamics. In a study in some districts of Zimbabwe, for example, Chivi and Umzingwane Zimbabwe, lack of adequate information about climate change has resulted in limited adoption of adaptation actions as farmers are unaware of what they need to do to address their plight.

The limited access to climate change information is worsened by the fact that the mode and format of dissemination do not suit the literacy prowess of most rural farmers, for example, in Umzingwane, Gwanda, and Tsholotsho districts (Soropa et al., 2015; Ndlovu et al., 2020). Furthermore, the seasonal forecasting information is too coarse and hence not suited to enable most rural farmers to adjust their cropping patterns Kupika et al. (2019). Moreover, most rural areas are characterized by inadequate extension services, which act as a barrier to climate change adaptation. Adequate and effective agricultural extension services may result in smallholder farmers being equipped with requisite knowledge of climate change and its impacts including the appropriate adaptation actions to be implemented in each area (Zvamasiya et al., 2017). This is also worsened by the lack of adequate institutions that support adaptation. These institutions, when available, are the backbone of the success of adaptation actions.

Another important factor that acts as a barrier to climate change adaptation in the agricultural sector is gender inequalities, which are highly prevalent in most developing countries (Nyahunda and Tirivangasi, 2020). In most developing countries with weak socioeconomic women's rights and gender, inequalities are important in determining the ability of communities to adapt to the impacts of climate change, particularly between men and women. There is a large disparity between men and women, and in most cases, women are at a disadvantage and find themselves more reliant on natural resources for food and income. This makes them more vulnerable to climate change-induced stresses (Bhatasara, 2018).

Although there are several barriers, as outlined above, the implementation of adaptation strategies in the agricultural sector may present opportunities across the value chain. For example, the implementation of some adaptation measures may result in agricultural systems that are more resilient to climate variability. For example, the development of new crop varieties that are drought-resistant may reduce crop losses during hot, dry summers (Nciizah et al., 2021). In addition, since climate change requires dynamic thinking, existing institutions and the public may need to change so that they are more flexible and able to respond to changing conditions (Nciizah et al., 2021) (Box 1).

4 Climate change mitigation in the agriculture sector

Climate change mitigation involves all activities aimed at reducing greenhouse gas emissions (Oguge et al., 2024). Agriculture contributes 10%–12% cent of the total global anthropogenic GHG emissions or about 6.8 Gt of CO_2 per year (Verchot et al., 2009). In most agricultural systems, deforestation and land use

BOX 1 Climate change adaptation strategies in Nyanga

Case studies highlighting successful adaptation strategies in agriculture

Nyanga is in the semi-arid climatic region of Zimbabwe; hence, adaptation and mitigation strategies for the benefit of communities. The rainfall season in most parts of Nyanga has become uncertain, making it difficult to decide on the start of the season and selection of the appropriate crop varieties (Mushore et al., 2017). The quality of the agricultural season is also declining through the increasing frequency of mid-season dry spells. Rising temperatures combined with declining rainfall led to an increase in evapotranspiration, resulting in a high soil moisture deficit, hence affecting the proper maturity of crops (Manatsa et al., 2017).

In some of the wards of Nyanga district, smallholder farmers rely on production of small grains which are drought-resistant, growing of short, seasoned varieties, multiple cropping, livelihood diversification, dry planting, and early planting. Most of the adaptation strategies are less costly and easily infused with indigenous knowledge systems hence they are favored by most farmers. The adaptation strategies are broadly divided into ethno-science and techno-science (Matanga and Jere, 2011).

Ethno-science techniques are based on local indigenous knowledge of their physical environment while techno-science involves the adoption of modern technologies. The Ethno-science adaptation measures include growing drought-tolerant crops such as small grains, multiple planting, early planting, and barter trade. The drought-tolerant crops grown include millet, sorghum, and rapoko. Multiple cropping involves growing a mix of crops such as pumpkins, maize and beans, cow peas, sorghum, and rapoko at the same time. This helps to improve soil fertility as legumes are nitrogen-fixing and create soil cover that retains moisture.

The local farmers have an in-depth knowledge of their agroecological conditions, and they use indigenous indicators to determine the planting dates for their crops. Farmers have also found it feasible to do dry planting to counter uncertainty at the start of the season. Farmers have prepared land during the dry season, around September and October. This is done to ensure that the crops sprout when the first rains are received. Once they do dry farming, farmers can then focus on off-farm activities such as petty trade. Another strategy is barter trade, particularly in exchange for food items in times of deficit or poor crop yields. Farmers in regions where small grains do well exchange with those in areas where maize does well or exchange small grains with basic groceries. In extreme cases, poultry or livestock are exchanged for crops. Farmers in some parts of the Nyanga district have also embraced livelihood diversification even during periods of low agricultural output. The diversification activities include providing casual labor and selling in areas around the Nyanga forests.

change are responsible for an estimated 17% of the total global emissions (Solomon et al., 2007). Across most parts of the world, emissions from agriculture are expected to grow as demand for food increases, driven by population and income growth. As a result, the agricultural sector needs to contribute to emission reduction by decoupling production through actions that reduce food losses (Wiebe et al., 2019). Mitigation in agriculture involves deliberate human intervention aimed at reducing the sources or enhancing the sinks of greenhouse gases (Solomon et al., 2007). In addition to land use change, agricultural activities that require intervention include deforestation and soil organic carbon depletion, livestock production, rice cultivation, fertilizer application, and manure management (Rosenzweig and Tubiello, 2007). There is, therefore, a need to modify the current management of agricultural systems to help mitigate global anthropogenic emissions (Rosenzweig and Tubiello, 2007).

In most agricultural systems, mitigation options include actions aimed at reducing deforestation and forest degradation while at the same time enhancing sequestration through increasing afforestation and reforestation (Reyer et al., 2009). In addition, there is a need for interventions that enhance forest management to maintain or increase forest carbon density. Mitigation options in agriculture that help to reduce climate change variability impacts are outlined in the following section.

4.1 Cropland emission mitigation

This includes crop diversification and cultivation of legume crops under crop rotation (Liu et al., 2022). Crops and residues from agriculture can be used as a source of fuel, either directly or after conversion to fuels such as ethanol or diesel. Emissions of GHGs, notably CO_2, can also be reduced by agricultural management practices that forestall the opening up of new areas for agriculture.

4.2 Enhancing soil carbon sequestration

This option offers more benefits for biodiversity, soil fertility, and productivity by increasing concentration of carbon in the soil through better soil management practices. The action results in soil stabilization and increases food production resulting from the optimization of synthetic fertilizer utilization and reversal of land degradation (Takimoto et al., 2008). The large reserves of carbon most found in soil organic matter can be removed from the atmosphere by increasing the photosynthetic input of carbon (Smith et al., 2008). In countries such as Zimbabwe, GHGs from agricultural soils are being reduced by adopting conservation agriculture. The adoption of conservation agriculture (CA) is part of climate-smart agriculture strategies for reducing GHG emissions. In fact, CA has the potential to sequester up to the equivalent of 32% of all man-made GHG emissions through the reduction in the use of compound fertilizers in favor of organic farm-made manure (Brazier, 2015; O'Dell et al., 2020).

4.3 Livestock management

There is a need to improve livestock waste management to reduce the emission of CH_4 and N_2O through ruminant livestock management (Montes et al., 2013).

Practices for reducing CH_4 emissions from enteric fermentation include improved feeding practices, management changes, and animal breeding. Methane emissions can be reduced by feeding using more concentrates and replacing forages. Other practices include improving pasture quality, especially in developing countries, as it improves animal productivity and reduces the proportion of energy lost as CH_4 (Petersen et al., 2013). In addition to reducing emissions from livestock, there is a need to reduce emissions through manure management. To mitigate the emission of methane from manure, the manure can also be digested anaerobically to maximize retrieval of CH_4 as an energy source. It is also possible to store and handle the manure in solid rather than liquid form, which is known to suppress CH_4 emissions, although it may increase N_2O formation.

In regions that produce rice on a large scale, methane emissions need to be reduced to mitigate climate change. To mitigate emissions in rice-producing regions, there is a need to use cultivars with low methane exudation rates, adjust the timing of residue incorporation, and produce biochar from rice straw and husk to avoid GHG during decomposition (Jiang et al., 2017). In most savanna ecosystems, wildfire is a major source of emissions; hence, there is a need for prescribed burning (Volkova et al., 2021). To reduce GHGs emitted through fire burning, there is a need to reduce the frequency or extent of fires through more effective fire suppression mechanisms. In addressing the problem of fires, Zimbabwe has heavily relied on law enforcement and suppression rather than adopting a holistic community-based approach to fire management.

5 Case studies showcasing mitigation strategies in agriculture

In most parts where agriculture is the back-bone of the economy, there is a high potential to mitigate greenhouse gas (GHG) emissions. This can be achieved through improved cropland and grazing land management including restoration of degraded lands (Smith et al., 2008). The opportunities for agriculture to implement mitigation strategies are based on the contribution of mitigation practices to food security and climate change adaptation (Wilkes et al., 2013). Mitigation strategies are faced with barriers that hinder the implementation of proposed actions due to a number of factors, which range from technological to behavioral.

Mitigation strategies in agriculture are aimed at addressing farmers' priority concerns, while taking into consideration diversity among farmers (Darnhofer et al., 2010). Food security, high income, and profitability are the primary concerns of farmers. Hence, there is a need to ensure that mitigation strategies address the primary concerns of farmers to have a higher chance of success. In addition, there is also a need to take equity considerations into account. In most developing countries, the focus is on promoting agricultural modernization, which includes the commercialization of agriculture as a means of increasing the sector's contribution to economic growth, rural employment, and income generation. The adoption of mitigation practices under such situations is possible for large-scale commercial operations (Wang et al., 2023). Mitigation strategies should target specific characteristics of individual farmers (e.g., wealth levels, gender, ethnicity, and age), which may limit their ability to adopt and benefit from the promoted practices. If mitigation plans are poorly designed or implemented, they may even increase inequality of opportunity and outcomes among farmers.

The analysis of the relationship between the specific financing needs in the agriculture sector and suitable forms of financial or policy support is another aspect of overcoming barriers to mitigate climate change. Some of the barriers to the successful implementation of mitigation strategies include obsolete, lack of technical expertise on certain technologies, lack of appropriate measuring/data capturing equipment, lack of willingness to share information, especially by the private sector, and heavy reliance on imported technologies.

6 Policy frameworks for climate change and agriculture

Policies and institutional frameworks that support adaptation and mitigation in agriculture are derived from the UNFCCC, where countries are obliged to be signatories (McCarl, 2007). The UN framework was formed in 1992 to facilitate national adaptation and mitigation strategies in various sectors including agriculture (Kojwang and Larwanou, 2015). This provides an international reference framework for different countries to address and mainstream climate change into national policies. In this way, developing countries would build the adaptive to reduce the impacts of climate change on their economies. Other related conventions such as the Convention on Biodiversity and the United Nations Convention on Combating Desertification provide policy and institutional support for climate change adaptation and mitigation in agriculture (Kutter, 2015).

The international institutional support framework presented under UNFCCC offers opportunities for different governments to develop climate change policies related to adaptation and mitigation in agriculture (Burton et al., 2007). Hence, countries can develop climate policies and national adaptation plans (NAPs) to address climate change issues in agriculture. The designing of policies for dealing with climate change in agriculture offers great opportunities to address adaptation and mitigation in the sector. Most developing countries developed NAPs that outline the requirements for a country to adapt (Kim et al., 2017). The different conventions are important for adaptation and mitigation as they address global problems and offer opportunities for developing countries to access finance for mitigation and adaptation.

7 Gaps and opportunities for policy and institutional support for climate change adaptation and mitigation in agriculture

Although there are many international conventions that offer opportunities for adaptation and mitigation in the agricultural sector, gaps exist that limit their full potential. At the national level, various policy gaps exist, which makes it difficult to mainstream climate change into development sectors. One of the gaps is the nonexistence of climate policies in some developing countries. This makes it difficult for different agencies involved in climate change adaptation and mitigation to coordinate in addressing the problem. This also results in most governments not dedicating government-wide support to facilitate adaptation and mitigation. Through the development of a climate policy, the government will be forced to make provisions for integrating climate change and mitigation into national budgetary allocations (Djoghlaf, 2009). This is because lack of finance is one of the challenges to implementing climate adaptation and mitigation measures. The lack of national climate policies partially explains why the greater part of the population in developing countries still heavily depends on agriculture despite its high vulnerability to climate change (Ribot, 2013). A coherent climate policy will aim to reduce dependency on agriculture through human capacity and skills development to allow the population to move to secondary economic sectors.

In addition, the uncertainties related to the projected impacts and vulnerabilities of climate change leave people in a dilemma, creating gaps in the development and implementation of climate change adaptation and mitigation strategies in agriculture. The lack of spatially detailed socioeconomic information to understand vulnerability at the subnational level makes it difficult for people to understand the impacts of climate change on their livelihoods (Parker et al., 2019). Moreover, the weak-research-policy linkages in most developing countries have created gaps in climate change adaptation and mitigation in agriculture as cross-cutting issues such as HIV/AIDS, poverty, gender inequality, and food insecurity are poorly addressed.

8 Conclusion

This chapter has explored the concept of climate variability and change and how it impacts the agricultural sector across several systems. It then highlights the basis of using remote sensing for the detection of climate variability impacts on agriculture, focusing on the soil moisture availability and vegetation conditions. The chapter evaluated the performance of univariate and multivariate indices for crop monitoring, focusing on the remote sensing approach used: optical, thermal, microwave, and integrated indices. We also evaluated the performance of each of the different indices that are used for monitoring crop conditions as influenced by climate variability. The chapter concludes by proffering mitigation options for the agricultural sector drawn from global experiences.

References

Adger, W.N., Dessai, S., Goulden, M., Hulme, M., Lorenzoni, I., Nelson, D.R., Wreford, A., 2009. Are there social limits to adaptation to climate change? Clim. Change 93, 335–354.

Agutu, N.O., Awange, J.L., Zerihun, A., Ndehedehe, C.E., Kuhn, M., Fukuda, Y., 2017. Assessing multi-satellite remote sensing, reanalysis, and land surface models' products in characterizing agricultural drought in East Africa. Remote Sens. Environ. 194, 287–302.

Ahmad, M.J., Iqbal, M.A., Choi, K.S., 2020a. Climate-driven constraints in sustaining future wheat yield and water productivity. Agric Water Manag 231, 105991.

Ahmad, I., Singh, A., Fahad, M., Waqas, M.M., 2020b. Remote sensing-based framework to predict and assess the interannual variability of maize yields in Pakistan using Landsat imagery. Comput. Electron. Agric. 178, 105732.

Amin, A.Q., Nurul Azam, M., et al., 2010. Policy challenges towards potential climate change impacts: in search of agro-climate stability. Sci. Res. Essays 5 (18), 2681–2685.

Asseng, S., Ewert, F., Martre, P., Rötter, R.P., Lobell, D.B., Cammarano, D., Kimball, B.A., Ottman, M.J., Wall, G.W., White, J.W., Reynolds, M.P., Alderman, P.D., Prasad, P.V.V., Aggarwal, P. K., Anothai, J., Basso, B., Biernath, C., Challinor, A.J., De Sanctis, G., Doltra, J., Fereres, E., Garcia-Vila, M., Gayler, S., Hoogenboom, G., Hunt, L.A., Izaurralde, R.C., Jabloun, M., Jones, C.D., Kersebaum, K.C., Koehler, A.-K., Müller, C., Naresh Kumar, S., Nendel, C., O'Leary, G., Olesen, J.E., Palosuo, T., Priesack, E., Eyshi Rezaei, E., Ruane, A.C., Semenov, M.A., Shcherbak, I., Stockle, C., Stratonovitch, P., Streck, T., Supit, I., Tao, F., Thorburn, P.J., Waha, K., Wang, E., Wallach, D., Wolf, J., Zhao, Z., Zhu, Y., 2015. Rising temperatures reduce global wheat production. Nat. Clim. Change 5 (2), 143–147. https://doi.org/10.1038/nclimate2470.

Badarinath, K.V.S., Chand, T.K., 2007. Analysis of apparent thermal inertia over different land use/ land cover types using ENVISAT AATSR data. J. Indian Soc. Remote Sens. 35, 185–191.

Bett, B., Kiunga, P., Gachohi, J., Sindato, C., Mbotha, D., Robinson, T., Grace, D., 2017. Effects of climate change on the occurrence and distribution of livestock diseases. Prev. Vet. Med. 137, 119–129.

Bhatasara, S., 2018. Understanding adaptation to climate variability in smallholder farming systems in eastern Zimbabwe: a sociological perspective. Rev. Agric. Food Environ. Stud. 99 (2), 149–166.

Bhavani, P., Roy, P.S., Chakravarthi, V., Kanawade, V.P., 2017. Satellite remote sensing for monitoring agriculture growth and agricultural drought vulnerability using long-term (1982–2015) climate variability and socio-economic data set. Proc. Natl. Acad. Sci. India Sect. A: Phys. Sci. 87, 733–750.

Biesbroek, R., Klostermann, J., et al., 2011. Barriers to climate change adaptation in the Netherlands. Clim. Law 2 (2), 181–199.

Boyer, J.S., 1982. Plant productivity and environment. Science 218, 443–448.

Brazier, A., 2015. Climate Change in Zimbabwe. A Guide for Planners and Decision-Makers. Konrad Adenauer Stiftung.

Brown, D., Rance Chanakira, R., et al., 2012. Climate Change Impacts, Vulnerability and Adaptation in Zimbabwe, IIED Climate Change Working Paper No. 3.

Burton, M.R., Mader, H.M., Polacci, M., 2007. The role of gas percolation in quiescent degassing of persistently active basaltic volcanoes. Earth Planet. Sci. Lett. 264 (1–2), 46–60.

Chang, S., Isaev, E., Chen, H., Wu, B., Meng, J., Yan, N., Ma, Z., 2024. Exploring the linkages between different types of drought and their impacts on crop production in Kyrgyzstan. IEEE J. Sel. Top. Appl. Earth Obs. Remote Sens. 17, 4566–4580. https://doi.org/10.1109/JSTARS.2024.3359429.

Chanza, N., 2015. Indigenous knowledge and climate change: insights from Muzarabani, Zimbabwe. Doctoral dissertation, Nelson Mandela Metropolitan University.

Chen, C., Pang, Y., 2020. Response of maize yield to climate change in Sichuan province, China. Glob. Ecol. Conserv. 22, e00893.

Chen, Z., Wang, W., Fu, J., 2020. Vegetation response to precipitation anomalies under different climatic and biogeographical conditions in China. Sci. Rep. 10 (1), 830.

Dai, A., 2011. Drought under global warming: a review. Rev. Clim. Change 2 (1), 45–65. https://doi.org/10.1002/wcc.81.

Darnhofer, I., Bellon, S., Dedieu, B., Milestad, R., 2010. Adaptiveness to enhance the sustainability of farming systems. A review. Agron. Sustain. Dev. 30, 545–555.

Dikshit, A., Pradhan, B., Huete, A., Park, H.-J., 2022. Spatial based drought assessment: where are we heading? A review on the current status and future. Sci. Total Environ. 844, 157239.

Delgado, J.A., Short Jr., N.M., Roberts, D.P., Vandenberg, B., 2019. Big data analysis for sustainable agriculture on a geospatial cloud framework. Front. Sustain. Food Syst. 3, 54.

Djoghlaf, A., 2009. Climate change and biodiversity loss: the great new threats to peace, security and development. In: Climate Change and Energy Insecurity. Routledge, p. 110.

Du, L., Tian, Q., Yu, T., Meng, Q., Jancso, T., Udvardy, P., Huang, Y., 2013. A comprehensive drought monitoring method integrating MODIS and TRMM data. Int. J. Appl. Earth Obs. Geoinf. 23, 245–253.

Dubois, P.C., van Zyl, J., Engman, T., 1995. Measuring Soil Moisture with Imaging Radars. IEEE Trans. Geosci. Remote Sens. 33, 915–926.

Ferrara, V., Lindberg, J., 2023. Climate and environmental change perceptions: a case from rural Sicily, Italy. In: Routledge Handbook of Climate Change Impacts on Indigenous Peoples and Local Communities. Routledge, pp. 109–123.

Frischen, J., Meza, I., Rupp, D., Wietler, K., Hagenlocher, M., 2020. Drought risk to agricultural systems in Zimbabwe: a spatial analysis of hazard, exposure, and vulnerability. Sustainability 12 (3), 752.

Gifford, R., Kormos, C., et al., 2011. Behavioral dimensions of climate change: drivers, responses, barriers, and interventions. Rev. Clim. Change 2 (6), 801e827.

Godde, C.M., Mason-D'Croz, D., Mayberry, D.E., Thornton, P.K., Herrero, M., 2021. Impacts of climate change on the livestock food supply chain; a review of the evidence. Glob. Food Sec. 28, 100488.

Habte, M., Eshetu, M., Maryo, M., Andualem, D., Legesse, A., 2022. Effects of climate variability on livestock productivity and pastoralists perception: the case of drought resilience in Southeastern Ethiopia. Vet. Anim. Sci. 16, 100240.

Hatfield, J.L., Antle, J., Garrett, K.A., Izaurralde, R.C., Mader, T., Marshall, E., Nearing, M., Philip Robertson, G., Ziska, L., 2020. Indicators of climate change in agricultural systems. Clim. Change 163, 1719–1732.

Hatfield, J.L., Boote, K.J., Kimball, B.A., Ziska, L.H., Izaurralde, R.C., Ort, D., Thomson, A.M., Wolfe, D., 2011. Climate impacts on agriculture: implications for crop production. Agron. J. 103 (2), 351–370.

Hazaymeh, K., Hassan, Q.K., 2016. Remote sensing of agricultural drought monitoring: a state of art review. AIMS Environ. Sci. 3 (4), 604–630.

IPCC, 2014. Summary for policymakers. Climate Change 2014: Impacts, Adaptation, and Vulnerability. Part A: Global and Sectoral Aspects. Contribution of Working Group II to the Fifth Assessment Report of the Intergovernmental Panel on Climate Change. Cambridge University Press, Cambridge, UK and New York, NY.

Izaurralde, R.C., Thomson, A.M., Morgan, J.A., Fay, P.A., Polley, H.W., Hatfield, J.L., 2011. Climate impacts on agriculture: implications for forage and rangeland production. Agron. J. 103 (2), 371–381.

Jiang, Y., van Groenigen, K.J., Huang, S., Hungate, B.A., van Kessel, C., Hu, S., Zhang, J., Wu, L., Yan, X., Wang, L., Chen, J., 2017. Higher yields and lower methane emissions with new rice cultivars. Glob. Chang. Biol. 23 (11), 4728–4738.

Jones, L., Boyd, E., 2011. Exploring social barriers to adaptation: insights from Western Nepal. Glob. Environ. Chang. 21 (4), 1262–1274.

Kang, Y., Khan, S., et al., 2009. Climate change impacts on crop yield, crop water productivity and food security, a review. Prog. Nat. Sci. 19 (12), 1665–1674.

Kim, Y., Smith, J.B., Mack, C., Cook, J., Furlow, J., Njinga, J.L., Cote, M., 2017. A perspective on climate-resilient development and national adaptation planning based on USAID's experience. Clim. Dev. 9 (2), 141–151.

Klein, R.J.T., Midgley, G.F., et al., 2014. Adaptation Opportunities, Constraints, and Limits. Cambridge University Press, Cambridge.

Kojwang, H.O., Larwanou, M., 2015. An overview of nationally appropriate mitigation actions (NAMAs) and national adaptation programmes of action (NAPAs) in Africa. Int. For. Rev. 17 (3), 103–113.

Kruse, J., Simon, J., et al., 2013. Soil respiration and soil organic matter decomposition in response to climate change. Dev. Environ. Sci. 13, 131–149.

Kumar, V., Chu, H.J., 2024. Seasonal drought severity identification using a modified multivariate index: a case study of Indo-Gangetic Plains in India. J. Hydrol. 629, 130632.

Kupika, O.L., Gandiwa, E., et al., 2019. Local ecological knowledge on climate change and ecosystem-based adaptation strategies promote resilience in the middle Zambezi biosphere reserve, Zimbabwe. Scientifica 1.

Kutter, A., 2015. The United Nations Convention to Combat Desertification: Policies and Programs for Implementation. EOLSS.

Li, J., Li, Y., Yin, L., Zhao, Q., 2024. A novel composite drought index combining precipitation, temperature and evapotranspiration used for drought monitoring in the Huang-Huai-Hai Plain. Agric. Water Manag. 291, 108626.

Liu, Z., Yang, X., Chen, F., Wang, E., 2013. The effects of past climate change on the northern limits of maize planting in Northeast China. Clim. Change 117, 891–902.

Liu, X., Zhu, X., Pan, Y., Li, S., Liu, Y., 2016. Ma Agricultural drought monitoring: progress, challenges, and prospects. J. Geogr. Sci. 26 (6), 750–767. https://doi.org/10.1007/s11442-016-1297-9.

Liu, C., Plaza-Bonilla, D., Coulter, J.A., Kutcher, H.R., Beckie, H.J., Wang, L., Floc'h, J.B., Hamel, C., Siddique, K.H., Li, L., Gan, Y., 2022. Diversifying crop rotations enhances agroecosystem services and resilience. Adv. Agron. 173, 299–335.

Lobell, D.B., Schlenker, W., Costa-Roberts, J., 2011. Climate trends and global crop production since 1980. Science 333, 616–620.

Oguge, O., Nyamondo, J., Adera, N., Okolla, L., Okoth, B., Anyango, S., Berhane, K., 2024. Fine particulate matter air pollution and health implications for Nairobi, Kenya. Environ. Epidemiol. 8 (3), e307.

Ortiz, A.M.D., Outhwaite, C.L., Dalin, C., Newbold, T., 2021. A review of the interactions between biodiversity, agriculture, climate change, and international trade: research and policy priorities. One Earth 4 (1), 88–101.

Maes, W.H., 2012. Estimating evapotranspiration and drought stress with ground-based thermal remote sensing in agriculture: a review. J. Exp. Bot. 63, 695–709. https://doi.org/10.1093/jxb/err313.

Malhi, G.S., Kaur, M., Kaushik, P., 2021. Impact of climate change on agriculture and its mitigation strategies: a review. Sustainability 13 (3), 1318.

Manatsa, D., Mushore, T., et al., 2017. Improved predictability of droughts over southern Africa using the standardized precipitation evapotranspiration index and ENSO. Theor. Appl. Climatol. 127, 259–274.

Matanga, E., Jere, S., 2011. The Effectiveness of Ethno-Science Based Strategies in Drought Mitigation in Mberengwa District of Southern Zimbabwe. University of Pennsylvania, Clarion.

McCarl, B.A., 2007. Adaptation Options for Agriculture, Forestry and Fisheries. A Report to the UNFCCC Secretariat Financial and Technical Support Division. United Nations Framework Convention on Climate Change.

Meesters, A.G.C.A., De Jeu, R.A.M., Owe, M., 2005. Analytical derivation of the vegetation optical depth from the microwave polarization difference index. IEEE Geosci. Remote Sens. Lett. 2, 121–123.

Montes, F., Meinen, R., Dell, C., Rotz, A., Hristov, A., Oh, J., Dijkstra, J., 2013. Mitigation of methane and nitrous oxide emissions from animal operations: II. A review of manure management mitigation options. J. Anim. Sci. 91 (11), 5070–5094.

Moran, M., 2004. Thermal infrared measurement as an indicator of planet ecosystem health. Therm. Remote Sens. Land Surf. Process. 103, 257–282.

Moussa Kourouma, J., Eze, E., Negash, E., Phiri, D., Vinya, R., Girma, A., Zenebe, A., 2021. Assessing the spatio-temporal variability of NDVI and VCI as indices of crops productivity in Ethiopia: a remote sensing approach. Geomat. Nat. Haz. Risk 12 (1), 2880–2903.

Mugambiwa, S., 2018. Adaptation measures to sustain indigenous practices and the use of indigenous knowledge systems to adapt to climate change in Mutoko rural district of Zimbabwe. Jamba 10 (1).

Mushore, T., Manatsa, D., et al., 2017. Investigating the implications of meteorological indicators of seasonal rainfall performance on maize yield in a rain-fed agricultural system: case study of Mt. Darwin District in Zimbabwe. Theor. Appl. Climatol. 129, 1167–1173.

Nciizah, T., Nciizah, E., et al., 2021. Role of Small Grains in Adapting to Climate Change: Zvishavane District. Zimbabwe, Geneva, USK Library.

Ndlovu, S., Mathe, B., et al., 2020. Factoring water harvesting into climate change adaptation: endogenous responses by smallholder farmers in Gwanda District, Zimbabwe. Cogent Soc. Sci. 6 (1), 1784652.

Nyahunda, L., Tirivangasi, H.M., 2020. Effects of climate change on rural women in Makhado municipality, Limpopo Province, South Africa. Gender Behav. 16 (4), 12013–12020.

O'Dell, D., Eash, N.S., et al., 2020. Conservation agriculture as a climate change mitigation strategy in Zimbabwe. Int. J. Agric. Sustain. 18 (3), 250–265.

Parker, L., Bourgoin, C., Martinez-Valle, A., Läderach, P., 2019. Vulnerability of the agricultural sector to climate change: the development of a pan-tropical Climate Risk Vulnerability Assessment to inform sub-national decision making. PLoS One 14 (3), e0213641.

Parry, M.L., 2007. Climate Change 2007 – Impacts, Adaptation and Vulnerability: Working Group II Contribution to the Fourth Assessment Report of the IPCC. Cambridge University Press, London and New York.

Petersen, S.O., Blanchard, M., Chadwick, D., Del Prado, A., Edouard, N., Mosquera, J., Sommer, S. G., 2013. Manure management for greenhouse gas mitigation. Animal 7 (s2), 266–282.

Pruski, F.F., Nearing, M.A., 2002. Climate-induced changes in erosion during the 21st century for eight US locations. Water Resour. Res. 38 (12), 34–41.

Qin, Q., Wu, Z., Zhang, T., Sagan, V., Zhang, Z., Zhang, Y., Zhang, C., Ren, H., Sun, Y., Xu, W., Zhao, C., 2021. Optical and thermal remote sensing for monitoring agricultural drought. Remote Sens. (Basel) 13 (24), 5092.

Quaye, F., Denis, N., Valentina, H., 2018. Climate change impacts on farmland values in the Southeast United States. Sustainability 10 (10), 346. https://doi.org/10.3390/su10103426.

Ray, D.K., West, P.C., Clark, M., Gerber, J.S., Prishchepov, A.V., Chatterjee, S., 2019. Climate change has likely already affected global food production. PLoS One 14 (5), e0217148.

Raza, A., Razzaq, A., Mehmood, S.S., Zou, X., Zhang, X., Lv, Y., Xu, J., 2019. Impact of climate change on crops adaptation and strategies to tackle its outcome: a review. Plants 8 (2), 34.

Reyer, C., Guericke, M., Ibisch, P.L., 2009. Climate change mitigation via afforestation, reforestation and deforestation avoidance: and what about adaptation to environmental change? New Forests 38, 15–34.

Rhee, J., Im, J., Carbone, G.J., 2010. Monitoring agricultural drought for arid and humid regions using multi-sensor remote sensing data. Remote Sens. Environ. 114 (12), 2875–2887.

Ribot, J., 2013. Vulnerability does not just fall from the sky: toward multi-scale pro-poor climate policy. In: Handbook on Climate Change and Human Security. Edward Elgar Publishing, pp. 164–199.

Rosenzweig, C., Tubiello, F.N., 2007. Adaptation and mitigation strategies in agriculture: an analysis of potential synergies. Mitig. Adapt. Strat. Glob. Chang. 12, 855–873.

Sánchez, N., González-Zamora, Á., Martínez-Fernández, J., Piles, M., Pablos, M., 2018. Integrated remote sensing approach to global agricultural drought monitoring. Agric. For. Meteorol. 259, 141–153.

Schlenker, W, Roberts, M.J., 2009. Nonlinear temperature effects indicate severe damages to US crop yields under climate change. Proc. Natl. Acad. Sci. U. S. A. 106 (37), 15594–15598.

Sharara, A., Shekede, M.D., Gwitira, I., Masocha, M., Dube, T., 2022. Fine-scale multi-temporal and spatial analysis of agricultural drought in agro-ecological regions of Zimbabwe. Geomat. Nat. Haz. Risk 13 (1), 1342–1365.

Silva, G., 2018. Feeding the World in 2050 and Beyond. Part 1: Productivity Challenges. Michigan. https://www.canr.msu.edu/news/feeding-the-world-in-2050-and-beyond-part-1.

Siswanto, S., Wardani, K.K., Purbantoro, B., Rustanto, A., Zulkarnain, F., Anggraheni, E., Dewanti, R., Nurlambang, T., Dimyati, M., 2022. Satellite-based meteorological drought indicator to support food security in Java Island. PLoS One 17 (6), e0260982. https://doi.org/10.1371/journal.pone.0260982.

Smith, P., Fang, C., et al., 2008. Impact of global warming on soil organic carbon. Adv. Agron. 97, 1–43.

Solomon, S., Qin, D., Manning, M., Chen, Z., Marquis, M., Averyt, K., Tignor, M., Miller, H., 2007. IPCC fourth assessment report (AR4). Clim. Change 374.

Soropa, G., Gwatibaya, S., et al., 2015. Indigenous knowledge system weather forecasts as a climate change adaptation strategy in smallholder farming systems of Zimbabwe: case study of Murehwa, Tsholotsho and Chiredzi districts. Afr. J. Agric. Res. 10 (10), 1067–1075.

Sultan, B., Defrance, D., Iizumi, T., 2019. Evidence of crop production losses in West Africa due to historical global warming in two crop models. Sci. Rep. 9 (1), 12834.

Takimoto, A., Nair, P.K.R., et al., 2008. Carbon stock and sequestration potential of traditional and improved agro-forestry systems in the West African Sahel. Agric. Ecosyst. Environ. 125, 159–166.

Tao, F., Yokozawa, M., Liu, J., Zhang, Z., 2008. Climate–crop yield relationships at provincial scales in China and the impacts of recent climate trends. Clim. Res. 38 (1), 83–94.

Team, C.W., Pachauri, R.K., et al., 2014. Climate change 2014: synthesis report. In: Contribution of Working Groups I. II and III to the Fifth Assessment Report of the Intergovernmental Panel on Climate Change. IPCC, Geneva.

Thornton, P., Nelson, G., Mayberry, D., Herrero, M., 2021. Increases in extreme heat stress in domesticated livestock species during the twenty-first century. Glob. Chang. Biol. 27 (22), 5762–5772.

Tirivarombo, S.O.D.E., Osupile, D., Eliasson, P., 2018. Drought monitoring and analysis: standardised precipitation evapotranspiration index (SPEI) and standardised precipitation index (SPI). Phys. Chem. Earth A/B/C 106, 1–10.

Ulrichs, M., Slater, R., Costella, C., 2019. Building resilience to climate risks through social protection: from individualised models to systemic transformation. Disasters 43, S368–S387.

Van Loon, A.F., Van Lanen, H.A., 2012. A process-based typology of hydrological drought. Hydrol. Earth Syst. Sci. 16 (7), 1915–1946.

Verchot, L.V., Petkova, E., et al., 2009. The State of REDD Negotiations: Consensus Points, Options for Moving Forward and Research Needs to Support the Process. Center for International Forestry Research.

Vermeulen, S.J., Campbell, B.M., Ingram, J.S., 2012. Climate change and food systems. Annu. Rev. Environ. Resour. 37, 195–222.

Vogel, J., 2022. Drivers of phenological changes in southern Europe. Int. J. Biometeorol. 66 (9), 1903–1914.

Vogel, E., Donat, M.G., Alexander, L.V., Meinshausen, M., Ray, D.K., Karoly, D., Meinshausen, N., Frieler, K., 2019. The effects of climate extremes on global agricultural yields. Environ. Res. Lett. 14 (5), 054010.

Volkova, L., Roxburgh, S.H., Weston, C.J., 2021. Effects of prescribed fire frequency on wildfire emissions and carbon sequestration in a fire adapted ecosystem using a comprehensive carbon model. J. Environ. Manage. 290, 112673.

Wang, S., Höhler, J., Ang, F., Lansink, A.O., 2023. Dutch dairy farmers' adoption of climate mitigation measures—the role of socio-psychological and socio-demographical factors. J. Clean. Prod. 427, 139187.

Wang, L., Qu, J.J., 2009. Satellite remote sensing applications for surface soil moisture monitoring: a review. Front. Earth Sci. China 3, 237–247.

Wako, G., Tadesse, M., Angassa, A., 2017. Camel management as an adaptive strategy to climate change by pastoralists in southern Ethiopia. Ecol. Process. 6, 1–12.

Wei, W., Wang, J., Ma, L., Wang, X., Xie, B., Zhou, J., Zhang, H., 2024. Global drought-wetness conditions monitoring based on multi-source remote sensing data. Land 13 (1), 95.

Wiebe, K., Robinson, S., et al., 2019. Climate change, agriculture and food security: impacts and the potential for adaptation and mitigation. Sustain. Food Agric., 55–74.

Whitcraft, A.K., Becker-Reshef, I., Killough, B.D., Justice, C.O., 2015. Meeting earth observation requirements for global agricultural monitoring: an evaluation of the revisit capabilities of current and planned moderate resolution optical earth observing missions. Remote Sens. (Basel) 7 (2), 1482–1503.

Wilhite, D.A., 1993. Planning for Drought: A Methodology. Drought Assessment, Management, and Planning: Theory and Case Studies. Springer US.

Wilhite, D.A., 2000. Drought as a Natural Hazard: Concepts and Definitions. Routledge.

Wilhite, D.A., Buchanan-Smith, M., 2005. Drought as hazard: understanding the natural and social context. Drought and Water Crises. Science, Technology, and Management Issues. CRC Press, Boca Raton.

Wilkes, A., Tennigkeit, T., et al., 2013. National Integrated Mitigation Planning in Agriculture: A Review Paper, MICCA/CGIAR/CCAFS/FAO.

Wu, B., Zhang, M., Zeng, H., Tian, F., Potgieter, A.B., Qin, X., Yan, N., Chang, S., Zhao, Y., Dong, Q., Boken, V., 2023. Challenges and opportunities in remote sensing-based crop monitoring: a review. Natl. Sci. Rev. 10 (4), nwac290.

Zhang, T., Huang, Y., 2013. Estimating the impacts of warming trends on wheat and maize in China from 1980 to 2008 based on county level data. Int. J. Climatol. 33 (3), 699–708.

Zhu, P., Burney, J., Chang, J., Jin, Z., Mueller, N.D., Xin, Q., Xu, J., Yu, L., Makowski, D., Ciais, P., 2022. Warming reduces global agricultural production by decreasing cropping frequency and yields. Nat. Clim. Chang. 12 (11), 1016–1023.

Zvamasiya, B., Nyikahadzoi, K., et al., 2017. Factors influencing smallholder farmers' behavioural intention towards adaptation to climate change in transitional climatic zones: a case study of Hwedza District in Zimbabwe. J. Environ. Manage. 198, 233–239.

Chapter 5

Climate variability and rangeland ecosystems

Kgabo Humphrey Thamaga[a], Mahlatse Kganyago[b], Mthunzi Mndela[c], and Katlego Mashiane[d]

[a]*Department of GIS and Remote Sensing, University of Fort Hare, Alice, South Africa,* [b]*Department of Geography, Environmental Management and Energy Studies, University of Johannesburg, Johannesburg, South Africa,* [c]*Department of Livestock and Pasture, University of Fort Hare, Alice, South Africa,* [d]*Department of Geography, University of the Free State, Phuthaditjhaba, South Africa*

Chapter outline

1 Introduction

Rangelands are naturally occurring ecosystems that are characterized by a diversity of grasses, forbs, shrubs, and trees. They account for 54% of the global terrestrial land surface and support a population of over 800 million people (ILRI et al., 2021; Gibson and Newman, 2019). The livelihoods and food security of these people are sustained by rangeland ecosystem goods and services. Among others, rangelands provide forage for livestock and wildlife and serve as

Remote Sensing of Climate. https://doi.org/10.1016/B978-0-443-21731-9.00008-9
Copyright © 2024 Elsevier Inc. All rights are reserved, including those for text and data mining, AI training, and similar technologies.

a habitat for diverse flora and mammal populations. In arid and semiarid environments, rangelands are essential for livestock grazing and for improving ecosystem services for pastoral societies. On the other hand, the amount and timing of rainfall patterns due to climate variability have a significant influence on rangeland productivity (vegetation abundance, species diversity, nutritional quality, and forage biomass). These put enormous pressure on rangeland ecosystems across the globe due to unsustainable anthropogenic activities and climate variabilities, thereby causing risks to diversity and productivity.

Rangelands are experiencing multiple threats due to climate variability and anthropogenic actions both locally and globally. These include changes in species composition, reduced biomass, loss of vegetation cover and biodiversity, and soil degradation (Moyo and Ravhuhali, 2022). Rising temperatures and shifting rainfall patterns affect rangeland condition across the globe, while livestock overgrazing leads to changes in vegetation abundance, population, structure, and diversity. To understand the impact of climate variability, we need effective tools for early, accurate detection and frequent monitoring of rangeland productivity and ecological implications to derive appropriate decision-making and sustainable management strategies.

Satellite remote sensing emerged to complement traditional ground-based field inventories through synoptic spatial coverage and reduced costs. Remote sensing emerged from aerial photographs, satellite-based remote sensing imagery, and unmanned aircraft/unmanned aerial vehicles to provide comprehensive information at varying geographic locations, multitemporal scales, and resolutions. Remote sensing technology provides numerous advantages for frequent rangeland monitoring, including spatial explicit information that in situ surveys cannot achieve, frequent data collection, and noninvasive and cost-effective acquisition. Since the early 1970s, satellite imagery such as Landsat data collection has provided opportunities to detect, map, and monitor the influence of climate variability and the status of rangeland productivity. Over time, recent sensor developments, including Sentinel datasets, can provide information about climate variability across large areas, even in remote or inaccessible areas, which can be particularly useful for monitoring rangelands at the regional or even global scale. Satellite images can measure multiple indicators of rangeland productivity decline, such as climate variability, impacts of land use land cover, grass/tree nutritive value, plant species succession, tree density, woody cover, and biomass, which can provide a more comprehensive picture of rangeland status.

The rest of this chapter is structured as follows: Section 2 provides an overview of the global distribution of rangelands, Section 3 discusses the effects of climate variability on rangelands, Section 4 provides an account of the applications of remote sensing in detecting, assessing, and monitoring the impacts of climate change and variability on rangeland productivity, Section 5 focuses on the integrated data analysis of climate and satellite-based products for modeling and projection of climate variability on rangelands, Section 6 provides some details of the advances in the big geospatial data analytics in rangelands ecosystems, and Section 7 provides a summary and conclusion.

2 Global rangeland spatial extent

Globally, rangelands cover about 79,509,420.64 km^2 (54%) of the Earth's total surface (148,326,000 km^2). Rangelands (Fig. 1) are categorized into seven biomes including deserts and xeric shrublands (27,984,644.64 km^2 [35%]), flooded grasslands and savannas (1,096,129.62 km^2 [1%]), Mediterranean forests, woodlands, and scrub (3,227,266.28 km^2 [6%]), montane grasslands and shrublands (5,203,411.00 km^2 [4%]), temperate grasslands, savannas, and shrublands (10,104,079.63 km^2 [13%]), tropical and subtropical grasslands, savannas, and shrublands (20,295,424.19 km^2 [26%]), and tundra (11,598,465.28 km^2 [15%]). These rangeland types possess numerous unique vegetation and plant associations that host a variety of different biota depending on the regional climate, soils, and anthropogenic influences.

3 Effects of climate change and variability on rangeland ecosystems

Over 40% of the Earth's surface is rangelands, which provide valuable goods and ecosystem services, thereby sustaining the livelihoods of billions of people globally (McCollum et al., 2017). Rangelands are reservoirs for biodiversity conservation (Mndela et al., 2022); they provide forage for wildlife and livestock and play a central role in energy balance, contributing significantly to global socioeconomic systems (McCollum et al., 2017). However, global climate change, characterized by erratic rainfall, global warming, rising emissions

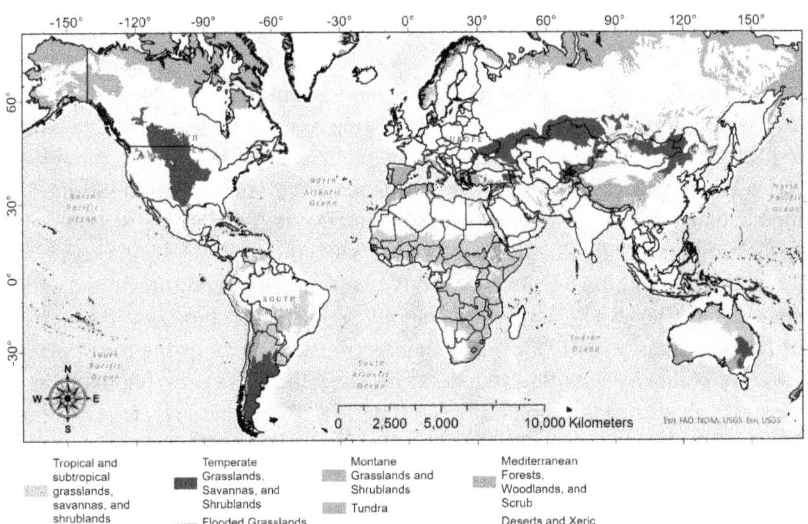

FIG. 1 Spatial distribution of global rangeland ecosystem biomes derived from Terrestrial Ecoregions of the World (TEOW, https://www.worldwildlife.org/publications/terrestrial-ecoregions-of-the-world) data (Olson et al., 2001).

of atmospheric gases, and predominantly elevated carbon dioxide (eCO_2), threatens biodiversity and productivity and disrupts rangeland ecosystem processes. The consequences of climate change are already witnessed in several regions across the globe. They are expected to worsen with time, given the future trends of increasing temperatures by 1–5°C and a decline in rainfall by 30%. The increase in temperatures and decline in rainfall in some regions will have profound negative effects on rangelands through evaporative soil moisture loss and, thus, plant moisture stress (Reeves et al., 2018). However, the impact of global climate change on rangeland ecosystems varies geographically, with some regions experiencing frequent and intensive droughts while others experience intensive rainstorms. These global climate changes drive shifts in vegetation at the community level and biome shifts, with many plant and animal species already shifting their original distributional range. However, given the complexity of the effects of climate change on a global scale, it is unclear how global rangeland systems respond to these complex changes. This is further complicated by current predictive models, which oversimplify vegetation responses to single climate factors, which is not an ideal scenario given that global climate factors interact at a large landscape scale. The adaptive responses of plants to global change vary considerably, depending on plant physiological traits, the extremes of climate change, and their interactions with atmospheric gases and local anthropogenic disturbances.

3.1 Effect of climate change on woody vegetation

It is widely documented that woody plant encroachment is limited largely by soil moisture. Thus, an increase in precipitation, together with eCO_2 in arid and semiarid areas, drives an increase in woody proliferation, biomass, and canopy cover. The climate-CO_2 aDGVM model simulations suggest that the world's grassy biomes, largely in semiarid areas, are likely to switch to a woodier plant-dominated state due to woody plant encroachment (Kumar et al., 2021). The woody cover is predicted to increase with precipitation in semiarid rangelands to an upper limit of 600 mm/year, beyond which the woody cover closes (Sankaran et al., 2005). Precipitation and atmospheric CO_2 interact globally, with precipitation enhancing woody plant seedling recruitment and establishment, while eCO_2 increases canopy cover and biomass through C fertilization (Briske, 2017). eCO_2 enhances the efficiency of the moisture use of woody plants by reducing transpiration rates and increasing photosynthetic rate and leaf area index, potentially counteracting drought effects (O'Connor et al., 2022). This, in turn, enhances the growth of woody plant encroachers even in water-limited drylands (Briske, 2017).

In frost-prone areas, an increase in temperatures offsets the lethal effects of frost on woody plant seedlings, thereby enhancing the survival and proliferation of woody plants (Venter et al., 2018); even in deserts where aridity is high, shrub cover and densities are increasing in the face of climate change

(Briske, 2017). Severe droughts reduce grass cover and increase shrub densities and cover in desert grasslands. Of the common shrub species responding positively to drought-induced decline in grass cover in deserts is *Larrea tridentata*. Unlike herbaceous plants, most encroaching shrubs have a deep root system facilitating deep water uptake.

3.2 Climate variability effects on herbaceous plant diversity

Dating back to historical times when temperate grasslands were predominant, a rise in temperatures and low CO_2 at the time facilitated an increase in the abundance of C4 tropical grasses at the expense of their C3 counterparts. This facilitated the expansion of tropical grasslands and savannas across the globe, which are dominated by C4 grasses. However, even in subtropical and tropical rangelands, many species are on the verge of extinction, as the current temperature extremes and drought prevalence are intensifying, causing shifts in growing seasonal patterns and length. Recently, many parts of the world have witnessed frequent longer dry seasons, which drives changes in plant species richness and abundance as other species go extinct or shift their distributional range. Grasslands are highly sensitive to temperature changes and variable interannual and seasonal rainfall distribution (Sun et al., 2021). The specialist herbaceous species with narrow niche breadth are particularly more vulnerable to climate extremes (Sintayehu, 2018) and are likely to undergo extinction. The future extreme climate may drive the extinction of more than a million species. The biodiversity-rich Cape flora of South Africa is projected to lose almost 11% of species in the face of future climate extremes (Sintayehu, 2018). This is due mainly to the fact that climate change fosters change not only by affecting the hydrological cycle but also the nutrient supply and other important biogeochemical processes. Nonetheless, species respond differently to climate extremes, depending on their phenology and physiology.

Temperate and tropical grassland species are likely to respond differentially to climate change, as C3 plants characterize the former while the latter is characterized by C4 plants. These plant photosynthetic pathways respond differently to changing temperatures and water supply, with C4 plants exhibiting superior adaptation to high temperatures and low soil moisture conditions relative to C3 plants. However, given that climate and rising CO_2 interact to influence terrestrial ecosystems, it is imperative to understand that plant responses to climate and elevated atmospheric gases. While C3 plants get saturated at higher CO_2 levels than C4 plants, the optimal temperatures at which they attain high C-assimilation ranges from 28°C to 32°C, whereas C4 plants range from 33°C to 37°C (Lara et al., 2011). Hence, increasing temperatures beyond these optimal levels may limit C3 grass adaptation to extreme climates despite their superior adaptation to higher atmospheric CO_2 scenarios. As reported in many temperate grasslands of the world, intense warming is taking place, thereby shifting C3 grass dominance to C4 grass dominance. The current atmospheric

CO_2 concentrations are still below a saturation point for C4 grass photosynthesis; hence, these grasses displace C3 plants in rangelands experiencing high temperatures, such as tropical rangelands. In the Western USA, for instance, C3 and C4 grasses utilize differential geographical spaces in the face of climate change, with C4 grasses increasing in abundance in higher latitudes with higher temperatures, whereas C3 plants dominate lowlands. A decline in C3 abundance in higher latitudes is associated with higher temperatures and low precipitation (Havrilla et al., 2023).

3.3 Effect of climate change on rangeland productivity

In ecological zones where climate change, particularly precipitation, favors woody plant encroachment, a decline in herbaceous productivity and cover is expected due to shading by woody plants (Sankaran et al., 2005). A decline in herbaceous productivity has serious implications for livestock production and wildlife. However, in climax grasslands at equilibrium, increased precipitation may increase herbaceous biomass without increasing the woody densities. Several studies conducted worldwide indicate that the annual net primary productivity (ANPP) of grasslands increases with precipitation and elevated CO_2. Global warming, however, reduces ANPP in grasslands. In nine grasslands of China, warming reduced ANPP by an average of 6.1%, but the effects were inconsistent across grasslands, with the Alpine and Lowland grasslands being at the receiving edge compared to temperate grasslands (Wu et al., 2021). In contrast, in New Zealand grasslands where C3 and C4 grasses co-occurred, simulated heat waves increased the abundance of C4 grasses and enhanced productivity, whereas productivity declined by 60% where C3 plants dominated (White et al., 2000). During warming, soil moisture declines, and plants tend to wilt and senesce due to moisture stress. This is exacerbated when C3 grasses dominate the ecosystem, as higher temperatures increase photorespiration in these plants. Wu et al. (2021) found that high temperatures reduced ANPP in all grasslands while precipitation increased ANPP, with the effects of radiation varying with grassland types, as it increased ANPP in temperate while it reduced it in Alpine grasslands (Wu et al., 2021). This suggests that the effects of climate vary from place to place depending on the prevailing climatic factor and grassland type. Drought simulation studies indicate that rainfall decline reduces the ANPP of grasses across the globe (Magandana et al., 2020; Sun et al., 2021).

4 Role of remote sensing in assessing the impacts of climate change and variability on rangelands

Remote sensing technology is renowned for its capability to gather a myriad of data about the dynamic and complex processes of the environment. Its advantage lies in the fact that sensors can be mounted on the ground (in situ sensors), near the ground (proximal sensors), on the aircraft (aerial sensors), and onboard a spacecraft (satellite or space-based sensors). Over the years, the technology has evolved in all sensor types mentioned. Still, the satellite sensors experienced the most rapid advances, which included various aspects such as improvements in the spatial resolutions (from coarse resolution to super-resolution sensors), revisits in high to very spatial resolution sensors through the establishment of satellite constellations and CubeSats, and radiometry. In tandem, techniques to process acquired satellite images have also developed significantly, providing opportunities for assessing rangelands and the impacts of external climate-related factors.

Time series datasets from wide-swath sensors such as Advanced Very High-Resolution Radiometer (AVHRR), Satellite Pourl' Observation de la Terre Vegetation (SPOT-VGT), Medium Resolution Imaging Spectrometer (MERIS), and MODerate Resolution Imaging Spectroradiometer (MODIS) have been invaluable in the assessment of rangeland health, the impact of droughts and Wildfires on the rangeland ecosystems (Diouf et al., 2015; Xie et al., 2022). Moreover, these sensors have been applied to develop various Essential Climate Variables (ECVs), such as Fraction of Photosynthetically Active Radiation (FPAR), Land Surface Temperature (LST), Leaf Area Index (LAI), aboveground biomass (Diouf et al., 2015), Fire Radiative Power, burned areas, land cover, Lake surface-water temperature, soil moisture, snow-covered areas and snow water equivalent, and water extents. Vegetation Indices (NDVI and EVI), Gross Primary Productivity (GPP)/Net Primary Productivity, Land Cover, Evapotranspiration, Soil Water Index, and Burned Area provided mainly by NASA (https://modis.gsfc.nasa.gov/) and Copernicus Global Land Service (https://land.copernicus.eu/). For instance, to address the urgent need for dynamic estimates of spatial distributions and temporal variations of C3 and C4 grasses in response to climate change, to inform policy-makers, land managers, and climate change mitigation measures, Xie et al. (2022) developed a phenology-based method to identify C3 and C4 grasses using MODIS Vegetation Indices product (MYD13A1), assess the presence and extent of grass compositional changes across 15 years (i.e., 2003–2017), and investigate the role of climate on these grass composition changes in southeast Australia. Their study found that interannual changes in the ratio of C4/C3 grass areas in southeast Australia were significantly associated with seasonal rainfall patterns rather than annual and seasonal temperatures or annual rainfall. Specifically, the (log) ratio of warm to cool season monthly average precipitation was identified as a key factor influencing the grassland composition. These shifts in grassland composition can cause extreme fire danger and changes in forage quality for livestock and wild animals. They noted that projected decreases in cool-season rainfall across southern Australia might cause a shift toward summer-dominant rainfall, favoring an increase in C4 grasses and a decrease in C3 grasses. Another study spanning the period 1999–2013 by Diouf et al. (2015) estimated the annual biomass production in the Sahelian rangelands using the

FaPAR-based phenological metrics from SPOT-VGT time series and in situ biomass data to R^2 values ranging from 0.64 to 0.77 and MAE from 625 to 489.21 kg·DM/ha.

4.1 Operational decision support systems for monitoring the rangeland changes

The global and regional efforts such as Famine Early Warning Systems Network (FEWS NET) and Group on Earth Observations (GEO-GLAM) Rangeland and Pasture Productivity (RAPP), and regional projects such as Enhancing Food Security in African Agricultural Systems with the Support of Remote Sensing (AfriCultuReS) project (Alexandridis et al., 2020) heavily rely on the coarse resolution products to develop value-added products and services. For example, the GEO-GLAM RAPP map (https://map.geo-rapp.org/) provides monthly fractional vegetation cover (FCover) at a 500 m spatial resolution based on MODIS Nadir BRDF-Adjusted Reflectance product (MCD43A4) for supporting strategic decisions about erosion management in rangelands. Similarly, the rangeland mask from the Anomaly Hotspots of Agricultural Production (ASAP, https://agricultural-production-hotspots.ec.europa.eu/) system by the Joint Research Council is provided at the spatial resolution of 0.0089286° (about 1 km^2) with the latest version (ver. 4) at 500 m spatial resolution (Fig. 2), where each pixel represents the area fraction of the rangeland

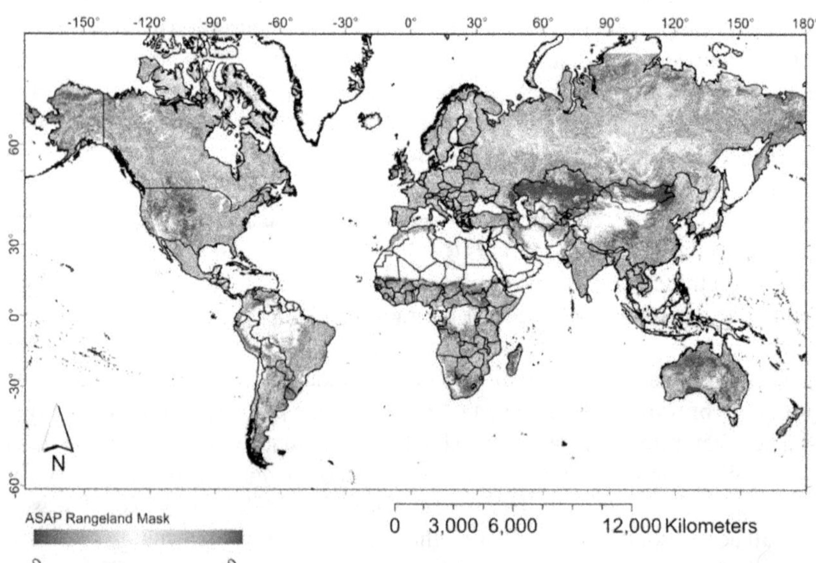

FIG. 2 A new ASAP Rangeland mask (rangeland area fraction) provided at 500 m spatial resolution (accessed from https://agricultural-production-hotspots.ec.europa.eu/).

(i.e., percentage of the pixel with rangeland) (Perez-Hoyos et al., 2017). On the other hand, national platforms such as the Rangeland Analysis Platform (https://rangelands.app/) in the United States of America incorporate the medium resolution data. The availability of online platforms allows rapid analysis of patterns in rangeland changes in response to climate factors. It provides intuitive graphics (i.e., maps and charts) and statistics for supporting decision-making.

Moreover, the data can be used for research purposes. A study by Smith et al. (2022) demonstrates the utility of these platforms for research. They used a remotely sensed vegetation cover product based on Landsat data from the Rangeland Analysis Platform (https://rangelands.app/) to quantify the expansion of annual grasses, that is, cheatgrass (*Bromus tectorum*), red brome (*B. rubens*), medusa head (*Taeniatherum caput-medusae*), and ventenata (*Ventenata dubia*) dominance in the Great Basin over the past three decades (i.e., 1990–2020) and tested for shifts in the distribution along topographic gradients. The study found that the expansion of annual grass dominance varied spatially (i.e., across the Great Basin) and temporally. It confirmed the widespread elevational ascent of annual grass dominance and identified shifts in distribution along topographic gradients, potentially facilitated by climate change. According to the study, warmer temperatures, earlier snowmelt, and increasing summer aridity are creating more favorable conditions for the establishment and growth of these exotic grasses at higher elevations. Their results showed that the median elevation of transition to annual grass dominance had increased by 138–459 m in 30 years. The study also observed upslope movement into cooler north-facing aspects, indicating the widespread phenomenon of elevational expansion of annual grass dominance. Extended fire seasons resulting from climate change may accelerate the transition to annual grass dominance.

The value of coarse resolution data for regional-scale monitoring is also shown by launching new sensors, such as the Visible and Infrared Imaging Spectroradiometer (VIIRS) and Sentinel-3, aimed at data continuity of MODIS and MERIS, respectively. Today, these sensors have been applied in diverse ecosystems and various applications, including wildfire assessment, vegetation health and drought assessment, and forest monitoring (Harris et al., 2011; Li et al., 2020).

In contrast, the medium- to high-resolution sensors, such as those from heritage missions such as Landsat and SPOT, are critical for landscape-scale analysis. Earlier sensors from these missions provided immersive opportunities to explore analytical methods at a reasonable spatial resolution (i.e., 20–30 m) to discern and monitor localized changes to rangeland and characterize the effect of climatic factors. For many decades, these sensors provided spatially and temporally consistent datasets, as well as adequate spectral properties for analyzing the patterns and dynamics of rangeland resources (Chen and Gillieson, 2009). Between 1996 and 1998, Chen and Gillieson (2009) estimated saltbush and total vegetation cover in semiarid rangeland environments using Landsat Thematic Mapper (TM). In hindsight, the archives provide many opportunities for long-term analysis of the effects of climate variability and change on rangelands.

One of the prospective limitations could be the lack of dense acquisitions over Africa due to commercial data licenses for Landsat Multispectral Scanner (MSS) and for Satellite Pourl' Observation de la Terre (SPOT)-1 to -5. Moreover, the ICT capabilities at the time also played a role, where data providers (i.e., Space agencies and private Earth Observation (EO) companies) could not store and rapidly process big datasets.

Nonetheless, the value of the datasets discussed here lies in their integrated use rather than the analysis of individual sensors. In support of this notion, Mardian et al. (2021) compared the performance of MODIS and Landsat datasets for detecting grassland to cropland conversions and found that MODIS outperformed Landsat, highlighting the importance of high-temporal resolution remote sensing data and advanced change detection methods, such as BFAST seasonal for accurately monitoring and managing grassland conversions. The authors suggested that future research could integrate additional data sources, such as phenological information, to improve the operational monitoring of grassland inventories. They also recommended exploring more sophisticated time series decomposition methods and incorporating multisensor data fusion techniques to enhance the accuracy and reliability of grassland conversion monitoring. Indeed, multisensor data fusion ensures seamless integration of coarse- to medium- to high-resolution sensor datasets through space and time, incorporating new sensors such as Landsat-8 and -9 and Sentinel-2 constellations. Such integrated analysis also allows us to overcome limitations of optical data, such as reduced image quality and utility caused by the changing atmospheric conditions, mainly the presence of opaque clouds and associated shadows (Boyte et al., 2018). With many EO sensor data used simultaneously, the chances of a cloudy pixel on one image (acquired by sensor *A*) becoming clear on another image acquired by sensor *B* are increased, thus also increasing the robustness and reliability of the analysis. Kganyago et al. (2022) used multisensor EO data and a Random Forest classifier to characterize the spatial distribution of African rangelands, aiming to support grazing and browsing capacity modeling, rangeland assessment, and management policy development. The results demonstrate that rangelands can be characterized with good accuracies exceeding 70% in most AfriCultuReS pilot countries using high-resolution land cover maps and MCD12Q1 products as training and testing data. These spatial distribution maps can aid in monitoring rangeland health, productivity, phenology, and changes. The study by Pringle et al. (2021) addressed the challenges faced by land managers, such as assessing forage quality accurately and promptly due to factors such as large grazing properties, widely dispersed herds, spatial heterogeneity of large paddocks, and the spatial variability of rainfall, by proposing using NDVI and EVI derived from Landsat series spanning 23 years, as proxies for forage quality to provide more accurate and timely assessments of forage quality (dietary crude protein and in vivo dry-matter digestibility [DMD]). A linear mixed foresting model was used to predict future forage quality for cattle in the dry savannas of northeast Australia. Another study (Kearney et al., 2022) focused on understanding the linkages between primary and secondary productivity in rangelands. It developed a model to predict weekly diet quality using phenological metrics derived from pseudo-daily satellite time series data and to predict the subseasonal performance of individual free-ranging yearling cattle based on diet quality and forage production predictions.

4.2 Multisource datasets and their capabilities in mapping and monitoring climate change variability in rangeland environments

The integration of various sensor data is referred to as data fusion, which entails integrating data from different sensors by applying mathematical models to harness beneficial features from each of those combined satellite imagery (Chang and Bai, 2018; Zhu et al., 2010). Multisensor data fusion can occur at pixel, feature, and decision levels. The pixel level involves the prediction of co-registered and preprocessed spectral data at pixel scale on days when there are no acquisitions (i.e., prediction day) of higher spatial resolution data. In such cases, high-temporal frequency sensors, such as MODIS and Sentinel-3, are employed to estimate daily surface reflectance of high-spatial-resolution data such as that from Landsat-8/-9 OLI or Sentinel-2 MSI, thus closing the data gaps of the 5-day or 8-day revisits of both sensors, respectively (Ovakoglou et al., 2020). To achieve high-temporal and high-spatial resolutions, studies mainly employed and adapted the Spatial Temporal Adaptive Reflectance Fusion Method algorithm by Gao et al. (2006). The algorithm predicts the surface reflectance on a prediction day at a high spatial resolution as a linearly weighted combination of the coarse temporal changes and available fine spatial reflectance (Fu et al., 2013). The re-projected and resampled high-temporal but low-spatial resolution images are slid over the high-spatial but low-temporal resolution images through a moving window at a determined pixel size, searching for spectrally similar pixels while calculating an optimal weight for each neighbor using spectral, temporal, and spatial differences, at the same time computing the surface reflectance of the central pixel of the moving window (Gao et al., 2006). Subsequently, other algorithms, such as the enhanced spatial and temporal adaptive reflectance fusion model, were developed to improve the temporal and spatial consistency of the simulated fine-resolution reflectance data, especially critical in heterogeneous landscapes (Zhu et al., 2010). This enhances our capacity to map, assess, and monitor rangelands at landscape scales and the external factors such as climate that affect them. Boyte et al. (2018) used (four regression models) to integrate 30 m Landsat data with 250 m eMODIS NDVI data, resulting in the production of four weekly, atmospherically corrected, and nearly cloud-free, downscaled 30-m synthetic MODIS-NDVI predictions with R^2 values of between 74% and 85%. Feature-level data fusion involves further processing of the pixel-level data to derive distinct features, such as through thresholding a vegetation index, image segmentation, or classification, which will be integrated for a more informative product. The decision level data fusion uses decision rules to fuse various extracted information; thus, it is highly dependent on the accuracy of the extracted features (Table 1).

TABLE 1 Applications of satellite data for rangeland mapping, assessment, and monitoring.

Reference	Satellite sensor/product	Climate variables	Analysis type	Study area	Accuracy
Smith et al. (2022)	Landsat MSS, TM, ETM+, OLI	Temperature, climatic water deficit, precipitation	Classification (*k*-Means clustering)	Great Basin, USA	OA: 88%
D'Adamo et al. (2021)	GIMMS-NDVI, CCI Soil Moisture, VOD, MODIS-land cover	CHIRPS Precipitation	Trend analysis (Spearman's rank trend test), Spearman's rank correlation	Africa	N/A[a]
Xie et al. (2022)	MODIS-EVI, ESA CCI land cover	Precipitation, Temperature	Trend analysis, classification (Gaussian distribution function)	Southeast Australia	N/A[a]
Ding et al. (2022)	MODIS-EVI, LST, NDSI, ESA CCI land cover	ERA5-land monthly precipitation	Trend analysis (Sen-Theil slope, Man-Kendall trend test), Spearman rank correlation analysis	Tianshan Mountains, Central Asia	N/A[a]
Mardian et al. (2021)	MODIS-NDVI and EVI, Landsat NDVI, EVI, SAVI, and MSAVI		Change detection (BFAST, BFAST seasonal, and BEAST)	Central Alberta, Canada	63%–76%
Rigge et al. (2021)	Landsat	PRISM-Precipitation, Temperature	Regression (Generalised Additive Model (GAM) and Cubist RT)	Western USA	*R*: 0.63
Wylie et al. (2004)	SPOT-VGT NDVI	NOAA Climate Prediction Center Precipitation and Temperature	Regression (RT)	Kazakh steppe, Kazakhstan	R^2: 0.48–0.72
Pringle et al. (2021)	Landsat TM, ETM+, OLI NDVI, and EVI		Spectral unmixing	Queensland, Australia	MAE: 0.86%–0.95%
Kganyago et al. (2022)	Landsat TM, ETM+, OLI		Classification (Random Forest)	Sub-Saharan Africa	OA: >70%
Diouf et al. (2015)	SPOT-VGT FaPAR		Regression (Partial Least Squares)	Sahel region	R^2: 64%–77%

[a]*Study did not explicitly report the accuracy.*

5 Integrated climate and satellite-based products for precision modeling and projection of climate variability on rangeland

Scientific predictions and knowledge on climate change and its effects are paramount for mitigation and adaptation strategies, as the consequences could be dire for these ecosystems. Grasslands are one of the important biomes of the world as they are home to the most rangelands. Rangeland vegetation dynamics are more likely to be altered in extent and composition due to climate change. Globally, grassland temperatures are expected to rise, impacting the ecological dynamics of rangelands. The ramifications of climate change and variability on rangelands are somewhat studied and documented for the global north; however, they remain scant for the global south. Fortunately, remote sensing can study rangelands in a more spatial explicit manner, thus providing the necessary comprehensive monitoring systems for managing and understanding rangelands under certain climatic projections. One of the major limitations in analyses involving climatic variables is the availability of the ground station data, which are unevenly distributed in various regions of the world, expensive to operate and maintain, and subject to human error in manual stations. Climate and meteorological reanalysis data combine satellite measurements with ground station data to overcome data gaps. Common variables used to assess climate variability include Precipitation (P), Temperature, Relative Humidity, Potential Evapotranspiration (PET), and Soil moisture (SM) from datasets such as Climate Hazards Group InfraRed Precipitation with Station Data (CHIRPS, https://data.chc.ucsb.edu/products/CHIRPS-2.0/), Global Land Evaporation Amsterdam Model (GLEAM), Multisource Weighted-Ensemble Precipitation (MSWEP), Tropical Rainfall Measuring Mission (https://climatedataguide.ucar.edu/climate-data/indices), Tropical Applications of Meteorology using Satellite data and ground-based observations (TAMSAT, https://www.tamsat.org.uk/), African Rainfall Climatology version 2 (ARC2, https://www.icpac.net/data-center/arc2/), Global Land Data Assimilation System (https://ldas.gsfc.nasa.gov/index.php/data), European Centre for Medium-range Weather Forecasts (https://climate.copernicus.eu), and ERA5 Reanalysis (https://climatedataguide.ucar.edu/climate-data/era5-atmospheric-reanalysis), among others. Gebrechorkos et al. (2023) developed four high-resolution, that is, 5 km, Standardized Precipitation Evaporation Indices (SPEI) using monthly precipitation from the CHIRPS and MSWEP, and PET from the GLEAM and hourly PET (hPET). The SPEI products agreed with observation-based measurements of SPEI, root zone SM, and vegetation health indices (Fig. 3).

Critically, integrating datasets from multiple sources, such as EO and meteorological and climate data (i.e., in situ and reanalysis), can aid multidisciplinary and comprehensive analyses to discern the impacts of climate variability and change on rangelands. Studies (D'Adamo et al., 2021; Ding et al., 2022) integrated the meteorological and bioclimatic variables into remotely sensed radiometric variables in mapping, assessment, and modeling in rangeland

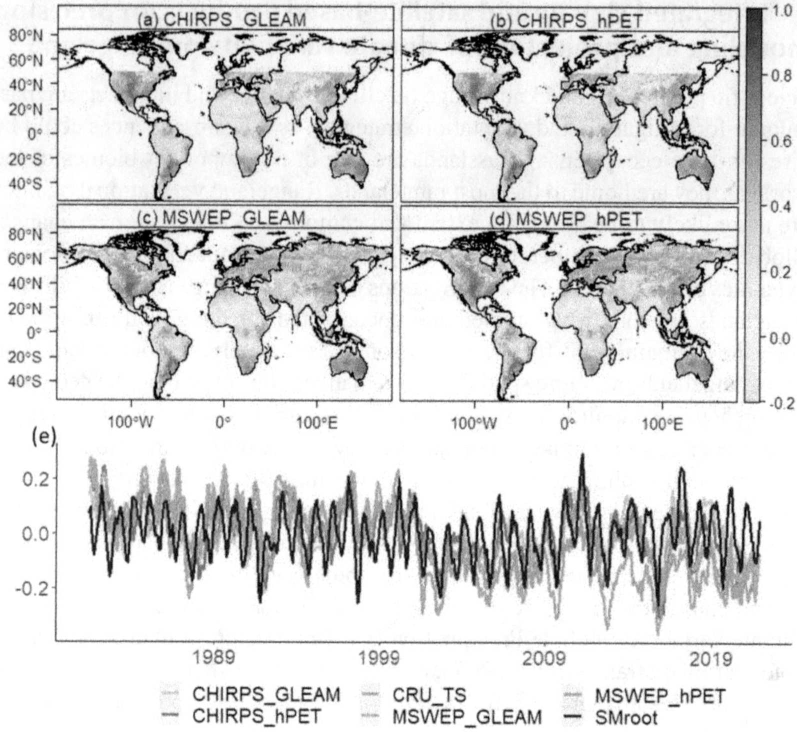

FIG. 3 Agreement between soil moisture at the root zone and the high-resolution SPEI over 6-month accumulation developed by Gebrechorkos et al. (2023) based on (A) CHIRPS_GLEAM, (B) CHIRPS_hPET, (C) MSWEP_GLEAM, and (D) MSWEP_hPET. The products are publicly distributed under a CC BY 4.0 license from the Centre for Environmental Data Analysis (https://doi.org/10.5285/ac43da11867243a1bb414e1637802dec).

environments. In mapping the annual exotic grasses in Western USA, Smith et al. (2022) combined the vegetation products from the Landsat series with climate variables from the TerraClimate dataset (https://www.climatologylab.org/terraclimate.html) such as annual minimum temperature, cumulative annual climatic water deficit, June to September (summer, dry season) precipitation, and October to June (growing season) precipitation, to an accuracy of 88%. However, the challenge is the relatively low resolution of the bioclimatic variables when compared to the remotely sensed variables. To account for differences in resolutions, studies such as D'Adamo et al. (2021) used harmonization methods to bring the datasets to a common resolution, that is, 25 km × 25 km, using a bilinear resampling algorithm. This allowed for meaningful comparison and analysis across different datasets. However, the individual data artifacts persist in the results. Consequently, super-resolving and downscaling techniques of some bioclimatic variables by building empirical relations with remotely sensed data have been attempted. Inarguably, integrating data from various sources is crucial to achieving robust modeling of climate impacts on natural resources such as rangelands.

Ding et al. (2022) sought to characterize changes in the spatial pattern of vegetation Green-up Date (GUD) along the elevation gradient across the Tianshan Mountains in the arid and semiarid central Asia for the period of 2001–2020 using MODIS-EVI and LST and ERA5-land monthly precipitation. The study found elevation-dependent changes in GUD across the Tianshan Mountains, with shifts toward higher elevations observed between 2001–2005 and 2016–2020. The study also found significant elevation dependence in the impacts of temperature and precipitation on GUD changes, indicating the important role of climate factors in driving these changes. Furthermore, the research highlighted the altered elevational patterns of GUD, which have implications for sustainable rangeland management and biodiversity in the region. In another study, D'Adamo et al. (2021) used the MODIS global land cover product (MCD12) and combined various spatial data products such as the NDVI (GIMMS NDVI3 g.v1 at 8 km × 8 km), Vegetation Continuous Fields (VCFs, 5 km × 5 km), Vegetation optical depth (VOD, 25 km × 25 km), CHIRPS (5 km × 5 km), ESA CCI v04.2 SM dataset (25 km × 25 km), ECV, and the simulated biomass carbon by the Dynamic Global Vegetation Model LPJ-GUESS, to identify rangelands where changes in vegetation greenness were mostly driven or unaffected by long-term climate change, and provide a long-term overview of how rangeland natural vegetation cover has changed across Africa in the last three decades. The study found long-term increases in vegetation greenness were observed in the western Sahel, Chad, and southern Africa. The associated changes in the vegetation structure included increases in trees, shrubs, and herbaceous vegetation, as well as a decrease in bare ground. Rigge et al. (2021) sought to apply models developed over a reference period (1985–2018) to future weather conditions under two emission scenarios, that is, Representative Concentration Pathway (RCP) 4.5 and 8.5, to evaluate changes in functional group abundance within the sagebrush biome. The primary objective was to inform rangeland managers of potential future vegetation composition, cover, and species distributions, which could improve the prioritization of conservation and restoration efforts. The two emission scenarios are RCP 4.5 and RCP 8.5, representing different trajectories of future greenhouse gas emissions and their potential impacts on vegetation. RCP 4.5 represents a lower greenhouse gas emissions scenario, while RCP 8.5 represents a higher emissions scenario. The study found that under RCP 4.5 and RCP 8.5 scenarios, there was an increasing dominance of nonsagebrush shrubs, annual herbaceous cover, and bare ground over herbaceous and sagebrush cover in the sagebrush encroached rangeland. Additionally, the results suggested that RCP 8.5 tended to result in more extreme changes in fractional component cover relative to the reference period than RCP 4.5.

6 Advancements in remote sensing applications and analysis in rangeland productivity and climate variability monitoring

Geospatial datasets have increased in volume over the years to the extent that the previous problem of lack of data has evolved into the problem of too much data. This necessitated research into new data formats, such as cloud-native GEOTIFF and data architectures that improve the storage and handling of multidimensional datasets. This has given birth to the concept of data cubes (Giuliani et al., 2017; Kopp et al., 2019), which refer to the arrangement of data in a spatially and temporally consistent manner to allow multitemporal analysis in a single shot (Egorov et al., 2019; Qiu et al., 2019). The Committee on Earth Observation Satellites (CEOS) has pioneered the concept and was piloted in Africa through the African Data Cubes, which were meant to be individual cubes hosting country-specific satellite datasets (Han et al., 2016). These efforts were recently scaled, thanks to the availability of cloud-based storage and computing platforms such as Amazon, Google Cloud, and Microsoft, and have evolved into a program called Digital Earth for Africa (Yuan et al., 2021, 2022). In fact, NASA and the United States Geological Survey adopted the cloud-based platform as their permanent hosting service for a myriad of sensors, including the widely utilized Landsat mission. At the same time, many data cubes were being developed, particularly in Europe, which resulted in platforms such as Sentinel Playground, EO Browser, and Sentinel Hub by Synergise (now owned by PlanetScope) (Dyke et al., 2021; Gomes et al., 2020). Indeed, one of the greatest planetary data analysis platforms to emerge is the Google Earth Engine (GEE) (Gorelick et al., 2017), which enabled research in various applications. Its strength, compared to its competitors, was the free access and myriad intuitive training materials on many subjects, including JavaScript coding exercises and applications. Sentinel Hub by Synergise, created as a spin-off company from the project funded by Horizon 2020, is also one of the ground-breaking platforms, providing access to petabytes of satellite data mainly from Copernicus/ESA missions and related products, while also allowing easy manipulation and intriguing visualizations. The cloud analysis platforms by Synergise also use JavaScript. Prior to this, other Horizon 2020 projects, such as EO AFRICA, offered Linux-based virtual machines, fully equipped with Free and Open-Source Software, to researchers worldwide. In addition, the private companies exploring the commercial potential of the product could access the systems for free for a certain period, that is, 3–6 months. Although limited information about the wide use of these initiatives, apart from GEE, is available in the literature, they played a key role in the success of the subsequent systems. The main thread around the successful adoption of any ground-breaking system is aggressive capacity building, without which adoption falls dramatically.

In the context of climate variability and rangelands mapping, assessment, and monitoring, the GEE platform has been widely utilized (Amani et al., 2020; Campos-Taberner et al., 2018; Long et al., 2019; Mutanga and Kumar, 2019).

Xie et al. (2019) used 30 years (1988–2017) of Landsat satellite imagery to map changes in vegetation cover amounts in large-scale rangelands of Queensland in Australia using GEE. The "big data" techniques and data repository applied in their study allowed efficient data access and processing on otherwise computationally intensive tasks in under 5 min. Bian et al. (2020) proposed a method for calculating the Mountain Green Cover Index (MGCI) based on Landsat-8 data and the GEE platform, addressing the limitations of the current MGCI calculation schemes and to characterize explicit mountain vegetation dynamics and vegetation-climate change interactions on a global scale. Another study, that is, Zhou et al. (2020), utilized the GEE platform and a machine learning algorithm, Random Forest, to identify the potential impact of different sampling times on the estimation of rangeland indicators. The findings indicate that temporally relevant training data improves predictions of rangeland indicators, but training data need not be from the same time for a prediction to be temporally relevant. Sentinel-2 and Landsat-8 data and cloud geo-processing technologies were used to develop a linear spectral mixture model for estimating the fractional vegetation cover in South African grasslands to address the detrimental impacts of grassland transformation and woody plant encroachment on grassland productivity for the livestock industry.

Inarguably, the previous decade has been dominated by big data analysis using contemporary shallow and deep machine learning algorithms such as Random Forest, Support Vector Machines, Artificial and Convoluted Neural Networks (A[*C*]NN), eXtreme Gradient Boosting (XGBoost), and Gaussian Process Regression and Classification (GPR[C]). Combined with the capability to analyze geometrical-aligned long-term datasets, the machine learning analytical approach presents a revolution in the EO-based analysis of climate-related phenomena and their impacts. Jones et al. (2018) utilized machine learning and cloud-based computing to produce annual, moderate-resolution percent land cover maps of rangelands across the Western United States. They integrated historical Landsat satellite data, gridded meteorology, abiotic land surface data, and over 30,000 field plots within a Random Forests model to predict per-pixel percent cover of annual forbs and grasses, perennial forbs and grasses, shrubs, and bare ground from 1984 to 2017.

7 Summary and conclusion

This chapter provided an overview of the impacts of climate variability and anthropogenic actions on rangeland ecosystems and the applications of remote sensing technology in detecting, assessing, and monitoring these impacts. Rangelands cover 54% of the global terrestrial land surface and support over 800 million people by providing forage for livestock, wildlife, and diverse flora. Climate variability affects rangeland productivity, leading to changes in species composition, reduced biomass, loss of vegetation and biodiversity, soil degradation, and other threats. Satellite remote sensing technology has emerged as a crucial tool for frequently monitoring rangeland productivity, providing spatial explicit information that traditional surveys cannot achieve. Various satellite imagery, such as Landsat and Sentinel data, have been utilized to detect, map, and monitor the influence of climate variability on rangeland productivity. Several papers reviewed in this chapter indicated that climate variability is found to have a profound impact on rangeland productivity, influencing the dynamics of vegetation cover and composition. The advancements in remote sensing applications and analysis in rangeland productivity and climate variability monitoring, including using cloud-based platforms such as GEE and the emergence of data cubes, machine learning techniques, and big geospatial data analytics, hold greater prospects for integrated multitemporal data analysis in understanding and projecting the effects of climate variability on rangelands. Overall, this chapter emphasizes the significance of remote sensing in understanding and monitoring the impacts of climate variability on rangeland ecosystems and addressing the challenges for future exploration.

References

Alexandridis, T.K., Ovakoglou, G., Cherif, I., Giménez, M.G., Laneve, G., Kasampalis, D., Moshou, D., Kartsios, S., Karypidou, M.C., Katragkou, E., García, S.H., Kganyago, M., Mashiyi, N., Pattnayak, K., Challinor, A., Pritchard, R., Brockington, D., Kagoyire, C., Beltran, J.S., 2020. Designing AfriCultuReS services to support food security in Africa. Trans. GIS 25, 692–720. https://doi.org/10.1111/tgis.12684.

Amani, M., Ghorbanian, A., Ahmadi, S.A., Kakooei, M., Moghimi, A., Mirmazloumi, S.M., Moghaddam, S.H.A., Mahdavi, S., Ghahremanloo, M., Parsian, S., Wu, Q., Brisco, B., 2020. Google Earth Engine cloud computing platform for remote sensing big data applications: a comprehensive review. IEEE J. Sel. Top. Appl. Earth Obs. Remote Sens. 13, 5326–5350. https://doi.org/10.1109/JSTARS.2020.3021052.

Bian, J., Li, A., Lei, G., Zhang, Z., Nan, X., 2020. Global high-resolution mountain green cover index mapping based on Landsat images and Google Earth Engine. ISPRS J. Photogramm. Remote Sens. 162, 63–76. https://doi.org/10.1016/j.isprsjprs.2020.02.011.

Boyte, S.P., Wylie, B.K., Rigge, M.B., Dahal, D., 2018. Fusing MODIS with Landsat 8 data to downscale weekly normalized difference vegetation index estimates for central Great Basin rangelands, USA. GIsci. Remote Sens. 55, 376–399. https://doi.org/10.1080/15481603.2017.1382065.

Briske, D.D. (Ed.), 2017. Rangeland systems. Walker, L.R., Howarth, R.W., Kaputska, L.A. (Eds.), In: Processes, Management and Challenges. Springer Nature, Texas, USA, pp. 25–58.

Campos-Taberner, M., Moreno-Martínez, Á., García-Haro, F.J., Camps-Valls, G., Robinson, N.P., Kattge, J., Running, S.W., 2018. Global estimation of biophysical variables from Google Earth Engine platform. Remote Sens. 10, 1–17. https://doi.org/10.3390/rs10081167.

Chang, N.B., Bai, K., 2018. Multisensor Data Fusion and Machine Learning for Environmental Remote Sensing. CRC Press.

Chen, Y., Gillieson, D., 2009. Evaluation of Landsat TM vegetation indices for estimating vegetation cover on semi-arid rangelands: a case study from Australia. Can. J. Remote. Sens. 35. https://doi.org/10.5589/m09-037.

D'Adamo, F., Ogutu, B., Brandt, M., Schurgers, G., Dash, J., 2021. Climatic and non-climatic vegetation cover changes in the rangelands of Africa. Glob. Planet Change 202. https://doi.org/10.1016/j.gloplacha.2021.103516.

Ding, C., Huang, W., Liu, M., Zhao, S., 2022. Change in the elevational pattern of vegetation greenup date across the Tianshan Mountains in Central Asia during 2001–2020. Ecol. Indic. 136. https://doi.org/10.1016/j.ecolind.2022.108684.

Diouf, A.A., Brandt, M., Verger, A., El Jarroudi, M., Djaby, B., Fensholt, R., Ndione, J.A., Tychon, B., 2015. Fodder biomass monitoring in Sahelian rangelands using phenological metrics from FAPAR time series. Remote Sens. 7, 9122–9148. https://doi.org/10.3390/rs70709122.

Dyke, G., Rosenqvist, A., Killough, B., Yuan, F., 2021. Intercomparison of Sentinel-1 datasets from Google Earth Engine and the Sinergise Sentinel Hub Card4L Tool. In: International Geoscience and Remote Sensing Symposium (IGARSS)., https://doi.org/10.1109/IGARSS47720.2021.9554039.

Egorov, A.V., Roy, D.P., Zhang, H.K., Li, Z., Yan, L., Huang, H., 2019. Landsat 4, 5 and 7 (1982 to 2017) Analysis Ready Data (ARD) observation coverage over the conterminous United States and implications for terrestrial monitoring. Remote Sens. 11. https://doi.org/10.3390/rs11040447.

Fu, D., Chen, B., Wang, J., Zhu, X., Hilker, T., 2013. An improved image fusion approach based on enhanced spatial and temporal the adaptive reflectance fusion model. Remote Sens. 5, 6346–6360. https://doi.org/10.3390/rs5126346.

Gao, F., Masek, J., Schwaller, M., Hall, F., 2006. On the blending of the Landsat and MODIS surface reflectance: predicting daily Landsat surface reflectance. IEEE Trans. Geosci. Remote Sens. 44. https://doi.org/10.1109/TGRS.2006.872081.

Gebrechorkos, S.H., Peng, J., Dyer, E., Miralles, D.G., Vicente-Serrano, S.M., Funk, C., Beck, H.E., Asfaw, D.T., Singer, M.B., Dadson, S.J., 2023. Global high-resolution drought indices for 1981–2022. Earth Syst. Sci. Data 15, 5449–5466. https://doi.org/10.5194/essd-15-5449-2023.

Gibson, D.J., Newman, J.A., 2019. Grasslands and Climate Change. Cambridge University Press., https://doi.org/10.1017/9781108163941.

Giuliani, G., Chatenoux, B., De Bono, A., Rodila, D., Richard, J.-P.P., Allenbach, K., Dao, H., Peduzzi, P., 2017. Building an Earth Observations Data Cube: lessons learned from the Swiss Data Cube (SDC) on generating Analysis Ready Data (ARD). Big Earth Data 1, 100–117. https://doi.org/10.1080/20964471.2017.1398903.

Gomes, V.C.F., Queiroz, G.R., Ferreira, K.R., 2020. An overview of platforms for big earth observation data management and analysis. Remote Sens. https://doi.org/10.3390/RS12081253.

Gorelick, N., Hancher, M., Dixon, M., Ilyushchenko, S., Thau, D., Moore, R., 2017. Google Earth Engine: planetary-scale geospatial analysis for everyone. Remote Sens. Environ. 202, 18–27. https://doi.org/10.1016/j.rse.2017.06.031.

Han, W., Di, L., Yu, G., Shao, Y., Kang, L., 2016. Investigating metrics of geospatial web services: the case of a CEOS federated catalog service for earth observation data. Comput. Geosci. 92, 1–8. https://doi.org/10.1016/j.cageo.2016.04.005.

Harris, S., Veraverbeke, S., Hook, S., 2011. Evaluating spectral indices for assessing fire severity in chaparral ecosystems (Southern California) using MODIS/ASTER (MASTER) airborne simulator data. Remote Sens. 3, 2403–2419. https://doi.org/10.3390/rs3112403.

Havrilla, C.A., Bradford, J.B., Yackulic, C.B., Munson, S.M., 2023. Divergent climate impacts on C3 versus C4 grasses imply widespread 21st century shifts in grassland functional composition. Divers. Distrib. 29, 379–394.

ILRI, IUCN, FAO, WWF, UNEP, ILC, 2021. Rangelands Atlas. ILRI, Nairobi Kenya.

Jones, M.O., Allred, B.W., Naugle, D.E., Maestas, J.D., Donnelly, P., Metz, L.J., Karl, J., Smith, R., Bestelmeyer, B., Boyd, C., Kerby, J.D., McIver, J.D., 2018. Innovation in rangeland monitoring: annual, 30 m, plant functional type percent cover maps for U.S. rangelands, 1984–2017. Ecosphere 9. https://doi.org/10.1002/ecs2.2430.

Kearney, S.P., Porensky, L.M., Augustine, D.J., Derner, J.D., Gao, F., 2022. Predicting spatial-temporal patterns of diet quality and large herbivore performance using satellite time series. Ecol. Appl. 32. https://doi.org/10.1002/eap.2503.

Kganyago, M., Ramoelo, A., Zoungrana, E., Nosiseko Mashiyi, I.G., 2022. Characterizing the spatial distribution of grazing and browsing resources in Africa using random forest classifier and multi-sensor data earth observation, South African National Space Agency, The Enterprise Building, Mark Shuttleworth School of Geography. In: 2022 IEEE International Geoscience and Remote Sensing Symposium. IEEE, pp. 4368–4371.

Kopp, S., Becker, P., Doshi, A., Wright, D.J., Zhang, K., Xu, H., 2019. Achieving the full vision of earth observation data cubes. Data (Basel) 4. https://doi.org/10.3390/data4030094.

Kumar, D., Pfeiffer, M., Gaillard, C., Langan, L., Scheiter, S., 2021. Climate change and elevated CO_2 favor forest over savanna under different future scenarios in South Asia. Biogeosci. Discuss. 18, 2957–2979. . https://doi.org/10.5194/bg-2020-169.

Lara, M.V., Andreo, C.S., Shanker, A., 2011. C4 plants adaptation to high levels of CO_2 and to drought environments. In: Abiotic Stress in Plants—Mechanisms and Adaptations. InTech, pp. 18–428. Available from: http://www.intechopen.com/books.

Li, F., Zhang, X., Kondragunta, S., Schmidt, C.C., Holmes, C.D., 2020. A preliminary evaluation of GOES-16 active fire product using Landsat-8 and VIIRS active fire data, and ground-based prescribed fire records. Remote Sens. Environ. 237, 111600. https://doi.org/10.1016/j.rse.2019.111600.

Long, T., Zhang, Z., He, G., Jiao, W., Tang, C., Wu, B., Zhang, X., Wang, G., Yin, R., 2019. 30m Resolution global annual burned area mapping based on Landsat images and Google Earth Engine. Remote Sens. 11, 1–30. https://doi.org/10.3390/rs11050489.

Magandana, T.P., Hassen, A., Tesfamariam, E.H., 2020. Seasonal herbaceous structure and biomass production response to rainfall reduction and resting period in the semi-arid grassland area of South Africa. Agronomy 10, 1807. https://doi.org/10.3390/agronomy10111807.

Mardian, J., Berg, A., Daneshfar, B., 2021. Evaluating the temporal accuracy of grassland to cropland change detection using multitemporal image analysis. Remote Sens. Environ. 255, 112292. https://doi.org/10.1016/j.rse.2021.112292.

McCollum, D.W., Tanaka, J.A., Morgan, J.A., Mitchell, J.E., Fox, W.E., Maczko, K.A., Hidinger, L., Duke, C.S., Kreuter, U.P., 2017. Climate change effects on rangelands and rangeland management: affirming the need for monitoring. Ecosyst. Health Sustain. 3, e01264. https://doi.org/10.1002/ehs2.1264.

Mndela, M., Madakadze, I.C., Nherera-Chokuda, F., Dube, S., Ramoelo, A., Mangwane, M., Tjelele, J., 2022. Short-term responses of herbaceous vegetation to bush clearing in semi-arid rangelands of South Africa. Pastoralism 12, 17. . https://doi.org/10.1186/s13570-022-00235-7.

Moyo, B., Ravhuhali, K.E., 2022. Abandoned croplands: drivers and secondary succession trajectories under livestock grazing in communal areas of South Africa. Sustainability 2022 (14), 6168.

Mutanga, O., Kumar, L., 2019. Google Earth Engine applications. Remote Sens. 11, 11–14. https://doi.org/10.3390/rs11050591.

O'Connor, R.C., Blumenthal, D.M., Ocheltree, T.W., Nippert, J.B., 2022. Elevated CO_2 counteracts effects of water stress on woody rangeland-encroaching species. Tree Physiol., 1–12. . https://doi.org/10.1093/treephys/tpac150.

Olson, D.M., Dinerstein, E., Wikramanayake, E.D., Burgess, N.D., Powell, G.V.N., Underwood, E. C., D'Amico, J.A., Itoua, I., Strand, H.E., Morrison, J.C., Loucks, C.J., Allnutt, T.F., Ricketts, T.H., Kura, Y., Lamoreux, J.F., Wettengel, W.W., Hedao, P., Kassem, K.R., 2001. Terrestrial ecoregions of the world: a new map of life on Earth. Bioscience 51 (11), 933–938.

Ovakoglou, G., Alexandridis, T.K., Clevers, J.G.P.W., Gitas, I.Z., 2020. Downscaling of MODIS leaf area index using landsat vegetation index. Geocarto Int., 1–24. https://doi.org/10.1080/10106049.2020.1750062.

Perez-Hoyos, A., Rembold, F., Kerdiles, H., Gallego, J., 2017. Comparison of global land cover datasets for cropland monitoring. Remote Sens. 9. https://doi.org/10.3390/rs9111118. The data sets were last updated on 12-03-2019 (version 03).

Pringle, M.J., O'Reagain, P.J., Stone, G.S., Carter, J.O., Orton, T.G., Bushell, J.J., 2021. Using remote sensing to forecast forage quality for cattle in the dry savannas of northeast Australia. Ecol. Indic. 133. https://doi.org/10.1016/j.ecolind.2021.108426.

Qiu, S., Lin, Y., Shang, R., Zhang, J., Ma, L., Zhu, Z., 2019. Making Landsat time series consistent: evaluating and improving Landsat analysis ready data. Remote Sens. 11. https://doi.org/10.3390/rs11010051.

Reeves, M.C., Manning, M.E., DiBenedetto, J.P., Palmquist, K.A., Lauenroth, W.K., Bradford, J.B., Schlaepfer, D.R., 2018. Effects of climate change on rangeland vegetation in the Northern Rockies Region. In: Halofsky, J.E., Peterson, D.L., Dante-Wood, S.K., Hoang, L., Ho, J.J., Joyce, L.A. (Eds.), Climate Change Vulnerability and Adaptation in the Northern Rocky Mountains—Part 2. Gen. Tech. Rep. RMRS-GTR-374. U.S. Department of Agriculture, Forest Service, Rocky Mountain Research Station, Fort Collins, CO, pp. 275–316 (Chapter 7).

Rigge, M., Shi, H., Postma, K., 2021. Projected change in rangeland fractional component cover across the sagebrush biome under climate change through 2085. Ecosphere 12. https://doi.org/10.1002/ecs2.3538.

Sankaran, M., Hanan, N.P., Scholes, R.J., et al., 2005. Determinants of woody cover in African savannas. Nature 438, 846–849.

Sintayehu, D.W., 2018. Impact of climate change on biodiversity and associated key ecosystem services in Africa: a systematic review. Ecosyst. Health Sustain. 4, 225–239. https://doi.org/10.1080/20964129.2018.1530054.

Smith, J.T., Allred, B.W., Boyd, C.S., Davies, K.W., Jones, M.O., Kleinhesselink, A.R., Maestas, J.D., Morford, S.L., Naugle, D.E., 2022. The elevational ascent and spread of exotic annual grass dominance in the Great Basin, USA. Divers. Distrib 28. https://doi.org/10.1111/ddi.13440.

Sun, Y., Sun, Y., Yao, S., Akram, M.A., Hu, W., Dong, L., Li, H., Wei, M., Gong, H., Xie, S., Aqeel, M., Ran, J., Degen, A.A., Guo, Q., Deng, J., 2021. Impact of climate change on plant species richness across drylands in China: from past to present and into the future. Ecol. Indic. 132, 108288. . https://doi.org/10.1016/j.ecolind.2021.108288.

Venter, Z.S., Cramer, M.D., Hawkins, H.J., 2018. Drivers of woody plant encroachment over Africa. Nat. Commun. 9, 2272. . https://doi.org/10.1038/s41467-018-04616-8.

White, T.A., Campbell, B.D., Kemp, P.D., Hunt, C.L., 2000. Sensitivity of three grassland communities to simulated extreme temperature and rainfall events. Glob. Chang. Biol. 6, 671–684.

Wu, G.L., Cheng, Z., Alatalo, J.M., Zhao, J., Liu, Y., 2021. Climate warming consistently reduces grassland ecosystem productivity. Earths Future 9, e2020EF001837. . https://doi.org/10.1029/2020EF001837.

Wylie, B.K., Gilmanov, T.G., Johnson, D.A., Saliendra, N.Z., Akshalov, K., Tieszen, L.L., Reed, B. C., Laca, E., 2004. Intra-seasonal mapping of CO2 flux in rangelands of northern Kazakhstan at one-kilometer resolution. Environ. Manag. 33. https://doi.org/10.1007/s00267-003-9156-8.

Xie, Z., Phinn, S.R., Game, E.T., Pannell, D.J., Hobbs, R.J., Briggs, P.R., McDonald-Madden, E., 2019. Using Landsat observations (1988–2017) and Google Earth Engine to detect vegetation cover changes in rangelands—a first step towards identifying degraded lands for conservation. Remote Sens. Environ. 232, 111317. https://doi.org/10.1016/j.rse.2019.111317.

Xie, Q., Huete, A., Hall, C.C., Medlyn, B.E., Power, S.A., Davies, J.M., Medek, D.E., Beggs, P.J., 2022. Satellite-observed shifts in C3/C4 abundance in Australian grasslands are associated with rainfall patterns. Remote Sens. Environ. 273. https://doi.org/10.1016/j.rse.2022.112983.

Yuan, F., Lewis, A., Leith, A., Dhar, T., Gavin, D., 2021. Analysis ready data for Africa. In: International Geoscience and Remote Sensing Symposium (IGARSS)., https://doi.org/10.1109/IGARSS47720.2021.9554019.

Yuan, F., Repse, M., Leith, A., Rosenqvist, A., Milcinski, G., Moghaddam, N.F., Dhar, T., Burton, C., Hall, L., Jorand, C., Lewis, A., 2022. An operational analysis ready radar backscatter dataset for the African continent. Remote Sens. 14. https://doi.org/10.3390/rs14020351.

Zhou, B., Okin, G.S., Zhang, J., 2020. Leveraging Google Earth Engine (GEE) and machine learning algorithms to incorporate in situ measurement from different times for rangelands monitoring. Remote Sens. Environ. 236. https://doi.org/10.1016/j.rse.2019.111521.

Zhu, X., Chen, J., Gao, F., Chen, X., Masek, J.G., 2010. An enhanced spatial and temporal adaptive reflectance fusion model for complex heterogeneous regions. Remote Sens. Environ. 114. https://doi.org/10.1016/j.rse.2010.05.032.

Chapter 6

Climate variability and natural ecosystems productivity

Onalenna Gwate[a] and Munyaradzi Davis Shekede[b,c]
[a]*Department of Geography and Geo-Information Sciences, Lupane State University, Lupane, Zimbabwe,* [b]*Department of Geography Geospatial Sciences and Earth Observation, University of Zimbabwe, Harare, Zimbabwe,* [c]*Department of Geospatial Sciences and Earth Observation, Zimbabwe National Geospatial and Space Agency, Harare, Zimbabwe*

Chapter outline

1 Introduction

Climate is highly variable spatially and temporally owing to both natural and anthropogenic activities. Many earth system processes are closely coupled with climate. Consequently, climate variability influences ecological functioning by mediating ecosystem productivity. Dynamics in climate influence photosynthesis, which typifies new biomass creation by autotrophs. Given the high variability in the climate system, it is critical to understand how and to what extent this could affect ecosystem productivity. Suffices to note that ecosystem productivity forms the basis of life on the earth system by supplying energy through food webs. This chapter seeks to explore mechanisms of climate variability and impacts on the carbon cycle, ecosystems, and ecosystem productivity. The chapter concludes

Remote Sensing of Climate. https://doi.org/10.1016/B978-0-443-21731-9.00003-X
Copyright © 2024 Elsevier Inc. All rights are reserved, including those for text and data mining, AI training, and similar technologies.

by examining strategies for mitigating the impact of climate variability on natural ecosystem productivity in pursuit of land degradation neutrality as espoused in the United Nations Framework on Combating Desertification.

2 Climate variability

Climate is the long-term statistical description of the atmospheric conditions, averaged over a specified period, of at least three decades. The climate system is very complex and comprises the atmosphere, ocean, land surface, ice and snow surfaces/cryosphere (both land and ocean areas), and biosphere (both terrestrial and marine). The interactions among these components are key in characterizing the climate of a particular area. Owing to this complexity, climate is bound to vary spatially and temporally (Ghil, 2002). Climate variability relates to oscillations in the mean state and other statistics of the climate on all temporal and spatial scales, beyond individual weather events (IPCC, 2012). These fluctuations can influence patterns of rainfall, temperature, and other variables on timescales anywhere from a few weeks to a few decades. Climate variability should be distinguished from climate change, which relates to the alterations in the state of the climate that can be identified (e.g., using statistical tests) by changes in the mean and/or the variability of its properties, and that persists for an extended period, typically decades or longer (IPCC, 2012).

Natural and anthropogenic factors modulate climate variability, which may be internal or external to the climate system. Natural forcings of climate variability include those directly driven by a purely periodic external force, like the diurnal or the seasonal cycle of insolation linked to the earth's rotation. Variations due to the nonlinear interplay of feedbacks within the climate system produce internal variability in the climate system (Salinger et al., 2005). For example, the evolution of the El Niño/Southern Oscillation, North Atlantic Oscillation (NAO), and changes in sea-surface temperature have significant influence on short-term temperature and precipitation dynamics in affected areas. Other natural causes include external forcings such as astronomical periodicities, sunspot cycle, vulcanicity, ocean currents, as well as perihelion and aphelion dynamics. Variations associated with random fluctuations in physical or chemical constituents of the atmosphere (internal or external) linked to natural phenomena and anthropogenic activities that alter atmospheric composition also cause short-term variability in climate (Ghil, 2002).

The effects of climate variability are largely like those of climate change. For example, both lead to extreme events such as flooding or drought (UNFCCC, 2007). The frequency of these climate extremes will change in response to shifts in both mean climate and climate variability. However, it is not clear to what extent each contributes to the magnitude of these events. For example, globally, model simulations have shown that precipitation variability in most climate models increased in over 66% of global land area in response to warming. Under climate change (RCP 8.5), the global, multimodel mean, precipitation variability is projected to increase (2%–5%) over land and ocean (Pendergrass et al., 2017). In addition, some studies have shown that the increased occurrence of monthly high-temperature events is a function of a warming mean climate while future changes in monthly heavy-precipitation events depend to a considerable degree on trends in climate variability (van der Wiel and Bintanja, 2021). Consequently, there is a high degree of uncertainty regarding the contribution of climate variability and mean climate to extreme events. Thus, there is a need to fully account for all sources of uncertainty such as scenarios, intermodal uncertainty, and internal variability in climate projections to enable better and more appropriate interpretation of results. For example, recent studies have shown that climate internal variability represents more than 50% of the total climate uncertainty in certain climate projections (Schwarzwald and Lenssen, 2022). This demonstrates the importance of studying and understanding climate variability because it is critical in climate change science.

Life on Earth is closely coupled with climate and thus climate variability inadvertently influences the biosphere. For example, agriculture is highly vulnerable to climate variability owing to dynamics in temperature and water for agricultural production. In southern Africa, climate variability linked to the El Niño/Southern Oscillation grossly influences food security dynamics. Both high and low precipitation may result in various forms of extreme events such as drought, which may adversely affect human beings and livelihoods. This when combined with other stressors like population increase may lead to food insecurity. For example, research has shown that climate variability explained >60% of yield variability in substantial areas of the global breadbaskets (Ray et al., 2015). Consequently, it is prudent to understand the full range of impacts of climate change on biological and food systems to effectively address the effects of climate variability and extreme events on human vulnerability and food security in the context of increasing population, particularly in developing countries (Thornton et al., 2014). On the contrary, extremely high precipitation events affect the hydrological cycle and often lead to flooding. For example, in East Africa, long rains have tended toward a drier state from the mid-1980s to 2010 while the short rains have become wetter since the mid-1980s (Palmer et al., 2023). There is consensus that the short rains will deliver more rainfall than the long rains by 2030–2040, with implications for sustaining agricultural yields and triggering climate-related public health emergencies. Consequently, mitigating the impacts of future eastern African climate requires continued investments in agriculture, clean water, medical and emergency infrastructures, and development and adoption of adaptation strategies, as well as targeted early-warning systems driven by improved meteorological observation (Palmer et al., 2023). To reduce vulnerability to climate change, understanding climate variability is key. Climate variability should be understood in the context of timing and interactions of different climatic stresses on plant growth and development, particularly at higher temperatures, and the impacts on crops, livestock, and farming systems of changes in climate variability and extreme events on pest-weed-disease complexes (Thornton et al., 2014).

Increased warming may result in sea level rises (Climate change widespread, rapid, and intensifying—IPCC). Sea levels have been rising globally by over 3 mm per year in recent decades, with an acceleration in this trend now observable, and a rise of 1.1 m projected by 2100 (Climate change: sea level rise to affect "three times more people"—BBC News). Such sea level rise would lead to the submersion of coastal areas. Countries like The Netherlands, which is below sea level, may be submerged. This could create ecological refugees and loss of property and life. Climate variability may also have serious geological effects. For example, Burroughs (2001) notes that climate events like the waxing and waning of ice sheets, the desiccation of continental interiors, or the drying up of oceans had major impacts on the geology of large parts of the earth. In addition, climate change helps to release stresses in the earth's crust, for example, during the last Ice Age, the eruption of volcanoes in the Mediterranean region appears to have occurred more often at times of rapid sea level changes. This changes the loading of the earth's crust by the build-up and collapse of ice. Dynamics in sea levels may also lead to human migrations, but further research is needed on the fundamental mechanisms underlying sea level rise migration, tipping points, thresholds and feedbacks, risk perception, and migration to fully understand migration responses to sea level rise (Hauer et al., 2020).

3 Climate variability effects on natural ecosystems in different regions of the world

An understanding of the influence of climate variability on ecosystem production is critical for mitigation planning. For example, at a global scale, it is predicted that climate change and variability will increase the global net ecosystem production substantially. However, the response of natural terrestrial ecosystems depends largely upon changes in individual factors and varies widely among climate-vegetation zones (Cao and Woodward, 1998). As a result, consideration of these other factors could enhance understanding of the close connection between carbon sequestration and climate variability, leading to more realistic predictions of both ecosystem responses and climate change.

Climate variability also influences plant productivity. In Africa, it has been demonstrated that climate change reduced plant production and soil carbon stocks and caused net CO_2 release. However, the fertilization effect of increasing atmospheric CO_2 on photosynthesis reversed the reduction and led to more carbon accumulation in vegetation. Cao et al. (2001) found that the mean rate of carbon sequestration in the period 1981–1995 was 0.34 Gt C y^{-1}. Overall, Africa was not necessarily a significant carbon sink, because a large part of the carbon sequestration was offset by the carbon release arising from land use changes as well as burning (Cao et al., 2001).

In a study in China, it has been demonstrated that warming stimulates net carbon uptake under wet conditions but depresses it under very dry conditions (Quan et al., 2019). In addition, future warmer conditions could lead to variable feedbacks of ecosystems to climate warming, owing to dynamics in moisture conditions (Quan et al., 2019). Climate variability in the form of droughts causes dynamics in carbon uptake via photosynthesis and release through respiration, and dynamics in these are modulated by plant survival strategies. Emerging evidence has revealed that consideration of plant survival strategies will lead to an improved understanding of the influence of climate variability on ecosystem production (Van der Molen et al., 2011). Hence, plant strategies must be given an adequate role in global vegetation models if the effects of climate variability such as droughts on the carbon cycle are to be described in a way that justifies the interacting processes (Van der Molen et al., 2011). Similarly, Gwate and Ndou (2022) reported a unimodal relationship between net primary production (NPP) and rainfall in a teak, Miombo, and *Vachellia* woodland in Zimbabwe. This relationship suggests that climate change and variability are likely to shift ecosystem productivity at any place. The integration of in situ observations and remote sensing has improved the observation of various components of ecosystem productivity in the context of climate variability (Schimel et al., 2015).

Ongoing programs to develop global carbon budgets are crucial in understanding the influence of climate variability on ecosystem production. For example, Le Quéré et al. (2013) provided a baseline to keep track of annual carbon budgets in the future. In a study to understand global carbon dynamics, an exceptionally high carbon sink was reported in 2011 and this was driven by the growth of semi-arid vegetation in the Southern Hemisphere, with almost 60% of carbon uptake attributed to Australian ecosystems, where prevalent La Nina conditions caused up to six consecutive seasons of increased precipitation (Poulter et al., 2014). In addition, since 1981, a 6% expansion of vegetation cover over Australia was associated with a fourfold increase in the sensitivity of continental net carbon uptake to precipitation. This indicates the importance of climate dynamics in modulating ecosystem production and that semi-arid ecosystems of the world are important drivers of global carbon cycle interannual variability. At the same time, a synthesis of global air temperature data reveals a greater rate of warming in winter than in summer in northern mid and high latitudes, and the inverse pattern in some tropical regions (Xia et al., 2014). The same study also revealed a decline in the diurnal temperature range over 51% of the global land area and an increase over only 13%, because night-time temperatures in most locations have risen faster than daytime temperatures. Analyses of satellite data, model simulations, and in situ observations suggest that the impact of seasonal warming varies between regions. Therefore, ascertaining the effects of non-uniform climate warming on terrestrial ecosystems is a key challenge in carbon cycle research (Xia et al., 2014).

Dynamics in climate also influence the ecology and biogeochemistry of the deep sea. It affects surface ocean processes and the production of organic carbon, which ultimately comprises the primary food supply to the deep-sea ecosystems. Climate change has enhanced ocean stratification, and this undermines vertical mixing. This will in turn affect oceanic communities throughout the ocean depth, entrenching variability in primary production and carbon export flux to the deep sea (Smith et al., 2009). Meanwhile, in a field experiment, Yvon-Durocher et al. (2010) found that the key components of the carbon cycle were strongly related to temperature in aquatic ecosystems. Their temperature dependence was typically constrained by the average activation energy of their metabolic pathway, and this increased progressively for gross primary production (GPP), ecosystem respiration (ER), and methane flux. The same study found that warming of 4°C decreased the sequestration of CO_2 by 13%, increased the fraction of primary production effluxing such as methane by 20%, and the fraction of ER as methane by 9%, in line with the offset in their respective activation energies. Because methane has 21 times the greenhouse gas (GHG) radiative potential of CO_2, these results suggest aquatic ecosystems could drive a previously unknown positive feedback between warming and the carbon cycle (Yvon-Durocher et al., 2010).

4 Natural ecosystem productivity and the carbon cycle

The carbon cycle entails the fluxes of carbon among four main reservoirs: fossil carbon, the atmosphere, the oceans, and the terrestrial biosphere over time scales of hours to centuries (Schimel, 1995). The unprecedented high anthropogenic CO_2 emissions cannot be absorbed by natural ecosystems and our understanding of the interactions between the components of the earth systems and the relationship between the carbon cycle and other biogeochemical and climatological processes is limited (Falkowski et al., 2000). Before the Industrial Era, atmospheric CO_2 concentration was 280 ± 10 ppm for several thousand years, rising to 367 ppm at the beginning of the 21st century owing to industrialization (Prentice et al., 2001). Schimel (1995) reported that during the 1980s fossil carbon emissions averaged 5.5 Gt y^{-1} and the atmosphere gained 3.2 Gt C y^{-1} while oceans are believed to have absorbed 2.0 Gt C y^{-1}. At the same time, the regrowing forests of the Northern Hemisphere could have sunk 0.5 Gt C y^{-1} over the same period. Tropical deforestation is thought to have released an average of 1.6 Gt C y^{-1} over the 1980s. In principle, fluxes between the four pools should balance but often they do not (Schimel, 1995). Carbon fluxes from wetlands particularly in the Southern Hemisphere are not well understood. A recent review concluded that there is an urgent need for systematic data on carbon dynamics in tropical wetlands to provide a robust understanding of how they differ from well-studied northern wetlands and allow the incorporation of tropical wetlands into global climate change models (Sofie et al., 2014).

During photosynthesis, plants amass new biomass as they release water in exchange for atmospheric carbon. This biomass moves up the food web when autotrophs are consumed by herbivores, which are in turn consumed by carnivores and other higher life forms up the food web. The process of respiration returns carbon to the atmosphere as CO_2. In addition, when organisms die, and decay carbon also returns to the atmosphere or is integrated into soil along with some of their waste. Burning of biomass also releases large amounts of carbon stored in plants back into the atmosphere. On longer timescales, significant amounts of carbon are transferred between rocks and the ocean and atmosphere through the process of weathering. The resulting sediments along with organic material can be transported from the land to enter the ocean where they sink to the bottom. This carbon from land, as well as carbon atoms in CO_2 absorbed by the ocean from the atmosphere, can become incorporated into $CaCO_3$ shells made by algae, plants, and animals. These shells become buried as the successive layers of sediment are compressed and cemented into limestone rock. Over millions of years, these carbon-bearing rocks can be exposed to sufficient heat and pressure to melt, causing them to release their carbon back into the atmosphere as CO_2 via volcanism. Some of these rocks will also be exposed at the surface of the earth through mountain building and weathering, and the cycling begins again. Carbon from the mantle is also released into the atmosphere as carbon dioxide through volcanic activity. Carbon is also transferred to rocks from the biosphere, via the formation of fossil fuels.

However, this natural carbon cycle has been interrupted by human activities. Before humans perturbed the environment, the carbon emitted by natural processes (like decay of organic matter) was approximately equal to the carbon absorbed (for instance, by vegetation growth), leading to a relatively stable atmospheric CO_2 concentration since the end of the last glaciation 10,000 years ago. This helped ensure that surface temperature remained stable. However, since the advent of the Industrial Revolution ($\sim$1750), atmospheric CO_2 has been increasing tremendously (Falkowski et al., 2000). This return of carbon back into the atmosphere as CO_2 is occurring at a rate that is hundreds to thousands of times faster than it took to bury it, and much faster than it can be removed by the carbon cycle (e.g., by weathering). Thus, the CO_2 released from the burning of fossil fuels is accumulating in the atmosphere, leading to increases in mean temperatures and it also dissolves in the ocean, resulting in acidification. This anthropogenic perturbation is the main driver of climate change and, therefore, it may be inconvenient to talk of natural ecosystem productivity. Although forests are essential in the carbon cycle, recently, it has been noted that in the context of climate change, tropical forests are likely to become a carbon source, owing to continued forest loss and the effect of climate change on the ability of the remaining forests to capture excess atmospheric CO_2 (Mitchard, 2018). There is some uncertainty regarding the influence of extreme events on the carbon cycle as increasing extreme events such as droughts or storms can lead to a decrease in regional ecosystem carbon stocks and, therefore, have the potential to attenuate an expected increase in terrestrial carbon uptake (Reichstein et al., 2013). Many models of the carbon cycle tend to ignore the invaluable role of inland water bodies in holding carbon. Recently, it has been argued that inland waters have a significant role in the sequestration, transport, and mineralization of organic carbon that is similar to terrestrial carbon sink for anthropogenic emissions (Battin et al., 2009). Therefore, integration of these fluxes into the traditional carbon cycle is needed for appropriate CO_2 management and climate change mitigation.

5 Climate variability effects on natural ecosystem productivity and the carbon cycle

The terrestrial biosphere is a key component of the global carbon cycle and its carbon balance is strongly influenced by climate. Natural ecosystem productivity is tightly coupled with climate variability. Understanding the variability of the carbon cycle is a daunting task due to interconnections between different components. For example, separating trends and induced variability from natural variability and random sampling errors. However, given the large datasets of CO_2 observation since the 1960s, variability in carbon can now be better explained (Baldocchi et al., 2016). The variations in the CO_2 growth rate are linked to climate factors related to heat and drought on the vegetation, leading to ecosystem carbon losses through decreased vegetation productivity and/or increased respiration (Heimann and Reichstein, 2008). Climate extremes such as droughts or storms can lead to a decrease in regional ecosystem carbon stocks and, therefore, have the potential to offset an expected increase in terrestrial carbon uptake. For example, in the *Caatinga* system, Mendes et al. (2020) demonstrated that the dynamics of the components of the CO_2 fluxes varied depending on the magnitude and distribution of rainfall and, as a consequence, on the variability of the vegetation state. This was because the length of the wet period and total accumulated rainfall modulated the behaviors of carbon fixation rates in the *Caatinga* biome. The magnitude of change and the timing and magnitude of climate variability were deemed critical in ecosystem productivity. For example, Craine et al. (2012) demonstrated that drought and high intensity of precipitation reduced grass productivity only during a 110-day period, whereas high temperatures reduced productivity only during 25 days in July in a temperate humid grassland. However, the response of ecosystems to climate factors such a precipitation variability is difficult to constrain due to myriad physiological and abiotic variables that limit terrestrial productivity.

Biodiversity dynamics are also critical for explaining variation in ecosystem productivity. In the Western Hemisphere, across multiple biomes, regions of greater plant diversity were not so sensitive to temperature variability at the interannual and seasonal scales compared to those of less plant diversity (Oliveira et al., 2022). Although these areas displayed lower sensitivity to interannual variability in precipitation, they were highly sensitive to precipitation seasonality. Therefore, conserving landscapes of greater diversity may help stabilize ecosystem functioning under climate change, possibly securing the continuous provision of productivity-related ecosystem service to people (Oliveira et al., 2022). This was also buttressed by Hisano et al. (2018) who observed that biodiversity itself may improve biodiversity, as more-diverse systems could be more resilient to climate change impacts, through the positive relationship between diversity and ecosystem functioning.

Ecosystem productivity is not only influenced by climate variability but other factors related to land management. For example, a terrestrial carbon cycle can be due to changes in constituent fluxes and losses as a result of fire, soil erosion, or harvesting (Baldocchi et al., 2016). It has been reported that global land vegetation can sequester more CO_2 if location-specific optimal land management practices are taken into consideration (Baldocchi et al., 2016; Sha et al., 2022). This suggests that optimizing land management could be critical in attenuating emissions to mitigate climate change. Overall, it has been demonstrated that the variability of the global carbon cycle is changing as the amplitude of seasonal CO_2 concentration and the net land sink are increasing (Baldocchi et al., 2016). The causes of these trends in the global carbon cycle remain uncertain and a challenge for future research (Baldocchi et al., 2016).

Climate variability is tightly coupled with ecosystem productivity. For example, model simulations estimate that over the past 95 years, a combination of increasing atmospheric CO_2 with historical temperature and precipitation variability caused a 4.2% (4.3 Pg C) decrease in the total carbon storage of potential vegetation in the conterminous United States, with vegetation carbon decreasing by 7.2% (3.2 Pg C) and soil organic carbon decreasing by 1.9% (1.1 Pg C) due to dynamics in dry periods (Tian et al., 1999). In addition, precipitation variability alone reduced total carbon storage by 9.5% while temperature variability alone did not significantly affect carbon storage. The effect of CO_2 fertilization alone increased total carbon storage by 4.4% (Tian et al., 1999). Similarly, droughts influenced carbon dynamics, for example, in the Amazon Basin, during drought years, the ecosystems acted as a source of carbon to the atmosphere and during years of normal rains, these ecosystems were a carbon sink (Tian et al., 1998). The same study reported that soil moisture influenced carbon storage dynamics in the Amazon Basin.

Climate variability also influences ecosystem productivity on maritime surfaces. For example, the spatial variability in the physical oceanographic responses to the NAO forcing led to spatial differences in biological responses (Straile et al., 2003), and this affected ecosystem productivity in oceans. In addition, in a simulation study, Le Mézo et al. (2016) showed that the temporal variability of marine pelagic organisms' abundance is not only influenced by natural climate fluctuations but also by the structure of the pelagic community. Consequently, the size- and community-dependent response of marine ecosystems to climate variability could impact overall maritime ecosystem productivity in the context of climate variability.

6 Climate variability and spatial variations in natural ecosystem productivity

Terrestrial global ecosystem productivity varies spatially. For example, in a modeling study, Li et al. (2015) showed that NPP had an increasing temporal trend in the past 150 years, which also indicated that the terrestrial ecosystem was positively responding to climate change in that time period. In terms of the global average, the simulated NPP varied from 21.5 to 69.3 Pg C y^{-1}, and the multimodel ensemble NPP was about 50.6, which was almost consistent with the International Geosphere Biosphere Program (IGBP) NPP result of 55.1 and Moderate Resolution Imaging Spectroradiometer (MODIS) NPP results of 60.5 Pg C y^{-1} (Li et al., 2015). The largest NPP distribution was found between 30°S and 0° for IGBP with 657.259 g C m^{-2} y^{-1} and MODIS with 687.042 g C m^{-2} y^{-1}, and the lowest NPP distribution was found between 60°N and 90°N for IGBP with 223.217 g C m^{-2} y^{-1} and MODIS with 255.243 g C m^{-2} y^{-1} (Li et al., 2015). However, identifying the spatio-temporal variations of global ecosystem productivity under future climate changes has not been described adequately. In a modeling study, Lu et al. (2024) predicted that the global GPP will increase under the future climate change in the 21st century. The annual GPP is expected to be 115.122, 116.537, 117.626, and 120.097 Pg C in 2100 under four future scenarios, and the predicted monthly GPP showed seasonal differences. Meanwhile, GPP tended to increase in the northern mid-high latitude regions and decrease in the lower latitudes (equator). For the climate zones form Köppen-Geiger classification, the arid, cold, and polar zones presented increased GPP, while GPP in the tropical zone will decrease in the future (Lu et al., 2024).

Plant traits are essentially a function of climate, and they are critical in predicting ecosystem production. Consequently, climate change or variability is likely to alter plant traits and this has serious implications on ecosystem productivity. Madani et al. (2018) reported that by 2070, tropical and arid ecosystems will be more suitable for plants with relatively lower canopy height, specific leaf area, and seed mass, while far northern latitudes will favor woody and taller plants than at present. The same study predicted that annual GPP in northern biomes (≥45°N) will increase by 31% (+8.1 ± 0.5 Pg C), but this will be offset by a 17.9% GPP decline in the tropics (−11.8 ± 0.84 Pg C).

Climate variability influences spatial variations in natural ecosystem productivity. For example, climate variability will have a positive effect on the Italian forests as GPP was likely to increase particularly over the evergreen forest types rainfall (Fibbi et al., 2019). However, this increasing trend should be reduced by the end of the century for all forest types except mountain conifers because of increasing temperature and decreasing rainfall (Fibbi et al., 2019). It has been demonstrated that the spatio-temporal patterns of vegetation respond to precipitation and meteorological variables. For example, vegetation, particularly in the wetter areas, does not always respond directly and proportionately to precipitation variation; either because of antecedent soil moisture conditions or because variations in temperature and humidity attenuate the vegetation responses to changes in water availability (Rishmawi et al., 2016). In addition, productivity was also influenced by intra-annual precipitation, regardless of changes in total precipitation. In the Yellowstone Basin, the annual mean temperature showed a positive correlation with NPP in 99.6% of the basin, 91.4% of which passed the 0.01 significant test. In addition, NPP and annual precipitation positively correlated in 87.1% of the basin, 41.7% of which passed the 0.01 significant test. In Brazil, rainfall variation explained as much as 50% of plant productivity, using the vegetation index as a proxy, and rainfall anomaly explained 80% of the vegetation productivity anomaly (Salimon and Anderson, 2018). In 75.2% of the basin, NPP was related negatively to annual sunshine hours, 17.6% of which passed the 0.01 significant test (Tian et al., 2022). In another global study, water availability was found to be a dominant driver of the local interannual variability in GPP and terrestrial ecosystem respiration and at a global stage the net ecosystem exchange (NEE), temporal NEE variability is mostly driven by temperature fluctuations (Jung et al., 2017). Overall, the same study revealed that spatial climate covariation drives the global carbon cycle response. These results indicate the close coupling between climate and ecosystem production.

Climate variability also influenced ecosystem productivity as a global study showed an increasing trend in potential NPP and actual net primary production, while there was a decreasing trend of the vegetation production gap in most parts of the world and this trend is more significant under climate change (Cao et al., 2023). At the same time, evaluating the climate potential productivity (CPP) of terrestrial vegetation is crucial to ascertain the threshold of vegetation productivity. In a global study, the CPP showed signs of productivity saturation and that future climate may not be conducive to the increases in vegetation productivity in tropical rainforests and arid/semi-arid areas due to decreased CPP. Under climate change, increases in average annual temperature, minimum temperature, and solar radiation will be beneficial to CPP increase in most parts of the globe. These results may help to inform targeted policies for future sustainable development, resource allocation, and vegetation management.

In a global study, using the vegetation sensitivity index, Seddon et al. (2016) found that the Arctic tundra, parts of the boreal forest belt, the tropical rainforest, alpine regions worldwide, steppe and prairie regions of Central Asia and North and South America, the Caatinga deciduous forest in eastern South America, and eastern areas of Australia were ecologically sensitive with amplified responses to climate variability. This suggests that ecosystems in drier conditions tend to show higher sensitivity to water variability, and ecosystems with colder climate are likely to be more sensitive to temperature variability (Seddon et al., 2016; You et al., 2018). This variation in sensitivity also has implications for ecosystem productivity and ecosystem well-being. In China (Heihe River Basin), You et al. (2018) found that the sensitivity and resilience of vegetation to climate variability were different in the upper, middle, and lower reaches of the basin. Temperature variability dominantly controlled vegetation anomalies in the upper reach, but water variability was the dominant climatic factor in the middle and lower reaches. The different responses of semi-arid to hyper-arid ecosystems to climate variability depended much on the distinct climatic conditions and diverse vegetation types. Understanding these differences is critical when planning climate change mitigation and adaptation interventions.

In a study in China, vegetation in the Inner Mongolia Plateau is strongly affected by precipitation anomalies and vegetation has a 1–2-month lag response to precipitation anomalies and is significantly correlated with 2–6 months cumulative precipitation anomalies. For example, the largest normalised vegetation difference index (NDVI) precipitation correlation appeared in areas with 150–500 mm of mean annual precipitation, 0.075–0.275 of a fraction of precipitation days, and 19–23 of precipitation concentration index (Chen et al., 2020). In a study covering 50% of global land and five biomes, the productivity of mesic grasslands and forests decreased by ~28% and ~7% (respectively) in response to more irregular rain within the year, while the sensitivity was halved in response to higher year-to-year variability. On the contrary, the productivity of boreal forests increased under higher variability on both timescales (Ritter et al., 2020).

Detection and identification of the impacts of climate change on ecosystems are critical. A comparison of trends in actual productivity (NDVI) with climate-induced potential productivity showed that the trends in average productivity in nearly 40% of global grassland areas have been significantly affected by climate change. The contribution of climate change to variability in grassland productivity was 15.2%–71.2% during 1982–2011 (Gao et al., 2016).

However, determining the effects of nonuniform climate warming on terrestrial ecosystems is a key challenge in carbon cycle research. For example, spring warming has largely stimulated ecosystem productivity at latitudes between 30° and 90°N, but suppressed productivity in other regions (Xia et al., 2014). This necessitates the need to evaluate ecosystem productivity resilience to climate variability. In this regard, Seddon et al. (2016) found that ecologically sensitive regions with amplified responses to climate variability were located in the Arctic tundra, parts of the boreal forest belt, the tropical rainforest, alpine regions worldwide, steppe and prairie regions of Central Asia and North and South America, the Caatinga deciduous forest in eastern South America, and eastern areas of Australia. This is important as it helps to give insights to why some regions appear to be more sensitive than others, and what impact this has on the resilience of ecosystem service provision and human well-being.

Across biomes, drought conditions during critical subannual climate periods could have a strong negative impact on ecosystem production. For example, in the southwest United States, the forests' gross ecosystem production (GEP) was driven by cool-season precipitation and constrained by warm-season maximum temperature. GEP in both shrublands and grasslands was driven by summer precipitation and constrained by high daily summer maximum temperatures. In grasslands, there was a negative relationship between temperature and GEP in July, but no relationship in August and September (Barnes et al., 2016). In a study in the Tibetan Plateau, climate variability greatly influenced alpine grass productivity. For example, increasing temperature and precipitation resulted in an increase in above-ground net primary production (ANPP). The ANPP positively correlated with annual temperature and precipitation, but negatively with grazing intensity for both periods, an indication that precipitation variability was the key factor for alpine grassland ANPP increase across Tibet (Li et al., 2021). In a model intercomparison study, O'Sullivan et al. (2020) found that tropical regions exhibit the largest interannual variability in GPP, with both rainforests and savannas contributing substantially. Variability in savanna GPP was likely predominantly driven by water availability, although temperature could play a role via soil moisture-atmosphere feedbacks. In addition, Gwate and Ndou (2022) reported a unimodal relationship between precipitation and NPP, an indication that beyond a critical threshold, ecosystem productivity declines despite increasing rainfall. This illustrates the vulnerability of ecosystem production to precipitation variability. Similarly, increased precipitation variability significantly reduced ecosystem primary production. For example, dominant plant functional types showed opposite responses as perennial-grass productivity decreased by 81%, whereas shrub productivity increased by 67% (Gherardi and Sala, 2015). This pattern was explained by different nonlinear responses to precipitation. Grass productivity presented a saturating response to precipitation where dry years had a larger negative effect than the positive effects of wet years. In contrast, shrubs showed an increasing response to precipitation that resulted in an increase in average productivity with increasing precipitation variability. This is so because under drought conditions, grasses reduce their abundance and their ability to absorb water is then transferred to deep soil layers that are exclusively explored by shrubs.

7 Climate change mitigation in natural ecosystems

Climate variability is affecting all aspects of life on Earth. For example, species are responding to climate change through changes in morphology and behavior, phenology, and geographic range shifts owing to plastic and evolutionary responses (Weiskopf et al., 2020). In addition, climate variability is leading to widespread changes in productivity, species interactions, vulnerability to biological invasions, and other emergent properties. As a result, nature's contribution to society is also grossly altered. Admittedly, not all impacts are negative, even positive climatic changes can require costly societal adjustments. Consequently, natural resources managers need proactive, flexible adaptation strategies that consider historical and future outlooks to minimize costs over the long term. Although such initiatives are emerging, implementation is not yet widespread and systematic due to a lack of information and institutional barriers (Weiskopf et al., 2020).

Consequently, industrialized and developing countries face policy adoption or outcome gaps. Hence, there is a need for better measurement of both components of the implementation gap, particularly when it comes to the policy outcome gap, and for analysis of the causes across countries. This will help devise strategies to close these gaps to ensure that tenets of the Paris Agreement to limit warming to 1.5°C are met (Fransen et al., 2023). However, there are considerable gaps and challenges in reporting ecosystem emissions and removals. Furthermore, confounding factors can influence the assessment of mitigation outcomes, including the slow rate of carbon sequestration in many ecosystems and the risk of stored carbon being released by wildfire or land-use change. Hence, long-term measurement is required, as patterns of carbon uptake change over time (Morecroft et al., 2019).

Climate change and climate variability may result in ecosystem degradation and hence ecosystem restoration could be critical in sustainably managing the earth system. This will require reducing net GHG emissions and adapting to changes that cannot be prevented (Morecroft et al., 2019). This could be achieved through managing carbon in biological systems, safeguarding existing stores of carbon, reducing emissions, and maximizing the potential of natural and agricultural areas for removing carbon from the atmosphere (Trumper et al., 2009). For instance, reducing deforestation rates by 50% by 2050 and then maintaining them at this level until 2100 would avoid the direct release of up to 50 Gt C this century, an equivalent of 12% of the emissions reductions needed to keep atmospheric concentrations of carbon dioxide below 450 ppm (Trumper et al., 2009). Consequently, protecting the natural resources base is fundamental to climate management. For example, intact forests are important for carbon storage, carbon sequestration, local climate regulation, water supply, and biodiversity (Morecroft et al., 2019). However, the widespread incorporation of climate change into natural resource management has remained elusive (Weiskopf et al., 2020).

Nature-based solutions (NBS) arguably provide the best way of delivering land-based options, climate mitigation through protection, and restoration and sustainable management of natural carbon sinks and reservoirs. For example, NBS implemented across all ecosystems can deliver emission reductions and removal of at least 5 Gt CO_2e per year, with a maximum estimate of 11.7 Gt CO_2e per year by 2030 (UNEP and IUCN, 2021). By 2050, this will rise to at least 10 Gt CO_2e per year, with a maximum estimate of 18 Gt CO_2e per year (UNEP and IUCN, 2021). NBS can contribute to making our cities carbon-neutral. For example, assuming maximum theoretical implementation, NBS in the residential, transport, and industrial sectors could reduce urban carbon emissions by up to 25% in Europe. Spatially prioritizing different types of NBS in 54 major European Union cities could reduce anthropogenic carbon emissions by an average of 17.4% (Pan et al., 2023). Coupling NBS with other existing measures under climate change could reduce total carbon emissions by 57.3% in 2030, with both indirect pathways and sequestration. The value and importance of nature need to be better reflected in economic and political decision-making and in a stronger integration between biodiversity, climate change, and development agendas. There is a need to more effectively communicate the evidence base that already exists so that scientific knowledge is comprehensible to decision-makers to generate political will. Isolated islands of success of NBS should serve as learning sites and existing information could be systematized and effectively (Pramova et al., 2012). It will also be vital to identify how key elements of the complexity that enhance resilience and adaptation can be supported and propagated. Maintaining long-term monitoring, which is the only way to fully understand trajectories in complex contexts and evaluate the success of management interventions will be important in implementing NBS (Malhi et al., 2020).

NBS should be carefully planned to achieve their full benefits. Within the armpit of NBS, tree planting in open ecosystems such as savannas is being encouraged in pursuit of restoration objectives. However, the major concern is tree planting in inappropriate places. For example, in grassy ecosystems of Africa, erroneously identified as degraded, 1 million km^2 of land has been targeted for restoration through tree planting under the assumption that they are degraded while this is often not the case (Bond et al., 2019). In addition, tree planting in such ecosystems also faces competition from other important land uses such as crop cultivation and has deleterious effects on the hydrological cycle. At the same time, it is believed that intentionally allowing or promoting invasion by nonnative trees into areas characterized by treeless vegetation could contribute to climate change mitigation by increasing carbon sequestration. For example, nonnative tree invasions can increase carbon sequestration in some ecosystems. However, the negative impacts of this far outweigh the positives as tree invasions can alter fire regimes, soil C sequestration, and light absorption, all of which influence the impact invaders may have on climate regulation (Nuñez et al., 2021). Hence, success in the deployment of NBS should be monitored and evaluated, to enjoy the full benefits. To enable this to happen, metrics are required for biodiversity, mitigation, and adaptation parameters, which reflect the needs of local communities as well as international reporting (Morecroft et al., 2019). Furthermore, to ensure successful implementation of NBS, it will be necessary to understand how ecological systems function, how they are changing and will continue to change under environmental conditions with no historical analog, and what interventions are needed to maintain and restore ecosystems. In addition, in terms of environmental economics, there is a need to understand the costs and benefits of any intervention, and how those costs and benefits are distributed across society (Malhi et al., 2020).

At the same time, ecosystem restoration will not be possible in all places. For example, many formerly forested and peatland areas have been cleared and drained for agriculture (Morecroft et al., 2019). Ecosystem restoration in such areas may reduce crop production or displace it to other locations where it may have an equally or even more negative impact. Such ecosystems may have been transformed beyond restoration, becoming novel systems (Gwate et al., 2016; Hobbs et al., 2014). Therefore, trying to restore such systems could be futile and society needs to accept that some degraded system as a result of climate variability may not be restored.

It is essential that climate mitigation policy is guided by the best available science concerning ecosystem carbon, and decisions should be informed by the overall costs and benefits of carbon management. Therefore, strengthening the science and policy interface for biodiversity and ecosystem services will contribute to the management and sustainable use of biodiversity and ecosystem, to sustainable development, and has a great role in carbon sequestration and reducing the GHG from the atmosphere to mitigate climate change (Daba and Dejene, 2018; Trumper et al., 2009). Priority research areas included general goals, policy and implementation strategies for conservation and resource use, integrated (ecosystem-based) management and adaptation, and managing threats and stressors (Williams et al., 2019). These research priorities provide a unified guide for policymakers, funding organizations and researchers to strategically direct resources, maximize stakeholder uptake of resulting knowledge, and minimize the impacts of climate change on natural ecosystems. Given the pace of climate change, it is imperative that we inform and accelerate adaptation progress in all regions around the world (Williams et al., 2019). Synergistic interactions between climate change and other human pressures likely pose the most important threats to global biodiversity. Understanding the impacts of such interactions across multiple scales will assist in allocating resources between ameliorating existing stressors and implementing new adaptive strategies that specifically address climate change (e.g., species translocations). The best solution may not be the optimum for any individual stressor but rather one that maximizes the benefits across multiple stressors and manages natural assets and dependent human systems. Furthermore, funding for technology research and development is also a very important aspect moving forward (Fawzy et al., 2020).

Energy-related CO_2 absorption and fixation are key solutions to global warming. In pursuit of this, economic instruments such as carbon tax will be key as they are likely to reduce GHG emissions (Trumper et al., 2009). Carbon markets are developing in industrialized countries. Consequently, the portfolios of policy instruments used by developed countries should be widened, increasing the portfolio coverage of those instruments to all sectors including in developing countries (Birkmann and Lauer, 2022). Negative carbon emissions contribute immensely to negative emissions. To aggressively drive negative emissions projects, policymakers and governments should devise appropriate policy instruments and support frameworks with a special focus on carbon pricing. Furthermore, the financial industry should provide enhanced financial support and efficiently utilize financial resources and policy support as most of the related technologies can be deployed immediately; however, efficient carbon pricing mechanisms that focus on carbon removal need to be aggressively developed and introduced (Fawzy et al., 2020). Currently, carbon pricing would be insufficient to economically sustain carbon removal projects, apart from the existing framework for afforestation and reforestation projects.

Biodiversity and ecosystem service play an important role in carbon sequestration in pursuit of climate mitigation. It is unequivocal that well managed and protected biodiversity can deliver vital ecosystem services for climate change adaptation and mitigation through carbon capture (Daba and Dejene, 2018). For example, decreasing biodiversity due to climate change may inadvertently undermine ecosystem services such as carbon sequestration and reduce the GHG from the atmosphere to mitigate climate change.

Decarbonization will be key to climate change mitigation. Therefore, ecosystems that contain "irrecoverable carbon" are vulnerable to release upon land use conversion and, once lost, may not be recoverable on timescales relevant to avoiding dangerous climate impacts. Such ecosystems include high densities in peatlands, mangroves, old-growth forests, and marshes and are believed to contain at least 260 Gt of irrecoverable carbon. To achieve climate goals, we must safeguard these irrecoverable carbon pools through an expanded set of policy and finance strategies (Goldstein et al., 2020). Related to this, vegetated marine systems are important in carbon sequestration. For example, marine vegetated habitats contribute 50% of carbon burial in marine sediments. Their canopies dissipate wave energy and high burial rates raise the seafloor, buffering the impacts of rising sea levels and wave action that are associated with climate change. The loss of a third of the global cover of these ecosystems involves a loss of CO_2 sinks and the emission of 1 Pg CO_2 annually. The conservation, restoration, and use of vegetated coastal habitats in eco-engineering solutions for coastal protection provide a promising strategy, delivering significant capacity for climate change mitigation and adaptation (Duarte et al., 2013).

8 Case studies of potentially viable mitigation strategies

Adapting to current climate variability rather than climate change will reduce vulnerability to climate change. It is important to consider the dynamics of vulnerability in a connected system to successfully adapt to climate variability. In addition, there is a need for greater engagement with various publics on the trade-offs involved in adaptation action and for improving communication about the complicated nature of the dynamics of vulnerability (Dilling et al., 2015).

Organized networks for data collection are critical in helping society to understand ecosystem's productivity to improve climate change mitigation. For example, the global ecosystems monitoring network is a network of sites where the productivity and carbon cycling of terrestrial ecosystems are tracked through a standard protocol, and frequently integrated with data on plant functional traits and broader environmental variables, such as tree species community, composition, soil, and climate (Malhi et al., 2021). This provides a repository for ecosystem productivity data to cover all tropical continents. Similarly, the Fluxnet network provides invaluable ecosystem production data to inform management (Schimel et al., 2015). Such organized networks collect important data to inform climate variability management.

Ecosystems play a pivotal role in climate mitigation owing to their ability to sequester carbon via the process of photosynthesis to offset emissions. Therefore, mitigation strategies in terrestrial ecosystems include protection of intact carbon stores, avoiding deforestation, reforestation of formerly forested land, and restoration of degraded peatlands (Morecroft et al., 2019). Using a global compilation of forest age observations, combined with a terrestrial biosphere model with explicit modeling of forest regrowth, to partition the global forest carbon sink between old-growth and regrowth stands over the period 1981–2010. Pugh et al. (2019) reported a carbon sink of 0.85 (0.66–0.96) Pg y^{-1} located in intact old-growth forest, primarily in the moist tropics and boreal Siberia, and 1.30 (1.03–1.96) Pg y^{-1} located in stands regrowing after past disturbance. This demonstrates successful rehabilitation as forests experienced autogenic regeneration. The same study estimated that forests will accumulate an additional 69 (44–131) Pg C in live biomass from changes in demography alone if natural disturbances, wood harvest, and reforestation continue at rates comparable to those during 1981–2010. Therefore, forest disturbance and allowing for regeneration could be critical in attenuating CO_2 emissions, contributing to climate change mitigation (Pugh et al., 2019). Similarly, protecting irrecoverable carbon stocks such as those in peatlands, mangroves, old-growth forests, and marshes could proffer successful mitigation to climate change within a broader context of mangroves, old-growth forests, and marshes (Goldstein et al., 2020). This may require further development of a set of responsive policy and finance strategies.

Managing and protecting marine vegetated habitats (seagrasses, saltmarshes, macroalgae, and mangroves) is a viable strategy for climate change mitigation. These habitats only occupy 0.2% of the ocean surface, but contribute 50% of carbon burial in marine sediments (Duarte et al., 2013). The loss of a third of the global cover of these ecosystems involves a loss of CO_2 sinks and the emission of 1 Pg CO_2 annually. Similarly, conservation of these landscapes will result in concomitant benefits in terms of climate mitigation. Rehabilitation work is going on in various parts of the world, including the reforestation of the Mekong Delta mangrove forests through the initiative of the Vietnamese government, large-scale mangrove afforestation programs have been successful in Thailand and India, and large-scale salt marsh restoration schemes are in place in China and the United States (Duarte et al., 2013). All these projects have a net impact of sequestering large quantities of carbon, resulting in climate change mitigation. Within the broader context of Blue Carbon initiatives, in Southeast Asia including the Andaman and Nicobar Islands of India, Stankovic et al. (2021) found that the average value of the total carbon storage within seagrass meadows was 121.95 ± 76.11 Mg ha^{-1} (average $\pm$ SD) and the total organic carbon, Corg stock of the seagrass meadows of this region was 429.11 ± 111.88 Tg, with the highest Corg stock in the Philippines (78%). The seagrass meadows of this region have the capacity to accumulate 5.85–6.80

Tg C y^{-1}, which accounts for USD 214.6–249.4 million. Under the current rate of decline of 2.82%, the seagrass meadows are emitting 1.65–2.08 Tg of CO_2 y^{-1} and the economic value of these losses accounts for USD 21.42–24.96 million (Stankovic et al., 2021). The potential of the seagrass meadows to offset current CO_2 emissions varies across the region, with the highest contribution to offset in the seagrass meadows of the Philippines (11.71%). Therefore, the Blue Carbon initiative provides a viable opportunity for successful mitigation and shows the invaluable role of biodiversity itself, in mitigating climate change as more-diverse systems could be more resilient to climate change impacts, and ecosystem functioning through the positive relationship between diversity and ecosystem functioning (Li et al., 2018).

Sustainable mitigation of climate change requires a holistic approach such as the water-food-energy nexus to better understand how some choices made in one part of the system might impact other valued outcomes or even create new vulnerabilities (Dilling et al., 2015). In addition, successful mitigation should take place with a broader understanding of threats and challenges to ecosystems, opportunities for improving ecosystem and societal resilience, the potential of NBS, and the invaluable role of academic research (Malhi et al., 2020). For example, Williams et al. (2019) noted that research priorities for natural ecosystems in a changing global climate should include general goals, policy and implementation strategies for conservation and resource use, integrated (ecosystem-based) management and adaptation across diverse and multiuse land, water and seascapes, managing threats and stressors as well as managing natural assets and dependent human system. Overall, greater engagement with various publics on the tradeoffs involved in adaptation action and improving communication about the complicated nature of the dynamics of vulnerability is crucial (Dilling et al., 2015). To better anticipate the adverse impacts of climate variability on ecosystem production, national and international programs will be critical to mobilize efforts to achieve necessary synergies (Fisher et al., 2017). Li et al. (2018) found that, under current sensitivity and resilience conditions, vulnerable areas were projected to develop in high Northern Hemisphere latitudes under climate change. Overall, the integration of all three aspects of vulnerability (exposure, sensitivity, and resilience) may offer more comprehensive and spatially explicit adaptation strategies to reduce the impacts of climate change on terrestrial ecosystems (Li et al., 2018).

9 Social security nets in Ethiopia

Government involvement was critical in successful adaptation. In 2006, the Ethiopian government and international partners established the Productive Safety Net Program Adapting to Climate Change—Three Success Stories (imf.org). The program provided cash and food to families that were unable to feed themselves due to droughts, delayed rains, and flooding that threatened food security. The aid was contingent on participation in local productivity-enhancing or environmental programs—for example, land rehabilitation, improvement of water sources, and construction of infrastructure such as roads and hospitals. Program beneficiaries experienced a 25% smaller drop in consumption after droughts relative to those not in the program. The program has also reduced soil loss by more than 40%, improved the availability of water, and increased land productivity. However, the sustainability of such initiatives is doubtful adaptation to Adapting to Climate Change—Three Success Stories (imf.org).

10 Climate change mitigation

This case study is based on Fekete et al. (2021). An exploration of climate change mitigation policies that lead to reduced GHG emissions, compared to scenarios without such policies revealed that in the past, Organisation of Economic Cooperation and Development (OECD) countries such as the EU Member States, Japan, Norway, and the United States, tended to implement policy packages that led to more rapid progress in improvements to energy intensity and GHG intensity, in comparison to emerging economies and developing countries. However, major emitting developing countries are now catching up and are increasingly implementing similar or even more aggressive packages than those of many OECD countries. The level of implementation of mitigation policies varies per sector as well. There are many policies that have led to clear deviation from business-as-usual trends in energy use and GHG emissions related to renewable electricity, passenger vehicles, and forestry. For industry and buildings, the comparable information available is too limited to allow any conclusions to be drawn. For the industry and freight transport sectors, no ambitious historical policies or forward-looking policies were identified in the five major emitting economies. Global replication of successful sector-level policies and their performance will reduce GHG emissions by 2030 compared to levels achieved under current policies. To fulfill the Paris goals, countries, therefore, need to expand their actions over very rapidly and beyond historical efforts.

To ensure successful mitigation at the global level, there is a need to develop comprehensive policy packages with financial incentives, which address user behavior and limit administrative barriers. Furthermore, we need to provide long-term security to actors in the sector, and we need to focus on underdeveloped sectors (in terms of technology or coverage by the policy framework), where countries have the capacity to do so. Building up transformative policy frameworks for those sectors can drive corresponding climate action around the world and is therefore also a responsibility of OECD countries (Fekete et al., 2021).

11 Conclusion

Climate variability is a natural phenomenon, but it has been amplified by anthropogenic activities since the industrial revolution. This affects ecosystems through modulating productivity by altering the process and magnitude of photosynthesis. Consequently, ecosystem services are affected and this in turn influences human well-being. Reducing greenhouse gas emissions will be key to the future. NBS are promising to reduce global warming through decarbonization and reversing environmental deterioration in pursuit of land degradation neutrality. Understanding dynamics in ecological systems' functionality under climate variability will be critical in successfully applying NBS. Research and development will be vital in propagating NBS through documenting successful case studies and communication. It will also be critical in uncovering priority research areas including general goals, policy and implementation strategies for conservation and resource use, integrated (ecosystem-based) management and adaptation, and managing threats and stressors. When considering ecosystem restoration, it is important to determine if restoration thresholds have not been surpassed, otherwise, society should embrace the system as "novel." The role of economic instruments such as carbon markets should be further developed and implicated throughout the globe to attenuate global warming. There is a need to further unroll NBS to promote land degradation neutrality.

References

Baldocchi, D., Ryu, Y., Keenan, T., 2016. Terrestrial carbon cycle variability. F1000Res. 5, 2371. https://doi.org/10.12688/f1000research.8962.1.

Barnes, M.L., Moran, S., Scott, R.L., Kolb, T.E., Ponce-Campos, G.E., Moore, D.J.P., Ross, M.A., Mitra, B., Dore, S., 2016. Vegetation productivity responds to sub-annual climate conditions across semiarid biomes. Ecosphere 7 (5), e01339.

Battin, T.J., Luyssaert, S., Kaplan, L.A., Aufdenkampe, A.K., Richter, A., Tranvik, L.J., 2009. The boundless carbon cycle. Nat. Geosci. 2 (9), 598–600. https://doi.org/10.1038/ngeo618.

Birkmann, J., Lauer, H., 2022. Climate change: mitigation and adaptation. In: Routledge Handbook of Environmental Hazards and Society. vol. 16, pp. 150–167, https://doi.org/10.4324/9780367854584-11.

Bond, W.J., Stevens, N., Midgley, G.F., Lehmann, C.E.R., 2019. The trouble with trees: afforestation plans for Africa. Trends Ecol. Evol. 34 (11), 963–965. https://doi.org/10.1016/j.tree.2019.08.003.

Burroughs, T.J., 2001. Climate Change: A Multidisciplinary Approach. Cambridge University Press, Cambridge, UK.

Cao, M., Woodward, F.I., 1998. Dynamic responses of terrestrial ecosystem carbon cycling to global climate change. Nature 393 (6682), 249–252. https://doi.org/10.1038/30460.

Cao, M., Zhang, Q., Shugart, H.H., 2001. Dynamic responses of African ecosystem carbon cycling to climate change. Clim. Res. 17 (2 Special 8), 183–193. https://doi.org/10.3354/cr017183.

Cao, D., Zhang, J., Zhang, T., Yao, F., Ji, R., Zi, S., 2023. Spatiotemporal variations and driving factors of global terrestrial vegetation productivity gap under the changing of climate, CO_2, landcover and N deposition. Sci. Total Environ. 880 (February), 162753. https://doi.org/10.1016/j.scitotenv.2023.162753.

Chen, Z., Wang, W., Fu, J., 2020. Vegetation response to precipitation anomalies under different climatic and biogeographical conditions in China. Sci. Rep. 10 (1), 830. https://doi.org/10.1038/s41598-020-57910-1.

Craine, J.M., Nippert, J.B., Elmore, A.J., Skibbe, A.M., Hutchinson, S.L., Brunsell, N.A., 2012. Timing of climate variability and grassland productivity. Proc. Natl. Acad. Sci. U. S. A. 109 (9), 3401–3405. https://doi.org/10.1073/pnas.1118438109.

Daba, M.H., Dejene, S.W., 2018. The role of biodiversity and ecosystem services in carbon sequestration and its implication for climate change mitigation. Int. J. Environ. Sci. Nat. Res. 11 (2), 555810. https://doi.org/10.19080/IJESNR.2018.10.5558.

Dilling, L., Daly, M.E., Travis, W.R., Wilhelmi, O.V., Klein, R.A., 2015. The dynamics of vulnerability: why adapting to climate variability will not always prepare us for climate change. Wiley Interdiscip. Rev. Clim. Change 6 (4), 413–425. https://doi.org/10.1002/wcc.341.

Duarte, C.M., Losada, I.J., Hendriks, I.E., Mazarrasa, I., Marbà, N., 2013. The role of coastal plant communities for climate change mitigation and adaptation. Nat. Clim. Chang. 3 (11), 961–968. https://doi.org/10.1038/nclimate1970.

Falkowski, P., Scholes, R.J., Boyle, E., Canadell, J., Canfield, D., Elser, J., Gruber, N., Hibbard, K., Hogberg, P., Linder, S., Mackenzie, F.T., Moore, B., Pedersen, T., Rosental, Y., Seitzinger, S., Smetacek, V., Steffen, W., 2000. The global carbon cycle: a test of our knowledge of earth as a system. Science 290 (5490), 291–296. https://doi.org/10.1126/science.290.5490.291.

Fawzy, S., Osman, A.I., Doran, J., Rooney, D.W., 2020. Strategies for mitigation of climate change : a review Intergovernmental Panel on Climate Change. Environ. Chem. Lett. 18 (6), 2069–2094. https://doi.org/10.1007/s10311-020-01059-w.

Fekete, H., Kuramochi, T., Roelfsema, M., den Elzen, M., Forsell, N., Höhne, N., Luna, L., Hans, F., Sterl, S., Olivier, J., van Soest, H., Frank, S., Gusti, M., 2021. A review of successful climate change mitigation policies in major emitting economies and the potential of global replication. Renew. Sustain. Energy Rev. 137 (December 2020). https://doi.org/10.1016/j.rser.2020.110602.

Fibbi, L., Moriondo, M., Chiesi, M., Bindi, M., Maselli, F., 2019. Impacts of climate change on the gross primary production of Italian forests. Ann. For. Sci. 76 (2). https://doi.org/10.1007/s13595-019-0843-x.

Fisher, J.B., Melton, F., Middleton, E., Hain, C., Anderson, M., Allen, R., Mccabe, M.F., Hook, S., Baldocchi, D., Townsend, P.A., Kilic, A., Tu, K., Miralles, D.D., Perret, J., Lagouarde, J.-P., Waliser, D., Purdy, A.J., French, A., Schimel, D., et al., 2017. The future of evapotranspiration: global requirements for ecosystem functioning, carbon and climate feedbacks, agricultural management, and water resources. Water Resour. Res. 53 (4), 2618–2626. https://doi.org/10.1002/2016WR020175.

Fransen, T., Meckling, J., Stünzi, A., Schmidt, T.S., Egli, F., Schmid, N., Beaton, C., 2023. Taking stock of the implementation gap in climate policy. Nat. Clim. Chang. 1–4. https://doi.org/10.1038/s41558-023-01755-9.

Gao, Q., Zhu, W., Schwartz, M.W., Ganjurjav, H., Wan, Y., Qin, X., Ma, X., Williamson, M.A., Li, Y., 2016. Climatic change controls productivity variation in global grasslands. Sci. Rep. 6 (October 2015), 1–10. https://doi.org/10.1038/srep26958.

Gherardi, L.A., Sala, O.E., 2015. Enhanced precipitation variability decreases grass- and increases shrub-productivity. Proc. Natl. Acad. Sci. U. S. A. 112 (41), 12735–12740. https://doi.org/10.1073/pnas.1506433112.

Ghil, M., 2002. Natural climate variability. In: Encyclopedia of Global Environmental Change. vol. 1, pp. 544–549.

Goldstein, A., Turner, W.R., Spawn, S.A., Anderson-Teixeira, K.J., Cook-Patton, S., Fargione, J., Gibbs, H.K., Griscom, B., Hewson, J.H., Howard, J.F., Ledezma, J.C., Page, S., Koh, L.P., Rockström, J., Sanderman, J., Hole, D.G., 2020. Protecting irrecoverable carbon in Earth's ecosystems. Nat. Clim. Chang. 10 (4), 287–295. https://doi.org/10.1038/s41558-020-0738-8.

Gwate, O., Ndou, G., 2022. Exploring water use and production dynamics of an indigenous African dry forest in south-western Zimbabwe. J. Arid Environ. 198, 104678. https://doi.org/10.1016/j.jaridenv.2021.104678.

Gwate, O., Mantel, S.K., Finca, A., Gibson, L.A., Munch, Z., Palmer, A.R., 2016. Exploring the invasion of rangelands by *Acacia mearnsii* (black wattle): biophysical characteristics and management implications. Afr. J. Range Forage Sci. 33 (4), 265–273. https://doi.org/10.2989/10220119.2016.1271013.

Hauer, M.E., Fussell, E., Mueller, V., Burkett, M., Call, M., Abel, K., McLeman, R., Wrathall, D., 2020. Sea-level rise and human migration. Nat. Rev. Earth Environ. 1 (1), 28–39. https://doi.org/10.1038/s43017-019-0002-9.

Heimann, M., Reichstein, M., 2008. Terrestrial ecosystem carbon dynamics and climate feedbacks. Nature 451 (7176), 289–292. https://doi.org/10.1038/nature06591.

Hisano, M., Searle, E.B., Chen, H.Y.H., 2018. Biodiversity as a solution to mitigate climate change impacts on the functioning of forest ecosystems. Biol. Rev. 93 (1), 439–456. https://doi.org/10.1111/brv.12351.

Hobbs, R.J., Higgs, E.S., Harris, J.A., 2014. Novel ecosystems: concept or inconvenient reality? A response to Murcia et al. Trends Ecol. Evol. 29 (12), 645–646. https://doi.org/10.1016/j.tree.2014.09.006.

IPCC, 2012. Annex 1: glossary of terms. In: Field, C.B., Barros, V., Stocker, T.F., Qin, D., Dokken, D.J., Ebi, K.L., Tignor, P.M.M.M. (Eds.), Managing the Risks of Extreme Events and Disasters to Advance Climate Change Adaptation. Cambridge University Press, UK, and New York, NY, pp. 555–564, https://doi.org/10.1002/9783527612024.oth1.

Jung, M., Reichstein, M., Schwalm, C.R., Huntingford, C., Sitch, S., Ahlström, A., Arneth, A., Camps-Valls, G., Ciais, P., Friedlingstein, P., Gans, F., Ichii, K., Jain, A.K., Kato, E., Papale, D., Poulter, B., Raduly, B., Rödenbeck, C., Tramontana, G., et al., 2017. Compensatory water effects link yearly global land CO_2 sink changes to temperature. Nature 541 (7638), 516–520. https://doi.org/10.1038/nature20780.

Le Mézo, P., Lefort, S., Séférian, R., Aumont, O., Maury, O., Murtugudde, R., Bopp, L., 2016. Natural variability of marine ecosystems inferred from a coupled climate to ecosystem simulation. J. Mar. Syst. 153 (September), 55–66. https://doi.org/10.1016/j.jmarsys.2015.09.004.

Le Quéré, C., Andres, R.J., Boden, T., Conway, T., Houghton, R.A., House, J.I., Marland, G., Peters, G.P., Van Der Werf, G.R., Ahlström, A., Andrew, R.M., Bopp, L., Canadell, J.G., Ciais, P., Doney, S.C., Enright, C., Friedlingstein, P., Huntingford, C., Jain, A.K., et al., 2013. The global carbon budget 1959-2011. Earth Syst. Sci. Data 5 (1), 165–185. https://doi.org/10.5194/essd-5-165-2013.

Li, S., Lü, S., Zhang, Y., Liu, Y., Gao, Y., Ao, Y., 2015. The change of global terrestrial ecosystem net primary productivity (NPP) and its response to climate change in CMIP5. Theor. Appl. Climatol. 121 (1–2), 319–335. https://doi.org/10.1007/s00704-014-1242-8.

Li, D., Wu, S., Liu, L., Zhang, Y., Li, S., 2018. Vulnerability of the global terrestrial ecosystems to climate change. Glob. Chang. Biol. 24 (9), 4095–4106. https://doi.org/10.1111/gcb.14327.

Li, M., Wu, J., Feng, Y., Niu, B., He, Y., Zhang, X., 2021. Climate variability rather than livestock grazing dominates changes in alpine grassland productivity across Tibet. Front. Ecol. Evol. 9 (March), 1–12. https://doi.org/10.3389/fevo.2021.631024.

Lu, Q., Liu, H., Wei, L., Zhong, Y., Zhou, Z., 2024. Global prediction of gross primary productivity under future climate change. Sci. Total Environ. 912 (December 2023), 169239. https://doi.org/10.1016/j.scitotenv.2023.169239.

Madani, N., Kimball, J.S., Ballantyne, A.P., Affleck, D.L.R., Van Bodegom, P.M., Reich, P.B., Kattge, J., Sala, A., Nazeri, M., Jones, M.O., Zhao, M., Running, S.W., 2018. Future global productivity will be affected by plant trait response to climate. Sci. Rep. 8 (1), 1–10. https://doi.org/10.1038/s41598-018-21172-9.

Malhi, Y., Franklin, J., Seddon, N., Solan, M., Turner, M.G., Field, C.B., Knowlton, N., 2020. Climate change and ecosystems: threats, opportunities and solutions. Philos. Trans. R. Soc. Lond. B Biol. Sci. 375 (1794). https://doi.org/10.1098/rstb.2019.0104.

Malhi, Y., Girardin, C., Metcalfe, D.B., Doughty, C.E., Aragão, L.E.O.C., Rifai, S.W., Oliveras, I., Shenkin, A., Aguirre-Gutiérrez, J., Dahlsjö, C.A.L., Riutta, T., Berenguer, E., Moore, S., Huasco, W.H., Salinas, N., da Costa, A.C.L., Bentley, L.P., Adu-Bredu, S., Marthews, T.R., et al., 2021. The Global Ecosystems Monitoring network: monitoring ecosystem productivity and carbon cycling across the tropics. Biol. Conserv. 253 (November 2020). https://doi.org/10.1016/j.biocon.2020.108889.

Mendes, K.R., Campos, S., da Silva, L.L., Mutti, P.R., Ferreira, R.R., Medeiros, S.S., Perez-Marin, A.M., Marques, T.V., Ramos, T.M., de Lima Vieira, M.M., Oliveira, C.P., Gonçalves, W.A., Costa, G.B., Antonino, A.C.D., Menezes, R.S.C., Bezerra, B.G., Santos E Silva, C.M., 2020. Seasonal variation in net ecosystem CO_2 exchange of a Brazilian seasonally dry tropical forest. Sci. Rep. 10 (1), 9454. https://doi.org/10.1038/s41598-020-66415-w.

Mitchard, E.T.A., 2018. The tropical forest carbon cycle and climate change. Nature 559 (7715), 527–534. https://doi.org/10.1038/s41586-018-0300-2.

Morecroft, M.D., Duffield, S., Harley, M., Pearce-Higgins, J.W., Stevens, N., Watts, O., Whitaker, J., 2019. Measuring the success of climate change adaptation and mitigation in terrestrial ecosystems. Science 366 (6471). https://doi.org/10.1126/science.aaw9256.

Nuñez, M.A., Davis, K.T., Dimarco, R.D., Peltzer, D.A., Paritsis, J., Maxwell, B.D., Pauchard, A., 2021. Should tree invasions be used in treeless ecosystems to mitigate climate change? Front. Ecol. Environ. 19 (6), 334–341. https://doi.org/10.1002/fee.2346.

Oliveira, B.F., Moore, F.C., Dong, X., 2022. Biodiversity mediates ecosystem sensitivity to climate variability. Commun. Biol. 5 (1). https://doi.org/10.1038/s42003-022-03573-9.

O'Sullivan, M., Smith, W., Sitch, S., Friedlingstein, P., Arora, V.K., Haverd, V., Jain, A.K., Kato, E., Kautz, M., Lombardozzi, D., Nabel, J.E.M.S., Tian, H., Vuichard, N., Wiltshire, A., Zhu, D., Buermann, W., 2020. Climate-driven variability and trends in plant productivity over recent decades based on three global products. Glob. Biogeochem. Cycles 34, e2020GB006613.

Palmer, P.I., Wainwright, C.M., Dong, B., Maidment, R.I., Wheeler, K.G., Gedney, N., Hickman, J. E., Madani, N., Folwell, S.S., Abdo, G., Allan, R.P., Black, E.C.L., Feng, L., Gudoshava, M., Haines, K., Huntingford, C., Kilavi, M., Lunt, M.F., Shaaban, A., Turner, A.G., 2023. Drivers and impacts of Eastern African rainfall variability. Nat. Rev. Earth Environ. 4 (4), 254–270. https://doi.org/10.1038/s43017-023-00397-x.

Pan, H., Page, J., Shi, R., Cong, C., Cai, Z., Barthel, S., Thollander, P., Colding, J., Kalantari, Z., 2023. Contribution of prioritized urban nature-based solutions allocation to carbon neutrality. Nat. Clim. Chang. 13 (August). https://doi.org/10.1038/s41558-023-01737-x.

Pendergrass, A.G., Knutti, R., Lehner, F., Deser, C., Sanderson, B.M., 2017. Precipitation variability increases in a warmer climate. Sci. Rep. 7 (1), 1–9. https://doi.org/10.1038/s41598-017-17966-y.

Poulter, B., Frank, D., Ciais, P., Myneni, R.B., Andela, N., Bi, J., Broquet, G., Canadell, J.G., Chevallier, F., Liu, Y.Y., Running, S.W., Sitch, S., Van Der Werf, G.R., 2014. Contribution of semi-arid ecosystems to interannual variability of the global carbon cycle. Nature 509 (7502), 600–603. https://doi.org/10.1038/nature13376.

Pramova, E., Locatelli, B., Djoudi, H., Somorin, O.A., 2012. Forests and trees for social adaptation to climate variability and change. Wiley Interdiscip. Rev. Clim. Change 3 (6), 581–596. https://doi.org/10.1002/wcc.195.

Prentice, I.C., Farquhar, G.D., Fasham, M.J.R., Goulden, M.L., Heimann, M., 2001. The carbon cycle and atmospheric carbon dioxide. In: Pitelka, L., Ramirez Rojas, A. (Eds.), Climate Change 2001: The Scientific Basis. Cambridge University Press, Cambridge, UK and New York, NY, pp. 183–237. https://www.ipcc..ch/site/assets/uploads/2018/02/tar-03.pdf. (Accessed 25 June 2024).

Pugh, T.A.M., Lindeskog, M., Smith, B., Poulter, B., Arneth, A., Haverd, V., Calle, L., 2019. Role of forest regrowth in global carbon sink dynamics. Proc. Natl. Acad. Sci. U. S. A. 116 (10), 4382–4387. https://doi.org/10.1073/pnas.1810512116.

Quan, Q., Tian, D., Luo, Y., Zhang, F., Crowther, T.W., Zhu, K., Chen, H.Y.H., Zhou, Q., Niu, S., 2019. Water scaling of ecosystem carbon cycle feedback to climate warming. Sci. Adv. 5 (8), 1–8. https://doi.org/10.1126/sciadv.aav1131.

Ray, D.K., Gerber, J.S., Macdonald, G.K., West, P.C., 2015. Climate variation explains a third of global crop yield variability. Nat. Commun. 6, 1–9. https://doi.org/10.1038/ncomms6989.

Reichstein, M., Bahn, M., Ciais, P., Frank, D., Mahecha, M.D., Seneviratne, S.I., Zscheischler, J., Beer, C., Buchmann, N., Frank, D.C., Papale, D., Rammig, A., Smith, P., Thonicke, K., Van Der Velde, M., Vicca, S., Walz, A., Wattenbach, M., 2013. Climate extremes and the carbon cycle. Nature 500 (7462), 287–295. https://doi.org/10.1038/nature12350.

Rishmawi, K., Prince, S.D., Xue, Y., 2016. Vegetation responses to climate variability in the northern arid to sub-humid zones of sub-Saharan Africa. Remote Sens. 8, 910. https://doi.org/10.3390/rs8110910.

Ritter, F., Berkelhammer, M., Garcia-Eidell, C., 2020. Distinct response of gross primary productivity in five terrestrial biomes to precipitation variability. Commun. Earth Environ. 1 (1), 1–8. https://doi.org/10.1038/s43247-020-00034-1.

Salimon, C., Anderson, L., 2018. How strong is the relationship between rainfall variability and caatinga productivity? A case study under a changing climate. An. Acad. Bras. Cienc. 90 (2), 2121–2127. https://doi.org/10.1590/0001-3765201720170143.

Salinger, J., Sivakumar, M.V.K., Motha, R.P., 2005. Increasing climate variability and change: reducing the vulnerability of agriculture and forestry. Clim. Chang. 70, 9–29. https://doi.org/10.1007/1-4020-4166-7.

Schimel, D.S., 1995. Terrestrial ecosystems & the carbon cycle. Glob. Chang. Biol. 1 (December 1994), 77–91.

Schimel, D., Pavlick, R., Fisher, J.B., Asner, G.P., Saatchi, S., Townsend, P., Miller, C., Frankenberg, C., Hibbard, K., Cox, P., 2015. Observing terrestrial ecosystems and the carbon cycle from space. Glob. Chang. Biol. 21 (5), 1762–1776. https://doi.org/10.1111/gcb.12822.

Schwarzwald, K., Lenssen, N., 2022. The importance of internal climate variability in climate impact projections. Proc. Natl. Acad. Sci. U. S. A. 119 (42), 1–8. https://doi.org/10.1073/pnas.2208095119.

Seddon, A.W.R., Macias-Fauria, M., Long, P.R., Benz, D., Willis, K.J., 2016. Sensitivity of global terrestrial ecosystems to climate variability. Nature 531 (7593), 229–232. https://doi.org/10.1038/nature16986.

Sha, Z., Bai, Y., Li, R., Lan, H., Zhang, X., Li, J., Liu, X., Chang, S., Xie, Y., 2022. The global carbon sink potential of terrestrial vegetation can be increased substantially by optimal land management. Commun. Earth Environ. 3 (1), 1–10. https://doi.org/10.1038/s43247-021-00333-1.

Smith, K.L., Ruhl, H.A., Bett, B.J., Billett, D.S.M., Lampitt, R.S., Kaufmann, R.S., 2009. Climate, carbon cycling, and deep-ocean ecosystems. Proc. Natl. Acad. Sci. U. S. A. 106 (46), 19211–19218. https://doi.org/10.1073/pnas.0908322106.

Sofie, S., Black, C.R., Evers, S., Hoyos-Santillan, J., Wright, E.L., Turner, B.L., 2014. Tropical wetlands: a missing link in the global carbon cycle? Tropical wetlands: a missing link in the global carbon cycle? Glob. Biogeochem. Cycles 28, 1371–1386. https://doi.org/10.1111/1462-2920.13280.

Stankovic, M., Ambo-Rappe, R., Carly, F., Dangan-Galon, F., Fortes, M.D., Hossain, M.S., Kiswara, W., Van Luong, C., Minh-Thu, P., Mishra, A.K., Noiraksar, T., Nurdin, N., Panyawai, J., Rattanachot, E., Rozaimi, M., Soe Htun, U., Prathep, A., 2021. Quantification of blue carbon in seagrass ecosystems of Southeast Asia and their potential for climate change mitigation. Sci. Total Environ. 783, 146858. https://doi.org/10.1016/j.scitotenv.2021.146858.

Straile, D., Livingstone, D.M., Weyhenmeyer, G.A., George, D.G., 2003. The response of freshwater ecosystems to climate variability associated with the North Atlantic oscillation. Geophys. Monogr. Ser. 134, 263–279. https://doi.org/10.1029/134GM12.

Thornton, P.K., Ericksen, P.J., Herrero, M., Challinor, A.J., 2014. Climate variability and vulnerability to climate change: a review. Glob. Chang. Biol. 20 (11), 3313–3328. https://doi.org/10.1111/gcb.12581.

Tian, H., Melillo, J.M., Kicklighter, D.W., 1998. Effect of interannual climate variability on carbon storage in Amazonian ecosystems. Nature 396 (December), 664–667.

Tian, H., Melillo, J.M., Kicklighter, D.W., Mcguire, A.D., Helfrich, J., 1999. The sensitivity of terrestrial carbon storage to historical climate variability and atmospheric CO_2 in the United States. Tellus Ser. B Chem. Phys. Meteorol. 51 (2), 414–452. https://doi.org/10.3402/tellusb.v51i2.16318.

Tian, H., Ji, X., Zhang, F., 2022. Spatiotemporal variations of vegetation net primary productivity and its response to meteorological factors across the Yellow River Basin during the period 1981–2020. Front. Environ. Sci. 10 (July), 1–10. https://doi.org/10.3389/fenvs.2022.949564.

Trumper, K., Bertzky, M., Dickson, B., Van Der Heijden, G., Jenkins, M., Manning, P., 2009. The natural fix? The role of ecosystems in climate mitigation. A UNEP rapid response assessment. In: Director. United Nations Environment Programme, UNEP WCMC. http://search.ebscohost.com/login.aspx?direct=true&db=lah&AN=20103015280&site=ehost-live.

UNEP, IUCN, 2021. Nature-Based Solutions for Climate Change Mitigation. United Nations Environment Programme. UNEP.

UNFCCC, 2007. Climate change: impacts, vulnerabilities and adaptation in developing countries. In: United Nations Framework Convention on Climate Change., https://doi.org/10.1029/2005JD006289.

Van der Molen, M.K., Dolman, A.J., Ciais, P., Eglin, T., Gobron, N., Law, B.E., Meir, P., Peters, W., Phillips, O.L., Reichstein, M., Chen, T., Dekker, S.C., Doubková, M., Friedl, M.A., Jung, M., van den Hurk, B.J.J.M., de Jeu, R.A.M., Kruijt, B., Ohta, T., et al., 2011. Drought and ecosystem carbon cycling. Agric. For. Meteorol. 151 (7), 765–773. https://doi.org/10.1016/j.agrformet.2011.01.018.

van der Wiel, K., Bintanja, R., 2021. Contribution of climatic changes in mean and variability to monthly temperature and precipitation extremes. Commun. Earth Environ. 2 (1), 1–11. https://doi.org/10.1038/s43247-020-00077-4.

Weiskopf, S.R., Rubenstein, M.A., Crozier, L.G., Gaichas, S., Griffis, R., Halofsky, J.E., Hyde, K.J.W., Morelli, T.L., Morisette, J.T., Muñoz, R.C., Pershing, A.J., Peterson, D.L., Poudel, R., Staudinger, M.D., Sutton-Grier, A.E., Thompson, L., Vose, J., Weltzin, J.F., Whyte, K.P., 2020. Climate change effects on biodiversity, ecosystems, ecosystem services, and natural resource management in the United States. Sci. Total Environ. 733. https://doi.org/10.1016/j.scitotenv.2020.137782.

Williams, S.E., Hobday, A.J., Falconi, L., Hero, J.M., Holbrook, N.J., Capon, S., Bond, N.R., Ling, S.D., Hughes, L., 2019. Research priorities for natural ecosystems in a changing global climate. Glob. Chang. Biol. 26 (2), 410–416. https://doi.org/10.1111/gcb.14856.

Xia, J., Chen, J., Piao, S., Ciais, P., Luo, Y., Wan, S., 2014. Terrestrial carbon cycle affected by non-uniform climate warming. Nat. Geosci. 7 (3), 173–180. https://doi.org/10.1038/ngeo2093.

You, N., Meng, J., Zhu, L., 2018. Sensitivity and resilience of ecosystems to climate variability in the semi-arid to hyper-arid areas of Northern China: a case study in the Heihe River Basin. Ecol. Res. 33 (1), 161–174. https://doi.org/10.1007/s11284-017-1543-3.

Yvon-Durocher, G., Allen, A.P., Montoya, J.M., Trimmer, M., Woodward, G., 2010. The temperature dependence of the carbon cycle in aquatic ecosystems. Adv. Ecol. Res. 43 (C). https://doi.org/10.1016/B978-0-12-385005-8.00007-1.

Chapter 7

Application of remote sensing techniques to monitor climate variability effects on groundwater-dependent ecosystems

Tatenda Dalu[a,b], Chantel Chiloane[c], Farai Dondofema[d], Timothy Dube[c], Elsie N. Leshaba[a], Fannie M. Masina[a], and Linton F. Munyai[a]

[a]*Aquatic Systems Research Group, School of Biology and Environmental Sciences, University of Mpumalanga, Nelspruit, South Africa,* [b]*South African Institute for Aquatic Biodiversity, Makhanda, South Africa,* [c]*Department of Earth Science, Institute for Water Studies, University of the Western Cape, Cape Town, South Africa,* [d]*Department of Geography and Environmental Sciences, University of Venda, Thohoyandou, South Africa*

Chapter outline

1 Introduction

Groundwater-dependent ecosystems (GDEs) can be defined as ecosystems that are partially or entirely dependent on a continuous supply of groundwater for their current composition, structure, and function (Dalu and Wasserman, 2022). While GDEs are recognized as a crucial component of the natural environment, they are not yet fully understood (Duran-Llacer et al., 2021; Liu et al., 2021). Groundwater plays a significant role in various aquatic and terrestrial

Remote Sensing of Climate. https://doi.org/10.1016/B978-0-443-21731-9.00002-8
Copyright © 2024 Elsevier Inc. All rights are reserved, including those for text and data mining, AI training, and similar technologies.

ecosystems, including rivers, lakes, hyporheic and riparian habitats, subterranean aquifers and caves, wetlands and springs, and estuarine and nearshore marine ecosystems (Kløve et al., 2011; Doody et al., 2017). GDEs are in global ecosystems, serving as habitats for unique species and playing a pivotal role in maintaining hydrological equilibrium (Gleeson et al., 2020; Masina et al., 2023). It is crucial to protect GDEs due to their numerous ecosystem services (Kumar and Kanaujia, 2014; Chaikumbung et al., 2016). These unique systems support biodiversity with many distinctive plant and animal species only found within these ecosystems (Huang et al., 2020; Mungenge et al., 2023). Biodiversity encompasses the entire range of life, including genetic, population, species, and ecosystem diversity. It further involves the relationships between various biological components and their ecological roles within specific environmental conditions such as physico-chemical conditions and hydrogeomorphological units (Boulton, 2005; Dalu and Wasserman, 2022). Consequently, when there is a seasonal change in water temperature and discharge regimes are minimal, groundwater provides a continuous flux of dissolved ions, nutrients, and organic matter critical for species inhabiting GDEs (Erostate et al., 2020). These unique ecosystems significantly contribute to the regional biodiversity of freshwater and wetland ecosystems by forming a link between groundwater and surface water, as well as between groundwater and terrestrial ecosystems (Barquín and Scarsbrook, 2008).

They also play a crucial role in water purification by locking up pollutants in their sediment and vegetation (Kumar and Kanaujia, 2014). Especially, high levels of nutrients such as phosphorous and nitrogen, commonly associated with agricultural runoff and sewage effluent, can be minimized by wetlands. This prevents the nutrients from reaching toxic levels in groundwater used for drinking purposes (Barquín and Scarsbrook, 2008). In addition, it helps to reduce the risk of eutrophication. This process occurs when a high nutrient level causes a massive boost in algal growth, depleting oxygen and blocking out the light that aquatic plants and animals need to survive (Eamus and Froend, 2006; Williams et al., 2022). Furthermore, these unique ecosystems support livelihoods through the provision of access to clean water and protection from natural disasters such as floods and droughts (Alaibakhsh et al., 2017). Moreover, they provide habitats for migratory birds, rare plant species, and invertebrates (Kumar and Kanaujia, 2014; Chaikumbung et al., 2016). Lastly, they also provide an ecological service such as carbon sequestration which is important in climate mitigation (Villa and Bernal, 2018). However, despite their ecological significance, GDEs are typically influenced by extreme climate variability, a phenomenon that has become more apparent because of ongoing global climate change (Kløve et al., 2014; Williams et al., 2022). Kløve et al. (2014) indicated that climate variability is expected to amplify global concerns through decreased precipitation and increased evapotranspiration. These changes will adversely affect recharge rates and potentially lead to higher groundwater abstraction rates (Kløve et al., 2014).

The reliance of ecosystems on groundwater varies from continuous to seasonal or occasional, such as only a few months to every few years. This reliance becomes evident when groundwater supply is interrupted for a significant period of time, leading to noticeable changes in plant function, often indicated by a decline in water-use rates (Danielopol and Pospisil, 2001). In arid climates, groundwater is often the primary water source for vegetation. Specific ecosystems, like springs, rely entirely on groundwater and would not exist without it (Orellana et al., 2012; Huntington et al., 2016). This reliance on groundwater is also reflected in the plant and animal species in these ecosystems, as they have adapted to thrive under these unique conditions (Mango et al., 2017; Liu et al., 2021). Due to their reliance on groundwater for sustenance, GDEs are exceptionally susceptible to climate variability and change (Gosling et al., 2015). Consequently, comprehending how these ecosystems react to shifting climate patterns is imperative for their preservation and effective management (Gorelick et al., 2015). In this regard, remote sensing techniques have emerged as indispensable tools for investigating various GDE types and their responses to climate variability (Brocca et al., 2014).

Geology and water residence time are essential factors that significantly influence natural water chemistry in GDEs. Groundwater contributes nutrients and electron acceptors such as sulfate and usually creates specific physicochemical conditions in GDEs (Duran-Llacer et al., 2021). Water pH plays a crucial role in the solubility and biological availability of nutrients and heavy metals. At lower pH, metals tend to occur in bioavailable forms, while different nutrients are best taken up at pH levels, offering suitable adsorption conditions. On the contrary, water parameters such as redox potential are essential in the reductive dissolution of iron oxyhydroxides and the state of redox-sensitive elements (Verlicchi and Grillini, 2020). In GDEs linked to surface waters, seasonal and daily variations in photosynthesis can be a significant natural cause of pH variations. Therefore, water chemical balance modification may result in irreversible changes in the entire GDE (Gronwall and Danert, 2020).

One particularly critical type of GDE is riparian zones, encompassing lands bordering rivers and streams. Riparian zones provide crucial support to diverse flora and fauna (Stromberg et al., 2010). By harnessing remote sensing capabilities, such as the use of satellite imagery and aerial photography, researchers have gained the capacity to closely monitor alterations in riparian vegetation, land cover, and hydrological dynamics (Pettorelli et al., 2014). An example is the Amazon Basin, where remote sensing has proven instrumental in evaluating the consequences of deforestation and shifting river flow on the health of riparian vegetation (Hess et al., 2019). Similarly, in the Murray-Darling Basin in Australia, remote sensing was used to facilitate the tracking of modifications in riparian vegetation in response to changes in water availability (Leblanc et al., 2007). Another vital category of GDEs is wetlands, which encompasses environments such as marshes, swamps, and floodplains. Wetlands play a pivotal role in supporting a myriad of species and delivering essential ecosystem services (Mitsch and Gosselink, 2015). Remote sensing has proven invaluable in monitoring wetland extent, water quality, and vegetation health (Murray et al., 2013). An illustrative case is found in the Everglades of Florida, USA, where remote sensing techniques were deployed to track water level fluctuations and vegetation composition (Suding et al., 2015). In the African context, the Okavango Delta offers an example where remote sensing contributed to assessing the impact of evolving flood patterns on wetland health and biodiversity (Papa et al., 2018).

Karst springs, existing in limestone landscapes characterized by underground drainage systems, represent another unique category of GDEs (Ford and Williams, 2013). Employing remote sensing methods such as light detection and ranging (LiDAR) and high-resolution satellite imagery, researchers have identified and monitored spring's locations and flows (Bonacci et al., 2018). In the Yunnan Province of China, remote sensing has played a pivotal role in locating karst springs and evaluating their flow rates, deepening our understanding of these distinctive ecosystems (Zhang et al., 2019). In the Balkans, remote sensing has been instrumental in scrutinizing the vulnerability of karst springs in the face of altering precipitation patterns and land use (Bonacci et al., 2018). In conclusion, GDEs are integral components of the world's ecosystems, magnified by their role in housing unique species and preserving hydrological equilibrium (Gleeson et al., 2020). Climate variability and change threaten these ecosystems, necessitating a comprehensive understanding of their responses to climatic shifts (Gosling et al., 2015). Remote sensing techniques have emerged as invaluable tools for studying various GDE types, including riparian zones, wetlands, and karst springs, and assessing their reactions to climate variability (Brocca et al., 2014). These technologies have been successfully applied globally, including in regions such as the Amazon Basin, the Murray-Darling Basin, the Everglades, the Okavango Delta, the Yunnan Province of China, and the Balkans, offering insights into the conservation and management of these ecosystems (Pettorelli et al., 2014; Hess et al., 2019; Leblanc et al., 2007; Murray et al., 2013; Papa et al., 2018; Zhang et al., 2019; Bonacci et al., 2018).

On a global scale, groundwater accounts for approximately one-third of all freshwater withdrawals; this valuable resource is responsible for supplying an estimated 36.0% of the water used for domestic purposes, 42.0% for agricultural purposes, and 27.0% for industrial purposes (Taylor et al., 2013). Studies conducted by Konikow and Kendy (2005), Kundzewicz and Doell (2009), Misra (2014), Foster et al. (2013), and Hamed et al. (2018) highlighted the decline in groundwater levels due to climate variation, which is a major problem globally. Climate changes have direct or indirect effects on groundwater resources in many ways (Havril et al., 2018; Morsy et al., 2017). Moreover, Havril et al. (2018) stated that climate variability directly impacts groundwater systems by altering recharge processes, flow dynamics, and fragmentation, affecting levels, flow dynamics, and flow hierarchy. Moreover, Basavarajappa et al. (2015)

indicated that the alteration of climatic circumstances directly impacts the hydrologic cycle, subsequently leading to a steady fluctuation in the groundwater table. Furthermore, it is evident that the impact of periodic climatic disturbances, such as El Niño/La Niña phenomena and their accompanying events, primarily manifests through both direct and indirect alterations to the operational dynamics of GDEs, particularly wetlands (House et al., 2016). In addition, Taylor et al. (2013) indicated that recharge is strongly influenced by climate variability over time, including climate extremes (i.e., droughts and floods) that are frequently associated with modes of climate variability such as the El Niño/Southern Oscillation at multiyear timescales and the Pacific Decadal Oscillation, Atlantic Multidecadal Oscillation, and others at longer timescales. Moreover, groundwater could become increasingly important under a rapidly changing climate, as more frequent and severe extremes, such as droughts and hotter temperatures increase the risk of plant mortality and reduce productivity (Marchionni et al., 2020).

Climate variability impacts groundwater resources through precipitation, temperature, and recharge conditions (Gintamo et al., 2021). Furthermore, Gintamo et al. (2021) highlighted that urban areas' hydrologic cycle may change due to reduced recharge rates and increased salinity due to reduced flushing. The continuation of this phenomenon may significantly alter global and local climate characteristics, including temperature and precipitation, leading to climate change that has profound effects on the hydrologic cycle through precipitation, evapotranspiration, and soil moisture with increasing temperatures (Kumar, 2012). Climate change is identified as a major threat to wetlands. It can influence a GDE by increasing temperature and changing hydrological patterns, which in turn can alter the biogeochemistry of the ecosystem (Salimi et al., 2021). Moreover, climate changes alter the existing hydrological regime to have the potential to significantly modify habitat suitability for many species and cause significant ecological changes (Salimi et al., 2021). Global warming and hydrological cycle changes cause erosion, salinization, altered tidal ranges, sediment inputs, nutrient loadings, and increased coastal flooding, affecting freshwater availability for humans and ecosystems (Lloréns, 2008).

Depending on each ecosystem's unique properties and flexibility, GDEs respond in different ways to changing hydrological dynamics. According to Jyväsjärvi et al. (2015), the climate-induced changes in GDEs have the potential to disrupt patterns of species distribution and result in the loss of habitats for endangered organisms or have specialized ecological requirements. Moreover, Kløve et al. (2011) indicated that climate change increases extreme weather events like floods and droughts, altering sedimentary and biological assemblages, affecting GDEs, and desiccating wetlands and springs. Kløve et al. (2014) stated that climate change stresses aquifers and GDEs, putting pressure on changes in groundwater levels and temporal patterns, endangering ecosystem services like arable land irrigation and ecosystem water needs, especially during droughts.

The inevitable results of climate change are increased climate variability and the frequency, intensity, duration, spatial extent, and timing of extreme weather and climate events (IPCC, 2014). Increased climate variability has ongoing adverse impacts on food and water security, society and economies, and environmental degradation (IPCC, 2021). For example, since the 1950s, hot extreme weather has become more frequent and intense across all regions, while cold extremes have become less frequent and severe. Ecological and agricultural droughts have become prevalent in some regions. Furthermore, the increasing frequency of extreme weather, such as heatwaves, drought, and floods, is projected for this century (IPCC, 2021). Impacts from these extreme climate events will have significant consequences on human and natural systems (Thornton et al., 2014). More specifically, natural systems such as GDEs are vulnerable to the effects of extreme climate variability. Thus, spatially and temporally explicit methods are required to investigate and understand the impacts of climate variability on GDEs to inform GDE conservation, management, and climate adaptation initiatives at both global and local scales. Remote sensing data offer valuable multiscale and long-time data sets for understanding the relationship between GDEs and climate variability. Such data sets are helpful in understanding historical trends, modeling current trends, and forecasting GDEs' response to changes in rainfall patterns and temperature. For example, Albano et al. (2020) utilized 35 years (1984–2018) of Landsat satellite imagery coupled with gridded climate data sets to quantify the status and trend of potential GDEs in response to climate variability and groundwater depth fluctuations.

Mapping and characterization of GDEs are crucial processes in environmental science, aquatic ecology, and nature conservation. These ecosystems rely on groundwater as a primary water source, and understanding their distribution and characteristics is vital for their protection and sustainable management. GDEs range from wetlands and springs to riparian zones, caves, and even some terrestrial habitats like forests and shrublands. These ecosystems are frequently found in locations with minimal surface water, dry or semi-arid climates, or where surface water is unpredictable seasonally or during droughts. Groundwater plays a crucial role, particularly in semi-arid environments, as a source of water for humans, and it maintains various ecosystems such as rivers, lakes, wetlands, and terrestrial systems (Rampheri et al., 2023a,b). Furthermore, GDEs have the potential to provide various ecological services such as carbon sequestration, water purification, drought relief, and flood prevention (Yang and Liu, 2020). To manage and distribute water resources sustainably, researchers have called for more studies to understand better the geographical position, distribution, and extent of GDEs (Kløve et al., 2011). This is due to the threats that GDEs face in the developing environment. This will help balance human water needs and environmental demands for sustainable management and GDE conservation. Additionally, it is critical to analyze and monitor GDEs' spatial extent and health condition considering climate change and water shortage, especially in dry environments with inadequate data, to improve their conservation. Therefore, this chapter disseminates the impacts of climate variability on GDEs where the damaging effects are projected to be the greatest. More specifically, literatures on the remote sensing of the spatial and temporal responses of GDEs to climate variability, such as rainfall patterns, drought, and temperature changes, are discussed.

2 Remote sensing-based indicators of GDE health

Remote sensing-derived indicators and indices can be used to assess the health and condition of GDEs, including vegetation health indicators, water stress indicators, and ecosystem function metrics. Remote sensing data has become critical in assessing landscapes at coarse and fine spatial scales because it makes it easier to identify possible GDEs, their groundwater dependence, and their health (Fig. 1). In conjunction with ancillary data, remote sensing is increasingly becoming a cost-effective and beneficial tool for diagnosing groundwater reliance (Doody et al., 2017). The spectral vegetation indicators acquired from remote sensing may be used to assess vegetation biomass, water utilization, plant stress, and plant health and identify GDE through distribution and temporal trend analysis. The correlation between plant density and water availability in arid and semi-arid areas provides scientists with remote sensing-based tools for detecting "green islands" (Akasheh et al., 2008; Eamus et al., 2015). Remote sensing indices such as Normalized Difference Vegetation Index (NDVI) (Rouse et al., 1974), Enhanced Vegetation Index (Huete et al., 2002), and Normalized Difference Water Index (NDWI) (Gao, 1996) have all been used to indirectly evaluate the greenness, wetness, and stress of GDE surface conditions in recent studies (Box et al., 2022). During the dry season, the indices identify plants that maintain continuous greenness, wetness, and high density (Lv et al., 2012).

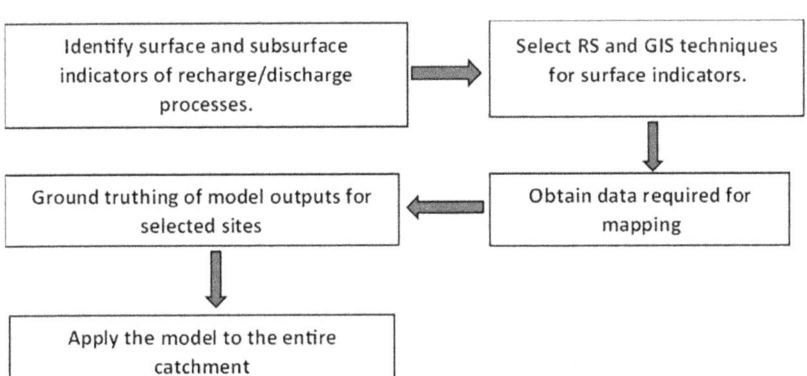

FIG. 1 A schematic of the methodology used by Tweed et al. (2007) in the use of remote sensing and geographical information system (GIS) to identify the location of GDEs in a landscape.

The capability of NDVI to provide information on the health and productivity of vegetation makes it an essential indicator for GDEs (Barron et al., 2014). The NDVI can be used to detect losses in vegetation that may be connected to changes in groundwater supply or quality and track changes in vegetation health over time. The NDVI can also improve the identification and mapping of vegetative stress zones, which might indicate changes in groundwater availability or quality (Fu and Burgher, 2015), making it one of the most influential indices for mapping GDEs. The NDVI product has the potential to generate a map that will help managers and scientists better understand the connections between groundwater and vegetation, allowing them to make more informed decisions about managing these resources. When assessing vegetation characteristics, vegetation indices overcome the impacts of soil background, zenith angle, and atmospheric composition while improving the vegetation signal (Thamaga et al., 2021).

Another commonly used index is the NDWI, which offers information on vegetation water content and provides an essential approach for monitoring vegetation and GDEs (Fontana and Collischonn, 2019). NDWI improves the identification of vegetation areas under water stress and, thus, can be used as an early warning sign of declining groundwater levels (Xu et al., 2022). The NDWI data can be effectively used in the management of groundwater resources, the protection of ecosystems, and the preservation of biodiversity. NDWI may also be used to track changes in land use land cover (LULC), which can directly influence GDEs, and to relate moisture conditions to GDEs, making GDE mapping easier (Barron et al., 2014). Remote sensing advancements have resulted in a revised conventional NDVI formulation incorporating evapotranspiration inside the GDE (Groeneveld and Baugh, 2007). Using meteorological data from grids such as the Food and Agriculture Organization (FAO)'s WaPOR (Water Productivity through Open access of Remotely sensed derived data) (Allen et al., 1998; FAO, 2019), the NDVI may be calibrated to measure accurate evapotranspiration (Eta) rates. This method is helpful in the vegetated desert and semi-arid locations with a shallow water table when rainfall is limited and often variable, but water availability to roots is reasonably steady.

The Leaf Area Index (LAI), which is the ratio of a canopy's total leaf area to the ground area covered by the canopy, is another remote sensing surrogate beneficial for GDE health monitoring. The LAI loss is expected to be slower in sites with access to groundwater than in sites without groundwater access (MacFarlane et al., 2007). Measuring LAI with remotely sensed (satellite-based) data is now feasible. The Moderate Resolution Imaging Spectroradiometer (MODIS) is a popular instrument. This sensor is onboard the TERRA and AQUA satellites, which circle the Earth from north to south (TERRA) or south to north (AQUA). These data are used to determine the LAI of locations worldwide, with historical data going back a decade or more available, as well as the variations in LAI of patches of vegetation during wet and dry times.

The NDVI and Normalized Difference Moisture Index (NDMI) are calculated using Landsat Operational Land Imager and Thermal Infrared Sensor (Landsat OLI and TIRS) data plus accessible meteorological data. The spatial resolution of Landsat remote sensing images is typically modest (25 m) (Glanville et al., 2016). As a result, the approach may fail to identify narrow riparian corridors and tiny wetland ecosystems. NDVI and NDMI can also identify regions unrelated to groundwater usage while ignoring delays in water availability and vegetation conditions (Glanville et al., 2016). Based on the premise that all GDEs can be recognized by this attribute alone (Glanville et al., 2016), this technique seeks to identify GDEs as a unique ecosystem type (i.e., places that are greener or wetter than surrounding areas).

The Soil Adjusted Vegetation Index (SAVI) considers senescent vegetation and background soil impacts, which produce estimate errors due to soil brightness and bare soil cover (Dube et al., 2019; Thamaga and Dube, 2018). As a result, the SAVI is more suited to predicting GDE cover when densities are low during dry seasons. Synthetic aperture radar (SAR) is an active sensor that gives accurate moisture inundation, seasonal changes in the water table, vegetation patterns, and the influence of natural and anthropogenic disturbance (Reschke et al., 2012). On the contrary, LiDAR is a remote sensing technique that may be used to create very precise vertical profiles of plants, which are critical for predicting potential evapotranspiration from vegetation. Thermal remote sensing techniques have also been utilized to detect regions inside GDEs that exhibit significant levels of evapotranspiration and/or saturation (Werstak et al., 2012).

However, groundwater dependency is only one feature of an ecosystem, which includes a wide range of ecosystem types. As a result, a single threshold value for a single indicator is unlikely to convey the complexities of an ecosystem's groundwater dependency within a landscape (Glanville et al., 2016). While remote sensing approaches for identifying GDEs can be beneficial in assessing vast regions, it is critical to recognize the limits of the related technology. GDE detection and monitoring can be improved using vegetation indices such as the NDVI and SAVI (Chiloane et al., 2023). Field assessments can help identify and map ecosystems on a finer geographic scale (Pérez Hoyos et al., 2016). Invasive vegetation depletes groundwater supplies and reduces groundwater recharge, limiting water availability to indigenous GDE. According to various research, conservation management should include the likelihood that invasive species would devour ecological groundwater supplies (Currie et al., 2009). As a result, invasive species control and restoration must be prioritized.

According to various studies, one of the significant limitations of remote sensing technology for mapping GDEs is that the size of specific ecosystems is frequently less than the size of a single pixel in the available remotely sensed images (Barron et al., 2014). Thus, remote sensing data sets do not clearly show the feature and, instead, the pixel value, representing a mixture of several features within the pixel (Werstak et al., 2012; Lewis and White, 2013). Another impediment to accurately detecting GDEs using remote sensing data sets is that some of these ecosystems have spectral features that are difficult to differentiate from nearby land cover, making delineation challenging.

3 Remote sensing of climate variability impacts on GDEs

A large proportion of the global population depends on ecosystem services provided by GDEs. GDEs are susceptible to impacts from increased rainfall variability that affect groundwater availability for maintaining GDEs permanently or intermittently. Changes in rainfall patterns, i.e., magnitude, duration, timing, and intensity due to global warming, are predicted to alter the groundwater regime, namely, the timing, duration magnitude, intensity, and groundwater level fluctuations to which GDEs are adapted (Kath et al., 2014). Groundwater dependency varies in space and time and between individuals, species, and communities (Eamus et al., 2015; Boulton, 2020). Increased temperatures may reduce groundwater availability due to increased evapotranspiration from soils, vegetation, and shallow surface water bodies. For instance, Bohn et al. (2016) studied climate variability and land use effects on shallow lakes in Argentina, using satellite imagery (MODIS-NDVI), Standard Precipitation Evapotranspiration Index (SPEI), the Oceanic Nina Index with in situ precipitation data. Their results indicated that the frequency of the lake area change was more linked with rainfall than land use activities. However, other hydrological balance variables accentuate or mitigate the climate effects. Climate indices such as SPEI determine drought magnitude and duration in diverse environments considering precipitation and atmospheric evaporative demand. Moreover, Meyers et al. (2021) utilized combined remote sensing (Landsat-NDVI) and groundwater residence time data (^{3}H, ^{36}Cl) to identify spring responses to the 2011–2017 California drought. Spring productivity was linked with surrounding healthy vegetation, thus, vegetation health indices prior to, during, and after the drought were compared with spring water residence time. Springs discharging modern groundwater ($<$100 years) showed evidence of increased drying and desiccation as the drought progressed, while spring-discharging old groundwater ($>$100 years) showed limited response to the drought effects. These observations support the concept that groundwater from deeper aquifers with stable groundwater flow paths buffer effects from short-term to long-term climate variability.

Kibler et al. (2021) studied drought effects on vegetation mortality by comparing the distribution of green vegetation and nonphreatophyte vegetation (brown vegetation) to groundwater depth fluctuations during the 2012–2019 California drought. They observed that riparian GDEs were more sensitive to groundwater variability during drought periods, and the propagation of the brown wave-vegetation dieback was associated with local changes in groundwater depth due to low rainfall. Similar studies using remote sensing data highlighted those changes in GDE surface water extent, vegetation vigor, and changes in species composition are linked to climate variability (Dwire et al., 2018; Kim et al., 2018; Albano et al., 2020; Majola et al., 2022).

While remote sensing provides spatial and long-term data for evaluating climate effects on GDEs, this method has inherent limitations. When using remote sensing data, it is not possible to differentiate plant cover by species or life form (Meyers et al., 2021; Elmore et al., 2006; Kibler et al., 2021). Remote-sensing data does not indicate whether plant community floristic composition is maintained through groundwater decline and recharge (Elmore et al., 2006). Vegetation response to drought is not homogenous but dependent on the vegetation species-rooting depth and degree of groundwater dependence consequently, vegetation species composition should be captured from field surveys. Moreover, hyperspectral imaging coupled with advanced tools such as whole genome sequencing, throughput phenotyping, and fluvial hydrologic/plant hydraulics modeling may provide opportunities to predict patterns of mortality in groundwater-dependent vegetation influenced by climate variability (Hultine et al., 2020). Information acquired from the combination of remote-sensing data analysis and targeted field monitoring can inform groundwater management toward protecting GDEs (Elmore et al., 2006).

4 Response of GDEs to climate variability

Understanding the response of GDEs to climate variability has remained a critical area of study, and remote sensing has played a pivotal role in understanding and monitoring these responses. GDEs rely on groundwater as their primary water source, making them highly sensitive to changes in climate factors such as temperature, precipitation, and hydrological patterns. Temperature variability has been a subject of extensive research in relation to GDEs, with remote sensing playing a pivotal role in elucidating this connection. Rising temperatures can profoundly affect GDEs, particularly in arid and semi-arid regions. Studies such as Allen et al. (2017) have employed thermal infrared remote sensing to monitor land surface temperatures in riparian areas, revealing that elevated temperatures correspond to increased evapotranspiration rates from riparian vegetation, potentially stressing GDEs by reducing groundwater availability. Similarly, research in the Murray-Darling Basin of Australia (Bretreger et al., 2021) employed remote sensing techniques to track temperature fluctuations, finding that prolonged periods of high temperatures during drought events adversely affected riparian ecosystems, leading to decreased vegetation health. Additionally, remote sensing has been utilized in the Amazon Basin to assess the impact of deforestation and changing river flow on riparian vegetation health, identifying areas where such changes had caused significant alterations in land surface temperatures, serving as early warning indicators of potential stress on riparian ecosystems (Hess et al., 2019). These studies (i.e., Allen et al., 2017; Hess et al., 2019; Bretreger et al., 2021) collectively underscore the critical role of remote sensing in monitoring temperature variability and its implications for GDEs, providing valuable insights and early warning capabilities for conservation and management efforts aimed at preserving these vital ecosystems.

In addition to temperature, precipitation patterns are fundamental to the dynamics of GDEs, and remote sensing techniques have played a pivotal role in understanding their influence. Changes in rainfall patterns, intensity, and distribution have been found to significantly impact GDEs, making monitoring precipitation patterns a critical aspect of GDE research, including in South Africa and other regions of Africa and beyond. Satellite-derived rainfall data, a product of remote sensing, enables the detection of anomalies in rainfall patterns, including periods of drought or increased precipitation. For example, a study in the Okavango Delta in Botswana, an important GDE in southern Africa, utilized remote sensing to analyze long-term precipitation data and its correlation with wetland dynamics. The research demonstrated how remote sensing can provide insights into the response of GDEs to shifts in precipitation patterns (Ashwini et al., 2023).

Moreover, hydrological dynamics, intricately linked to GDEs, are another focal point of remote sensing application, and this is particularly relevant in African regions where GDEs are prevalent. GDEs in South Africa, such as those in the Western Cape province, rely on groundwater, and alterations in hydrological patterns, such as fluctuations in groundwater levels, river flow variations, and changes in wetland extent, can have profound implications for their well-being. Remote sensing techniques, including satellite imagery and groundwater monitoring through remote sensing, facilitate monitoring of hydrological changes within GDEs. For instance, a study in the Western Cape of South Africa used remote sensing to assess wetland dynamics and changes in water levels, shedding light on the response of local GDEs to hydrological alterations (Le Maitre et al., 2016). In Africa, where GDEs are diverse and critical for biodiversity and water resources, remote sensing has offered valuable insights into their response to climate variability. It allowed researchers to monitor shifts in vegetation cover, wetland extent, and water availability, contributing to a better understanding of how these ecosystems adapt to changing environmental conditions (Weigand et al., 2019; Beck, 2019).

Overall, remote sensing offered a comprehensive and effective means of investigating how GDEs respond to climate variability. It enabled tracking temperature, precipitation, and hydrological changes in GDE areas over time. By analyzing long-term remote sensing data, researchers gain valuable insights into the dynamics of these ecosystems and understand their health and resilience in the face of changing climatic conditions. This knowledge has been essential for the conservation and sustainable management of GDEs, particularly in regions vulnerable to climate change's impacts.

5 Remote sensing approaches for mapping and characterizing GDEs

Remote sensing data combined with GIS methods have been shown to offer geographical and temporal data in GDE studies successfully. This aids in the robust, rapid, and efficient delineation and mapping of GDEs (Barron et al., 2014; Pérez Hoyos et al., 2016; Wu et al., 2018). Passive optical sensors, including multispectral and hyperspectral remote sensing satellite images such as Landsat, MODIS, Satellite Pour l'Observation de la Terre (SPOT), and Sentinel data sets, as well as active sensors such as LiDAR and Radio Detection and Ranging (RADAR) imagery, have been used to investigate the spatial distribution of GDEs. Other data-gathering methods, such as field surveys or ground-based measurements, have been deemed inadequate. Several studies have demonstrated the remarkable progress in the application of remote sensing in GDEs over the years using various satellite data to assess and examine these ecosystems. The progress is attributed to the technological advances that resulted in well-organized data processing. The increase in technology breakthroughs in remotely sensed data (e.g., Sentinel, which was launched in 2016) indicates the potential for expansion in the application of remote sensing in GDE evaluation and monitoring in the coming years.

In the past decade, much research in GDEs has been undertaken in arid and semi-arid regions utilizing remote sensing techniques. The research used satellite data sets with varying spatial, spectral, and temporal resolutions for a variety of objectives to examine various features of GDEs (Münch and Conrad, 2007; Barron et al., 2014; Pérez Hoyos et al., 2016; Box et al., 2022). The majority of GDE studies across the globe made use of medium to coarse multispectral data sets (Landsat and Sentinel) because these are freely, quickly, and easily accessible data sets with spatial, spectral, and temporal resolution that are technologically advanced (McAllister et al., 2022; Thamaga et al., 2022). Even though zero to low-cost satellite data provide medium to coarse resolution, they are also capable of providing credible findings in the evaluation and monitoring of GDEs.

6 Identification, delineation, mapping, and monitoring of GDEs using remote sensing capabilities

Remote sensing offers rapid and spatially comprehensive tools for assessing many aspects of GDEs, such as vegetation cover, water dynamics (i.e., recharge, discharge, quality, and quantity), and interactions between them and climatic factors. Various types of satellite sensors, including active sensors such as RADAR (Hong et al., 2010; Dabboor and Brisco, 2018) and LiDAR (Lang et al., 2012), and passive sensors such as Landsat (Doody et al., 2017;

Collados-Lara et al., 2021), Sentinel (Kaplan and Avdan, 2017; Slagter et al., 2020), MODIS (Gou et al., 2015; Xu et al., 2021), and SPOT across arid and semi-arid environments have been used previously to map and understand GDE dynamics. Several multispectral and hyperspectral satellite images, such as Landsat, Sentinel, MODIS, and SPOT, have been utilized to map GDEs. Many publications on GDE mapping, from local to worldwide, have been produced due to the repeated global coverage provided by satellite sensors such as Landsat imagery (Münch and Conrad, 2007; Li and Roy, 2017; Thamaga et al., 2021). According to Doody et al. (2017), the importance of using Landsat data for GDE assessment and monitoring is that they are able to detect landscapes that are wetter or greener than their surroundings, suggesting that they have access to more water, such as groundwater based on their band strength. Therefore, Landsat is the best tool for assessing and monitoring GDEs at various spatial scales. Most of the GDEs are affected by climate change, which leads them to deterioration and results in unsustainable ecosystem services. However, remote sensing has the potential to identify such ecosystems and measures can be put in place to deal with ecosystem degradation before further damage can be incurred. Fig. 2 provides a conceptual framework for the impacts of climate change on GDEs.

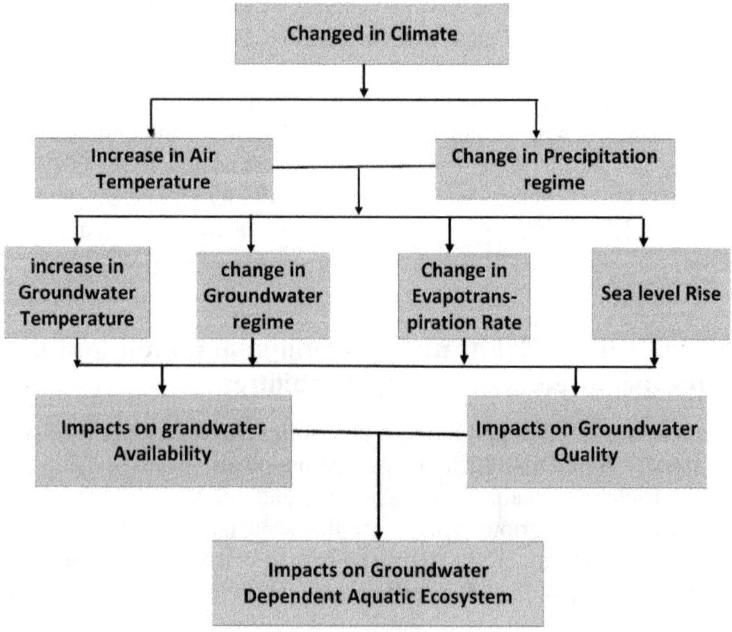

FIG. 2 A conceptual framework on impacts of climate change on GDE.

7 Remotely sensed algorithms on GDE assessment and monitoring

Many studies (i.e., Münch and Conrad, 2007; Pérez Hoyos et al., 2016; Martínez-Santos et al., 2021; Chiloane et al., 2022) have commonly utilized techniques such as image classification coupled with machine-learning algorithms to characterize, assess, and monitor GDEs. Image classification algorithms include supervised and unsupervised classification, object-based classification, principal component analysis, and hybrid classification (Mpakairi et al., 2022a,b). Among these algorithms, supervised classification algorithms outperform unsupervised classification techniques and are thus more widely utilized and robust in GDE studies. Table 1 highlights some of the algorithms reviewed and applicable to GDEs.

TABLE 1 Available machine-learning algorithms that have been used in GDE studies.

Algorithm	Remote sensing data	Performance (accuracy assessment)	References
Random forest	Sentinel	Wetland classification in Cousen watershed, France (87.0%)	Rapinel et al. (2019)
	WorldView-2	High-density biomass estimation for wetlands in Isimangaliso Wetland Park, South Africa	Mutanga et al. (2012)
	SAR data	Wetland classification, Canada (94.0%)	Mahdavi et al. (2018)
	Landsat	Wetland classification along Etrix River in North Xinjiang, China (93.0%)	Tian et al. (2016)
	RapidEye	Wetland mapping In Peninsula, Newfoundland and Labrador, Canada (76.1%)	Mahdianpari et al. (2018)
Classification and regression trees	Landsat	Land cover and land use in Segara Anakan, Indonesia (97.0%)	Farda (2017)
Support vector machine	Landsat	Analyze spatial and temporal variation of wetlands in the Paarl River Delta (94.9%)	Han et al. (2018)
	LiDAR	Classified wetland vegetation (97.7%)	Bian et al. (2021)

Continued

TABLE 1 Available machine-learning algorithms that have been used in GDE studies—cont'd

Algorithm	Remote sensing data	Performance (accuracy assessment)	References
	Landsat	Identifying the risk zones of East Kolkata wetlands (91.1%)	Ghosh and Das (2020)
	Landsat	Assessing the extent of waterbodies within wetlands along the Klip River, South Africa (84.0%)	Zwedzi (2019)
Maximum likelihood classification	Landsat	Mapping wetlands in the Mekong Basin (ranges from 77.0% to 94.0%)	MacAlister and Mahaxay (2009)
	IKONOS	Mapping invasive wetland plants in the Hudson River National Estuarine Research Reserve in New York (ranges from 45.0% to 77.7%)	Laba et al. (2010)
Artificial neural network	Landsat	Classifying and mapping coastal wetlands of Cukurova Delta, Turkey (90.2%)	Berberoglu et al. (2004)

Depending on the objectives of their study, several studies (e.g., Laba et al., 2010; Farda, 2017; Mahdavi et al., 2018; Han et al., 2018; Rapinel et al., 2019; Ghosh and Das, 2020) have employed machine-learning algorithms to elucidate GDE dynamics. The most often utilized methods in the evaluation and monitoring of ecosystems such as wetlands are support vector machine (SVM) and random forest (RF) (Table 1). Han et al. (2018) used SVM to analyze the spatial and temporal variation of wetlands in the Paarl River Delta using Landsat, and the SVM model had a high classification accuracy of 95.0% (Table 1). Ghosh and Das (2020) used Landsat to identify the risk areas of East Kolkata Wetlands, with SVM classification scores of 91.1%. Because it can overcome problems connected with locations with complex land cover and types, the SVM classification approach produces high accuracies. It solves these problems by allowing users to alter parameter values and add training groups to collected data. Therefore, to get better results, the availability of enhanced remote sensing generation (high-resolution imagery) coupled with other characteristics such as the created machine-learning algorithm and in situ remotely sensed data might provide a significant possibility for GDE mapping (Mpakairi et al., 2022a,b).

8 Advanced indices for GDE mapping

Since supervised classification is prone to analyst errors, spectral derivation of distant sensed data remains the most often used method for identification, delineating, mapping, classification, and monitoring GDE characteristics (Barron et al., 2014; Gou et al., 2015; Pérez Hoyos et al., 2016; Chiloane et al., 2020). Water indices such as NDWI, MNDWI, Automated Water Extraction Index for Shadow, Water Ratio Index, Land Surface Water Index, and Topographic Wetness Index have been noted as the best to study GDEs. However, assessing GDE characteristics using indices is time-consuming and prone to human error (Bian et al., 2021). This is because indices require the preparation of training data which is often limited in terms of GDE occurrence. This presents difficulties for precise GDE studies such as delineation since these data may further require field validation.

9 Conclusion and recommendations

Remote sensing techniques provide a powerful tool for monitoring the effects of climate variability on GDEs. By using satellite imagery to track changes in vegetation cover, water surface area, and other surface features, we can gain insights into how GDEs are responding to drought, rising temperatures, and other climate-related stresses. Remote sensing is particularly useful for monitoring GDEs because it can be used to cover large areas at a relatively low cost. This makes it possible to track changes in GDEs over time and to identify trends that may not be apparent from ground-based observations. In addition, remote sensing can be used to monitor GDEs in remote and inaccessible areas. This is important because many GDEs are located in arid and semi-arid regions, where ground-based monitoring can be difficult and expensive. Thus, remote sensing techniques are still under development, but they have the potential to revolutionize the way we monitor and manage GDEs in the face of climate change. By combining remote sensing data with other information sources, such as climate models and hydrological data, we can develop a more comprehensive understanding of how GDEs are responding to climate variability and identify strategies for protecting these important ecosystems.

For a better understanding of GDEs, we recommend an integrated approach to using remote sensing tools. This involves the utilization of spectrally derived indices and supervised image classification systems coupled with machine-learning algorithms, technologically advanced satellite images, and ancillary data (i.e., topographic data like DEM) as well as integrating with expert knowledge. This will offer a more conceptually comprehensive assessment of GDEs than assessments based solely on one method. Despite the availability of improved robust remote sensing methods and modeling algorithms, assessing GDEs at various scales remains difficult. This is primarily due to the variability of GDEs (e.g., ecological water requirements, species diversity), which is difficult to capture when utilizing the medium to coarse spatial resolution sensors (Kaplan and Avdan, 2017; Gxokwe et al., 2020; Thamaga et al., 2021). Therefore, there is still a severe need for researchers to identify specific remote sensing tools for a specific characterization of GDE components that may not be extensively studied using the existing indices and models. One of these limitations is the discrepancy between the resolution of remote sensing data and the scale of the ecological features being identified (Glanville et al., 2016).

References

Akasheh, O.Z., Neale, C.M.U., Jayanthi, H., 2008. Detailed mapping of riparian vegetation in the middle Rio Grande River using high-resolution multi-spectral airborne remote sensing. J. Arid Environ. 72, 1734–1744.

Alaibakhsh, M., Emelyanova, I., Barron, O., Khiadani, M., Warren, G., 2017. Large-scale regional delineation of riparian vegetation in the arid and semi-arid Pilbara region, WA. Hydrol. Process. 31, 4269–4281.

Albano, C.M., Minor, B., Freed, Z., Huntington, J.L., 2020. Status and trends of groundwater dependent vegetation in relation to climate and shallow groundwater in the Harney Basin, Oregon. In: Technical Report, Publication. Desert Research Institute, p. 41280.

Allen, R.G., Pereira, L.S., Raes, D., Smith, M., 1998. Crop evapotranspiration—guidelines for computing crop water requirements. In: FAO Irrigation and Drainage Paper 56. vol. 300. FAO, Rome, p. 6541.

Allen, M.A., Voogt, J.A., Christen, A., 2017. Time-continuous hemispherical urban surface temperatures. Remote Sens. 10 (1), 3.

Ashwini, K., Verma, R.K., Sriharsha, S., Chourasiya, S., Singh, A., 2023. Delineation of groundwater potential zone for sustainable water resources management using remote sensing-GIS and analytic hierarchy approach in the state of Jharkhand, India. Groundw. Sustain. Dev. 21, 100908.

Barquín, J., Scarsbrook, M., 2008. Management and conservation strategies for coldwater springs. Aquat. Conserv.: Mar. Freshw Ecosyst. 18 (5), 580–591.

Barron, O.V., Emelyanova, I., Van Niel, T.G., Pollock, D., Hodgson, G., 2014. Mapping groundwater-dependent ecosystems using remote sensing measures of vegetation and moisture dynamics. Hydrol. Process. 28 (2), 372–385.

Basavarajappa, H.T., Pushpavathi, K.N., Manjunatha, M.C., 2015. Climate change and its impact on groundwater table fluctuation in Precambrian rocks of Chamarajanagara district, Karnataka, India using geomatics technique. Int. J. Geomat. Geosci. 5 (4), 510–524.

Beck, E., 2019. Ecology: ecosystems and biodiversity. In: Emergence and Modularity in Life Sciences. Springer, Cham, pp. 195–213.

Berberoglu, S., Yilmaz, K.T., Özkan, C., 2004. Mapping and monitoring of coastal wetlands of Cukurova Delta in the Eastern Mediterranean region. Biodivers. Conserv. 13, 615–633.

Bian, L., Melesse, A.M., Leon, A.S., Verma, V., Yin, Z., 2021. A deterministic topographic wetland index based on LiDAR-derived DEM for delineating open-water wetlands. Water 13 (18), 2487.

Bohn, V.Y., Delgado, A.L., Piccolo, M.C., Perillo, G.M., 2016. Assessment of climate variability and land use effect on shallow lakes in temperate plains of Argentina. Environ. Earth Sci. 75, 1–15.

Bonacci, O., Gabrovšek, F., Stevanović, Z., 2018. Karst Springs: a global perspective on their origin, ecology, and conservation. Water 10 (12), 1887.

Boulton, A.J., 2005. Chances and challenges in the conservation of groundwaters and their dependent ecosystems. Aquat. Conserv. 15, 319–323.

Boulton, A.J., 2020. Editorial: conservation of groundwaters and their dependent ecosystems: integrating molecular taxonomy, systematic reserve planning and cultural values. Aquat. Conserv. Mar. Freshwat. Ecosyst. 30 (1), 1–7.

Box, J.B., Leiper, I., Nano, C., Stokeld, D., Jobson, P., Tomlinson, A., Cobban, D., Bond, T., Randall, D., Box, P., 2022. Mapping terrestrial groundwater-dependent ecosystems in arid Australia using Landsat-8 time-series data and singular value decomposition. Remote Sens. Ecol. Conserv. 8 (4), 464–476.

Bretreger, D., Yeo, I.Y., Kuczera, G., Hancock, G., 2021. Remote sensing's role in improving transboundary water regulation and compliance: the Murray-Darling Basin, Australia. J. Hydrol. X 13, 100112.

Brocca, L., Melone, F., Moramarco, T., Wagner, W., Hasenauer, S., Hahn, S., 2014. Assimilation of surface- and root-zone ASCAT soil moisture products into rainfall-runoff modeling. IEEE Trans. Geosci. Remote Sens. 52 (11), 6866–6878.

Chaikumbung, M., Doucouliagos, H., Scarborough, H., 2016. The economic value of wetlands in developing countries: a meta-regression analysis. Ecol. Econ. 124, 164–174.

Chiloane, C., Dube, T., Shoko, C., 2020. Monitoring and assessment of the seasonal and inter-annual pan inundation dynamics in the Kgalagadi Transfrontier Park, Southern Africa. Phys. Chem. Earth, A/B/C 118, 102905.

Chiloane, C., Dube, T., Shoko, C., 2022. Impacts of groundwater and climate variability on terrestrial groundwater dependent ecosystems: a review of geospatial assessment approaches and challenges and possible future research directions. Geocarto Int. 37 (23), 6755–6779.

Chiloane, C., Dube, T., Shoko, C., 2023. Multispectral remote sensing of potential groundwater dependent vegetation in the greater Floristic region of the Western Cape, South Africa. S. Afr. Geogr. J., 1–19.

Collados-Lara, A.J., Pardo-Igúzquiza, E., Pulido-Velazquez, D., Baena-Ruiz, L., 2021. Estimation of the monthly dynamics of surface water in wetlands from satellite and secondary hydroclimatological data. Remote Sens. 13 (12), 2380.

Currie, B., Milton, S.J., Steenkamp, J.C., 2009. Cost-benefit analysis of alien vegetation clearing for water yield and tourism in a mountain catchment in the Western Cape of South Africa. Ecol. Econ. 68 (10), 2574–2579.

Dabboor, M., Brisco, B., 2018. Wetland monitoring and mapping using synthetic aperture radar. Wetl. Manag. Assess. Risk Sustain. Solut. 1, 13.

Dalu, T., Wasserman, R.J. (Eds.), 2022. Fundamentals of Tropical Freshwater Wetlands: From Ecology to Conservation Management. Elsevier, Oxford.

Danielopol, D.L., Pospisil, P., 2001. Hidden biodiversity in the groundwater of the Danube flood plain national park (Austria). Biodivers. Conserv. 10, 1711–1721.

Doody, T.M., Barron, O.V., Dowsley, K., Emelyanova, I., Fawcett, J., Overton, I.C., Pritchard, J.L., Van Dijk, A.I., Warren, G., 2017. Continental mapping of groundwater dependent ecosystems: a methodological framework to integrate diverse data and expert opinion. J. Hydrol. Reg. Stud. 10, 61–81.

Dube, T., Pandit, S., Shoko, C., Ramoelo, A., Mazvimavi, D., Dalu, T., 2019. Numerical assessments of leaf area index in tropical savanna rangelands, South Africa using Landsat 8 OLI derived metrics and in-situ measurements. Remote Sens. 11 (7), 11. https://doi.org/10.3390/rs11070829.

Duran-Llacer, I., Arumı, J.L., Arriagada, L., Aguayo, M., Rojas, O., Gonzalez-Rodrıguez, L., Rodrıguez-Lopez, L., Martınez-Retureta, R., Oyarzun, R., Singh, S.K., 2021. A new method to map groundwater-dependent ecosystem zones in semi-arid environments: a case study in Chile. Sci. Total Environ., 151528.

Dwire, K.A., Mellmann-Brown, S., Gurrieri, J.T., 2018. Potential effects of climate change on riparian areas, wetlands, and groundwater-dependent ecosystems in the Blue Mountains, Oregon, USA. Clim. Serv. 10, 44–52.

Eamus, D., Froend, R., 2006. Groundwater-dependent ecosystems: the where, what and why of GDEs. Aust. J. Bot. 54 (2), 91–96.

Eamus, D., Zolfaghar, S., Villalobos-Vega, R., Cleverly, J., Huete, A., 2015. Groundwater-dependent ecosystems: recent insights from satellite and field-based studies. Hydrol. Earth Syst. Sci. 19 (10), 4229–4256.

Elmore, A.J., Manning, S.J., Mustard, J.F., Craine, J.M., 2006. Decline in alkali meadow vegetation cover in California: the effects of groundwater extraction and drought. J. Appl. Ecol. 43 (4), 770–779.

Erostate, M., Huneau, F., Garel, E., Ghiotti, S., Vystavna, Y., Garrido, M., Pasqualini, V., 2020. Groundwater dependent ecosystems in coastal Mediterranean regions: characterization, challenges and management for their protection. Water Res. 172, 115461.

FAO, 2019. WaPOR Database Methodology: Level 3 Data—Using Remote Sensing to Support Solutions to Reduce Agricultural Water Productivity Gaps, Rome. (68 pp. Licence: CC BY-NC-SA 3.0 IGO).

Farda, N.M., 2017. Multi-temporal land use mapping of coastal wetlands area using machine learning in Google Earth Engine. In: IOP Conference Series: Earth and Environmental Science. vol. 98 (1). IOP Publishing, p. 012042.

Fontana, R.B., Collischonn, W., 2019. Remote Sensing and GIS Based Analysis of Groundwater Dependent Ecosystems in the Brazilian Cerrado. Simpósio Brasileiro de Recursos Hídricos (23.: Foz do Iguaçu, 2019). Anais [recurso eletrônico]. ABRH, Porto Alegre, p. 2019.

Ford, D., Williams, P., 2013. Karst Hydrogeology and Geomorphology. John Wiley and Sons.

Foster, S., Chilton, J., Nijsten, G.J., Richts, A., 2013. Groundwater—a global focus on the 'local resource'. Curr. Opin. Environ. Sustain. 5 (6), 685–695.

Fu, B., Burgher, I., 2015. Riparian vegetation NDVI dynamics and its relationship with climate, surface water and groundwater. J. Arid Environ. 113, 59–68.

Gao, B., 1996. NDWI—A normalized difference water index for remote sensing of vegetation liquid water from space. Remote Sens. Environ. 58 (3), 257–266.

Ghosh, S., Das, A., 2020. Wetland conversion risk assessment of East Kolkata wetland: a Ramsar site using random forest and support vector machine model. J. Clean. Prod. 275, 123475.

Gintamo, T.T., Mengistu, H., Kanyerere, T., 2021. GIS-based modelling of climate variability impacts on groundwater quality: Cape Flats aquifer, Cape Town, South Africa. Groundw. Sustain. Dev. 15, 100663.

Glanville, K., Ryan, T., Tomlinson, M., Muriuki, G., Ronan, M., Pollett, A., 2016. A method for catchment scale mapping of groundwater-dependent ecosystems to support natural resource management (Queensland, Australia). Environ. Manag. 57, 432–449.

Gleeson, T., Befus, K.M., Jasechko, S., Luijendijk, E., Cardenas, M.B., 2020. The global volume and distribution of modern groundwater. Nat. Geosci. 13 (5), 392–397.

Gorelick, S.M., Zheng, C., Gao, H., 2015. Global change and groundwater management. Annu. Rev. Environ. Resour. 40, 23–47.

Gosling, S.N., Arnell, N.W., Todd, M.C., 2015. A global assessment of the impact of climate change on water scarcity. Clim. Change 134 (3), 371–385.

Gou, S., Gonzales, S., Miller, G.R., 2015. Mapping potential groundwater-dependent ecosystems for sustainable management. Groundwater 53 (1), 99–110.

Groeneveld, D.P., Baugh, W.M., 2007. Correcting satellite data to detect vegetation signal for eco-hydrologic analyses. J. Hydrol. 344 (1), 135–145.

Gronwall, J., Danert, K., 2020. Regarding groundwater and drinking water access through a human rights lens: self-supply as a norm. Water 12, 419.

Gxokwe, S., Dube, T., Mazvimavi, D., 2020. Multispectral remote sensing of wetlands in semi-arid and arid areas: a review on applications, challenges and possible future research directions. Remote Sens. 12 (24), 4190.

Hamed, Y., Hadji, R., Redhaounia, B., Zighmi, K., Bâali, F., El Gayar, A., 2018. Climate impact on surface and groundwater in North Africa: a global synthesis of findings and recommendations. Euro-Mediterr. J. Environ. Integr. 3, 1–15.

Han, X., Pan, J., Devlin, A.T., 2018. Remote sensing study of wetlands in the Pearl River Delta during 1995–2015 with the support vector machine method. Front. Earth Sci. 12, 521–531.

Havril, T., Tóth, Á., Molson, J.W., Galsa, A., Mádl-Szőnyi, J., 2018. Impacts of predicted climate change on groundwater flow systems: can wetlands disappear due to recharge reduction? J. Hydrol. 563, 1169–1180.

Hess, L.L., Melack, J.M., Filho, B.S., Wang, Y., 2019. Amazonian waters: Trends and variability in an era of deforestation and hydroelectric dams. In: Land Use and the Carbon Cycle. Springer, pp. 15–34.

Hong, S.H., Wdowinski, S., Kim, S.W., Won, J.S., 2010. Multi-temporal monitoring of wetland water levels in the Florida Everglades using interferometric synthetic aperture radar (InSAR). Remote Sens. Environ. 114 (11), 2436–2447.

House, A., Thompson, J., Acreman, M., 2016. Projecting impacts of climate change on hydrological conditions and biotic responses in a chalk valley riparian wetland. J. Hydrol. 534, 178–192.

Huang, J., Wang, W., Cui, X., Wang, D., Liu, W., Liu, X., Wang, S., 2020. Environmental risk-based hydroeconomic evaluation for alluvial aquifer management in arid river basin. Sci. Total Environ. 711.

Huete, A., Didan, K., Miura, T., Rodriguez, E., Gao, X., Ferreira, L., 2002. Overview of the radiometric and biophysical performance of the MODIS vegetation indices. Remote Sens. Environ. 83 (1), 195–213.

Hultine, K.R., Froend, R., Blasini, D., Bush, S.E., Karlinski, M., Koepke, D.F., 2020. Hydraulic traits that buffer deep-rooted plants from changes in hydrology and climate. Hydrol. Process. 34 (2), 209–222.

Huntington, J., McGwire, K., Morton, C., Snyder, K., Peterson, S., Erickson, T., Niswonger, R., Carroll, R., Smith, G., Allen, R., 2016. Assessing the role of climate and resource management on groundwater dependent ecosystem changes in arid environments with the Landsat archive. Remote Sens. Environ. 185, 186–197.

IPCC, 2014. Climate change 2014: impacts, adaptation, and vulnerability. Summary for policymakers. In: Contributions of the Working Group II to the Fifth Assessment Report. Available at: https://doi.org/10.1016/j.renene.2009.11.012.

IPCC, 2021. Synthesis Report of the IPCC Sixth Assessment Report (AR6). European University Institute, pp. 2–5. Available at: https://eur-lex.europa.eu/legal-content/PT/TXT/PDF/?uri=CELEX:32016R0679andfrom=PT%0Ahttp://eur-lex.europa.eu/LexUriServ/LexUriServ.do?uri=CELEX:52012PC0011:pt:NOT.

Jyväsjärvi, J., Marttila, H., Rossi, P.M., Ala-Aho, P., Olofsson, B.O., Nisell, J., Backman, B., Ilmonen, J., Virtanen, R., Paasivirta, L., Britschgi, R., 2015. Climate-induced warming imposes a threat to North European spring ecosystems. Glob. Chang. Biol. 21 (12), 4561–4569.

Kaplan, G., Avdan, U., 2017. Mapping and monitoring wetlands using Sentinel-2 satellite imagery. ISPRS Ann. Photogramm. Remote Sens. Spat. Inf. Sci. 4, 271–277.

Kath, J., Reardon-Smith, K., Le Brocque, A.F., Dyer, F.J., Dafny, E., Fritz, L., Batterham, M., 2014. Groundwater decline and tree change in floodplain landscapes: identifying non-linear threshold responses in canopy condition. Glob. Ecol. Conserv. 2, 148–160.

Kibler, C.L., Schmidt, E.C., Roberts, D.A., Stella, J.C., Kui, L., Lambert, A.M., Singer, M.B., 2021. A brown wave of riparian woodland mortality following groundwater declines during the 2012–2019 California drought. Environ. Res. Lett. 16 (8), 084030.

Kim, J.B., Kerns, B.K., Drapek, R.J., Pitts, G.S., Halofsky, J.E., 2018. Simulating vegetation response to climate change in the Blue Mountains with MC2 dynamic global vegetation model. Clim. Serv. 10, 20–32.

Kløve, B., Allan, A., Bertrand, G., Druzynska, E., Ertürk, A., Goldscheider, N., Henry, S., Karakaya, N., Karjalainen, T.P., Koundouri, P., Kupfersberger, H., 2011. Groundwater dependent ecosystems. Part II. Ecosystem services and management in Europe under risk of climate change and land use intensification. Environ. Sci. Policy 14 (7), 782–793.

Kløve, B., Ala-Aho, P., Bertrand, G., Gurdak, J.J., Kupfersberger, H., Kværner, J., Muotka, T., Mykrä, H., Preda, E., Rossi, P., Uvo, C.B., 2014. Climate change impacts on groundwater and dependent ecosystems. J. Hydrol. 518, 250–266.

Konikow, L.F., Kendy, E., 2005. Groundwater depletion: a global problem. Hydrogeol. J. 13, 317–320.

Kumar, C.P., 2012. Climate change and its impact on groundwater resources. Int. J. Eng. Sci. 1 (5), 43–60.

Kumar, A., Kanaujia, A., 2014. Wetlands: significance. Threats and their conservation. Environment 7, 1–24.

Kundzewicz, Z.W., Doell, P., 2009. Will groundwater ease freshwater stress under climate change? Hydrol. Sci. J. 54 (4), 665–675.

Laba, M., Blair, B., Downs, R., Monger, B., Philpot, W., Smith, S., Sullivan, P., Baveye, P.C., 2010. Use of textural measurements to map invasive wetland plants in the Hudson River National Estuarine Research Reserve with IKONOS satellite imagery. Remote Sens. Environ. 114 (4), 876–886.

Lang, M., McDonough, O., McCarty, G., Oesterling, R., Wilen, B., 2012. Enhanced detection of wetland-stream connectivity using LiDAR. Wetlands 32, 461–473.

Le Maitre, D., Dzikiti, S., Gush, M., 2016. Remote sensing data on evaporation for spatial and temporal extrapolation of sitespecific data to landscapes and catchments. Report No. CSIR/NRE/ECOS/ER/2016/0019/B, Natural Resources and the Environment. CSIR, Stellenbosch.

Leblanc, M., Tweed, S.O., van Dijk, A.I., Timbal, B., 2007. A review of historic and future hydrological changes in the Murray-Darling Basin. Glob. Planet. Change 57 (3–4), 267–294.

Lewis, M.M., White, D., 2013. Evaluation of remote sensing approaches. In: Allocating Water and Maintaining Springs in the Great Artesian Basin. Volume IV. Spatial Survey and Remote Sensing of Artesian Springs of the Western Great Artesian Basin. vol. IV. National Water Commission, Kingston, Jamaica.

Li, J., Roy, D.P., 2017. A global analysis of Sentinel-2A, Sentinel-2B and Landsat-8 data revisit intervals and implications for terrestrial monitoring. Remote Sens. 9 (9), 902.

Liu, C., Liu, H., Yu, Y., Zhao, W., Zhang, Z., Guo, L., Yetemen, O., 2021. Mapping groundwater-dependent ecosystems in arid Central Asia: implications for controlling regional land degradation. Sci. Total Environ. 797, 149027.

Lloréns, J.L.P., 2008. Impacts of climate change on wetland ecosystems. Water Supply 7 (2.117), 8.

Lv, J., Wang, X.S., Zhou, Y., Qian, K., Wan, L., Eamus, D., Tao, Z., 2012. Groundwater-dependent distribution of vegetation in Hailiutu River catchment, a semi-arid region in China. Ecohydrology 6, 142–149.

MacAlister, C., Mahaxay, M., 2009. Mapping wetlands in the Lower Mekong Basin for wetland resource and conservation management using Landsat ETM images and field survey data. J. Environ. Manag. 90 (7), 2130–2137.

MacFarlane, C., Hoffman, M., Eamus, D., Kerp, N., Adams, M.A., Higginson, S., McMurtrie, R., 2007. Estimation of leaf area index in eucalypt forest using digital photography. Agric. For. Meteorol. 143, 176–188.

Mahdavi, S., Salehi, B., Granger, J., Amani, M., Brisco, B., Huang, W., 2018. Remote sensing for wetland classification: a comprehensive review. GISci. Remote Sens. 55 (5), 623–658.

Mahdianpari, M., Salehi, B., Rezaee, M., Mohammadimanesh, F., Zhang, Y., 2018. Very deep convolutional neural networks for complex land cover mapping using multispectral remote sensing imagery. Remote Sens. 10 (7), 1119.

Majola, K., Xu, Y., Kanyerere, T., 2022. Review: assessment of climate change impacts on groundwater-dependent ecosystems in transboundary aquifer settings with reference to the Tuli-Karoo transboundary aquifer. Ecohydrol. Hydrobiol. 22 (1), 126–140.

Mango, N., Siziba, S., Makate, C., 2017. The impact of adoption of conservation agriculture on smallholder farmers' food security in semi-arid zones of southern Africa. Agric. Food Secur. 6 (1), 1–8.

Marchionni, V., Daly, E., Manoli, G., Tapper, N.J., Walker, J.P., Fatichi, S., 2020. Groundwater buffers drought effects and climate variability in urban reserves. Water Resour. Res. 56 (5), e2019WR026192.

Martínez-Santos, P., Díaz-Alcaide, S., De la Hera-Portillo, A., Gómez-Escalonilla, V., 2021. Mapping groundwater-dependent ecosystems by means of multi-layer supervised classification. J. Hydrol. 603, 126873.

Masina, F.M., Wasserman, R.J., Wu, N., Mungenge, C.P., Dondofema, F., Keates, C., Shikwambana, P., Dalu, T., 2023. Macroinvertebrate diversity in relation to limnochemistry in an Austral semi-arid transboundary aquifer region pan system. Sci. Total Environ. 878, 163161.

McAllister, E., Payo, A., Novellino, A., Dolphin, T., Medina-Lopez, E., 2022. Multispectral satellite imagery and machine learning for the extraction of shoreline indicators. Coast. Eng. 174, 104102.

Meyers, Z.P., Frisbee, M.D., Rademacher, L.K., Stewart-Maddox, N.S., 2021. Old groundwater buffers the effects of a major drought in groundwater-dependent ecosystems of the eastern Sierra Nevada (CA). Environ. Res. Lett. 16 (4), 044044.

Misra, A.K., 2014. Climate change and challenges of water and food security. Int. J. Sustain. Built Environ. 3 (1), 153–165.

Mitsch, W.J., Gosselink, J.G., 2015. Wetlands. John Wiley and Sons.

Morsy, K.M., Alenezi, A., AlRukaibi, D.S., 2017. Groundwater and dependent ecosystems: revealing the impacts of climate change. Int. J. Appl. Eng. Res. 12 (13), 3919–3926.

Mpakairi, K.S., Dube, T., Dondofema, F., Dalu, T., 2022a. Spatio-temporal variation of vegetation heterogeneity in groundwater dependent ecosystems within arid environments. Ecol. Inform. 69, 101667.

Mpakairi, K.S., Dube, T., Dondofema, F., Dalu, T., 2022b. Spatial characterisation of vegetation diversity in groundwater-dependent ecosystems using in-situ and Sentinel-2 MSI satellite data. Remote Sens. 14 (13), 2995.

Münch, Z., Conrad, J., 2007. Remote sensing and GIS based determination of groundwater dependent ecosystems in the Western Cape, South Africa. Hydrogeol. J. 15, 19–28.

Mungenge, C.P., Wasserman, R.J., Dondofema, F., Keates, C., Masina, F.M., Dalu, T., 2023. Assessing chlorophyll-a and water quality dynamics in arid-zone temporary pan systems along a disturbance gradient. Sci. Total Environ. 873, 162272.

Murray, N.J., Keith, D.A., Bland, L.M., Nicholson, E., McAlpine, C.A., 2013. Threat to the eastern Australian coastal zone from the 2011 Tohoku tsunami. PLoS One 8 (5), e62742.

Mutanga, O., Adam, E., Cho, M.A., 2012. High density biomass estimation for wetland vegetation using WorldView-2 imagery and random forest regression algorithm. Int. J. Appl. Earth Obs. Geoinf. 18, 399–406.

Orellana, F., Verma, P., Loheide, S.P., Daly, E., 2012. Monitoring and modeling water-vegetation interactions in groundwater-dependent ecosystems. Rev. Geophys. 50 (3).

Papa, F., Prigent, C., Aires, F., Jiménez, C., Rossow, W.B., Rossow, W.B., 2018. Intercomparison of a large ensemble of satellite-based humidity products and implications for climate studies. Environ. Res. Lett. 13 (11), 115007.

Pérez Hoyos, I.C., Krakauer, N.Y., Khanbilvardi, R., Armstrong, R.A., 2016. A review of advances in the identification and characterization of groundwater dependent ecosystems using geospatial technologies. Geosciences 6 (2), 17.

Pettorelli, N., Laurance, W.F., O'Brien, T.G., Wegmann, M., Nagendra, H., Turner, W., 2014. Satellite remote sensing for applied ecologists: opportunities and challenges. J. Appl. Ecol. 51 (4), 839–848.

Rampheri, M.B., Dube, T., Dondofema, F., Dalu, T., 2023a. Progress in the remote sensing of groundwater-dependent ecosystems in semi-arid environments. Phys. Chem. Earth Parts A/B/C 130, 103359.

Rampheri, M.B., Dube, T., Dondofema, F., Dalu, T., 2023b. Identification and delineation of groundwater-dependent ecosystems (GDEs) in the Khakea-Bray transboundary aquifer using geospatial techniques. Geocarto Int. 38, 2172217.

Rapinel, S., Fabre, E., Dufour, S., Arvor, D., Mony, C., Hubert-Moy, L., 2019. Mapping potential, existing and efficient wetlands using free remote sensing data. J. Environ. Manag. 247, 829–839.

Reschke, J., Bartsch, A., Schlaffer, S., Schepaschenko, D., 2012. Capability of C-band SAR for operational wetland monitoring at high latitudes. Remote Sens. 4, 2923–2943.

Rouse, J.W., Haas, R.H., Schell, J.A., Deering, D.W., Harlan, J.C., 1974. Monitoring the vernal advancement of retrogradation of natural vegetation. In: NASA/GSFC, Type III, Final Report, 371, p. 371.

Salimi, S., Almuktar, S.A., Scholz, M., 2021. Impact of climate change on wetland ecosystems: a critical review of experimental wetlands. J. Environ. Manag. 286, 112160.

Slagter, B., Tsendbazar, N.E., Vollrath, A., Reiche, J., 2020. Mapping wetland characteristics using temporally dense Sentinel-1 and Sentinel-2 data: a case study in the St. Lucia wetlands, South Africa. Int. J. Appl. Earth Obs. Geoinf. 86, 102009.

Stromberg, J.C., Lite, S.J., Dixon, M.D., Paradzick, C., 2010. Comparative water use by native and invasive riparian plants. Am. J. Bot. 97 (7), 1079–1092.

Suding, K.N., Farrer, E.C., King, A.J., Kueppers, L., Spasojevic, M.J., 2015. Vegetation change at high elevation: scale dependence and interactive effects on Niwot Ridge. Plant Ecol. Divers. 8 (5–6), 713–725.

Taylor, R.G., Scanlon, B., Döll, P., Rodell, M., Van Beek, R., Wada, Y., Longuevergne, L., Leblanc, M., Famiglietti, J.S., Edmunds, M., Konikow, L., 2013. Ground water and climate change. Nat. Clim. Chang. 3 (4), 322–329.

Taylor, R.G., Todd, M.C., Kongola, L., Maurice, L., Nahozya, E., Sanga, H., MacDonald, A.M., 2013. Evidence of the dependence of groundwater resources on extreme rainfall in East Africa. Nat. Clim. Change 3 (4), 374–378.

Thamaga, K.H., Dube, T., 2018. Remote sensing of invasive water hyacinth (Eichhornia crassipes): a review on applications and challenges. Remote Sens. Appl. 10, 36–46. https://doi.org/10.1016/j.rsase.2018.02.005.

Thamaga, K.H., Dube, T., Shoko, C., 2021. Advances in satellite remote sensing of the wetland ecosystems in sub-Saharan Africa advances in satellite remote sensing of the wetland. Geocarto Int. 37 (20), 1–22.

Thamaga, K.H., Dube, T., Shoko, C., 2022. Advances in satellite remote sensing of the wetland ecosystems in sub-Saharan Africa. Geocarto Int. 37 (20), 5891–5913.

Thornton, P.K., Ericksen, P.J., Herrero, M., Challinor, A.J., 2014. Climate variability and vulnerability to climate change: a review. Glob. Chang. Biol. 20 (11), 3313–3328.

Tian, S., Zhang, X., Tian, J., Sun, Q., 2016. Random forest classification of wetland landcovers from multi-sensor data in the arid region of Xinjiang, China. Remote Sens. 8 (11), 954.

Tweed, S.O., Leblanc, M., Webb, J.A., Lubczynski, M.W., 2007. Remote sensing and GIS for mapping groundwater recharge and discharge areas in salinity prone catchments, southeastern Australia. Hydrogeol. J. 15, 75–96.

Verlicchi, P., Grillini, V., 2020. Surface and groundwater quality in South Africa and Mozambique—analysis of the most critical pollutants for drinking purposes and challenges in water treatment. Water 12, 305.

Villa, J.A., Bernal, B., 2018. Carbon sequestration in wetlands, from science to practice: an overview of the biogeochemical process, measurement methods, and policy framework. Ecol. Eng. 114, 115–128.

Weigand, M., Wurm, M., Dech, S., Taubenböck, H., 2019. Remote sensing in environmental justice research—a review. ISPRS Int. J. Geo-Inf. 8 (1), 20.

Werstak, C.E., Housman, I., Maus, P., Fisk, H., Gurrieri, J., Carlson, C.P., Johnston, B.C., Stratton, B., Hurja, J.C., 2012. Groundwater-Dependent Ecosystem Inventory Using Remote Sensing: RSAC-10011-RPT1. Remote Sensing Applications Center, Salt Lake City, UT, USA.

Williams, J., Stella, J.C., Voelker, S.L., Lambert, A.M., Pelletier, L.M., Drake, J.E., Friedman, J.M., Roberts, D.A., Singer, M.B., 2022. Local groundwater decline exacerbates response of dryland riparian woodlands to climatic drought. Glob. Chang. Biol. 28 (22), 6771–6788.

Xu, D., Wang, C., Chen, J., Shen, M., Shen, B., Yan, R., Li, Z., Karnieli, A., Chen, J., Yan, Y., Wang, X., 2021. The superiority of the normalized difference phenology index (NDPI) for estimating grassland aboveground fresh biomass. Remote Sens. Environ. 264, 112578.

Wu, J., Kobayashi, H., Stark, S.C., Meng, R., Guan, K., Tran, N.N., Gao, S., Yang, W., Restrepo-Coupe, N., Miura, T., Oliviera, R.C., 2018. Biological processes dominate seasonality of remotely sensed canopy greenness in an Amazon evergreen forest. New Phytol. 217 (4), 1507–1520.

Xu, W., Kong, F., Mao, R., Song, J., Sun, H., Wu, Q., et al., 2022. Identifying and mapping potential groundwater-dependent ecosystems for a semi-arid and semi-humid area in the Weihe River, China. J. Hydrol. 609, 127789.

Yang, X., Liu, J., 2020. Assessment and valuation of groundwater ecosystem services: a case study of Handan City, China. Water 12 (5), 1455.

Zhang, X., Gao, Y., Long, X., Brierley, G., 2019. Karst spring discharge and its sensitivity to climatic variation in Yunnan, SW China. J. Hydrol. 578, 124114.

Zwedzi, L., 2019. Delineating Wetland Waterbodies of Wide Spatial Variation Using Remote Sensing Techniques. University of Johannesburg, South Africa.

Chapter 8

Climate variability and water resources

Timothy Dube and Trisha D. Bhaga
Department of Earth Science, Institute for Water Studies, University of the Western Cape, Cape Town, South Africa

Chapter outline

Remote Sensing of Climate. https://doi.org/10.1016/B978-0-443-21731-9.00001-6
Copyright © 2024 Elsevier Inc. All rights are reserved, including those for text and data mining, AI training, and similar technologies.

1 Introduction

1.1 Background and significance

Climate variability is a complex phenomenon that involves natural fluctuations and changes in patterns and climatic elements, such as temperature, precipitation, wind patterns, and ocean currents (Huang et al., 2017; Verner et al., 2018; Varghese et al., 2021). Climate variability occurs due to anthropogenic activities, such as the burning of fossil fuels, deforestation, land use changes, and aerosol emissions, as well as natural factors such as solar radiation, volcanic activity, and El Nino and La Nina (Seager et al., 2003; Varghese et al., 2021; Affoh et al., 2022; Ahmed et al., 2023). Climate variability can have significant impacts on ecosystems, agriculture, water resources, economies, and human societies; therefore, studying and understanding climate variability and its impact on water resources is vital (Bhaga et al., 2020; Hagenlocher et al., 2019; Frischen et al., 2020). Water is a finite resource and, therefore, climate variability can significantly influence the quantity and distribution of water resources, as water resources and climate are intimately connected (Bhaga et al., 2023). The effects of climate variability on water resources will lead to negative impacts on agriculture, especially in regions dependent on rainfed agriculture, which causes food insecurity (Gohar and Cashman, 2016; Sharara et al., 2022). Drought caused by climate variability leads to an economic loss of approximately US \$6–8 billion per year (Zeng et al., 2019; Saha et al., 2022). Drought affected 2 million people globally in 1972 and 14 million people in 2002/2003 (Bogale and Erena, 2022). The Earth's temperature has increased by approximately 0.08°C per decade since 1880 and based on the National Oceanic and Atmospheric Administration's (NOAA) temperature data, 2022 was the sixth warmest year (NOAA, 2023). Many regions across the globe are experiencing water stress; therefore, effective water resource management and planning requires an understanding of how climate variability affects water resources to assist policymakers and water managers in developing efficient adaptive strategies to mitigate the impacts of climate variability. The monitoring of water resources needs to be conducted in a timely manner and continuously to notice any changes; therefore, remotely sensed data provides the opportunity to monitor climate variability and water resources in a spatially explicit and timely manner. By understanding the role climate variability plays in water resource changes, researchers can better understand and predict how these events might evolve in the future and address the challenges posed by climate variability. This knowledge is crucial for informed decision-making, sustainable development, the preservation of natural ecosystems, and human well-being. Moreover, by studying and understanding climate variability, several sustainable development goals (SDGs) will be addressed, namely, Goal 1: "No poverty," Goal 2: "Zero hunger," Goal 3: "Good health and well-being," Goal 6: "Clean water and sanitation," Goal 11: "Sustainable cities and communities," Goal 12: "Responsible production and consumption," Goal 13: "Climate action," and Goal 15: "Life on land" (Zhang et al., 2019; Bhaga et al., 2020).

1.2 Role of remote sensing in assessing climate variability impacts on water resources

Remote sensing involves the use of satellite, aircraft, drone, and other platforms to collect information on the Earth's surface, atmosphere, and oceans from a distance (USGS, 2023). Remote sensing plays a crucial role in assessing the impacts of climate variability on water resources by providing valuable data and insights that are difficult to obtain through traditional in situ-based methods (Huang et al., 2018). Remote sensing methods provide the ability to monitor changes on the Earth's surface, atmosphere, and oceans in a timely, cost-effective manner and are able to collect data in a repeatable manner from regions that are inaccessible (AghaKouchak et al., 2015; Bhaga et al., 2023). Remote sensing allows for the collection of large-scale, repeatable data and long-term monitoring of various aspects of the Earth's environment, such as temperature, precipitation, soil moisture, snow cover, and vegetation (USGS, 2023). These data provide a comprehensive view of how climate variability is affecting water resources over time. Long-term monitoring using remote sensing also aids researchers in studying trends and changes in water resources over extended periods (Mishra and Singh, 2010). Remote sensing also enables the monitoring of water resources in remote and inaccessible areas, which provides a comprehensive understanding of the impacts of climate variability on water availability and quality across the globe (Huang et al., 2018). Satellite data and other remote sensing platforms provide frequent and consistent measurements over time, which enables researchers to observe short-term and long-term changes in water resources due to climate variability (Govender et al., 2022; Sharara et al., 2022). The spatial resolution of satellite data provides high-resolution data that is vital to analyze local and regional variations in water resources, which is important for understanding how climate variability affects different areas (Huang et al., 2018). Remote sensing instruments are able to estimate precipitation levels, and runoff and monitor changes in river and stream flow, which aids researchers in understanding how climate variability impacts water availability and the hydrological cycle (Cancelliere et al., 2007; Chawira et al., 2013). Remote sensing data can be used to identify drought conditions by monitoring changes in soil moisture, vegetation health, and water levels and is able to provide early warnings of potential flood events by tracking changes in river levels and precipitation patterns (Huang et al., 2018; NOAA, 2023). Climate variability can lead to changes in water quality by influencing turbidity, suspended sediment concentrations, and water temperature, which impact aquatic ecosystems and human water usage (Ahmed et al., 2023); however, water quality assessments can also be conducted using remotely sensed data by providing information about turbidity, suspended sediment concentrations, and water temperature (Janke et al., 2006; Gxokwe et al., 2022). Remotely sensed data is also used to monitor changes in glacial extent and volume, which is important for understanding the impacts of climate variability on freshwater resources supplied by glacier melt (Mishra and Singh, 2010). Remote sensing data can be integrated with climate models and hydrological models to improve the understanding of the interaction between climate variability and water resources (Frischen et al., 2020; Khan and Gilani, 2021; Ahmed et al., 2023). Analysis of remote sensing data provides vital information for water resource management, which enables policymakers and planners to make informed decisions regarding water allocation, infrastructure development, and disaster preparedness. Remotely sensed data is valuable in assessing the impacts of climate variability on water resources, due to its ability to provide consistent, comprehensive, and high-resolution data to improve the understanding of how climate variability affects water availability, quality, and distribution, aiding in sustainable water resource management strategies (Mishra and Singh, 2010; Govender et al., 2022; Dube et al., 2023).

1.3 Importance of understanding the interactions between climate variability and water availability

Climate variability influences precipitation patterns and intensity, duration, and frequency and has an impact on surface water and groundwater availability and, therefore, affects water resources (Gohar and Cashman, 2016). Understanding the relationship between climate and water availability is vital due to their profound impacts on ecosystems, societies, and economies. This understanding is crucial for ecosystem health, agriculture and food security, human health, water scarcity and stress, energy production, economic impacts, water management and planning, disaster preparedness, international relations, and sustainable development. Ecosystems rely on a consistent supply of water to support their biodiversity, structure, and functions; therefore, changes in water availability due to climate variability disrupt aquatic and terrestrial ecosystems, which can cause habitat loss, altered species distributions, and the collapse of fragile ecosystems (Zhou et al., 2017). Agriculture depends on reliable water sources for irrigation and livestock (Frischen et al., 2020). Climate variability can lead to irregular rainfall patterns, droughts, and floods, impacting crop yields, livestock productivity, and overall food production, which can cause food shortages and price fluctuations (Wilhite et al., 2007; West et al., 2019). Water scarcity can be exacerbated by climate variability; therefore, understanding the dynamics between climate variability and water availability is vital to develop strategies to mitigate water scarcity's adverse effects on communities and industries (Sharara et al., 2022; Bhaga et al., 2023). Energy generation such as hydropower and cooling water for thermal power plants rely on adequate water resources and climate variability can affect water availability (World Health Organization, 2013). Industries such as manufacturing, mining, agriculture, and tourism that rely on stable water resources can be severely affected by changes in water availability (Huang et al., 2017). Climate variability can disrupt production processes, increase operational costs, and lead to economic losses (Saha et al., 2022). Accurate predictions of water availability and demand are essential for efficient water resource management and planning, and understanding the relationship between climate variability and water availability assists policymakers and water managers develop adaptive strategies and infrastructure to mitigate the impacts of extreme weather events such as floods and droughts (Gohar and Cashman, 2016; Dube et al., 2023). Understanding the relationship between natural disasters such as floods and droughts and climate variability will lead to better disaster preparedness and response planning, which will aid communities in minimizing the impacts of such events (Bhaga et al., 2023). Shared water resources often cross national boundaries, which can make them potential sources of conflict; therefore, understanding how climate variability influences water availability in transboundary regions is essential for promoting cooperation and sustainable management practices among nations (Wilhite et al., 2007). Achieving the SDGs requires a comprehensive understanding of the impacts of climate variability, which can guide policies and strategies that ensure the SDGs are met and that present and future generations have equitable access to water resources (Griggs et al., 2013; United Nations, 2019; Moyer and Hedden, 2020). These factors highlight the importance of understanding the relationship between climate variability and water availability for addressing the challenges posed by changing climate patterns. This knowledge informs decision-making, aids adaptation strategies, and promotes the sustainable management of water resources.

2 Remote sensing principles and techniques

2.1 Overview of remote sensing in water resources monitoring

Remote sensing has become a powerful tool for monitoring and managing water resources on regional and global scales (Bhaga et al., 2020; Dube et al., 2023). It involves the collection of information about the Earth's surface and atmosphere by using sensors aboard satellites, aircraft, drones, and ground-based platforms and operates across the electromagnetic spectrum, including the visible, infrared, microwave, and thermal wavelengths (Huang et al., 2018; USGS, 2023). These sensors detect reflected, emitted, or transmitted radiation that allows for the extraction of valuable information-related water resources. The application of remote sensing in water resource monitoring provides a holistic and timely understanding of various hydrological components and processes (Zhou et al., 2017; Seaton et al., 2020) and provides insights into surface water dynamics, groundwater levels, soil moisture, water quality, and climate-driven changes that hinder water availability and distribution (Mishra and Singh, 2011; Zhou et al., 2017; Bhaga et al., 2020; Govender et al., 2022). Surface water bodies such as lakes, rivers, reservoirs, wetlands, and dams can be mapped and monitored by using remote sensing, and changes in water quantity, fluctuations, and trends are tracked over time to understand water availability dynamics (Zhou et al., 2017). Satellite-based techniques aid in estimating groundwater storage changes in regions where in situ measurements are challenging (Mishra and Singh, 2010). To understand water availability, microwave sensors are utilized to estimate soil moisture content. Remote sensing data can also be used in conjunction with hydrological models to improve the accuracy of water balance estimations, flood predictions, and drought assessments (Bhaga et al., 2020). The ability of remote sensing to provide multidimensional, spatially explicit, and timely data proves to be crucial for hydrological modeling, irrigation management, and water resource management. This shows the vital role remote sensing plays in addressing challenges posed by water scarcity, climate variability, and sustainable water management, which aids decision-makers, researchers, and policymakers in creating effective strategies and mitigation.

2.2 Sensors and platforms used for remote sensing of climate variability

The advancement of technology in remote sensing techniques allows for data to be collected at different temporal, spatial, and spectral resolutions, which can be used for assessing climate variability, drought, floods, and monitoring water resources. There are a number of satellite systems and sensors used to collect data on climate variability, namely the Moderate Resolution Imaging Spectroradiometer (MODIS), Advanced Very High-Resolution Radiometer (AVHRR), Advanced Microwave Scanning Radiometer (AMSR), Visible Infrared Imaging Radiometer Suite, Sentinel Satellites, Tropical Rainfall Measuring Mission (TRMM), Landsat missions, IKONOS, MERIS, QuickBird, RapidEye, Satellite Pour l'Observation de la Terre (SPOT) and Worldview (Table 1). The MODIS is onboard NASA's TERRA and AQUA satellites, captures data in multiple spectral bands, and provides information on land surface temperature cloud cover and vegetation health (Gxokwe et al., 2020).

TABLE 1 Satellite systems and sensors used to collect data on climate variability.

Satellite name	Bands	Spatial resolution	Temporal resolution (day)	Data availability
AVHRR	5	Low	1 day	Freely available
AMSR	6	Medium	2	Free
IKONOS	5	Medium	1–2	Expensive
Landsat missions	4–11	Low-medium	16–18	Free
MERIS	15	Medium	3	Free
MODIS	7	Low	1	Free
QuickBird	5	High	1–3.5	Expensive
RapidEye	5	High	5.5	Expensive
Sentinel Satellites	13	Medium-high	5	Free
SPOT	4	High	26	Free

2.3 Data acquisition and preprocessing techniques

The effective use of remote sensing to monitor climate variability and water resources is based on a structured approach that includes data collection and rigorous preprocessing techniques (Henchiri et al., 2020). First, a data collection strategy, which involves the selection of appropriate data sources suited to the research objectives, needs to be planned. Subsequent data collection, whether by satellite, drone, or aircraft requires careful consideration of the sensor characteristics, such as the frequency and spatial resolution to ensure the data collected is relevant (AghaKouchak et al., 2015). After data collection, a series of preprocessing steps are required to turn raw data into accurate and reliable information. Geometric correction corrects distortions caused by viewing angle and platform movement and radiometric correction normalizes radiometric values for variations in sensor sensitivity, atmospheric conditions, and solar angle (Zhou et al., 2017). Atmospheric correction is essential for optical remote sensing because it removes atmospheric noise and reveals accurate surface reflectance values (Seaton et al., 2020). Cloud and shadow removal techniques improve optical data usability by reducing the impact of obstructing elements. Data consolidation integrates multiple datasets, which leads to better quality and holistic information through the combination of multispectral, hyperspectral, and radar data (Gxokwe et al., 2020). Feature extraction methods identify and delineate specific target features such as water bodies and wetlands, facilitating further analysis. Change detection techniques highlight changes in land cover/land use, water bodies, and other related features, which allows for a deeper understanding of the impacts of climate variability (Bhaga et al., 2020). Data integration combines remote sensing data with in situ measurements, hydrological models, and climate data, which creates a comprehensive perspective on the interaction between climatic factors and water resources. Rigorous validation procedures involve basic ground-truthing data to confirm the accuracy of the results obtained from the remotely sensed data. Quality checks or accuracy assessments verify the reliability of the processed data, ensuring a strong basis for further analysis. This culmination of preprocessing and data collection processes allows researchers to gather precise information about the complex relationships between climate variability and water resources, which aids management strategies and decision-making.

2.4 Remote sensing data types relevant to water resources analysis

In the field of water resource analysis, remote sensing provides a variety of data that play a vital role in understanding the complexities of aquatic systems, water bodies, climate impacts, and hydrological processes (AghaKouchak et al., 2015). The key remotely sensed data types for water resource analysis include optical imagery, thermal infrared imagery, microwave imagery, hyperspectral imagery, light detection and ranging (LiDAR) data, gravimetry data, synthetic aperture radar (SAR) data, vegetation indices, meteorological data, and bathymetric data (Huang et al., 2018; Feng et al., 2019; Henchiri et al., 2020). Optical imagery is able to capture visible and near-infrared light, which aids in identifying water bodies and monitoring changes in land cover while highlighting water quality fluctuations through the analysis of spectral reflectance patterns (Yu and Chen, 2019). Thermal infrared imagery detects emitted thermal radiation from surfaces, which enables the estimation of water surface temperatures and provides information regarding evaporation rates. On the contrary, microwave imagery penetrates clouds and vegetation, which provides important information about soil moisture and water content changes (Mishra and Singh, 2011). Hyperspectral data, having a high spectral resolution, reveals specific indicators of water quality, which assists in understanding water composition and pollution levels (Govender et al., 2022). LiDAR data emits laser pulses, measures surface interaction times, and maps complex river and floodplain topographies (West et al., 2019). Satellite missions such as the Gravity Recovery and Climate Experiment (GRACE) provide information about gravity and elevation, which assists in analyzing groundwater reserves and river flows (Frappart, 2020). SAR data is vital for mapping floods, and monitoring water level fluctuations and changes on the Earth's surface. Vegetation indices obtained from optical data provide information on vegetation health and water content and are able to detect soil-water interactions (Zhou et al., 2017). Meteorological data from weather satellites provide the meteorological parameters needed to understand climate-induced changes to water resources (Henchiri et al., 2020). Bathymetry data obtained from airborne LiDAR and satellite-derived bathymetry collects information on water depths, which allows for the assessment of reservoir capacity and sediment dynamics (West et al., 2019). By utilizing these types of multifaceted remotely sensed data, researchers get a holistic understanding of water availability, quality, and dynamics, which leads to informed decision-making to improve the management of water resources and climate adaptation strategies.

3 Climate variability and its impacts on water resources

3.1 Understanding climate variability patterns and drivers

Understanding climate variability patterns and their drivers is vital for understanding the complex dynamics of the Earth's climate system (Wilhite et al., 2007; Affoh et al., 2022; Bogale and Erena, 2022; Ahmed et al., 2023). As mentioned earlier, climate variability refers to the natural fluctuations in climatic factors like temperature, precipitation, and atmospheric circulation over short and long periods. Understanding these patterns needs the integration of observational data, satellite measurements, and models (Frappart, 2020). By analyzing and understanding these drivers and their interactions, researchers can discern the factors that influence climate variability on a regional and global scale (Ahmed et al., 2023). These insights will improve the understanding of climate variability and provide information for predictions and adaptation strategies to address challenges posed by climate variability.

3.2 Key climate variables affecting water resources

Several factors impact the availability, distribution, and quality of water resources, namely temperature, precipitation, evapotranspiration (ET), and wind patterns (Mishra and Singh, 2010; Varghese et al., 2021). An increase in temperature can cause accelerated snowmelt in mountainous areas, which changes the timing and magnitude of river flows (Ahmed et al., 2023). Changes in temperature cause a rise in ET rates, which affect water levels in lakes, dams, and reservoirs (Bhaga et al., 2023). Rainfall patterns, such as intensity, frequency, and distribution, have a direct impact on water availability and rainfall variability can lead to droughts during extended dry periods or flooding during extended rainfall events (Asfaw et al., 2018). ET, driven by temperature and humidity, influences the amount of water lost from surface water bodies (Marshall et al., 2012). Wind patterns affect the distribution of rainfall and can lead to an increase in ET rates from open water bodies. Moreover, wind patterns influence ocean currents which have dire effects on regional climates and rainfall patterns (Deivanayagam et al., 2022). Understanding the dynamics between these factors is pivotal for efficient water resource management, the development of adaptive strategies, and the formulation of policies that ensure sustainable development in the face of evolving climate conditions.

3.3 Impacts of climate variability on hydrological processes and water availability

Climate variability has dire impacts on hydrological processes and water availability, which causes complex changes in the distribution and availability of water, which leads to changes in rainfall patterns, causing droughts and changing snowmelt dynamics (Park et al., 2016; Dzurume et al., 2021). On the contrary, changes in rainfall patterns can cause reduced snow accumulation, which affects the natural storage and gradual release of water (Valadao et al., 2021). Moreover, climate variability influences the occurrence of monsoons and seasonal rains, which has direct impacts on the agricultural sector and water availability for irrigation, particularly in regions dependent on rainfed crops (Masocha et al., 2018; Bhuiyan, 2004; Udmale et al., 2014). Extended dry periods caused by climate variability cause a decrease in soil moisture, groundwater recharge, and reduced surface water levels, leading to water scarcity (Affoh et al., 2022; Deivanayagam et al., 2022; Saha et al., 2022; Ahmed et al., 2023). Saltwater intrusion into freshwater systems can arise due to an increase in sea levels, propagated by climate variability, which impacts water quality and availability in coastal areas (Mishra and Singh, 2011). Variability of sea surface temperatures influences the frequency and strength of tropical cyclones, causing heavy rainfall and flooding risks. These effects have dire consequences on ecosystems, industries dependent on water, human populations, and water resource management strategies (Bhaga et al., 2020). Understanding these impacts is vital for adaptive planning and water resource management. This indicates the need to develop holistic strategies that consider the various effects of climate variability. By understanding the relationship between climate variability and hydrological processes, researchers and decision-makers can develop and implement early warning systems to mitigate negative impacts, enhance water security, and encourage resilience.

4 Remote sensing applications for water resources assessment

4.1 Estimation of precipitation using remote sensing data

There is a variety of methods that leverage different sensors and techniques to estimate rainfall with increasing accuracy in areas where ground-based methods are challenging due to complex terrains or vandalism (Du et al., 2013; Henchiri et al., 2020). The Weather Surveillance Radar is a radar-based method, whereas the TRMM and Global Precipitation Measurement are examples of spaceborne radar systems, which use the backscattering of radar signals off rainfall particles to estimate rainfall intensity and accumulation (Du et al., 2013; Zhou et al., 2017). Microwave radiometry, such as the Special Sensor Microwave Imager and the Advanced Microwave Scanning Radiometer, utilizes the interaction between microwave radiation and raindrops (Park et al., 2016). Passive microwave radiometers aboard satellites measure the natural microwave radiation that is emitted from water vapor and raindrops. Precipitation rates can be inferred through various algorithms that consider atmospheric conditions by analyzing the microwave brightness temperature (Jamali et al., 2011; Seyama et al., 2019). Radar and radiometer data can be combined to create hybrid approaches, which improves the accuracy and coverage of precipitation estimation (Du et al., 2013). Although remote sensing methods to estimate precipitation offer beneficial insights, it is not without challenges. To reduce uncertainties related to precipitation type and intensity, results obtained from satellite data need to be calibrated and validated (Zeng et al., 2019; Henchiri et al., 2020). Combining remotely sensed estimates with ground-based results and hydrological models has the potential to create robust, holistic, and accurate results, which will aid researchers and decision-makers in understanding precipitation patterns.

4.2 Remote sensing-based monitoring of ET and soil moisture

ET and soil moisture are two vital factors of the hydrological cycle and have critical implications for water resource management. ET is the combination of surface evaporation and plant transpiration and can be detected by remote sensing sensors using infrared imagery (Marshall et al., 2012; Alemayehu et al., 2017). Multispectral data from satellites such as MODIS and Landsat can be used in conjunction with meteorological factors to estimate sensible and latent heat fluxes by using models such as the two-source energy balance model (Alemayehu et al., 2017). ET will then be able to be detected on various scales, which can be used to understand water use patterns and land-atmosphere interactions. Soil moisture monitoring is a vital determinant of water availability and land productivity. There is an array of sensors used for soil moisture monitoring, such as AMSR-E, Sentinel-1, SMAP, and SMOS. These sensors detect microwave radiation that is emitted by the Earth's surface. Sentinel-2 and Landsat satellites are able to estimate soil moisture indirectly by assessing vegetation health (Jamali et al., 2011; Fontanet et al., 2018). The results obtained from these satellites can be integrated into hydrological models such as the soil and water assessment tool (SWAT) and variable infiltration capacity (VIC) to improve the accuracy of the ET and soil moisture estimations (Guug et al., 2020; University of Washington Computational Hydrology Group, 2021). While there are challenges regarding calibration and validation, remote sensing methods to monitor ET and soil moisture are still able to provide vital information and able to pave the way for informed water resource management.

4.3 Mapping and monitoring of surface water resources using remote sensing

The mapping and monitoring of surface water resources by utilizing remote sensing is vital for up-to-date and holistic information about water bodies (AghaKouchak et al., 2015; Masocha et al., 2018; West et al., 2019; Seaton et al., 2020; Govender et al., 2022; Bhaga et al., 2023). The development of indices to identify surface waterbodies by using satellite data has led to the mapping of surface waterbodies (Zhou et al., 2017). An array of indices exists, however, each has its own advantages and disadvantages (Jiao et al., 2019; Bhaga et al., 2020). Satellite data first need to be preprocessed and atmospherically corrected to ensure accurate water body detection (Seaton et al., 2020). Thereafter thresholding and spectral indices can be computed to detect and classify water bodies from satellite imagery, and these indices separate water pixels from other features. These spectral indices use the reflectance characteristics of water and its surrounding land features to detect surface water bodies by using remotely sensed images (Zhou et al., 2017; Huang et al., 2018). The most commonly used indices include the Normalized Difference Water Index (NDWI), Modified Normalized Difference Water Index (MNDWI), Water Index (WI), Tasseled Cap Wetness (TCW), and the Modified Soil-Adjusted Vegetation Index (MSAVI). By computing these indices, researchers can identify and monitor water bodies, which aids in water resource management, hydrological studies, and monitoring.

4.4 Remote sensing techniques for groundwater assessment

Remote sensing methods provide a noninvasive and spatially extensive method to obtain information about subsurface water and assess groundwater resources (Yeh et al., 2006; Bonsor et al., 2018; Frappart, 2020). The GRACE satellites are able to estimate groundwater variations by measuring changes in the Earth's gravitational field (Bonsor et al., 2018). SAR sensors aboard Sentinel-1 use interferometric methods to detect ground deformation caused by changes in groundwater levels, which indicates whether groundwater levels are increasing or decreasing (Zhou et al., 2017). Electromagnetic induction sensors aboard aircraft or satellites are able to detect electromagnetic responses to measure moisture, salinity, and the presence of groundwater (AghaKouchak et al., 2015). Thermal infrared imaging is able to locate groundwater discharge areas by analyzing temperature anomalies due to underground runoff (Yeh et al., 2006). Aerial and satellite data such as LiDAR and satellite-derived bathymetry are able to provide topographic mapping in underwater environments, which can be used to assess groundwater storage (Bhaga et al., 2020). Vegetation health can be assessed from remotely sensed data such as the Normalized Difference

Vegetation Index, which can be suggestive of the potential presence of groundwater (Jamali et al., 2011; Masocha et al., 2018; Seaton et al., 2020). The integration of remote sensing with ground-based measurements in hydrogeophysical methods improves the accuracy of groundwater assessments and these techniques provide researchers and policymakers with invaluable data and assist with the improvement of groundwater models, thereby advancing water management strategies.

5 Methodologies for analyzing climate variability impacts

5.1 Time-series analysis of remote sensing data for climate trend detection

Time-series analysis of remotely sensed data is a powerful tool for detecting and understanding long-term climate trends (Janke et al., 2006; Asfaw et al., 2018). Changes influenced by climate variability can be monitored and detected by analyzing the temporal evolution of features on the Earth's surface (Mishra and Singh, 2011; Masocha et al., 2018; Ahmed et al., 2023). By harnessing satellite sensors, such as Landsat, MODIS, and Sentinel, with consistent observations over time, researchers can monitor land cover/land use changes, land surface temperature, and sea ice extent (AghaKouchak et al., 2015; Zhou et al., 2017; Huang et al., 2018). Trends, periodic patterns, and anomalies can be distinguished by using techniques such as linear regression, Mann-Kendall analysis, and seasonal decomposition (Malik et al., 2020). This helps identify changes in land cover/land use and other indicators that reflect climate-induced changes. The relationship between the Earth's surface dynamics and climatic factors can be analyzed and assessed by combining remotely sensed data with climate records (Ahmed et al., 2023; Bhaga et al., 2023). Overall, time-series analysis is vital to detect climate trends and improve the understanding of long-term climatic effects to aid adaptation and mitigation strategies.

5.2 Modeling approaches for understanding climate-water interactions

Modeling methods are essential tools to understand the complex interactions that link climate and hydrological processes. There is a wide range of modeling techniques that enable researchers to simulate and predict the dynamics that govern these dynamics (Michaelides et al., 2009; Seager et al., 2009; Mishra and Singh, 2011). Models such as the SWAT or the VIC analyze the movement of water in the hydrological cycle and consider factors such as precipitation and temperature to understand flow patterns, runoff patterns, and groundwater recharge (Guug et al., 2020; University of Washington Computational Hydrology Group, 2021). Global and regional climate models provide insight into climatic patterns and changes and indicate the influence of climate factors on regional hydrological conditions. Land surface models consider soil-atmosphere dynamics and allow for the analyses of energy, water, and carbon fluxes. Integrated water resources models provide a holistic overview by integrating hydrological, socioeconomic, and environmental aspects, which indicates how policies and interventions are related to climate impacts (Herndon et al., 2020). Aquifer models consider groundwater movement, recharge, and storage and indicate climate-induced changes. Data-driven, machine learning, and agent-based models help to indicate nonlinear relationships and predictions and analyze individual factors influenced by climate variability. Scenario modeling allows researchers to simulate different climate and hydrological scenarios, which predicts potential impacts and aids with adaptation strategies. With these various modeling methods, researchers can improve the understanding of the interactions between climate variability and water, which assists in predictions and the development of strategies and plans to improve water resource management.

5.3 Integration of remote sensing with hydrological models for water resources assessment

The integration of remotely sensed data with hydrological models allows for accurate, spatially explicit, and timely information, which improves water resource monitoring and assessment (Mishra and Singh, 2011). By combining the monitoring strengths of remote sensing with the prediction capabilities of hydrological models, a comprehensive understanding of climate variability and its influence on water resources is provided. Remotely sensed data provide spatial information on land cover, vegetation, soil moisture, and meteorological conditions, which are vital inputs for models and lead to more accurate model predictions and simulations, especially in data-scarce regions (AghaKouchak et al., 2015; Hagenlocher et al., 2019; Bogale and Erena, 2022; Govender et al., 2022; Dube et al., 2023). The validation of the alignment between model outputs and remote sensing ensures that the results are highly accurate. The continuous observations allow for effective and timely decision-making in water resource management (Bhaga et al., 2023). These integrated models also provide information regarding the impacts of climate variability on water resources. This provides a comprehensive framework for adaptative management strategies and policymaking.

6 Case studies and examples

6.1 Case study 1: Remote sensing of climate impacts on river flow and streamflow patterns

A study by Immerzeel et al. (2009) used remotely sensed data to analyze the impacts of climate variability on river flow in two river basins, namely the upper Indus Basin and the upper Ganges Basin, in the Tibetan-Qinghai plateau and the Hindu-Kush, Pamir, Karakoram, and Himalaya mountain ranges, known for its complex terrain and limited in situ observations. A hydrological model integrated with MODIS snow cover data, TRMM precipitation, and temperature data obtained from the Climate Research Unit dataset. The runoff modeling was based on the Snowmelt Runoff Model (SRM), which was used to simulate daily runoff from snowmelt and rainfall in mountainous regions. The model was calibrated by using daily discharge rates from 2000 to 2005. The results of the study indicate the applicability of the integration between remotely sensed data and modeling. The remotely sensed data correlated will the in situ measurements and was able to track changes in snow extent, especially during the snowmelt period. The application of the SRM model with remotely sensed data was able to capture the seasonal flow patterns in the river basins. This demonstrates the value of remote sensing in data-scarce regions and highlights the importance of remote sensing for large-scale snow cover monitoring and runoff predictions.

Akhtar et al. (2021) conducted a study that assessed the challenge of assessing the impacts of climate variability on streamflow in data-scarce regions by combining remote sensing data and hydrological monitoring to predict changes in streamflow patterns in the Kabul River Basin (KRB) in Afghanistan. The SWAT model was used to assess the potential impacts of climate variability on streamflow. The SUFI-2 technique was used to calibrate and validate the model with streamflow data. The coefficient of determination (R^2) and Nash-Sutcliffe Efficiency (NSE) were used for calibration and achieved a value of 0.75–0.86 and 0.62–0.86, respectively. The Regional Climate Model (RegCM4-4) was leveraged to extract data for the simulated climate change scenarios and indicated the KRB would decrease by 5% under the RCP 4.5 and by 8.5% under the RCP 8.5. The results from this study highlight the importance of utilizing remote sensing in data-scarce regions and the potential it has to fill data gaps and improve predictions.

Another study by Wufu et al. (2021) investigated the changes in glacial runoff in the Tianshan region in Central Asia. The river discharge for 19 rivers was calculated by integrating cross-sections obtained by unmanned aerial vehicles (UAV) and remotely sensed images from Landsat and Sentinel-2 satellites. River discharge variations under climate variability were analyzed and in situ measured discharges were used as a reference. The UAV-derived results achieved an NSE of 0.98 and a root mean square error (RMSE) of 8.49 m^3/s. The results of this study indicate that major changes in glacial meltwater runoff and the combination of UAV and satellite data provide a holistic understanding of these changes. The authors were able to identify the influence of temperature and rainfall on glacial meltwater, highlighting the relationship between climate variability and glacial hydrology in the region. These results provided vital information for policymakers and water resource managers.

6.2 Case study 2: Monitoring the impact of climate variability on reservoir storage

A study by Awange et al. (2014) investigated the changes in water storage and their correlation with climate variability in the Nie Basin from 2002 to 2011.

Satellite data obtained from GRACE was used to estimate changes in terrestrial water storage and the independent component analysis was used to estimate changes in smaller subbasins. The Global Land Data Assimilation System model was used, along with monthly precipitation data from the TRMM. This allowed for an in-depth analysis of changes in water storage and to assess the influence of anthropogenic and climate variability caused by El Nino and the Indian Ocean Dipole. The results from the study show the changes in water storage in the Nile Basin and highlight the interannual and seasonal variability in water storage, which emphasizes the importance of considering these dynamics in water management. There was a 0.56 RMSE correlation between water storage and climate variability during the study period. This study indicates the relationship between water storage changes and climate variability within the Nile Basin and that by combining GRACE data with climate indices, a holistic understanding of how water storage is influenced by climate variability.

Luo et al. (2021) analyzed the changes in lake variations in the Altai Mountains in Asia by using altimetry data from TOPEX/Poseidon, Jason-1, Jason-2, and Sentinel-3 to monitor lake water levels. Landsat and Sentinel-2 data was used to characterize changes in lake surface extent in 43 lakes around the Altai Mountains from 2001 to 2018. The results indicate that lake surface extent decreased from 9835 km^2 in 2001 to 9652 km^2 in 2009 and rose to 9714 km^2 in 2018. This indicates the responses to changing climate conditions. The results from this study indicate the applicability of altimetry data for accurate and continuous monitoring of lake water levels. The combination of altimetry data and optical satellite data improves the understanding of how climate variability influences lake water levels and lake areas.

Donchyts et al. (2022) used multisensor satellite data combined with cloud analytics to monitor small- to medium-sized water reservoirs. There were 71,208 small- to medium-sized reservoirs identified by using Landsat and Sentinel data. The results were validated by using 768 daily in situ water-level measurements. The results indicated that intraannual variability is much higher for small reservoirs compared to medium-sized reservoirs. The results from this study prove that remote sensing techniques have the potential to provide vital information on water scarcity conditions.

6.3 Case study 3: Remote sensing-based assessment of drought and its effects on water resources

Tarnavsky et al. (2018) assessed drought patterns and severity in the East Shewa Zone in Ethiopia, using the WRSI. AVHRR and rainfall estimate data was used in this study and data was collected during the rainy season (June to September) from 1996 to 2008. The results indicated that the region experienced severe drought during 2002 and 2000. WRSI achieved a 76% anomaly of variability of grain yield and, therefore, proved to be a good agricultural drought indicator. The results achieved by these 20 indices illustrate that drought risk mapping is able to aid decision-makers and reduce the risk of drought with proper detection, planning, and mitigation strategies. The results from this study indicate the vital role remote sensing plays in drought monitoring and detection.

A study by Orimoloye et al. (2021) conducted in the Free State province in South Africa used regression-based models to identify drought conditions using MODIS and the Vegetation Condition Index (VCI). Google Earth Engine (GEE) was also used to perform a land use classification during 2001, 2015, and 2020, using Landsat-7 and Landsat-8 data. Multiple regression, principal component regression, partial least square regression (PLSR), random forest regression, support vector regression, and multivariate adaptive regression splines were computed for this study. The VCI was able to detect drought conditions in the area. The study achieved OA of 0.86, 0.95, and 0.93 for 2001, 2015, and 2020, respectively. The land use classification map indicated that grasslands and barren surfaces had the highest land cover and that cultivated lands decreased drastically in size due to drought occurrence. The RMSE results indicated that the best model for VCI prediction in the study area is the PLSR.

A study by Alahacoon and Edirisinghe (2022) evaluated and compared the performance of 111 drought monitoring indices on a global and regional scale. The study proposed the use of remote-sensing-based indices obtained from satellite data to provide a more holistic view of drought conditions. The study found that remote-sensing-based indices can accurately capture the spatiotemporal variations of drought and remote sensing indices that integrate climatic data such as temperature, precipitation, and vegetation health, achieve higher accuracies in drought detection. This study demonstrates the advantages of remote sensing techniques in capturing the dynamics of drought occurrence.

7 Challenges and future directions

7.1 Limitations and uncertainties in remote sensing-based analysis of climate variability and water resources

Despite the advantages of using remotely sensed data to analyze climate variability and its impact on water resources, it is not without inherent limitations and uncertainties that must be considered, as they can influence the accuracy, reliability, and interpretation of results (West et al., 2019; Dube et al., 2023). Atmospheric interference caused by aerosols, clouds, and water vapor can lead to inaccuracies in measurements by obscuring remote sensing signals (Gxokwe et al., 2020; Sharara et al., 2022). Cloud cover hinders data collection, which affects the temporal continuity of observations (AghaKouchak et al., 2015). Spatial, spectral, and temporal resolutions of sensors impose limitations on the level of detail and frequency of observations (Govender et al., 2022). Calibration of sensors is essential for accurate geometry and radiation characteristics, as calibration errors and sensor degradation can cause bias (Varghese et al., 2021). Mixed pixels in remotely sensed imagery caused by the heterogeneous nature of landscapes can lead to inaccuracies in the identification and classification of objects (Seaton et al., 2020; Huang et al., 2018; West et al., 2019; Zhou et al., 2017). Topographic effects affect the interaction of electromagnetic radiation with the terrain, causing different reflection patterns (Gxokwe et al., 2022). Validation of field data is vital, however, obtaining accurate baseline data is problematic. Integrating data from different satellites requires harmonization to minimize inconsistencies. Spectral confusion occurs when similar materials display similar spectral signatures. Uncertainties arise in parameters obtained from complex algorithms, which affects the accuracy of the results. The difference in scale between remotely sensed data and local decision-making can affect direct applicability and the assumptions inherent in the models used to estimate parameters contribute to the uncertainty (Feng et al., 2019; Hagenlocher et al., 2019). Acknowledging and addressing these limitations and uncertainties warrant that information derived from remotely sensed data is analyzed correctly, thus improving its application and usefulness for informed water resource management strategies.

7.2 Data availability and quality considerations

Availability and quality of data are fundamental considerations in climate variability studies, where accurate and complete datasets are essential to obtain reliable and relevant information (Frischen et al., 2020). Long-term and consistent observations are vital for capturing changes and patterns (Wilhite et al., 2007). Although high-resolution data provides localized detail, global analyses require large spatial datasets. Various data sources such as satellites, ground stations, and climate models supply a multidimensional understanding of the complex dynamics of climate variability (Gxokwe et al., 2022). To maintain data accuracy and integrity, standardized data formats, quality control, and validation based on ground truth measurements need to be maintained (Mishra and Singh, 2011). Truthful communication about uncertainty assists in understanding the results, and it is essential to fill in data gaps and missing values to avoid misinterpretation of the results (Gxokwe et al., 2022). Collaboration between experts in different fields and participation in open data initiatives fosters an accessible database. Data availability and quality are vital for creating accurate insights into climate variability and researchers should, therefore, aim to access, evaluate, and integrate datasets that align with the objectives of their study while being mindful of data limitations and uncertainties.

7.3 Advancements and future prospects in remote sensing technology for water resources assessment

With the improvement in technology, the field of remote sensing has advanced tremendously in recent years, thereby improving water resource assessment and management. These advancements together with emerging trends create a promising future for the application of remote sensing technologies. The development of satellites with improved spatial and temporal resolutions will allow for more detailed and frequent monitoring of water resources and the fusion of hyperspectral and multispectral data for a holistic analysis of water availability will lead to more accurate and insightful observations (Gxokwe et al., 2020; Bhaga et al., 2023). The use of UAVs enhances localized studies, whereas big data analytics and machine learning programs help in data processing and analysis (Khan and Gilani, 2021). Improvements in radar technology such as SAR allow continuous monitoring in all types of weather conditions (West et al., 2019). Moreover, various satellite missions demonstrate a global commitment to advance the understanding of climate variability and water resources. As innovations continue to develop, the future of water resource monitoring looks promising to leverage the potential of remote sensing for sustainable management.

8 Conclusion

8.1 Summary of key findings

Remote sensing plays a significant role in climate variability detection and monitoring and in our understanding of the impacts of climate variability on water resources. The relationship between climate variability and water resources has implications for ecosystems, societies, and economies. Climate variability caused by natural fluctuations and anthropogenic influences causes changes in climatic factors like temperature, precipitation, and ocean currents. This variability is often attributed to fossil fuel consumption and solar radiation, which has significant impacts on water availability and quality. Rainfed agricultural areas are particularly vulnerable to droughts and often experience food shortages. Remote sensing techniques are a powerful tool that can provide a holistic assessment of the impacts of climate variability on water resources. Satellite data offers timely, high-resolution data on temperature, precipitation, soil moisture, and glacial extent. This information and understanding are essential for effective water resource management, disaster preparedness, informed policymaking, and achieving the SDGs.

8.2 Importance of remote sensing in understanding climate variability impacts on water resources

The application of remote sensing in the field of climate variability and water resource analysis is vital to understanding the complex relationship between these components. Remote sensing plays a vital role in improving our understanding of the impacts of climate variability on water resources, as it provides holistic and timely information at various spatial and temporal scales. Its application in facilitating the timely detection of extreme weather events, such as droughts and floods, allows for preparedness and targeted interventions. Furthermore, remote sensing can be utilized to assess water quality, monitor ecosystem health, and calibrate hydrological and climate models. The combination of data-driven information improves our understanding of climate-related changes and aids in the development of adaptation strategies for sustainable water management, which will assist in achieving SDGs. Moreover, remote sensing closes the gap between scientific knowledge and practical solutions by highlighting the transboundary nature of water and assisting in policy support. Therefore, this chapter illuminates the importance of remote sensing for understanding the impact of climate variability on water resources and steering the trajectory toward drought resilience and informed resource management planning.

8.3 Recommendations for further research and applications

With climate variability having dire impacts on water resources, there is a need for water resource monitoring and climate variability studies to evolve, and with the continuous improvement of technology, there are various promising approaches for further research and development. Leveraging advanced data fusion techniques that integrate remotely sensed data with in situ measurements and models will offer a more comprehensive understanding of climate-induced impacts on water resources. The use of high-resolution monitoring for small and intermittent water bodies can provide patterns and changes in water availability that were previously undetected. Furthermore, the integration of machine learning methods for data analysis has the potential to classify water bodies, detect anomalies, and model predictions, which can enhance the detection of changes and help understand the relationships in complex datasets. Blending remote sensing with social and hydrological data can provide a more holistic assessment of the socio-economic impacts of climate variability on water resources, which can aid in creating adaptive strategies. Adaptive strategies can lead to the development of early warning systems for extreme weather events, which subsequently lead to the improvement of timely disaster response and mitigation. A multidisciplinary approach that encompasses water quality monitoring, ecosystem services assessment, and cross-border water management will help understand the impacts of climate variability on water resources and highlight the potential of remote sensing as an essential tool for sustainable water management. Combining remote sensing with citizen science initiatives can aid in data collection and monitoring of water resources while engaging with local communities which can enhance data coverage and accuracy. These recommendations reflect emerging areas of research and potential applications that can aid our understanding of the impacts of climate variability on water resources, advance efficient management strategies, and highlight the opportunities remote sensing methods provide to improve the sustainability of water resources.

References

Affoh, R., Zheng, H., Dangui, K., Dissani, B.M., 2022. The impact of climate variability and change on food security in sub-Saharan Africa: Perspective from panel data analysis. Sustainability 14 (2).

AghaKouchak, A., et al., 2015. Remote sensing of drought: progress, challenges and opportunities. Rev. Geophys. 53, 452–480.

Ahmed, M., Ahmad, S., Kheir, A.M., 2023. Climate change: an overview. In: Global Agricultural Production: Resilience to Climate, pp. 1–30.

Akhtar, F., Awan, U.K., Borgemeister, C., Tischbein, B., 2021. Coupling remote sensing and hydrological model for evaluating the impacts of climate change on streamflow in data-scarce environment. Sustainability 13, 14025.

Alahacoon, N., Edirisinghe, M., 2022. A comprehensive assessment of remote sensing and traditional based drought monitoring indices at global and regional scale. Geomat. Nat. Hazards Risk 13 (1), 762–799.

Alemayehu, T., van Griensven, A., Senay, G.B., Bauwens, W., 2017. Evapotranspiration mapping in a heterogeneous landscape using remote sensing and global weather datasets: application to the Mara Basin, East. Remote Sens. 9 (390), 1–24.

Asfaw, A., Simane, B., Hassen, A., Bantider, A., 2018. Variability and time series trend analysis of rainfall and temperature in northcentral Ethiopia: a case study in Woleka sub-basin. Weather Clim. Extrem. 19, 29–41.

Awange, J.L., et al., 2014. Water storage changes and climate variability within the Nile Basin between 2002 and 2011. Adv. Water Resour. 73, 1–15.

Bhaga, T.D., Dube, T., Shekede, M.D., Shoko, C., 2020. Impacts of climate variability and drought on surface water resources in sub-Saharan Africa using remote sensing: a review. Remote Sens. 12, 4184.

Bhaga, T.D., Dube, T., Shekede, M.D., Shoko, C., 2023. Investigating the effectiveness of Landsat-8 OLI and Sentinel-2 MSI satellite data in monitoring the effects of drought on surface water resources in the Western Cape Province, South Africa. Remote Sens. Appl. 32.

Bhuiyan, C., 2004. Various Drought Indices for Monitoring Drought Condition in Aravalli Terrain of India. Indian Institute of Technology Kanpur, Kanpur.

Bogale, G.A., Erena, Z.B., 2022. Drought vulnerability and impacts of climate change on livestock production and productivity in different agro-ecological zones of Ethiopia. J. Appl. Anim. Res. 50 (1), 471–489.

Bonsor, H.C., et al., 2018. Seasonal and decadal groundwater changes in African sedimentary aquifers estimated using GRACE products and LSMs. Remote Sens. 10 (904), 1–20.

Cancelliere, A., Di Mauro, G., Bonaccorso, B., Rossi, G., 2007. Drought forecasting using the standardized precipitation index. Water Resour. Manag. 21, 801–809.

Chawira, M., Dube, T., Gumindoga, W., 2013. Remote sensing based water quality monitoring in Chivero and Manyame lakes of Zimbabwe. Phys. Chem. Earth 66, 38–44.

Deivanayagam, A., Sarangi, R.K., Palanisamy, M., 2022. Evaluating the influence of El Nino–Southern Oscillation (ENSO) patterns on the spatio-temporal variations of drought over southern Peninsular Indian region. J. Indian Soc. Remote Sens.

Donchyts, G., et al., 2022. High-resolution surface water dynamics in Earth's small and medium-sized reservoirs. Sci. Rep. 12.

Du, L., et al., 2013. A comprehensive drought monitoring method integrating MODIS and TRMM data. Int. J. Appl. Earth Obs. Geoinf. 23, 245–253.

Dube, T., Seaton, D., Shoko, C., Mbow, C., 2023. Advancements in earth observation for water resources monitoring and management in Africa: a comprehensive review. J. Hydrol. 623.

Dzurume, T., et al., 2021. Use of multispectral satellite data to assess impacts of land management practices on wetlands in the Limpopo Transfrontier River Basin, South Africa. S. Afr. Geogr. J.

Feng, P., Wang, B., Liu, D.L., Yu, Q., 2019. Machine learning-based integration of remotely-sensed drought factors can improve the estimation of agricultural drought in South-Eastern Australia. Agric. Syst. 173, 303–316.

Fontanet, M., Fernandez-Garcia, D., Ferrer, F., 2018. The value of satellite remote sensing soil moisture data and the DISPATCH algorithm in irrigation fields. Hydrol. Earth Syst. Sci. 22, 5889–5900.

Frappart, F., 2020. Groundwater storage changes in the major North African transboundary aquifer systems during the GRACE era (2003–2016). Water 2669.

Frischen, J., et al., 2020. Drought risk to agricultural systems in Zimbabwe: a spatial analysis of hazard, exposure and vulnerability. Sustainability 12 (752), 1–23.

Gohar, A.A., Cashman, A., 2016. A methodology to assess the impact of climate variability and change on water resources, food security and economic welfare. Agric. Syst. 147, 51–64.

Govender, T., Dube, T., Shoko, C., 2022. Remote sensing of land use-land cover change and climate variability on hydrological processes in sub-Saharan Africa: key scientific strides and challenges. Geocarto Int.

Griggs, D., et al., 2013. Sustainable development goals for planet and people. Nature 495 (305).

Guug, S.S., Abdul-Ganiyu, S., Kasei, R.A., 2020. Application of SWAT hydrological model for assessing water availability at the Sherigu catchment of Ghana and Southern Burkina Faso. HydroResearch 3, 124–133.

Gxokwe, S., Dube, T., Mazvimavi, D., 2020. Multispectral remote sensing of wetlands in semi-arid and arid areas: a review on applications, challenges and possible future research directions. Remote Sens. 12 (24).

Gxokwe, S., Dube, T., Mazvimavi, D., Grenfell, M., 2022. Using cloud computing techniques to monitor long-term variations in ecohydrological dynamics of small seasonally-flooded wetlands in semi-arid South Africa. J. Hydrol. 612.

Hagenlocher, M., et al., 2019. Drought vulnerability and risk assessments: state of the art, persistent gaps, and research agenda. Environ. Res. Lett. 14, 1–14.

Henchiri, M., et al., 2020. Identification of drought and performance evaluation of MODIS and TRMM through remote sensing: a case study in North and West Africa during 2002-2018. Preprints 1, 1–16.

Herndon, K., Muench, R., Cherrington, E., Griffin, R., 2020. An assessment of surface water detection methods for water resource management in the Nigerien Sahel. Sensors 431, 1–14.

Huang, S., et al., 2017. The propagation from meteorological to hydrological drought and its potential influence factors. J. Hydrol. 547, 184–195.

Huang, C., Chen, Y., Zhang, S., Wu, J., 2018. Detecting, extracting and monitoring surface water from space using optical sensors: a review. Rev. Geophys. 56, 333–360.

Immerzeel, W.W., Droogers, P., de Jong, S.M., Bierkens, M.F., 2009. Large-scale monitoring of snow cover and runoff simulation in Himalayan river basins using remote sensing. Remote Sens. Environ. 113 (1), 40–49.

Jamali, S., Seaquist, J.W., Ardo, J., Eklundh, L., 2011. Investigating temporal relationships between rainfall, soil moisture and MODIS-derived NDVI and EVI for six sites in Africa. In: International Symposium on Remote Sensing of Environment—The GEOSS Era: Towards Operational Environmental Monitoring.

Janke, R., Murray, R., Uber, J., Taxon, T., 2006. Comparison of physical sampling and real-time monitoring strategies for designing a contamination warning system in a drinking water distribution system. J. Water Resour. Plan. Manag. 132 (4), 1–16.

Jiao, W., et al., 2019. A new multi-sensor integrated index for drought monitoring. Agric. For. Meteorol. 268, 74–85.

Khan, R., Gilani, H., 2021. Global drought monitoring with drought severity index (DSI) using Google Earth Engine. Theor. Appl. Climatol. 427 (146), 411.

Luo, R., Yuan, Q., Yue, L., Shi, X., 2021. Monitoring recent lake variations under climate change around the Altai Mountains using multimission satellite data. IEEE J. Sel. Top. Appl. Earth Obs. Remote Sens. 14.

Malik, A., et al., 2020. Identification of EDI trend using Mann-Kendall and Şen-innovative trend methods (Uttarakhand, India). Arab. J. Geosci. 13, 951.

Marshall, M., Funk, C., Michaelson, J., 2012. Examining evapotranspiration trends in Africa. Clim. Dyn. 28, 1849–1865.

Masocha, M., et al., 2018. Surface water bodies mapping in Zimbabwe using Landsat 8 OLI multispectral imagery: a comparison of multiple water indices. Phys. Chem. Earth Parts A/B/C 106, 63–67.

Michaelides, S., et al., 2009. Precipitation: measurement, remote sensing, climatology and modeling. Atmos. Res. 98 (4), 512–533.

Mishra, A.K., Singh, V.P., 2010. A review of drought concepts. J. Hydrol. 391, 202–216.

Mishra, A.K., Singh, V.P., 2011. Drought modeling—a review. J. Hydrol. 403, 157–175.

Moyer, J.D., Hedden, S., 2020. Are we on the right path to achieve the sustainable development goals? World Dev. 127, 1–13.

NOAA, 2023. Climate Change: Global Temperature. Available at: https://www.climate.gov/news-features/understanding-climate/climate-change-global-temperature. (Accessed 14 August 2023).

Orimoloye, I.R., Ololade, O.O., Belle, J.A., 2021. Satellite-based application in drought disaster assessment using terra MOD13Q1 data across free state province, South Africa. J. Environ. Manag. 285.

Park, S., Im, J., Jang, E., Rhee, J., 2016. Drought assessment and monitoring through blending of multi-sensor indices using machine learning approaches for different climate regions. Agric. For. Meteorol. 216, 157–169.

Saha, A., et al., 2022. Vulnerability assessment of drought in India: insights from meteorological, hydrological, agricultural and socio-economic perspectives. Gondwana Res.

Seager, R., et al., 2003. Mechanisms of hemispherically symmetric climate variability. J. Clim. 16 (18), 2960–2978.

Seager, R., et al., 2009. Mexican drought: an observational modeling and tree ring study of variability and climate change. Atmosfera 22 (1), 1–31.

Seaton, D., Dube, T., Mazvimavi, D., 2020. Use of multi-temporal satellite data for monitoring pool surface areas occurring in non-perennial rivers in semi-arid environments of the Western Cape, South Africa. ISPRS J. Photogramm. Remote Sens. 167, 375–384.

Seyama, E.S., Masocha, M., Dube, T., 2019. Evaluation of TAMSAT satellite rainfall estimates for southern Africa: a comparative approach. Phys. Chem. Earth 112, 141–153.

Sharara, A., et al., 2022. Fine-scale multi-temporal and spatial analysis of agricultural drought in agro-ecological regions of Zimbabwe. Geomat. Nat. Hazards Risk 13 (1), 1342–1365.

Tarnavsky, E., Chavez, E., Boogaard, H., 2018. Agro-meteorological risks to maize production in Tanzania: sensitivity of an adapted Water Requirements Satisfaction Index (WRSI) model to rainfall. Int. J. Appl. Earth Obs. Geoinf. 73, 77–87.

Udmale, P.D., Ichikawa, Y., Kiem, A.S., Panda, S.N., 2014. Drought impacts and adaptation strategies for agriculture and rural livelihood in the Maharashtra State of India. Open Agric. J. 8, 41–47.

United Nations, 2019. Sustainable Development Goals. Available at: https://www.un.org/sustainabledevelopment/sustainable-development-goals/. (Accessed 13 March 2020).

University of Washington Computational Hydrology Group, 2021. VIC Model Overview. Available at: https://vic.readthedocs.io/en/latest/Overview/ModelOverview/. (Accessed 19 August 2023).

USGS, 2023. What Is Remote Sensing and What Is It Used for? Available at: https://www.usgs.gov/faqs/what-remote-sensing-and-what-it-used#:~:text=Remote%20sensing%20is%20the%20process,sense%22%20things%20about%20the%20Earth. (Accessed 14 August 2023).

Valadao, L.V., et al., 2021. Reservoir metrics estimated by remote sensors based on the Google Earth Engine platform. Remote Sens. Appl. 24, 100652.

Varghese, D., Rahulovic, M., Stojkovic, S., Crnojevic, V., 2021. Reviewing the potential of Sentinel-2 in assessing the drought. Remote Sens. 13, 3355.

Verner, D., et al., 2018. Climate change, variability and droughts in Morocco. In: Climate Variability, Drought, and Drought Management in Morocco's Agricultural Sector. Washington, DC, World Bank Group, pp. 1–4.

West, H., Quinn, N., Horswell, M., 2019. Remote sensing for drought monitoring and impact assessment: progress, past challenges and future opportunities. Remote Sens. Environ. 232, 1–14.

Wilhite, D.A., Svoboda, M., Hayes, M.J., 2007. Understanding the complex impacts of drought: a key to enhancing drought mitigation and preparedness. Water Resour. Manag. 21, 763–774.

World Health Organization, 2013. Environmental Health Challenges in Mauritania. Available at: https://www.who.int/features/2013/mauritania_environmental_health/en/. (Accessed 31 March 2020).

Wufu, A., et al., 2021. Changes in glacial meltwater runoff and its response to climate change in the Tianshan region detected using unmanned aerial vehicles (UAVs) and satellite remote sensing. Water 13, 1753.

Yeh, P.J., Swenson, S.C., Famiglietti, J.S., Rodell, M., 2006. Remote sensing of groundwater storage changes in Illinois using the Gravity Recovery and Climate Experiment (GRACE). Water Resour. Res. 42.

Yu, F., Chen, J., 2019. Examination progress about the application of remote sensing data in agricultural drought monitoring. Agron. J. 36 (6), 2170–2180.

Zeng, Z., et al., 2019. Agricultural drought risk assessment in Southwest China. Water 11 (5).

Zhang, X., et al., 2019. Urban drought challenge to 2030 sustainable development goals. Sci. Total Environ. 693, 1–11.

Zhou, Y., et al., 2017. Open surface water mapping algorithms: a comparison of water-related spectral indices and sensors. Water 9 (256), 1–16.

Chapter 9

Remote sensing of climate variability and flooding

Cletah Shoko[a], Mark Matsa[b], Timothy Dube[c], Iqra Atif[a], and Mwazvita Sachikonye Dalu[d]

[a]*Division of Geography, School of Geography, Archaeology and Environmental Studies, University of Witwatersrand, Johannesburg, South Africa,* [b]*Department of Geography, Environmental Sustainability and Resilience Building, Midlands State University, Gweru, Zimbabwe,* [c]*Department of Earth Science, Institute for Water Studies, University of the Western Cape, Cape Town, South Africa,* [d]*School of Biology and Environmental Sciences, University of Mpumalanga, Nelspruit, South Africa*

Chapter outline

1 Introduction

Climate variability is one of the global environmental concerns that influence the occurrence of floods (Galvin et al., 2001; Ficchì and Stephens, 2019). Although flooding has been occurring in previous years, climate variability events such as El Niño-Southern Oscillations (ENSO) and the North Atlantic Oscillations (NAO) have been exacerbating it in recent years (Twumasi et al., 2017). Ward et al. (2014) reported that excluding Greenland and Antarctica, significant anomalies in flood volumes have occurred in more than a third of the Earth's land surface with 34% during El Niño years, whereas 38% during La Niña years. The studies by Sun et al. (2022) further reported El Niño-induced flooding occurring in Australia, Europe, North and South America, South and East Asia, as well as South Africa. The occurrence of these floods remains a threat as it extends from lakes, dams, rivers, as well as oceans spreading inland to low-land areas, thereby imposing significant threats to humanity.

Remote Sensing of Climate. https://doi.org/10.1016/B978-0-443-21731-9.00015-6
Copyright © 2024 Elsevier Inc. All rights are reserved, including those for text and data mining, AI training, and similar technologies.

The impacts of climate variability-related flooding are enormous, from local to global scale. These impacts include environmental damages, such as land degradation, death, property destruction, and the rise of diseases (Anyamba et al., 2019; Zurqani et al., 2022). For example, the study by Twumasi et al. (2017) reported that La Niña-induced flooding in 2000 caused severe damage to property and displacement in southern African parts of Mozambique, Malawi, and Zimbabwe, whereas, in the years 2009–2011 and 2015, severe floods resulted in more than 200 fatalities and over 50,000 homeless in West Africa. Sajjad et al. (2022) highlighted serious life losses and economic damages in Asian countries, namely, India, Pakistan, and Bangladesh.

In addition, research on projected climate variability indicates high frequency and variability in El Niño events and severity associated with more floods across the globe (Ward et al., 2014; Muis et al., 2018). Similarly, considerable studies have reported an increase in the number of El Niño events and associated flooding in Asia (Coenraads, 2006; Loo et al., 2015; Hooshyaripor et al., 2020), United States (Munoz and Dee, 2017; Corringham and Cayan, 2019), Africa (Mashao et al., 2023; Gannon et al., 2018), Australia (Gu et al., 2020; Marren et al., 2011), Europe (Burgess and Klingaman, 2015; De Moel et al., 2009), and Latin America (Burgess and Klingaman, 2015). For example, Coenraads (2006) and Loo et al. (2015) noted that although floods have been common and frequent in Asia, they have been more frequent and intensified during the 20th century, due to climate variability. In addition, floods are also extending toward inland areas, and more flooding in coastal areas, as never before (Anyamba et al., 2019). These projections and associated impacts of climate variability-induced flooding emphasized the need for continuous monitoring of flooding, mapping flood extent and depth, identifying areas vulnerable to flooding or risk areas, as well as identifying the impacts of these floods due to climate variability.

Remote sensing plays an important role in climate variability and flood monitoring. This is achieved using remote sensing datasets in visualizing and mapping inundation (Twumasi et al., 2017, 2022), damage assessment (Atif et al., 2015; Mouratidis and Sarti, 2012; Irimescu et al., 2009), model areas vulnerable to flooding (Wu, 2021), assessing the impact of climate variability on flooding events, as well as linking climate variability and environmental factors in modeling flooding (Bhaga et al., 2020; Hooshyaripor et al., 2020). The availability of different remote sensing data and products allows for a better understanding of climate variability and flood monitoring. The datasets range from low-borne drones or unmanned aerial vehicles (UAVs) to spaceborne multispectral sensors (Twumasi et al., 2022). Remotely sensed observations can analyze flooding trends and related factors such as climate variability and land use land cover changes. These observations can also be used during the event for visualizing flood extent and affected areas (Twumasi et al., 2017), mapping affected areas or inundated areas, mapping flood-prone areas as well as deriving flood vulnerability indicators.

Similarly, advanced machine algorithms allow for remote sensing big data analytics for climate variability (Bhaga et al., 2020; Dijkstra et al., 2019; Ben-Romdhane et al., 2018) and analyzing flood trends and monitoring (Islam et al., 2021; Shahabi et al., 2020). These machine learning algorithms vary from simple spectral analysis, use of indices, and digital image classification for understanding climate variability as well as flood monitoring. In addition, the approaches allow for the integration of different remote sensing data (e.g., radar and optical datasets) (Shahabi et al., 2020) and products. In addition, advanced machine learning also allows the inclusion of environmental factors, such as land use land cover, as well as climatic factors in modeling flood hazard and risk areas, to predict areas susceptible to flooding due to climate variability.

The aim of this chapter, therefore, is to provide an overview of the remote sensing of climate variability and flooding. This includes remote sensing data that can be used for climate variability and flood monitoring. The chapter further highlights modeling/analytical approaches available to monitor climate variability and flooding using remote sensing data, and the strengths and weaknesses of these approaches. In addition, the global trends in the remote sensing of climate variability and flooding are also provided, drawing case studies of the remote sensing of climate variability within the flooding monitoring context, including the impacts of climate variability on flooding across the globe. Lastly, the chapter also highlights future directions in the remote sensing of climate variability and flooding.

2 Remote sensing applications for climate variability and flood monitoring

The integration of remote sensing data into hydrological modeling has emerged as a powerful tool for unraveling the complex relationship between climate variability and flood occurrence (Twumasi et al., 2022). By assimilating critical variables such as elevation, land cover, and precipitation from remote sensing, hydrological models create a dynamic framework for simulating and forecasting flood scenarios (Li and Chen, 2017). These models enable researchers and stakeholders to explore how changing extreme events interact with local landscape characteristics to influence flood dynamics. For example, the Soil and Water Assessment Tool (SWAT) model is a common hydrological model that effectively incorporates remote sensing data for flood prediction. SWAT integrates topographical information, land use data, and rainfall estimates obtained from remote sensing platforms to simulate hydrological processes such as runoff, infiltration, and streamflow (Abbaspour et al., 2007). This enables the assessment of various climate-induced scenarios on flooding patterns, allowing decision-makers to proactively design strategies for flood mitigation and disaster preparedness.

Recent advancements in remote sensing technologies have further enhanced the accuracy and sophistication of hydrological models. For instance, high-resolution elevation data obtained from satellite-based Light Detection and Ranging (LiDAR) enable more precise modeling of flow paths and floodplain inundation areas (Bhardwaj et al., 2020). Additionally, the integration of remote sensing-derived evapotranspiration data into hydrological models enhances their ability to capture water fluxes within the watershed (Guan et al., 2020). The study by Hoang et al. (2020) in the Mekong River Basin utilized a combination of satellite-based rainfall data, hydrological modeling, and historical flood records to assess the impact of climate variability on flood frequency and severity. This study demonstrated how hydrological models enriched with remote sensing data enable a comprehensive understanding of the intricate relationships between climate patterns and flooding dynamics. Overall, hydrological modeling empowered by remote sensing data provides an essential tool for predicting and comprehending the effects of climate variability on flood occurrence. These models serve as proactive frameworks that aid in flood risk assessment, mitigation, and preparedness, ultimately contributing to more effective disaster management strategies.

In addition, the capability of remote sensing to detect surface water variations is a cornerstone in flood mapping and monitoring. Through the utilization of multispectral (e.g., Landsat and Sentinel-2) data, remote sensing detects water extent, offering essential information for disaster response and recovery efforts (Brown and Ward, 2019). Previous studies (Mazhar et al., 2021; Chormanski et al., 2011) proved the potential of Landsat and Sentinel-2 in mapping the extent of floods. Similarly, satellite-based precipitation data, exemplified by the Global Precipitation Measurement (GPM) mission, revolutionized flood monitoring (Woods et al., 2023). These datasets offer accurate and expansive rainfall estimates, facilitating real-time flood forecasting and effective disaster management (Adler et al., 2017). The comparative analysis of imagery captured before and after flood events unveils alterations in land cover, river patterns, and floodplain configurations. This intelligence guides postflood impact assessments and the formulation of targeted mitigation strategies (Gonzalez et al., 2020).

In recent years, the utilization of advanced remote sensing techniques, particularly synthetic aperture radar (SAR) data, has revolutionized flood mapping by providing unprecedented capabilities to overcome challenging environmental conditions (DeVries et al., 2020; Chini et al., 2021). SAR technology emits microwave signals that penetrate through clouds, rain, and darkness, making it an invaluable tool for flood mapping during adverse weather conditions (Kumar et al., 2020). For instance, during the 2019 Cyclone Idai in Mozambique, SAR data from the Sentinel-1 satellite played a pivotal role in mapping the extensive flood extents caused by the cyclone's heavy rainfall (Chini et al., 2021). The rapid acquisition and processing of Sentinel-1 SAR imagery enabled the generation of accurate flood maps within hours of the disaster (Anusha and Bharathi, 2020). This information was indispensable for identifying the affected areas, assessing the extent of damage, and coordinating rescue and relief operations. Similarly, the European Space Agency's (ESA) Copernicus Emergency Management Service has been harnessing SAR data for timely flood mapping in various regions globally (ESA, 2021). Furthermore, the fusion of SAR data with optical imagery enhances the precision of flood mapping by providing additional contextual information. The synergy between Sentinel-1's SAR and Sentinel-2's optical data enables a comprehensive assessment of flood-affected areas, accounting for factors such as land cover, infrastructure, and human settlements (Anusha and Bharathi, 2020; ESA, 2021). The integration of SAR data with other remote sensing modalities enhances the accuracy and utility of flood maps, making them invaluable tools for disaster response and resource allocation (e.g., DeVries et al., 2020).

The convergence of remote sensing technologies and early warning systems has significantly bolstered the capacity to anticipate and mitigate the catastrophic impacts of floods, a phenomenon exacerbated by climate variability. The seamless fusion of these tools hinges on their ability to furnish real-time and precise data for flood prediction, thereby amplifying the effectiveness of preparedness and response strategies (Mendoza and Farias, 2020; Rahaman et al., 2021). For example, during the 2020 monsoon season in Bangladesh, a combination of satellite-based rainfall estimates and ground-based weather observations facilitated accurate flood forecasting and timely issuance of warnings (Rahaman et al., 2021). By assimilating high-resolution precipitation data from the GPM mission, the early warning system effectively captured the intricacies of the monsoonal rains, enabling authorities to take preemptive measures to mitigate the impending flood disaster. Furthermore, the combination of satellite-derived hydrological information and hydraulic modeling has led to advanced early warning systems. In Colombia, for instance, Mendoza and Farias (2020) harnessed remote sensing data on river discharge and water levels, for flood prediction. This comprehensive approach facilitated accurate and localized flood forecasts, enabling authorities to evacuate vulnerable communities and allocate resources efficiently. The utilization of remote sensing in early warning systems has also transcended regional boundaries. In Europe, the European Flood Awareness System leverages satellite-based data, including soil moisture measurements, precipitation estimates, and river levels, to predict potential flood events (Gosling et al., 2021).

With the availability of remote sensing satellites, equipped with an array of sensors, instruments such as the Moderate Resolution Imaging Spectroradiometer (MODIS), Visible Infrared Imaging Radiometer Suite, and Advanced Microwave Scanning Radiometer—Earth Observing System have become pivotal in capturing key climate variability parameters (Rodriguez and Lee, 2022). These encompass temperature, precipitation, humidity, and vegetation cover, enabling a comprehensive analysis of climatic shifts and trends (Rodriguez and Lee, 2022). The monitoring of sea surface temperature using remote sensing remains a cornerstone of oceanic conditions influencing global climate patterns. This contributes to forecasting climate variability phenomena, such as

El Niño and La Niña, both of which exhibit strong correlations with amplified flood occurrences (Chen et al., 2019; Smith and Johnson, 2021).

In addition, UAVs, commonly known as drones, in combination with satellite remote sensing, have demonstrated great potential in flood assessment and monitoring. UAVs offer more accurate detection of flood extent and damage assessment (Hashemi-Beni and Gebrehiwot, 2021). Compared with conventional satellite remote sensing platforms, UAVs offer several advantages. These airborne systems are associated with relatively low cost; they acquire multiview high-resolution imagery with minimal cloud shadows (Langhammer and Vacková, 2018). Most importantly, the use of UAVs is flexible in terms of time and frequency of data acquisition (Hashemi-Beni and Gebrehiwot, 2021). In this regard, UAVs offer a critical potential in remote sensing of flood mapping. Furthermore, UAV-derived data can also be integrated with satellite-based data for improved flood assessment. This integration allows for a multiscale approach, combining high-resolution UAV data with the global coverage of satellites, providing more comprehensive flood information.

3 Analytical approaches to the remote sensing of climate variability and flood monitoring

Several techniques have been developed and used with remote sensing data for climate variability and flood monitoring. These methods have become increasingly popular and important in recent years, due to the incredible amount of remote sensing data currently available. Some of the analytical approaches that are relevant and have been commonly used for climate variability and flood monitoring are classification, regression, and dimension reduction.

Different classification approaches exist; however, advanced supervised machine learning classifiers are becoming more prominent, compared with the traditional classifiers or unsupervised classifiers. For instance, support vector machines (SVMs) and random forests (RF) are gaining popularity as robust techniques for flooding extent and prediction using remote sensing data (Thanh Son et al., 2022). These advanced machine learning models are well-known for their capability to handle high-dimensional data, limited samples, and multisource data, making them more suited to remote sensing of climate variability and flooding over time (Liu et al., 2021; Ha and Kang, 2022; Langhammer, 2023). Therefore, these models have been successfully used in mapping flooded areas and predicting flood risk (Ha and Kang, 2022; Langhammer, 2023). Langhammer (2023) uses an SVM model coupled with a hydrometeorological sensor network to simulate floods in a montane basin, with high accuracy. The model accurately captures flood trends, timing, peaks, and volumes, and shows better performance on a daily timeframe. Ha and Kang (2022) applied an RF model for flood risk assessment in Korea. The model, evaluated using several metrics, creates an assessment map that helps in devising preventive measures against urban flooding and improving flood control and response capacities.

On the contrary, researchers have attempted to use a subpixel classification approach to handle the problem of "mixed" pixels, which is a common issue with coarse (e.g., MODIS) or moderate spatial resolution (e.g., Landsat) remotely sensed data. Spectral mixture analysis and soft, or fuzzy, classifiers are common methods in this regard. Giraldo Osorio and Gabriela García Galiano (2012) implemented a subpixel analysis methodology for identifying flooded areas, using digital elevation models (DEMs) and moderate-resolution satellite imagery. The results showed that it achieved 80% accurate flood mapping, significantly surpassing traditional methods' 53% performance. Although supervised classification approach has become more prominent, clustering unsupervised classification technique remains useful. K-means and agglomerative clustering are the most common clustering techniques applied to remote sensing analytical approaches. These approaches cluster identical pixels based on spectral values or other derived parameters such as vegetation indices or texture metrics. Sun et al. (2022) applied the K-means clustering and urban functional areas in Zhengzhou City, to improve flood modeling. The model, validated against three rainfall runoffs, achieved consistent simulated runoffs and accurate flood node prediction, aiding in flood forecasting and prevention.

Advanced regression models have also been used for climate variability and flooding monitoring. These models predict continuous variables using spectral data from satellite imagery which can be used to assess the extent of flooding and in the prediction of flood and associated impacts (Rodriguez and Lee, 2022; Abedi et al., 2022). In the context of climatic variability and flood monitoring, logistic and multinomial logistic regression can provide several advantages (Abedi et al., 2022). First, these algorithms typically provide a probability for each class, which might provide extra information regarding the classification's certainty. Second, these approaches can deal with nonlinear relationships between spectral values and land cover type that can occur in remote sensing data and may identify which spectral bands or features are most significant for categorization. Linear regression, RF, neural networks, boosted regression tree (BRT), and logistic regression are some of the prediction models that have been commonly used in the remote sensing of climate variability and flood monitoring. For instance, Eingrüber and Korres (2022) applied the multiple linear regression model to predict the frequency of extreme rainfall, achieving an increase in rainfall by 33%–51% and flood magnitudes by 31%–36% by 2099, using National Centers for Environmental Prediction, Global Climate Models, Statistical Downscaling Model, and SWAT models. Abedi et al. (2022) evaluated the classification and regression tree, RF, BRT, and extreme gradient boosting (XGBoost) to model the susceptibility to flash floods. The results showed that slope was the key factor for triggering floods, and the BRT model performed well with an accuracy of 0.899. On the contrary, Aryal et al. (2023) used artificial neural networks to establish a framework for predicting flood-prone zones in ungauged urban watersheds using satellite rainfall data and floodplain maps. The method was validated for both dry and wet seasons and generated quantifiable inundation extent percentages for various flooding events. Recently, Tella et al. (2023) successfully applied logistic regression to predict flood-prone areas in Kuala Lumpur. The study achieved an overall accuracy of 0.84 and emphasized the need for urban planning and improved drainage systems.

Dimension reduction is another statistical approach for reducing the number of variables in a dataset while retaining the important information. Dimension reduction approaches, such as principal component analysis (PCA), minimum noise fraction, and T-distributed stochastic neighbor embedding (t-SNE), are frequently used in remote sensing for multispectral or hyperspectral data, where each pixel has multiple linked spectral bands. Wu (2021) proposes an assessment framework, using PCA, to evaluate flood vulnerability in coastal regions of Connecticut, USA. The study underscores varying vulnerabilities across urban, suburban, and rural areas due to differing dominant factors. However, studies conducted in hydrological modeling using t-SNE have indicated that the approach is computationally demanding, especially with large datasets (Tang and Carey, 2021). This makes it difficult to apply for continuous monitoring of climate or flood assessment over time. Similarly, functional data analysis (FDA) is another approach that analyzes large datasets of observations as curves, surfaces, or general functions rather than discrete data points. In remote sensing, the FDA captures the data's fundamental functional nature by considering the spectral reflectance values of each pixel over a range of wavelengths, or the values of an area over time, as a continuous curve and attempts to evaluate these curves directly. One major limitation of the FDA to remote sensing data, particularly in the field of climate variability and flood monitoring, is the requirement for measurements taken continuously over a range of values (such as wavelengths or time), which is not always available. Nevertheless, Ghumman et al. (2020) used the FDA to compare global climate change models in assessing and predicting climate variability impact on streamflow in Pakistan's Upper Indus River Basin. Results suggested increasing streamflow till 2070, followed by a decline.

4 Case studies on the remote sensing of climate variability and flood monitoring

Global efforts toward the remote sensing of climate variability and flood monitoring are recognized (Hou et al., 2022; Muis et al., 2018). Considerable studies have been conducted in mapping flood extent, depth, areas vulnerable to flooding, environmental factors that contribute to flooding, the effect of climate variability on flooding, as well as the impacts of flooding, using remote sensing. Similarly, studies were conducted to understand and link flooding with climate variability such as El Niño or La Niña events (e.g., Muis et al., 2018; Hou et al., 2022; Jongman et al., 2012). These studies were conducted using different remote sensing datasets such as Landsat, Sentinel, and MODIS, as well as remote sensing products such as DEM, temperature, and rainfall. The use of approaches is also variable from traditional classifiers and regression analysis to advanced machine learning. Similarly, an integration of various datasets including remotely sensed data, products, meteorological (e.g., rainfall, temperature), hydrological (e.g., inflow, evaporation data) as well as environmental (e.g., land use land cover) datasets. Some of the studies focused on climate variability and flooding, whereas other studies focused on these two in isolation. At a global scale, Hou et al. (2022) determined the monthly water storage of global reservoirs and the effect of climate variability using Landsat images, geostatistical models, and satellite altimetry from 1984 to 2015. The study reported significant correlations between reservoir storages with rainfall, inflow, and evaporation. Also, Jongman et al. (2012) projected global river and coastal flood exposure using population, DEM, and land use land cover changes datasets. The study observed that 73% of the total global population exposed to river flooding is in Asia, followed by Europe and sub-Saharan Africa. In terms of future projections, the exposed population at risk is highest in sub-Saharan Africa, whereas a decline of 13% was projected for Europe in 2050.

Within the African continent, the use of remote sensing for climate variability (e.g., Komolafe et al., 2015) and flood extent, vulnerability, and damage assessment (e.g., Twumasi et al., 2017; Asante et al., 2007; Artan et al., 2002; De Groeve, 2010; Skakun et al., 2014) have also received considerable attention. Studies on remote sensing of floods in Africa focused mainly on coastal areas, large, recognized basins, and informal settlements of inland areas. For example, the review by Komolafe et al. (2015) provides detailed studies conducted in the coastal areas of Nigeria using different remote sensing data, products, meteorological data, and geographic information system techniques. The review concluded that there is a need for an integrated flood model, preassessment of flood impacts before the flood event, as well as determining the health impact of flooding in the country. In southern Africa, the studies on remote sensing of flood were conducted in South Africa (Cape Town and Durban), Mozambique, and Namibia. On the contrary, the issue of the remote sensing of climate variability is an emerging research field in Africa; limited studies have been conducted. Studies were conducted in assessing climate variability and flooding; however, the studies are conducted in isolation, thereby either focusing on flooding or climate variability, using remote sensing.

In Asia, considerable studies have been conducted to map flooding during extreme rainfall events using remote sensing and machine learning. Most studies were conducted in India (e.g., Saha et al., 2020; Ajin et al., 2013; Sowmya et al., 2015; Kundu et al., 2015; Sansare and Mhaske, 2020), China (e.g., Wei et al., 2023; Zhang et al., 2002; Wang et al., 2002; Na et al., 2013), and Pakistan (Atif et al., 2015; Uddin et al., 2013; Haq et al., 2012; Sajjad et al., 2020, 2022). Overall, these studies mainly focused on mapping flood extent or inundation extent, damage assessment, and risk areas. In addition, the majority of the studies used various passive optical satellite datasets, SARs, and limited UAVs datasets. For example, Atif et al. (2015) determined the spatial and temporal extent of flooding and damages associated with the 2014 flood in the Punjab province of Pakistan. The study used MODIS and Landsat remotely sensed data as well as field-based surveys and flow data, with multicriteria decision analysis. The results showed the potential of MODIS and in situ data in assessing flood damage and identifying areas with high damage.

In Australia, studies such as those by Mohammadi et al. (2017), Kalantar et al. (2021), Hess and Melack (2003), Jain et al. (2005), Tulbure et al. (2022), Ticehurst et al. (2015), and Li et al. (2016) have reported the potential of MODIS, Landsat, and Sentinel-2 passive multispectral datasets, as well as SAR data, in mapping flood extents and assessing vulnerable areas. These studies also used advanced machine learning approaches, such as neural networks, RF, and SVMs. For example, Tulbure et al. (2022) used a combination of Landsat-8, PlanetScope, and Sentinel-2 with RF to detect flood extent within the Murray-Darling Basin of Australia. The study reported over 90% accuracy, as well as the potential of water indices and shortwave infrared bands in detecting flooded areas. On the contrary, studies that focused on the remote sensing of climate variability (Xie et al., 2019) have reported the potential of climate indices, gravity recovery and climate experiments, and precipitation data tropical rainfall measuring mission (TRMM) in assessing the hydrological impacts of the Indian Ocean dipole, Southern Annular Mode, and ENSO in Australia. Studies which combined climate variability and flooding using remote sensing are rare within this continent.

In Europe, studies are conducted mainly focused on remote sensing of flood extent, compared with depth and risk analysis. This was reported by the review of De Moel et al. (2009), which was conducted to assess flooding in Europe, as well as the availability of methods and their use. It was noted that 80% of Europe focused on flood extent, with limited research on flood depth or risk assessment. It was also noted that the issue of incorporating climate change or extreme events in flood risk analysis and prediction is rare. Other studies that focused on flood assessment using remote sensing were those of Irimescu et al. (2009), Mourad et al. (2022), and Mouratidis and Sarti (2012), just to mention a few. The integration of climate variability in assessing flood is quite very rare, and most studies are done separately.

At the national level, Qiang et al. (2017) conducted the spatial and temporal changes in flood exposure in the United States. The study used different datasets which included census data, 100-year flood maps, and land cover data, with local Moran's index and weighted regression geostatistical analyses. The study observed that 8.3% of the total national population lived in high-risk flood zones in 2011, whereas the Federal Emergency Management Agency report estimated that rising seas and increasingly severe weather are anticipated to increase flood-risk areas by approximately 45% by 2100 (Qiang et al.,

2017). In addition, it was noted that developed countries have more established techniques and data archives for identifying flood zones. Recently, Mazza and Chen (2023) assessed the impacts of tropical cyclones on flooding in the United States at a continental scale. The study used various remotely sensed data, products, and field-based datasets from 2002 to 2019. These included Integrated Multi-satellitE Retrievals for GPM (IMERG), CPC MORPHing technique (CMORPH), and the TRMM-tropical multisatellite precipitation analysis products, as well as the high-resolution reanalysis (ERA5), ground-radar- and rain-gauge-based datasets. The study reported an underestimation of tropical cyclone rainfall and its flooding potential and the misplacement of cell values by satellite products within the coastal areas. This has been regarded as one of the uncertainties associated with satellite products. In a different study, Atashi et al. (2023) assessed the impacts of climate variability on water availability using Landsat, precipitation, and temperature data from ground-based measurements, as well as the global surface water dataset. The analyses were conducted for the Red River Basin of the United States and Canada, within the Google Earth Engine cloudy computing platform, from 1990 to 2019. Findings highlighted an increasing trend in rainfall and permanent water bodies area within the central parts of the basin, with high-frequency flooding. The study further recommends a comprehensive use of water indices and an assessment of water quality.

In Latin America, different studies were conducted to assess flooding using remote sensing, especially within the Amazon River Basin, which is the world's largest river basin (Arnesen et al., 2013; Santos et al., 2017; Schöngart and Junk, 2007; de Andrade and Szlafsztein, 2018). For example, Schöngart and Junk, (2007) reported Amazon River floods due to La Niña, whereas the study by Burgess and Klingaman (2015) in Jamaica predicted flood-prone areas using elevation models and meteorological data. The study predicted more frequent floods associated with extreme events. They further highlighted that flooding in Jamaica and Europe is becoming more frequent due to extreme events induced by climate variability. Although studies on extreme climate variability events and flood assessment were conducted (e.g., Marengo et al., 2012), the majority of studies were conducted in isolation. Studies on El Niño or La Niña were conducted using remote sensing data or products, to understand the spatial and temporal variations in rainfall (e.g., Bookhagen and Strecker, 2010; Moura et al., 2019; de Souza et al., 2020).

5 Strengths and weaknesses of the remote sensing of climate variability and flooding

The remote sensing approach presents an important for climate variability and flood monitoring. This has been facilitated by various factors. Compared with conventional methods of field-based data collection which required both more human labor and time to collect data about flood susceptibility, damages, and planning for recovery initiatives, remote sensing depends on the capacity of satellites to continually collect data for change analysis as well as planning purposes (Rahman and Di, 2020; Kucharczyk and Hugenholtz, 2021; Sadiq et al., 2022).

Flood monitoring in recent times has been enabled by the availability of free-of-charge remote sensing data which has enabled low-cost flood risk assessment (Turner et al., 2015; Li et al., 2016; Klein et al., 2017). These datasets include the Sentinel satellite constellation which provides SAR and multispectral data at a global scale, as well as the Landsat constellation and MODIS (Khanal et al., 2020; Pandey et al., 2021; Hansen and Loveland, 2012; Notti et al., 2019). The availability of these resources at no cost and archive data, which goes back to 40 years of Earth imaging, is crucial for continuous monitoring and management of flood events. These datasets have allowed the detection of flood-prone areas, flood damages, and flooded areas (Chen et al., 2013; Ticehurst et al., 2014; Chignell et al., 2015).

Furthermore, remote sensing data or products can be integrated with hydrological, meteorological, topographical, as well as socio-economic data for climate variability and flood monitoring (Farhadi and Najafzadeh, 2021). Remote sensing products such as DEMs from advanced spaceborne thermal emission and reflection radiometer (ASTER) and shuttle radar topography mission (SRTM) have been used with multispectral data (e.g., Landsat, Sentinel-2, and SPOT) and meteorological data to predict areas that are prone to flooding. These systems can be as complex as catchment scale modeled predictions, down to low-technology river gauges to give a specific community advanced warning of rising river levels. Change detection techniques based on high- and medium-resolution satellite data have the capacity to detect flood-induced property loss (Apostolopoulos and Nikolakopoulos, 2021; Sharma et al., 2022) at a large scale. This makes worldwide, regional, and local level flood monitoring and impact assessment easy and more accurate with the help of a few ground truthing data.

Simulation of flood extent using satellite data acquired through real-time active and passive satellite sensors makes flood extent monitoring easy, especially in coastal areas that are prone to flooding (Cohen et al., 2019). The availability of optical, thermal, and SAR sensors is an advantage for flood disaster management as flood prediction, susceptibility, and damage analysis (Hashemi-Beni and Gebrehiwot, 2021; Devrani et al., 2022). This makes flood disaster preparedness easy and accurate as high temporal and spatial resolution RADAR and LiDAR sensors are available to monitor floods under all weather conditions, especially for high-income countries who can afford to purchase this high-quality satellite data (Zeng et al., 2020). SARs from Sentinel-1 at a 12-day revisit and 10 m spatial resolution have been applied to determine flood extents during flooding events (Zeng et al., 2020). SAR sensors, especially the recently launched Sentinel-1 in combination with optical imagery (multisensor and multitemporal data), are capable of detecting flood-related damage and determining the extent of agricultural recovery postflooding. Passive optical sensors also have the power to detect water-inundated areas, vegetation loss, or growth which can be done based on various spectral indices like the normalized difference vegetation index, enhanced vegetation index, and other indices for built-up areas and water.

However, although the remote sensing approach is valuable in climate variability and flood monitoring, it has considerable limitations or challenges associated with its application (Jiménez-Jiménez et al., 2020). For example, the high acquisition cost of high-resolution data like IKONOS, GeoEye, and WorldView for fine-scale monitoring of floods makes remote sensing theoretical in poor regions of the world where people have access to only freely available satellite data like MODIS (250m resolution) and Landsat (30m resolution). In this regard, only limited studies have used them. For example, Farhadi and Najafzadeh (2021) reported that the high cost of acquisition is a major limitation for flooding monitoring and associated impacts in developing countries, such as Zimbabwe. Similarly, drones or UAVs are increasingly becoming popular in flood monitoring in countries like China (e.g., Feng et al., 2015; Li et al., 2021), the United States (e.g., Hashemi-Beni and Gebrehiwot, 2021), and Europe (e.g., Langhammer et al., 2018). In Africa, the use of UAVs is growing in monitoring flood extent during flood events. This, however, is very limited due to drone acquisition costs and the need for skilled personnel. On the contrary, hyperspectral remote sensing also requires specialized enhancement to provide more real-time flood monitoring results, large storage capacity, and heavy processing configuration which remains a major hurdle for processing digital flood event images.

The acquisition mode of multispectral remote sensing is also a major limitation in climate variability and flood monitoring (Lin et al., 2016). Optical passive remote sensing which remains one of the best approaches for water and inundation mapping is limited to day-light and under cloud-free conditions, which makes real-time monitoring of flood events a difficult task (Lin et al., 2016). More so, temporal resolutions of long-serving medium multispectral remote sensing satellites, such as Landsat at 16-day revisit frequency do not coincide with flood events which makes observation of real-time inundation difficult to accomplish. In addition, the low revisit time makes it difficult to acquire cloud-free images during the rainy season. On the contrary, MODIS has a daily revisit frequency of 250–1km spatial resolution. The research by Ticehurst et al. (2015) in Australia which aimed at improving the accuracy of MODIS in detecting flood extent noted its weakness in mapping flooded small rivers due to its coarse spatial resolution. Nevertheless, Ticehurst et al. (2015) commends the daily frequency and global coverage with archive data from MODIS for the year 2000, especially when integrated with other remote sensing datasets. This leaves real-time monitoring of inundation only possible for those flooding events that tend to coincide with the revisit time of some satellite sensors, under clear skies. This was also highlighted by the recent by Tulbure et al. (2022). The study tested the potential of harmonized Landsat-8 and Sentinel-2 in detecting ephemeral floods in Australia.

Although Sentinel-1 was launched to cover this loophole, there is still a lack of experience in processing radar data which makes remote sensing of flood events difficult to conduct in some areas. When it comes to monitoring post-flood recovery and even flood damage on both human-built structures and natural vegetation, remote sensing is limited to observing general patterns in mapped-out areas and lacks spectral detail except for developed countries where they have access to images acquired through hyperspectral sensors (Lin et al., 2016). This makes it difficult to distinguish between phenomena with narrow differences and spectral signatures. In this case, there is always a need for ground-based accuracy assessment to complement remote sensing. The disaster risk reduction sector has not been able to fully embrace remote sensing for flood monitoring from experimental to real-world settings because of the techno-economic dynamics of remote sensing data collection. The cost of high spatial and temporal resolution remote sensing data limits its effectiveness and applicability in poor regions of the world which are usually the worst affected by flood events (Munawar et al., 2022). Although there are several platforms to access remote sensing data on floods such as Landsat, advanced very high-resolution radiometer, ASTER, SRTM, and the ESA's Sentinel satellite series being available for free, these open-access data streams only offer medium-resolution spectral data (i.e., >10 pixel size) (Ajmar et al., 2017). For effective real-time flood monitoring, it is important to have high-resolution spectral and temporal data (<5 m) (Sharma et al., 2022) from commercial satellites such as Worldview and RapidEye.

Another factor that has proved to limit the applicability of remote sensing to flood monitoring is uncertainty. This issue was also highlighted by the review of De Moel et al. (2009), which focused on flood assessment methods and their availability in Europe. In addition, while there is an increasing number of satellites, increased frequency and rapid processing chains, such as Copernicus Emergency Management Services and the Global Flood Detection Systems (GFDS), allow for automatic and quasi-real-time flood mapping, these services are not always validated and are muddled with uncertainties (Copernicus Emergency Management Service, 2023). The inherent spaces between satellites combined with orbital cycles of satellites make it difficult to continuously monitor flood progression consistently over a given space of time (Nikam et al., 2017). These uncertainties have been linked to hydraulic models or the resolution of DEMs (De Moel et al., 2009; Bales and Wagner, 2009; Notti et al., 2019). Hence, hydrological and numerical models seldom provide imprecise results and fail to give reliable predictions regarding floods. In addition, these models are case-specific; they depend on the physical properties and climate of the specific study area (Zha et al., 2015; Cenci et al., 2017; Zhou et al., 2015). In such instances, maps that are based on field surveys become more reliable despite being time-consuming and geographically limited.

6 Conclusion

Climate variability is one of the global environmental concerns which influence the occurrence of floods. Although flooding has been occurring in the previous years, climate variability events such as ENSO and the NAO have been exacerbating it. The occurrence of climate variability-induced flooding remains a concern and emphasizes the need for continuous flood mapping, inundation extent, and duration, identify vulnerable areas, and assess the damage. Remotely sensed data and products provide useful information to understand flood occurrence, as well as relate flood occurrence to climate variability. The review found that although climate variability influences flooding, the majority of studies are conducted separately. In addition, most flooding assessments using remote sensing focused mainly on international or transboundary basins, thereby underestimating the influence of these floods in other small catchment or inland areas, at a local scale. This emphasized the need for future studies to expand flood assessment beyond mapping inundation, and to integrate climate variability in flood modeling.

References

Abbaspour, K.C., Yang, J., Maximov, I., Siber, R., Bogner, K., 2007. Modelling hydrology and water quality in the pre-alpine/alpine Thur watershed using SWAT. J. Hydrol. 333 (2–4), 413–430.

Abedi, R., Costache, R., Shafizadeh-Moghadam, H., Pham, Q.B., 2022. Flash-flood susceptibility mapping based on XGBoost, random forest and boosted regression trees. Geocarto Int. 37 (19), 5479–5496.

Adler, R.F., Gu, G., Huffman, G.J., Curtis, S., Bolvin, D., 2017. Satellite-based global precipitation measurement (GPM): methods, enhancements and applications. Atmos. Res. 197, 105–106.

Ajin, R.S., Krishnamurthy, R., Jayaprakash, M., Vinod, P.G., 2013. Flood hazard assessment of Vamanapuram River basin, Kerala, India: an approach using remote sensing & GIS techniques. Adv. Appl. Sci. Res. 4 (3), 263–274.

Ajmar, A., Boccardo, P., Broglia, M., Kucera, J., Giulio-Tonolo, F., Wania, A., 2017. Response to flood events: the role of satellite-based emergency mapping and the experience of the Copernicus Emergency Management Service. In: Flood Damage Survey and Assessment: New Insights from Research and Practice. Wiley, pp. 211–228.

Anusha, N., Bharathi, B., 2020. Flood detection and flood mapping using multi-temporal synthetic aperture radar and optical data. Egypt. J. Remote Sens. Space Sci. 23 (2), 207–219.

Anyamba, A., Chretien, J.P., Britch, S.C., Soebiyanto, R.P., Small, J.L., Jepsen, R., 2019. Global disease outbreaks associated with the 2015–2016 El Niño event. Sci. Rep. 9 (1), 1930.

Apostolopoulos, D., Nikolakopoulos, K., 2021. A review and meta-analysis of remote sensing data, GIS methods, materials and indices used for monitoring the coastline evolution over the last twenty years. Eur. J. Remote Sens. 54 (1), 240–265.

Arnesen, A.S., Silva, T.S., Hess, L.L., Novo, E.M., Rudorff, C.M., Chapman, B.D., McDonald, K. C., 2013. Monitoring flood extent in the lower Amazon River floodplain using ALOS/PALSAR ScanSAR images. Remote Sens. Environ. 130, 51–61.

Artan, G.A., Restrepo, M., Asante, K., Verdin, J., 2002. A flood early warning system for Southern Africa. In: Pecora 15 and Land Satellite Information 4th Conference Proceedings.

Aryal, A., Bhusal, A., Kalra, A., 2023. Evaluating dry and wet season precipitation from remotely sensed data using artificial neural networks for floodplain mapping in an ungauged watershed. Environ. Protect. Res. 29, 150–165.

Asante, K.O., Macuacua, R.D., Artan, G.A., et al., 2007. Developing a flood monitoring system from remotely sensed data for the Limpopo basin. IEEE Trans. Geosci. Remote Sens. 45, 1709–1714.

Atashi, V., Mahmood, T.H., Rasouli, K., 2023. Impacts of climatic variability on surface water area observed by remotely sensed imageries in the Red River Basin. Geocarto Int. 38 (1), 2209541.

Atif, I., Mahboob, M.A., Waheed, A., 2015. Spatio-temporal mapping and multi-sector damage assessment of 2014 flood in Pakistan using remote sensing and GIS. Indian J. Sci. Technol. 8 (35), 1–18.

Bales, J.D., Wagner, C.R., 2009. Sources of uncertainty in flood inundation maps. J. Flood Risk Manag. 2 (2), 139–147.

Ben-Romdhane, H., Al-Musallami, M., Marpu, P.R., Ouarda, T.B., Ghedira, H., 2018. Change detection using remote sensing in a reef environment of the UAE during the extreme event of El Niño 2015–2016. Int. J. Remote Sens. 39 (19), 6358–6382.

Bhaga, T.D., Dube, T., Shekede, M.D., Shoko, C., 2020. Impacts of climate variability and drought on surface water resources in sub-Saharan Africa using remote sensing: a review. Remote Sens. (Basel) 12 (24), 4184.

Bhardwaj, A.K., Muthuwatta, L.P., Ghosh, S., 2020. Assessment of future flood hazard in the Ganges-Brahmaputra-Meghna basin using a coupled hydrodynamic-hydrological model. Sci. Total Environ. 701, 134829.

Bookhagen, B., Strecker, M.R., 2010. Modern Andean rainfall variation during ENSO cycles and its impact on the Amazon drainage basin. In: Hoorn, C., Vonhof, H., Wesselingh, F. (Eds.), Neogene History of Western Amazonia and Its Significance for Modern Diversity. Blackwell Publishing, Oxford, England, pp. 223–241.

Brown, R.D., Ward, K.M., 2019. Remote sensing applications in flood monitoring: a comprehensive overview. Remote Sens. Appl. 33, 100798.

Burgess, M.L., Klingaman, N.P., 2015. Atmospheric circulation patterns associated with extreme cold winters in the UK. Weather 70 (7).

Cenci, L., Pulvirenti, L., Boni, G., Chini, M., Matgen, P., Gabellani, S., Squicciarino, G., Pierdicca, N., 2017. An evaluation of the potential of Sentinel 1 for improving flash flood predictions via soil moisture-data assimilation. Adv. Geosci. 44, 89–100.

Chen, X., Chen, Y., Smith, A.B., 2019. Sea surface temperature monitoring using remote sensing: recent developments and applications. Prog. Oceanogr. 180, 102234.

Chen, Y., Huang, C., Ticehurst, C., Merrin, L., Thew, P., 2013. An evaluation of MODIS daily and 8-day composite products for floodplain and wetland inundation mapping. Wetlands 33, 823–835.

Chignell, S.M., Anderson, R.S., Evangelista, P.H., Laituri, M.J., Merritt, D.M., 2015. Multitemporal independent component analysis and Landsat 8 for delineating maximum extent of the 2013 Colorado front range flood. Remote Sens. 7 (8), 9822–9843.

Chini, M., Chini, C.M., Olaya-Molina, L.F., 2021. Synthetic aperture radar for flood monitoring: a review. Rev. Geophys. 59 (2). e2020RG000727.

Chormanski, J., Okruszko, T., Ignar, S., Batelaan, O., Rebel, K.T., Wassen, M.J., 2011. Flood mapping with remote sensing and hydrochemistry: a new method to distinguish the origin of flood water during floods. Ecol. Eng. 37 (9), 1334–1349.

Coenraads, R., 2006. Natural Disasters and How We Cope. Millennium House Pty Ltd, Elanora Heights, Australia.

Cohen, S., Raney, A., Munasinghe, D., Loftis, J.D., Molthan, A., Bell, J., et al., 2019. The Floodwater Depth Estimation Tool (FwDET v2. 0) for improved remote sensing analysis of coastal flooding. Nat. Hazards Earth Syst. Sci. 19 (9), 2053–2065.

Copernicus Emergency Management Service, 2023. Directorate Space, Security and Migration, European Commission Joint Research Centre (EC JRC). https://emergency.copernicus.eu/. (Accessed July 2023).

Corringham, T.W., Cayan, D.R., 2019. The effect of El Niño on flood damages in the western United States. Weather Clim. Soc. 11 (3), 489–504.

de Andrade, M.M.N., Szlafsztein, C.F., 2018. Vulnerability assessment including tangible and intangible components in the index composition: an Amazon case study of flooding and flash flooding. Sci. Total Environ. 630, 903–912.

De Groeve, T., 2010. Flood monitoring and mapping using passive microwave remote sensing in Namibia. Geomat. Nat. Haz. Risk 1 (1), 19–35.

De Moel, H.D., Van Alphen, J., Aerts, J.C., 2009. Flood maps in Europe—methods, availability and use. Nat. Hazards Earth Syst. Sci. 9 (2), 289–301.

de Souza, V.A.S., Moreira, D.M., Rotunno Filho, O.C., Rudke, A.P., 2020. Extreme rainfall events in Amazonia: The Madeira river basin. Remote Sens. Appl. Soc. Environ. 18, 100316.

Devrani, R., Srivastava, P., Kumar, R., Kasana, P., 2022. Characterization and assessment of flood inundated areas of lower Brahmaputra River Basin using multitemporal Synthetic Aperture Radar data: a case study from NE India. Geol. J. 57 (2), 622–646.

DeVries, B., Huang, C., Armston, J., Huang, W., Jones, J.W., Lang, M.W., 2020. Rapid and robust monitoring of flood events using Sentinel-1 and Landsat data on the Google Earth Engine. Remote Sens. Environ. 240, 111664.

Dijkstra, H.A., Petersik, P., Hernández-García, E., López, C., 2019. The application of machine learning techniques to improve El Niño prediction skill. Front. Phys. 7, 153.

Eingrüber, N., Korres, W., 2022. Climate change simulation and trend analysis of extreme precipitation and floods in the mesoscale Rur catchment in western Germany until 2099 using Statistical Downscaling Model (SDSM) and the Soil & Water Assessment Tool (SWAT model). Sci. Total Environ. 838, 155775.

European Space Agency (ESA), 2021. Sentinel-1 and Sentinel-2 Working Together for Flood Mapping. Retrieved from: https://www.esa.int/Applications/Observing_the_Earth/Copernicus/Sentinel-1/Sentinel-1_and_Sentinel-2_working_together_for_flood_mapping.

Farhadi, H., Najafzadeh, M., 2021. Flood risk mapping by remote sensing data and random forest technique. Water 13 (21), 3115.

Feng, Q., Liu, J., Gong, J., 2015. Urban flood mapping based on unmanned aerial vehicle remote sensing and random forest classifier—a case of Yuyao, China. Water 7 (4), 1437–1455.

Ficchì, A., Stephens, L., 2019. Climate variability alters flood timing across Africa. Geophys. Res. Lett. 46 (15), 8809–8819.

Galvin, K.A., Boone, R.B., Smith, N.M., Lynn, S.J., 2001. Impacts of climate variability on East African pastoralists: linking social science and remote sensing. Clim. Res. 19, 161–172.

Gannon, K.E., Conway, D., Pardoe, J., Ndiyoi, M., Batisani, N., Odada, E., et al., 2018. Business experience of floods and drought-related water and electricity supply disruption in three cities in sub-Saharan Africa during the 2015/2016 El Niño. Glob. Sustain. 1, e14.

Ghumman, A.R., Alodah, A., Haider, H., Shafiquzzaman, M., 2020. Evaluating the impact of climate change on stream flow: integrating GCM, hydraulic modelling and functional data analysis. Arab. J. Geosci. 13, 1–15.

Giraldo Osorio, J.D., Gabriela García Galiano, S., 2012. Development of a sub-pixel analysis method applied to dynamic monitoring of floods. Int. J. Remote Sens. 33 (7), 2277–2295.

Gonzalez, G., Smith, L.C., Li, Y., 2020. Change detection analysis in flood impact assessment: a remote sensing approach. Nat. Hazards 98 (2), 625–647.

Gosling, S.N., Arnell, N.W., Best, M., Fung, F., Hattermann, F.F., Liu, X., et al., 2021. A global assessment of the impact of climate change on water resilience. Nat. Sustain. 4 (4), 263–273.

Gu, X., Zhang, Q., Li, J., Liu, J., Xu, C.Y., Sun, P., 2020. The changing nature and projection of floods across Australia. J. Hydrol. 584, 124703.

Guan, H., Wang, Q., Yuan, C., Wang, H., 2020. Improved remote sensing evapotranspiration for hydrological modeling in the Upper Mississippi River Basin. Remote Sens. (Basel) 12 (12), 1946.

Ha, J., Kang, J.E., 2022. Assessment of flood-risk areas using random forest techniques: Busan Metropolitan City. Nat. Hazards 1, 1–23.

Hansen, M.C., Loveland, T.R., 2012. A review of large area monitoring of land cover change using Landsat data. Remote Sens. Environ. 122, 66–74.

Haq, M., Akhtar, M., Muhammad, S., Paras, S., Rahmatullah, J., 2012. Techniques of remote sensing and GIS for flood monitoring and damage assessment: a case study of Sindh province, Pakistan. Egypt. J. Remote Sens. Space Sci. 15 (2), 135–141.

Hashemi-Beni, L., Gebrehiwot, A.A., 2021. Flood extent mapping: an integrated method using deep learning and region growing using UAV optical data. IEEE J. Sel. Top. Appl. Earth Obs. Remote Sens. 14, 2127–2135.

Hess, L.L., Melack, J.M., 2003. Remote sensing of vegetation and flooding on Magela Creek Floodplain (Northern Territory, Australia) with the SIR-C synthetic aperture radar. Hydrobiologia 500, 65–82.

Hoang, L., Kummu, M., Lauri, H., Koponen, J., 2020. Impacts of climate change and hydropower development on the hydrology of the Mekong River Basin. Sci. Total Environ. 717, 137164.

Hooshyaripor, F., Faraji-Ashkavar, S., Koohyian, F., Tang, Q., Noori, R., 2020. Annual flood damage influenced by El Niño in the Kan River basin, Iran. Nat. Hazards Earth Syst. Sci. 20 (10), 2739–2751.

Hou, J., Van Dijk, A.I., Beck, H.E., Renzullo, L.J., Wada, Y., 2022. Remotely sensed reservoir water storage dynamics (1984–2015) and the influence of climate variability and management at a global scale. Hydrol. Earth Syst. Sci. 26 (14), 3785–3803.

Irimescu, A., Stancalie, G.H., Craciunescu, V., Flueraru, C., Anderson, E., 2009. The use of remote sensing and GIS techniques in flood monitoring and damage assessment: a study case in Romania. In: Threats to Global Water Security. Springer, Netherlands, pp. 167–177.

Islam, A.R.M.T., Talukdar, S., Mahato, S., Kundu, S., Eibek, K.U., Pham, Q.B., et al., 2021. Flood susceptibility modelling using advanced ensemble machine learning models. Geosci. Front. 12 (3), 101075.

Jain, S.K., Singh, R.D., Jain, M.K., Lohani, A.K., 2005. Delineation of flood-prone areas using remote sensing techniques. Water Resour. Manage. 19, 333–347.

Jiménez-Jiménez, S.I., Ojeda-Bustamante, W., Ontiveros-Capurata, R.E., Marcial-Pablo, M.D.J., 2020. Rapid urban flood damage assessment using high resolution remote sensing data and an object-based approach. Geomat. Nat. Haz. Risk 11 (1), 906–927.

Jongman, B., Ward, P.J., Aerts, J.C., 2012. Global exposure to river and coastal flooding: long term trends and changes. Glob. Environ. Chang. 22 (4), 823–835.

Kalantar, B., Ueda, N., Saeidi, V., Janizadeh, S., Shabani, F., Ahmadi, K., Shabani, F., 2021. Deep neural network utilizing remote sensing datasets for flood hazard susceptibility mapping in Brisbane, Australia. Remote Sens. 13 (13), 2638.

Khanal, N., Matin, M.A., Uddin, K., Poortinga, A., Chishtie, F., Tenneson, K., Saah, D., 2020. A comparison of three temporal smoothing algorithms to improve land cover classification: a case study from NEPAL. Remote Sens. 12 (18), 2888.

Klein, I., Gessner, U., Dietz, A.J., Kuenzer, C., 2017. Global WaterPack – a 250 m resolution dataset revealing the daily dynamics of global inland water bodies. Remote Sens. Environ. 198, 345–362.

Komolafe, A.A., Adegboyega, S.A.A., Akinluyi, F.O., 2015. A review of flood risk analysis in Nigeria. Am. J. Environ. Sci. 11 (3), 157.

Kucharczyk, M., Hugenholtz, C.H., 2021. Remote sensing of natural hazard-related disasters with small drones: Global trends, biases, and research opportunities. Remote Sens. Environ. 264, 112577.

Kumar, D., Singh, V., Kumar, A., Roy, P.S., 2020. Synthetic aperture radar remote sensing for flood monitoring: a comprehensive review. J. Hydrol. 588, 125036.

Kundu, S., Aggarwal, S.P., Kingma, N., Mondal, A., Khare, D., 2015. Flood monitoring using microwave remote sensing in a part of Nuna river basin, Odisha, India. Nat. Hazards 76, 123–138.

Langhammer, J., 2023. Flood simulations using a sensor network and support vector machine model. Water 15 (11), 2004.

Langhammer, J., Janský, B., Kocum, J., Minařík, R., 2018. 3-D reconstruction of an abandoned montane reservoir using UAV photogrammetry, aerial LiDAR and field survey. Appl. Geogr. 98, 9–21.

Langhammer, J., Vacková, T., 2018. Detection and mapping of the geomorphic effects of flooding using UAV photogrammetry. Pure Appl. Geophys. 175, 3223–3245.

Li, X., Chen, Y., 2017. Integration of remote sensing data and hydrological modeling for flood risk assessment. J. Hydrol. 555, 520–532.

Li, L., Xu, T., Chen, Y., 2016. Improved urban flooding mapping from remote sensing images using generalized regression neural network-based super-resolution algorithm. Remote Sens. (Basel) 8 (8), 625.

Li, B., Hou, J., Li, D., Yang, D., Han, H., Bi, X., et al., 2021. Application of LiDAR UAV for high-resolution flood modelling. Water Resour. Manage. 35, 1433–1447.

Lin, L., Di, L., Yu, E.G., Kang, L., Shrestha, R., Rahman, M.S., et al., 2016. A review of remote sensing in flood assessment. In: 2016 Fifth International Conference on Agro-Geoinformatics (Agro-Geoinformatics). IEEE, pp. 1–4.

Liu, J., Xiong, J., Cheng, W., Li, Y., Cao, Y., He, Y., Duan, Y., He, W., Yang, G., 2021. Assessment of flood susceptibility using support vector machine in the belt and road region. Nat. Hazards Earth Syst. Sci. Discuss., 1–37.

Loo, Y.Y., Billa, L., Singh, A., 2015. Effect of climate change on seasonal monsoon in Asia and its impact on the variability of monsoon rainfall in Southeast Asia. Geosci. Front. 6 (6), 817–823.

Marengo, J.A., Tomasella, J., Soares, W.R., Alves, L.M., Nobre, C.A., 2012. Extreme climatic events in the Amazon basin: climatological and hydrological context of recent floods. Theor. Appl. Climatol. 107 (1–2), 73–85.

Marren, P.M., Woods, K.L., Abesser, C., Nützmann, G., Hill, M.C., Blöschl, G., Lakshmanan, E., 2011. Inundation of anabranching river flood plain wetlands: the Ovens River, Victoria, Australia. In: Conceptual and ModelIng Studies of Integrated GroundWater, Surface Water and Ecological Systems. IAHS Press, Wallingford, pp. 229–234.

Mashao, F.M., Mothapo, M.C., Munyai, R.B., Letsoalo, J.M., Mbokodo, I.L., Muofhe, T.P., et al., 2023. Extreme rainfall and flood risk prediction over the East Coast of South Africa. Water 15 (1), 50.

Mazhar, S., Sun, G., Wang, Z., Liang, H., Zhang, H., Li, Y., 2021. Flood mapping and classification jointly using MuWI and machine learning techniques. In: 2021 International Conference on Control, Automation and Information Sciences (ICCAIS). IEEE, pp. 662–667.

Mazza, E., Chen, S.S., 2023. Tropical cyclone rainfall climatology, extremes, and flooding potential from remote sensing and reanalysis datasets over the continental United States. J. Hydrometeorol. 24 (9), 1549–1562.

Mendoza, P.A., Farias, D., 2020. Integrating remote sensing and hydraulic modeling for improved early warning systems: a case study in Colombia. Nat. Hazards 104 (1), 671–689.

Mohammadi, A., Costelloe, J.F., Ryu, D., 2017. Application of time series of remotely sensed normalized difference water, vegetation and moisture indices in characterizing flood dynamics of large-scale arid zone floodplains. Remote Sens. Environ. 190, 70–82.

Moura, M.M., Dos Santos, A.R., Pezzopane, J.E.M., Alexandre, R.S., da Silva, S.F., Pimentel, S.M., et al., 2019. Relation of El Niño and La Niña phenomena to precipitation, evapotranspiration and temperature in the Amazon basin. Sci. Total Environ. 651, 1639–1651.

Mourad, K.A., Nordin, L., Andersson-Sköld, Y., 2022. Assessing flooding and possible adaptation measures using remote sensing data and hydrological modeling in Sweden. Clim. Risk Manage. 38, 100464.

Mouratidis, A., Sarti, F., 2012. Flash-flood monitoring and damage assessment with SAR data: issues and future challenges for earth observation from space sustained by case studies from the Balkans and Eastern Europe. In: Earth Observation of Global Changes (EOGC). Springer Berlin Heidelberg, Berlin, Heidelberg, pp. 125–136.

Muis, S., Haigh, I.D., Guimarães Nobre, G., Aerts, J.C., Ward, P.J., 2018. Influence of El Niño-Southern Oscillation on global coastal flooding. Earth's Future 6 (9), 1311–1322.

Munawar, H.S., Hammad, A.W., Waller, S.T., 2022. Remote sensing methods for flood prediction: a review. Sensors 22 (3), 960.

Munoz, S.E., Dee, S.G., 2017. El Niño increases the risk of lower Mississippi River flooding. Sci. Rep. 7 (1), 1772.

Na, X., Zang, S., Zhang, Y., Liu, L., 2013. Wetland mapping and flood extent monitoring using optical and radar remotely sensed data and ancillary topographical data in the Zhalong National Natural Reserve, China. In: Earth Resources and Environmental Remote Sensing/GIS Applications IV. vol. 8893. SPIE, pp. 390–396.

Nikam, B.R., Garg, V., Gupta, P.K., Thakur, P.K., Kumar, A.S., Chouksey, A., Aggarwal, S.P., Dhote, P., Purohit, S., 2017. Satellite-based mapping and monitoring of heavy snowfall in North Western Himalaya and its hydrologic consequences. Curr. Sci. 113 (12), 2328–2334.

Notti, D., Giordan, D., Calò, F., Pepe, A., Zucca, F., Galve, J.P., 2019. Low-cost flood mapping using free satellite data from Sentinels constellation. In: EGU General Assembly Conference Abstracts. EGU, p. 14773.

Pandey, P., Chauhan, P., Bhatt, C.M., Thakur, P.K., Kannaujia, S., Dhote, P.R., et al., 2021. Cause and process mechanism of rockslide triggered flood event in Rishiganga and Dhauliganga River Valleys, Chamoli, Uttarakhand, India using satellite remote sensing and in situ observations. J. Indian Soc. Remote Sens. 49, 1011–1024.

Qiang, Y., Lam, N.S., Cai, H., Zou, L., 2017. Changes in exposure to flood hazards in the United States. Ann. Am. Assoc. Geogr. 107 (6), 1332–1350.

Rahaman, S.A., Rahman, M.A., Ameen, M.A., 2021. Real-time flood forecasting and early warning system using satellite-based rainfall estimates: a case study from Bangladesh. J. Flood Risk Manage. 14 (1), e12588.

Rahman, M.S., Di, L., 2020. A systematic review on case studies of remote-sensing-based flood crop loss assessment. Agriculture 10 (4), 131.

Rodriguez, M.N., Lee, S.J., 2022. Remote sensing advances in climate variability assessment: a comprehensive review. Remote Sens. Rev. 43 (2), 128–150.

Sadiq, R., Akhtar, Z., Imran, M., Ofli, F., 2022. Integrating remote sensing and social sensing for flood mapping. Remote Sens. Appl.: Soc. Environ. 25, 100697.

Sajjad, A., Lu, J., Chen, X., Chisenga, C., Saleem, N., Hassan, H., 2020. Operational monitoring and damage assessment of riverine flood-2014 in the lower Chenab plain, Punjab, Pakistan, using remote sensing and GIS techniques. Remote Sens. (Basel) 12 (4), 714.

Sajjad, A., Lu, J., Chen, X., Chisenga, C., Mazhar, N., Nadeem, B., 2022. Riverine flood mapping and impact assessment using remote sensing technique: a case study of Chenab flood-2014 in Multan district, Punjab, Pakistan. Nat. Hazards, 1–20.

Saha, A., Salim, S.M., Sudheesan, D., Suresh, V.R., Nag, S.K., Panikkar, P., Paul, T.T., Palaniswamy, R., Das, B.K., Chowdhury, A.R., 2020. Geochemistry, mineralogy and nutrient concentrations of sediment of River Pampa in India during a massive flood event. Arab. J. Geosci. 13, 1–18.

Sansare, D.A., Mhaske, S.Y., 2020. Natural hazard assessment and mapping using remote sensing and QGIS tools for Mumbai city, India. Nat. Hazards 100, 1117–1136.

Santos, L.B., Carvalho, T., Anderson, L.O., Rudorff, C.M., Marchezini, V., Londe, L.R., Saito, S.M., 2017. An RS-GIS-based comprehensive impact assessment of floods—a case study in Madeira River, Western Brazilian Amazon. IEEE Geosci. Remote Sens. Lett. 14 (9), 1614–1617.

Schöngart, J., Junk, W.J., 2007. Forecasting the flood-pulse in Central Amazonia by ENSO-indices. J. Hydrol. 335 (1–2), 124–132.

Shahabi, H., Shirzadi, A., Ghaderi, K., Omidvar, E., Al-Ansari, N., Clague, J.J., et al., 2020. Flood detection and susceptibility mapping using sentinel-1 remote sensing data and a machine learning approach: hybrid intelligence of bagging ensemble based on k-nearest neighbor classifier. Remote Sens. (Basel) 12 (2), 266.

Sharma, V.K., Azad, R.K., Chowdary, V.M., Jha, C.S., 2022. Delineation of frequently flooded areas using remote sensing: a case study in part of Indo-Gangetic basin. In: Geospatial Technologies for Land and Water Resources Management. Springer International Publishing, Cham, Switzerland, pp. 505–530.

Skakun, S., Kussul, N., Shelestov, A., Kussul, O., 2014. Flood hazard and flood risk assessment using a time series of satellite images: a case study in Namibia. Risk Anal. 34 (8), 1521–1537.

Smith, J.K., Johnson, C.D., 2021. Atmospheric remote sensing and climate change: implications for flooding patterns. Climate Dynam. 56 (8), 3549–3571.

Sowmya, K., John, C.M., Shrivasthava, N.K., 2015. Urban flood vulnerability zoning of Cochin City, southwest coast of India, using remote sensing and GIS. Nat. Hazards 75, 1271–1286.

Sun, Y., Liu, C., Du, X., Yang, F., Yao, Y., Soomro, S., e. h., & Hu, C., 2022. Urban storm flood simulation using improved SWMM based on K-means clustering of parameter samples. J. Flood Risk Manage. 15 (4), e12826.

Tang, W., Carey, S., 2021. Classifying Annual Daily Hydrographs in Western North America Using t-Distributed Stochastic Neighbor Embedding (t-SNE). Authorea Preprints.

Tella, A., Mustafa, M.R.U., Balogun, A.O., Okolie, C.J., Bello Yamusa, I., Ibrahim, M., 2023. Spatial prediction of flood in Kuala Lumpur City of Malaysia using logistic regression. Int. Arch. Photogramm. Remote Sens. Spat. Inf. Sci. 48, 363–369.

Thanh Son, N., Thi Thu Trang, N., Bui, X.T., Thi Da, C., 2022. Remote sensing and GIS for urbanization and flood risk assessment in Phnom Penh, Cambodia. Geocarto Int. 37 (22), 6625–6642.

Ticehurst, C., Dutta, D., Karim, F., Petheram, C., Guerschman, J.P., 2015. Improving the accuracy of daily MODIS OWL flood inundation mapping using hydrodynamic modelling. Nat. Hazards 78, 803–820.

Ticehurst, C., Guerschman, J.P., Chen, Y., 2014. The strengths and limitations in using the daily MODIS open water likelihood algorithm for identifying flood events. Remote Sens. 6 (12), 11791–11809.

Tulbure, M.G., Broich, M., Perin, V., Gaines, M., Ju, J., Stehman, S.V., et al., 2022. Can we detect more ephemeral floods with higher density harmonized Landsat Sentinel 2 data compared to Landsat 8 alone? ISPRS J. Photogramm. Remote Sens. 185, 232–246.

Turner, R.J., Mansour, M.M., Dearden, R., O Dochartaigh, B.E., Hughes, A.G., 2015. Improved understanding of groundwater flow in complex superficial deposits using three-dimensional geological-framework and groundwater models: an example from Glasgow, Scotland (UK). Hydrogeol. J. 23 (3), 493–506.

Twumasi, Y.A., Merem, E.C., Ayala-Silva, T., Osei, A., Petja, B.M., Alexander, K., 2017. Techniques of remote sensing and GIS as tools for visualizing impact of climate change-induced flood in the Southern African region. Am. J. Clim. Change 6 (2), 306–327.

Twumasi, Y.A., Merem, E.C., Namwamba, J.B., Asare-Ansah, A.B., Annan, J.B., Ning, Z.H., et al., 2022. Flood mapping in Mozambique using Copernicus Sentinel-2 satellite data. Adv. Remote Sens. 11 (3), 80.

Uddin, K., Gurung, D.R., Giriraj, A., Shrestha, B., 2013. Application of remote sensing and GIS for flood hazard management: a case study from Sindh Province, Pakistan. Am. J. Geogr. Inf. Sys. 2 (1), 1–5.

Wang, S.X., Liu, Y.L., Yi, Z., Wei, C.J., 2002. Study on the method of establishment of normal water extent database for flood monitoring using remote sensing. In: IEEE International Geoscience and Remote Sensing Symposium. vol. 4, pp. 2048–2050.

Ward, P.J., Jongman, B., Kummu, M., Dettinger, M.D., Sperna Weiland, F.C., Winsemius, H.C., 2014. Strong influence of El Niño Southern Oscillation on flood risk around the world. Proc. Natl. Acad. Sci. U. S. A. 111, 15659–15664.

Wei, H., Yu, T., Tu, J., Ke, F., 2023. Detection and evaluation of flood inundation using CYGNSS data during extreme precipitation in 2022 in Guangdong Province, China. Remote Sens. 15 (2), 297.

Woods, D., Kirstetter, P.E., Vergara, H., Duarte, J.A., Basara, J., 2023. Hydrologic evaluation of the global precipitation measurement mission over the U.S.: flood peak discharge and duration. J. Hydrol. 617, 129124.

Wu, T., 2021. Quantifying coastal flood vulnerability for climate adaptation policy using principal component analysis. Ecol. Indic. 129, 108006.

Xie, J., Xu, Y.P., Wang, Y., Gu, H., Wang, F., Pan, S., 2019. Influences of climatic variability and human activities on terrestrial water storage variations across the Yellow River basin in the recent decade. J. Hydrol. 579, 124218.

Zeng, Z., Gan, Y., Kettner, A.J., Yang, Q., Zeng, C., Brakenridge, G.R., Hong, Y., 2020. Towards high resolution flood monitoring: an integrated methodology using passive microwave brightness temperatures and Sentinel synthetic aperture radar imagery. J. Hydrol. 582, 124377.

Zha, X., Huang, C., Pang, J., Liu, J., Xue, X., 2015. Reconstructing the palaeoflood events from slackwater deposits in the upper reaches of Hanjiang River, China. Quat. Int. 380, 358–367.

Zhang, J., Zhou, C., Xu, K., Watanabe, M., 2002. Flood disaster monitoring and evaluation in China. Global Environ. Change B. Environ. Hazard 4 (2), 33–43.

Zhou, Y., Lau, W.K., Huffman, G.J., 2015. Mapping TRMM TMPA into average recurrence interval for monitoring extreme precipitation events. J. Appl. Meteorol. Climatol. 54 (5), 979–995.

Zurqani, H.A., Al-Bukhari, A., Aldaikh, A.O., Elfadli, K.I., Bataw, A.A., 2022. Geospatial mapping and analysis of the 2019 Flood Disaster extent and impact in the city of Ghat in Southwestern Libya using Google Earth engine and deep learning technique. In: Environmental Applications of Remote Sensing and GIS in Libya. Springer International Publishing, Cham, pp. 205–226.

Chapter 10

Climate variability and drought

Samuel Kusangaya[a] and Munyaradzi Davis Shekede[a,b]
[a]*Department of Geography Geospatial Sciences and Earth Observation, University of Zimbabwe, Harare, Zimbabwe,* [b]*Department of Geospatial Sciences and Earth Observation, Zimbabwe National Geospatial and Space Agency, Harare, Zimbabwe*

Chapter outline

1 Introduction

Climate variability denotes natural fluctuations in weather patterns over different time scales, ranging from short-term fluctuations to long-term changes. Drought, on the contrary, refers to a prolonged period of below-normal precipitation leading to water shortages which, in turn, significantly impacts agriculture, ecosystems, and human populations. It is, therefore, a natural but temporary imbalance of water availability, consisting of persistent lower-than-average precipitation, of uncertain frequency, duration, and severity of unpredictable occurrence, resulting in diminished water resources availability and reduced carrying capacity of the ecosystems (Kunedzimwe et al., 2021). Climate variability and drought are closely related as changes in climate patterns influence the frequency, intensity, and duration of drought events.

Remote Sensing of Climate. https://doi.org/10.1016/B978-0-443-21731-9.00011-9
Copyright © 2024 Elsevier Inc. All rights are reserved, including those for text and data mining, AI training, and similar technologies.

Generally, drought occurrence is linked to periods when there is an imbalance between water demand and supply. The shortage of water, windy conditions, increasing temperatures, and consequent evapotranspiration all contribute to the magnitude and severity of drought conditions. Consequently, drought results in crop and general plant failure and has a major impact on agriculture, the economy, and the ecosystems of the affected areas. Furthermore, droughts often become highly visible when they are associated, rightly or wrongly, with famine (Sharara et al., 2022; Shekede et al., 2023). Thus, drought is regarded as one of the most harmful natural events with concomitant effects on economic and social systems (Van Loon, 2015; Wilhite, 2005; Wilhite and Glantz, 1985). Fig. 1 illustrates locations that have been affected by droughts across the globe historically (e.g., the drought of 1276–1299 in North America and the Millennium drought of 2001–2009 in Australia) or on an on-going basis as the case in central and northern Africa.

Drought is a recurring climate phenomenon whose frequency, impact, and intensity have increased over the years due to climate change (Mukherjee et al., 2018). The most popular perception of a drought is a meteorological

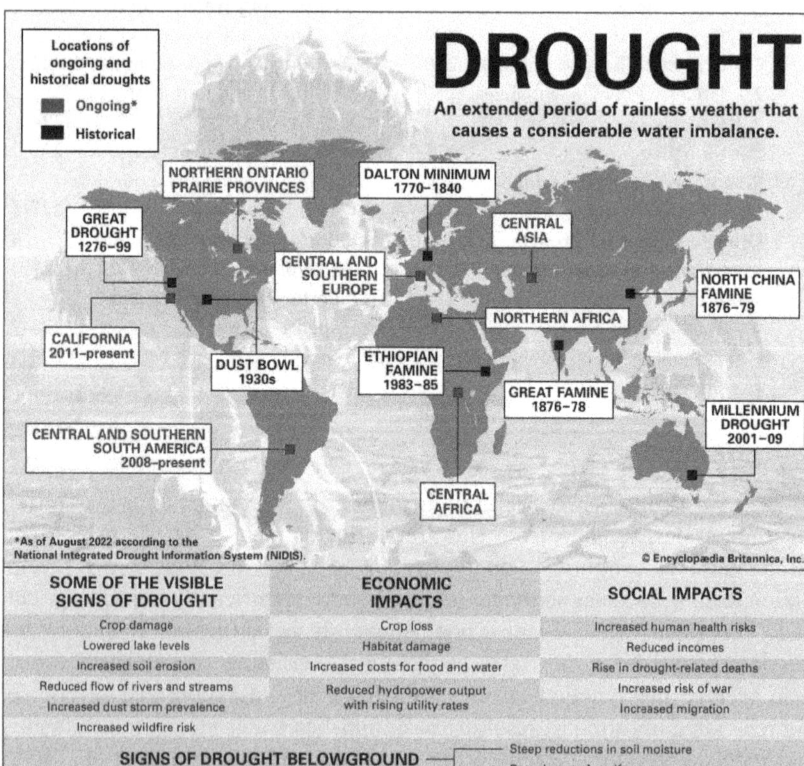

FIG. 1 Locations of significant present-day droughts and some major drought episodes throughout history. https://cdn.britannica.com/52/190352-050-F1D2DF24/Infographic-droughts-locations-drought-episodes-history.jpg.

phenomenon characterized by a less-than-expected average amount for a given period. Drought has been defined using the percentile of rainfall in relation to the long-term average. For instance, studies use thresholds of between 50% and 75% of the long-term average to define a drought (Castano, 2013; Bagheri, 2016). Wilhite and Glantz (1985) posit that drought can be defined from either a conceptual or operational framework. The conceptual approach focuses on describing droughts based on the physical processes related to precipitation, limited soil moisture, shortage of water in lakes and streams, and limited supply of water to society relative to demand. Thus through this approach, the following types of droughts have been identified: (1) hydrological, (2) meteorological, (3) agricultural, and (4) socioeconomic drought. Meteorological drought happens when dry weather patterns dominate an area. Hydrological drought occurs when low water supply becomes evident, especially in streams, reservoirs, and groundwater levels, usually after many months of meteorological drought. According to the National Centre of Environmental Information, meteorological drought can begin and end rapidly, while hydrological drought takes much longer to develop and then recover. Agricultural drought happens when crops are affected, and lastly, socioeconomic drought relates the supply and demand of various commodities to drought.

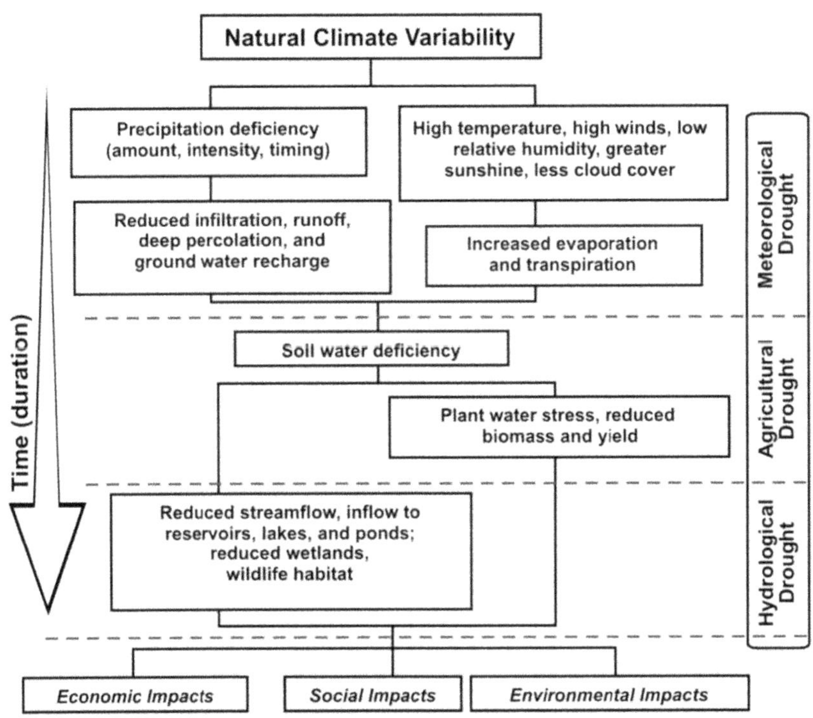

Drought characterization. National Drought Mitigation Center.

The operational definition of drought emphasizes detecting drought parameters, such as the onset, duration, termination, and even its severity (Mukherjee et al., 2018). Whichever approach is adopted, the common thread is that all drought types emanate from precipitation deficiency, which induces other types of drought depending on the duration of the drought episode. Droughts besides the above definitions or categories can also be characterized on a temporal scale. In this regard, some droughts can be considered regular and predictable as observed in subtropical regions characterized by a pronounced rainy season followed by a dry season. Thus each year, these seasonal droughts are a regular, normal feature of climate. On the contrary, droughts can be considered as the extreme end, are sporadic, and are oftentimes prolonged and acute (Tesfamariam et al., 2019, 2022). The following section adapted from Belayneh, (2012) lists and briefly describes the main forms of drought according to time-scale:

1.1 Seasonal drought

Seasonal drought occurs due to large-scale seasonal circulations within areas with well-defined dry and wet climates, such as subtropical or monsoon climates. Within sub-Saharan Africa, regular droughts are within these areas, with the bulk of the rain falling in season, however, within the summer followed by a virtually dry winter. Conversely, the Mediterranean climate is characterized by a rainy winter season and relatively warm and dry summers.

1.2 Irregular drought

Irregular droughts result from rainfall unreliability and are synonymous with. This type of drought is characterized by intermittent, erratic, sporadic, and unpredictable rainfall. This irregularity emanates from unreliable rainfall and is among the most damaging and complex droughts to manage. Despite the irregularity of this type of drought, it is still a normal and an integral part of a region's climate and often occurs in subhumid regions, including arid and semi-arid regions, grasslands, savannahs, and Mediterranean landscapes, encompassing approximately 47% of the Earth's terrestrial area (Wilhite et al., 2014).

1.3 Partial drought

Partial droughts are brief and irregular and involve an abnormal rainfall failure which can and may occur almost anywhere but is mostly experienced and is prevalent in humid and subhumid climates. Although partial droughts are brief and irregular, they often affect a relatively small area. This form of drought is insidious and damaging and hence can be difficult to detect (Wilhite, 2017). Partial droughts are characterized by the soil moisture deficit emanating from the drought and are not sufficiently severe to cause the symptoms of drought, but are nevertheless sufficient to reduce the uptake of nutrients from the soil. Thus, crop growth and yield will usually be below the optimum levels.

1.4 Permanent drought

Most droughts have a specified duration and will come to an end. However, when conditions do not return to normal, the drought becomes permanent (White, 2010; Wilhite, 2017; Wilhite and Pulwarty, 2017b). This results in areas experiencing drought over a long period, resulting in those affected areas turning into deserts.

2 Climate variability and drought

Assessments of the Intergovernmental Panel on Climate Change (IPCC) unequivocally demonstrated that the Earth's climate is changing due to increasing greenhouse gases (GHGs) (Gebrechorkos et al., 2020). It is projected that the future will likely result in being characterized by higher sea levels, including the occurrence of intense storms and heavy rainfall events, as well as the expansion of drought areas (McBean and Ajibade, 2009). These climate hazards are already occurring and are causing loss of life, social disruption, and economic hardships (Hulme et al., 2001; Desanker et al., 2005). Such hardships are already being heavily felt by the poor in most African countries (Magadza, 1996; Simms, 2015). Climate has limited adaptive capacity while there are diverse views regarding the magnitude and direction of change of climatic characteristics and the possible impacts at the scale at which land and water resources are managed. Several scholars have, however, noted that there is a certainty that climate change will thus affect the availability and use of water resources (Desanker et al., 2005; Matondo et al., 2005; Schulze, 2000; Schulze and Bulcock, 2011; Yamba et al., 2011).

Globally, approximately 4 billion people—more than half of the people on Earth—experience severe water scarcity during at least part of the year (Zhao et al., 2022). The availability of fresh water is likely to be further compromised with the expected increase in pressures from land degradation, pollution, population growth, and urbanization (Kusangaya et al., 2014; Gwatida et al., 2023). In southern Africa, climate change has the potential to impose additional pressures (Magadza, 1996; Battersby and Watson, 2018). Kusangaya et al. (2021) showed that 14 African countries are already experiencing water stress, while 11 others will be stressed by the year 2025, likely to be exacerbated by climate change. By 2020, an additional 75–250 million people in Africa are projected to be exposed to increased water stress (Gebrechorkos et al., 2020). According to Vörösmarty et al. (2005), water stress is already high for 25% of Africa's population, with a further 13% experiencing water stress from recurrent droughts. For the Southern Africa Development Community (SADC) countries, the

Falkenmark indicator of water stress (Falkenmark, 1989) shows that Lesotho, Malawi, South Africa, and Zimbabwe are already water stressed. Based on projected population growth from 2010 to 2050 by the United Nations Department of Economic and Social Affairs, Population Division (Yanda et al., 2010) predicts that Tanzania will fall within the moderate water stress category by 2020, and that by the year 2050, Malawi, South Africa, Tanzania, and Zimbabwe will be in the high stress category, all even without accounting for climate change. Thus, climate change will increase pressure on water resources which are already stressed. The population at risk of increased water stress in Africa is estimated to increase to 350–600 million people relative to the 1960–1990 baseline, up to 10% of the SADC population, by the 2050s (IPCC, 2023). The water sector and related sectors will be required to adapt to scarce water resources, which can be achieved through conservation, improved efficiency, technological improvements, and adoption of integrated water resource management (Matondo et al., 2005; Schulze and Bulcock, 2011; Alemaw and Matondo, 2020). Long-term trends analysis of atmospheric variables, such as temperature, rainfall, and evapotranspiration, has been used extensively as proxies for detecting changes in the climate (e.g., rainfall by Dore, 2005; Warburton et al., 2012), temperature (by Kruger, 2006; Wilk et al., 2013; Ziervogel et al., 2014), and evaporation (and soil moisture by Malisawa and De Rautenbach, 2012). The increased frequency of occurrence of extreme events, especially droughts, has been attributed to the changing climate. Most scientists (e.g., Nicholson, 2001; Rosenzweig et al., 2001; Mirza, 2003; Warburton et al., 2012) are of the view that the increased frequency of extreme events may be attributable to increasing GHG emissions.

2.1 Drivers of climate variability

In general, climate variability can increase the risk of drought, as well as the severity of drought events. This is because changes in weather patterns can lead to a reduction in rainfall, which, in turn, can lead to drought conditions. There is also evidence that climate variability can increase the duration and frequency of droughts, as well as the severity of these events. Droughts are mostly associated with the El Niño Southern Oscillation (ENSO) (Bhaga et al., 2020), which is an irregularly periodic variation in winds and sea surface temperatures (SSTs) over the tropical eastern Pacific Ocean, affecting the climate of much of the tropics and subtropics (Wang et al., 2018). ENSO can change the global atmospheric circulation, which, in turn, influences temperature and precipitation across the globe, including drought occurrence and frequency. Rainfall patterns are closely linked to atmospheric circulation patterns, and when these patterns change, it can lead to a decrease in rainfall. The warming phase of the sea temperature is known as El Niño and the cooling phase as La Niña. The Southern Oscillation is the accompanying atmospheric component, coupled with the sea temperature change. El Niño is accompanied by high air surface pressure in the tropical western Pacific and La Niña with low air surface pressure. In general, the warmer the ocean temperature anomalies, the stronger the El Niño (and vice versa). On the contrary, La Niña, which is the cooling of the ocean surface, is accompanied by below-average SSTs in the central and eastern tropical Pacific Ocean (Manatsa et al., 2011). In Indonesia, rainfall tends to increase while rainfall decreases over the central and eastern tropical Pacific Ocean. The normal easterly winds along the equator become even stronger. In general, the cooler the ocean temperature anomalies, the stronger the La Niña (and vice versa) (Bhaga et al., 2020). ENSO's climatic effects are global, such as droughts in Australia, Indonesia, and southern Africa, stormy weather in part of North America, and a weakening of the Indian monsoon. The risk of drought in southern Africa increases by 120% in an El Niño year.

Conversely, during El Niño events, drier and hotter weather occurs in parts of the Amazon River Basin, Colombia, and Central America. Winters during the El Niño are warmer and drier than average conditions in the Northwest, northern Midwest, and northern Mideast United States, so those regions experience reduced snowfalls. Conditions are also drier than normal from December to February in south-central Africa, mainly in Zambia, Zimbabwe, Mozambique, and Botswana. Direct effects of El Niño resulting in drier conditions occur in parts of Southeast Asia and Northern Australia, increasing bushfires, worsening haze, and decreasing air quality dramatically. Drier-than-normal conditions are also, in general, observed in Queensland, inland Victoria, inland New South Wales, and eastern Tasmania from June to August. As warm water spreads from the West Pacific and the Indian Ocean to the East Pacific, it causes extensive drought in the western Pacific. Singapore experienced the driest February in 2014 since records began in 1869, with only 6.3 mm of rain falling in the month and temperatures hitting as high as 35°C on 26 February. The years 1968 and 2005 had the next driest February when 8.4 mm of rain fell.

Climate change can increase disaster risk in a variety of ways by altering the frequency and intensity of hazard events, affecting vulnerability to hazards, and changing exposure patterns. Climate change is already modifying the frequency and intensity of many weather-related hazards as well as steadily increasing the vulnerability and eroding the resilience of exposed populations that depend on arable land, access to water, and stable mean temperatures and rainfall. Climate change could also lead to changes in the geographic distribution of weather-related hazards, which may lead to new patterns of risk.

2.2 Concept of drought and drought types

Drought is a complex phenomenon whose dimensions are still poorly understood (Frischen et al., 2020). Climate variability plays a significant role in the occurrence and intensity of droughts. Droughts are often influenced by natural climate phenomena that lead to irregular patterns in temperature and precipitation (Kusangaya et al., 2014). While climate variability refers to natural fluctuations, human-induced climate change is altering long-term climate patterns as well (Henchiri et al., 2021). Rising global temperatures, changes in precipitation patterns, and increased frequency of extreme weather events can influence drought occurrence and severity (Bhaga et al., 2020; Han et al., 2023). Climate change may lead to more intense and prolonged droughts in some regions, disrupting ecosystems, agriculture, and water supplies (Kusangaya et al., 2014). However, all droughts emanate from a precipitation deficiency over a specified period, resulting in an imbalance between water supply and societal or environmental water demands (Wilhite, 2005, 2017).

Due to the complex nature of drought, four main categories, as stated before, are used to understand drought occurrence, and these are meteorological, hydrological, agricultural, and socioeconomic drought (Wilhite, 2016). The operational definition of drought thus depends on the component of the biophysical or socioeconomic system affected (Winkler et al., 2017). Since droughts are triggered by precipitation deficiency (Wilhite, 2005), four dimensions, i.e., duration, geographical extent, intensity, and severity, are used to determine whether and to what extent water supply fails to meet the demand (Rossi, 2003). Although the four dimensions are critical for characterizing droughts, intensity is regarded as the most important (Tsakiris et al., 2007). Drought intensity is the ratio of drought severity divided by duration (Rossi et al., 1992). It is measured using a variety of indices, but the most commonly used one is the Palmer Drought Severity Index (PDSI) (Nagarajan, 2009; Winkler et al., 2018). The PDSI takes into account a number of factors, including precipitation, temperature, and soil moisture, to calculate a drought intensity value (Jacobi et al., 2013; Bogachev et al., 2023). PDSI ranges from −4 (extreme drought) to +4 (extreme wetness), with 0 being the baseline value (Jacobi et al., 2013). Drought intensity values can also be used to determine the impacts of a drought, with more severe droughts having more severe impacts. The severity of a drought depends on the duration, season of occurrence, spatial extent of the affected area as well as the impacts on socioeconomic activities and duration of the drought and the area affected (Wilhite, 2016). The severity of a drought is also determined by the social and economic impacts of drought. For example, a drought that lasts for a long time and affects a large area will have a more severe impact than a short-term drought that affects a small area (Masroor et al., 2020). In addition, the type of crops grown and the level of preparedness in a region can also affect the severity of a drought.

Globally, droughts have resulted in significant loss of life. For example, about 11 million deaths have been reported in the past, while approximately 2 billion people have been directly affected by drought-induced food shortages (Shabelle, 2011). The number of drought-related deaths is highest in the African continent, which has experienced 800,000 deaths since 1900, with about 262 million directly affected. The worst drought in Africa occurred during the 1991/1992 rainfall season when southern Africa was worst hit (Sharara et al., 2022). The 1992 drought resulted in countries such as Zimbabwe experiencing about 11% shrinkage in gross domestic product and food shortages (Maphosa,

1994; Sharara et al., 2022). In Zimbabwe, the 1991/1992 drought resulted in more than 1 million cattle deaths, and 6 million people were directly affected (Maphosa, 1994). Besides, recent droughts in Zimbabwe have increased the country's food insecurity where in 2019 alone, about 3.5 million people were considered food insecure by the United Nations World Food Programme (Sharara et al., 2022). In addition, the high frequency of droughts directly threatens agricultural-dependent economies and livelihoods and impedes the attainment of the United Nations Sustainable Development Goals, particularly, goals number one, two, and three, that is, no poverty, zero hunger, and good health and wellbeing, respectively (Frischen et al., 2020). In this regard, it is important to monitor droughts routinely and timeously to understand their spatial and temporal evolution as it is critical for identifying areas at risk and developing early warnings and disaster preparedness strategies (Magadza, 1996; Sharara et al., 2022).

Extreme hydrological events are increasing in frequency and magnitude (Nicholson, 2001; Reason and Keibel, 2004; Yanda et al., 2010; Harding et al., 2011) and the ENSO are becoming more intense as a result of climate change (Bunyavanich et al., 2003; Collins, 2011; Manatsa et al., 2011). For example, the frequency of droughts is projected to increase in southern Africa, most likely increasing the frequency of extremely low flows and low depletion of storage (Desanker and Justice, 2001; Desanker et al., 2005). These will inevitably affect aquatic ecosystems, water supply, irrigation, leisure, and hydropower generation. Field et al. (2012) showed that globally (a) Africa has the second highest number of extreme climate events after Asia, and (b) Africa has the highest number of hydrological extreme events. A significant number of these extreme events, hydrological (e.g., floods) and climatological (e.g., droughts) events, are occurring in southern Africa. Furthermore, southern African countries have undergone an increase of (a) occurrence of natural disasters in the last two decades (1991–2000 and 2001–2010) that is by far higher than in previous decades, with the highest occurrence rate in 2001–2010, and (b) occurrence of these water-related disasters, which has been increasing at an exponential rate. The number of natural disasters in the last decade is in most cases equivalent to the total number of disasters in the previous three decades. It is thus conclusive that there is an increasing trend of droughts and flood events in southern Africa (Kabat et al., 2002; Ahmed et al., 2018).

The frequency and magnitude of droughts can vary widely depending on geographic location, climate patterns, and local environmental conditions. In addition, the frequency and magnitude of droughts will continue to be affected by climate change as a result of changes in the timing, intensity, and frequency of rainfall. Climate change can influence the frequency of droughts since changes in climate patterns can lead to shifts in precipitation patterns, resulting in more or less frequent drought occurrences in different areas. In addition, human activities such as land use changes, deforestation, and water management practices can affect local and regional water cycles, potentially exacerbating drought conditions (Vörösmarty et al., 2005). Drought events cause considerable damage to infrastructure. Droughts also cause indirect adverse effects, such as epidemics (Kundzewicz et al., 2008; Knox et al., 2010). In addition to the direct impacts of drought on the environment, there are also a number of indirect adverse effects, including health impacts, economic impacts, and social impacts (Nhemachena and Mano, 2007; Orimoloye et al., 2022). In terms of health impacts, drought can lead to a number of issues, such as an increase in respiratory illnesses, heat-related illnesses, and mental health problems (Shabelle, 2011). The economic impacts of drought can be significant, as it can lead to a loss of agricultural productivity, decreased income, and increased food prices (Nhemachena and Mano, 2007). Social impacts can include displacement, food insecurity, and conflict (Ahmad et al., 2022). Droughts in southern Africa frequently affect Malawi, Mozambique, and Zimbabwe. Recent drought events have caused over a thousand deaths and loss of livestock, particularly for rural communities and poor urban households. Drought's occurrence in Africa has increased in frequency in recent years, causing massive damage and loss of life in Zimbabwe, Zambia, Malawi, Botswana, and Angola (Reason and Keibel, 2004; Reason, 2007). Tadross et al. (2005) and New et al. (2006) show evidence of changing rainfall seasonality and extreme events. These observations ultimately signify a significant shift in climate variables from the recent past.

Droughts cause food insecurity and considerable damage to economic production (Kabat et al., 2002). Kusangaya et al. (2021) pointed out that the six warmest years on record for Zimbabwe have occurred since 1987, with corresponding increased frequency of droughts in southern Africa since 1990, causing massive drops in crop yields in Zimbabwe, Malawi, and Zambia. A study by Kuri et al. (2020) revealed differential impacts of drought across the season with drought periods occurring at the vegetative stage having more impacts on crop yield than at the senescence stage. The increase in frequency and severity of droughts are related to climate change in most parts of the world, particularly in southern Africa (Kusangaya et al., 2014). In fact, studies have indicated that most parts of southern Africa will experience a decrease in rainfall and an increase in temperature, which is likely to lead to high soil-moisture deficiency (Hulme et al., 2001; Kabat et al., 2002; Kusangaya et al., 2014, 2021; Kunedzimwe et al., 2021; Sharara et al., 2022). With the advent of climate change, which is likely to see droughts becoming more frequent, there is a need to utilize all available tools to enhance knowledge on drought occurrence. Remotely sensed data thus becomes one of the key primary data sources as it has a large footprint with high temporal resolution and a long history of archival data, spanning over 40 years (Bhaga et al., 2020).

3 Remote sensing of drought: Drought monitoring

Historically, droughts have been monitored and investigated using ground-based point observations or interpolated grids (Hayes et al., 2011), primarily from meteorological (Hayes et al., 2005; Tesfamariam et al., 2019) and agricultural perspectives (Maphosa, 1994; Ojara et al., 2022). Globally, however, many areas still do not have weather measuring stations (e.g., at least one climate station per 5000 km^2) to provide ground-based observations of hydroclimatological variables over the long term (AghaKouchak et al., 2015). In many other regions, the available observations are not sufficient to capture the spatiotemporal variability of drought-related variables such as precipitation (Easterling et al., 2016). Furthermore, observations from different meteorological stations often have different record lengths and variable data quality (Wuebbles et al., 2016), which makes consistent global drought analysis using ground-based observations challenging (AghaKouchak et al., 2015).

Satellite remote sensing provides a synoptic view of the land and a spatial context for measuring drought impacts providing spatially continuous data with improved information on monitoring vegetation and land dynamics over large areas (Anteneh, 2012). Satellite remote sensing of the Earth's weather began with the Television and Infrared Observation Satellite (TIROS-1) mission in 1960. The success of this mission led to a series of additional weather- and climate-oriented satellite remote sensing missions. Remote sensing satellites observations have been used extensively for drought monitoring and impact assessment (AghaKouchak et al., 2015; Wang et al., 2018) through monitoring of drought-related variables and also to assess and quantify drought impacts (Marumbwa et al., 2020). Satellite multispectral, thermal infrared (IR), or microwave data are used to retrieve drought-related variables, including precipitation (West et al., 2019; Marumbwa et al., 2021), soil moisture (Thierfelder et al., 2013; Bhavani et al., 2017), or evapotranspiration (Kusangaya et al., 2021; Marumbwa et al., 2021). The specific hydrologic variable is then converted into a drought indicator by calculating the extent of an anomaly or departure from the longer-term environmental baseline. Those data are used to quantitatively assess and categorize drought severity. Satellite observations have also been used to assess drought ecosystem impacts, including vegetation health and growth by assessing the photosynthetic capacity of plants (Anyamba et al., 2003; Kusangaya et al., 2021; Gwatida et al., 2023). Precipitation deficits can lead to reduced photosynthetic capacity and changes in the absorption of solar radiation in photosynthetically active wavelengths by plants (Nikulin et al., 2012). Combinations of satellite visible and IR images have been widely used to monitor plant changes and water stress (Popova et al., 2014; Kusangaya et al., 2021).

New and advanced remote sensing data sets are increasingly becoming available (Wardlow et al., 2016). Satellite data has opened new avenues for global drought monitoring from different perspectives (e.g., meteorological, agricultural, hydrological, and ecological). The advantages of satellite-based sensors relative to traditional ground-based observations include global, near-real-time observations, consistent data records, and improved spatial resolution (Shekede et al., 2023; Sofia et al., 2023; Adou et al., 2024). The increasing volume of satellite observations and data products has led the science community into the era of big data (Ganguly et al., 2014; Gxokwe et al., 2022) and provided unique opportunities to develop advanced drought monitoring capabilities using multiple data sources. However, the abundance of data also presents major scientific challenges, including uncertainty assessment, managing data volumes, merging or fusion of multiple data sources, and ensuring consistency between different observations and data sets (Barrows et al., 2010; AghaKouchak et al., 2015).

The use of remote sensing-based drought monitoring and impact assessment presents opportunities as well as limitations, as outlined by Bhaga et al. (2020). Drought and surface water indices, namely, the Standardized Precipitation Index (SPI), PDSI, Normalized Difference Vegetation Index (NDVI), Vegetation Condition Index (VCI), and Water Requirement Satisfaction Index (WRSI), Normalized Difference Water Index (NDWI), Modified Normalized Difference Water Index (MNDWI), Land Surface Water Index (LSWI+5), Modified Normalized Difference Water Index (MNDWI+5), Automated Water Extraction Index (shadow) (AWEIsh), and Automated Water Extraction Index (nonshadow) (AWEInsh) are widely used for drought monitoring (Capra et al., 2013; Feyisa et al., 2014; Winkler et al., 2017; Ahmed et al., 2018; Zhao et al., 2022; Ali et al., 2023). Landsat-8 and Sentinel-2 satellite data are likely to be best suited to monitor climate variability, hydrological drought, and surface water bodies, due to their availability at a relatively low cost, impressive spectral, spatial, and temporal characteristics (Kuhn et al., 2019; Bhaga et al., 2020; Sofia et al., 2023; Adou et al., 2024). The most effective drought and water indices are SPI, PDSI, NDVI, VCI, NDWI, MNDWI, MNDWI+5, AWEIsh, and AWEInsh (Nagarajan, 2009; Núñez et al., 2014; Suryabhagavan, 2017; Winkler et al., 2017; Yihdego et al., 2019; Araneda-Cabrera et al., 2021; Ndayiragije and Li, 2022; Zhou et al., 2023). The NDVI of Rouse and Space (1978) is the most frequently used vegetation index (Bhavani et al., 2017; Mutanga et al., 2020; Gxokwe et al., 2022; Pacheco et al., 2022; Ali et al., 2023) and the first remote sensing-based measure used to monitor agricultural drought. The increasing role and potential of remote sensing in generating spatially explicit information on drought and climate variability affects surface water resources. Although progress has been made in advancing the application of remote sensing in drought monitoring (Sharara et al., 2022), this review indicates the need for further studies on assessing drought and climate variability impacts on water resources, especially in the context of climate change and increased water demand.

4 Drought indices for drought monitoring and early warning systems

A drought index is a prime variable for assessing the effect of drought and defining different drought parameters, which include intensity, duration, severity, and spatial extent (Özger et al., 2009; Gond et al., 2023). A drought index is a numerical measure used to quantify and characterize the severity and extent of drought conditions in a specific region (Guttman, 1998). These indices help in monitoring and assessing droughts over time, aiding in drought early warning systems and water resource management (Ali et al., 2023). A drought index should quantify the drought for different time scales. The most commonly used time scale for drought analysis is a year, followed by a month (Özger et al., 2009). The annual time scale is used to get information on the regional behavior of droughts, whereas the monthly time scale is more appropriate for monitoring the effects of drought in situations related to agriculture, water supply, and groundwater abstractions (Anteneh, 2012). A time series of drought indices provides a framework for evaluating drought parameters of interest (Özger et al., 2009). Several drought indices have been developed to quantify droughts, each having its own strengths and weaknesses. The most common drought indices include the PDSI, the commonly used Percent of Normal (PN), Rainfall Anomaly Index, deciles Crop Moisture Index (CMI), Bhalme and Mooly Drought Index, Surface Water Supply Index, National Rainfall Index, SPI, and Reclamation Drought Index (RDI) (Hayes et al., 2005; Suryabhagavan, 2017; Winkler et al., 2017; Hailesilassie et al., 2023). Drought indices can be classified into two groups: (a) satellite data-driven drought indices and (b) data-driven drought indices.

4.1 Satellite data-driven drought indices

4.1.1 Vegetation Condition Index

The VCI is a drought index that measures drought severity based on the health of vegetation. The VCI uses satellite data to measure the NDVI. When the VCI is below a certain threshold, it indicates that drought is likely occurring in the area. VCI is computed from satellite advanced very high-resolution radiometer radiance data adjusted for land, climate, ecology, and weather conditions (Marumbwa et al., 2021). The VCI is calculated by comparing the current NDVI value to the average NDVI value for the same area during a reference period, such as the average NDVI value for the past 30 years (Marumbwa et al., 2020). The VCI ranges from −1 (extreme drought) to +1 (extremely wet). The VCI can be used for drought detection and tracking (Marumbwa et al., 2021; Ali et al., 2023). The VCI allows the detection of drought and measurement of the time of its onset and its intensity, duration, and impact on vegetation (Özger et al., 2009; Marumbwa et al., 2020). However, since the VCI is based on NDVI, it is primarily useful for the summer growing season only (Bhaga et al., 2020).

4.1.2 Normalized Difference Vegetation Index

NDVI is a satellite data-driven index widely used to monitor vegetation and drought conditions (Bhavani et al., 2017; Ali et al., 2023; Adou et al., 2024).

It is expressed as the difference between the near-infrared (NIR) and red reflectance normalized by their sum (NDVI = [NIR − RED]/[NIR + RED]). The NDVI is the most commonly used vegetation index because its measurement scale ranges from −1 to 1, with zero being the approximate value for no vegetation. Negative values represent nonvegetative surfaces, whereas values closer to 1 are representative of dense vegetation (Measho et al., 2019; Mutanga et al., 2020). The NDVI is one of the widely used drought monitoring indices. A significant relationship has been reported between NDVI and precipitation and soil moisture (Bhavani et al., 2017; Measho et al., 2019; Bhaga et al., 2020; Ali et al., 2023), and thus, NDVI (or its derivatives) has been widely used for drought assessment and vegetation health monitoring (Nicholson, 2001; Anyamba et al., 2003; Núñez et al., 2014; West et al., 2019; Seaton et al., 2020).

4.2 Data-driven drought indices

Data-driven drought indices are calculated using measurements from satellites, weather stations, or other sources of data. These indices are often used together to provide a more complete picture of drought conditions. Data-driven drought indices use precipitation, either as the sole input or in combination with other meteorological elements such as precipitation and temperature or precipitation and soil moisture (Zaniolo et al., 2019; Zhao et al., 2022). Examples of data-driven drought indices include the Standardized Precipitation Evapotranspiration Index, the Soil Moisture Deficit Index, and the US Drought Monitor (USDM). The following section discusses commonly used data-driven drought indices, their strengths, and their limitations.

4.2.1 *Percent of Normal*

PN is a drought index that requires precipitation as its only input, usually over a minimum 30-year historical period (Bonaccorso et al., 2003). It is calculated by dividing actual precipitation by normal precipitation, typically considered to be a 30-year mean, and multiplying this by 100% (Bonaccorso et al., 2003; Bordi and Sutera, 2007; Bhaga et al., 2020). The PN can be computed for a variety of time scales such as a single month to a group of months within a season, annual or wet year (Bonaccorso et al., 2003; Morid et al., 2006). Normal precipitation is considered to be 100% for a specific location. However, the PN is limited as a tool for a comparison of climatic conditions of different areas (Sridhara et al., 2021) because precipitation on monthly or seasonal scales does not have a normal distribution. As a result, the mean precipitation is not the same as the median precipitation. Other standardized drought indices such as PDSI and SPI have been developed to overcome this limitation with respect to comparing different climatic regions.

4.2.2 *Palmer Drought Severity Index*

PDSI was developed by Palmer (1965) and it uses a combination of precipitation and temperature data to calculate drought conditions (Hayes et al., 2005;

Özger et al., 2009; Mishra and Singh, 2010). The index is based on the water balance, using a concept of supply and demand over a two-layer soil model. The basis of PDSI is the difference between the amount of precipitation required to retain a normal water balance level and the actual precipitation (Bordi and Sutera, 2007; Gutzler and Robbins, 2011; Mishra and Singh, 2011; Wang et al., 2018). Several coefficients are calculated, which define local hydrological norms related to temperature and precipitation. The calculation of the coefficients depends on the soil water capacity of the underlying layer (Bordi and Sutera, 2007; Mishra and Singh, 2011). The PDSI can be used to estimate short-term (less than a year), intermediate-term (1–5 years), and long-term (10–20 years) drought conditions. There are several limitations of the PDSI, which include an inherent time scale making PDSI more suitable for agricultural impacts and less so for hydrologic droughts. According to Mishra and Singh (2011), the PDSI assumes that all precipitation is rain, making values obtained during winter months and high elevations often questionable. In addition, the PDSI also assumes that runoff only occurs after all soil layers have been saturated, leading to an underestimation of runoff.

4.2.3 Crop Moisture Index

CMI was developed by Palmer (1968) (Hayes et al., 2005) to estimate drought conditions, especially for agricultural areas (Malisawa and De Rautenbach, 2012; Bhavani et al., 2017). The CMI takes into account factors such as precipitation, soil moisture, temperature, and crop growth stage. This makes it a particularly useful index for estimating drought conditions that may affect crops. The CMI is calculated weekly and is often used in conjunction with other drought indices, such as the PDSI or the USDM. Computation of CMI involves the use of precipitation, temperature, soil moisture, and crop growth stage to compute a simple moisture budget. First, the potential evapotranspiration (PET) is calculated using temperature and other variables. This is then compared to the actual evapotranspiration (AET), which is estimated based on precipitation and soil moisture. The difference between the PET and the AET is then used to calculate the CMI, which can range from −9 (extremely dry) to +9 (extremely wet). The CMI responds rapidly to changing conditions and this rapid response to changing short-term conditions may provide misleading information about long-term conditions. Another limitation of the CMI is its sensitivity to PET. An increase in CMI may occur with an increase in PET. An increase in the CMI value indicates wetter moisture conditions. The second limitation of the CMI is that it is not a long-term drought monitoring tool, due to its rapid response to changing short-term conditions (Malisawa and De Rautenbach, 2012).

4.2.4 The deciles index

The deciles index was developed by Gibbs and Maher (1967) and uses historical rainfall data to determine the likelihood of a certain amount of rainfall occurring by first classifying rainfall deciles (or rankings), from 1 to 10, with 1 being the lowest rank and 10 being the highest (Tsakiris et al., 2007; Kenabatho et al., 2012; Salehpour Jam et al., 2018). When computing the deciles index, monthly precipitation totals from a long-term record are first ranked from highest to lowest to construct a cumulative frequency distribution. The cumulative frequency distribution is then split into 10 parts (tenths of distribution or deciles). The first decile is the precipitation value not exceeded by the lowest 10% of all precipitation values in a record. The second decile is between the lowest 10% and 20% and so forth. By comparing the amount of precipitation in a month (or during several months) with the long-term cumulative distribution of precipitation amounts in that period, the severity of drought can be assessed (Morid et al., 2006). If precipitation falls into the lowest 20% (deciles 1 and 2), it is classified as "much below normal." Deciles 3 and 4 (20%–40%) indicate "below normal" precipitation, deciles 5 and 6 (40%–60%) indicate "near-normal" precipitation, 7 and 8 (60%–80%) indicate "above normal" precipitation and 9 and 10 (80%–100%) indicate "much above normal" precipitation (Morid et al., 2006). In the aforementioned method, monthly rainfall time series are normalized using the Box-Cox transformation. However, the deciles index requires a long climatological record to calculate the deciles accurately (Keyantash and Dracup, 2002).

4.2.5 Standardized Precipitation Index

SPI was developed by McKee et al. (1993) and is based on precipitation alone, making its evaluation relatively simple (Capra et al., 2013; Núñez et al., 2014; Cheval, 2015; Ahmed et al., 2018; Ngwenya and Simatele, 2024). The SPI is based on an equi-probability transformation of aggregated monthly precipitation into a standard normal variable (Svoboda et al., 2012; Cheval, 2015). The computation of the index requires fitting a probability distribution to aggregated monthly precipitation series (3, 6, 12, 24, and 48 months). The probability density function is then transformed into a normal standardized index whose values classify the category of drought characterizing each place and time scale (Guttman, 1991, 1998). The SPI can only be computed when sufficiently long (at least 30 years) and possibly continuous time series of monthly precipitation data are available (Guttman, 1998; Cacciamani et al., 2007). The SPI makes it possible to describe the drought on time scales that typically describe the four types of droughts: (1) hydrological, (2) meteorological, (3) agricultural, and (4) socioeconomic drought. The SPI enables the forecast of drought on both short and long-term time scales. However, indices developed over short-term time scales are useful in monitoring meteorological and agricultural drought. Likewise, indices developed over long-term time scales are useful for monitoring hydrologic and socioeconomic drought (Cacciamani et al., 2007). Additionally, the standardization of the SPI makes it particularly suited to compare drought conditions among different time periods and regions with different climates (Guttman, 1998; Cacciamani et al., 2007). The SPI may be used for monitoring both dry and wet conditions, whereby positive SPI values indicate greater than median precipitation and negative values indicate less than median precipitation (Morid et al., 2006; Akhtari et al., 2007). The "drought" parts of the SPI range are arbitrarily split into categories of near-normal, moderately dry, severely dry, and extremely dry of which the near-normal category is defined by a range of SPI values from 0.5 to −0.5, whereas the severely dry category is defined by a range of SPI values from −1.5 to −2 (Morid et al., 2006; Akhtari et al., 2007). A more negative SPI value is indicative of a more severe drought condition. A disadvantage of the SPI is that it is not always easy to find a probability distribution that models the raw precipitation data. Another disadvantage is that it is not always possible to access reliable time series data to produce a robust estimate of the distribution parameters (Akhtari et al., 2007).

5 Drought impacts on agriculture, ecosystems, human health and social systems, water resources

Wilhite and Pulwarty (2017a) posit that drought ranks highly among climate hazards that are pervasive, most damaging, and impacts a multitude of sectors, including agriculture, water availability, livestock breeding, ecosystems, and food supply. Most of these drought impacts have compounding, interactive, and cascading effects, thereby exacerbating morbidity and mortality risks, particularly in vulnerable populations (Salvador et al., 2023). For example, droughts influence agricultural systems and water resources, which in turn affects human health. The following section provides an overview of the effects of climate change and variability-induced droughts on the following sectors: water resources, human health, agriculture, and ecosystems.

5.1 Drought and water resources

According to Kusangaya et al. (2014) and Bhaga et al. (2020), climate variability and recurrent droughts have caused remarkable strain on water resources in most regions across the globe, with the arid and semi-arid areas being the hardest hit. The impacts of climate change have been notable on surface water resources, which are already under threat from massive abstractions due to increased demand, as well as poor conservation and unsustainable land management practices (Kusangaya et al., 2021). There has been an increasing need to enhance mitigation and adaptation mechanisms as the increasing impacts of drought and climate variability on water resources have gained increased attention in recent decades. According to Bhaga et al. (2020) and Sharara et al. (2022), even though the use of satellite technologies has gained prominence in generating spatially explicit information on drought and climate variability impacts across the globe, they are somewhat hampered by difficulties in detecting drought evolution due to its complex nature, varying scales, the magnitude of its occurrence, and inherent data gaps. A number of studies have been conducted to monitor and assess the impacts of climate variability and droughts on water resources using different remotely sensed and in situ datasets (AghaKouchak et al., 2015).

During the last two decades, the impacts of droughts have increased significantly with an increased number of droughts as well as an increase in their severity (Wilhite et al., 2007; Ifejika Speranza et al., 2008; Marumbwa et al., 2021; Orimoloye et al., 2022). For example, the drought of 1988 in the United States had an economic cost of approximately $40 billion on the US economy. Nearly 10% of the total land area of the United States experienced either severe or extreme droughts at any given time during the 20th century (Wilhite and Svoboda, 2005; Mishra and Singh, 2010). From 1980 to 2003 in the United States, drought accounted for 10 of the 58 weather-related disasters and accounted for $144 billion of the estimated $349 billion total cost of all weather-related disasters in the United States (Rossi et al., 1992; Ross and Lott, 2004). Hence, in economic terms alone, droughts are the costliest natural disasters in the United States. On the contrary, in Europe, major drought events have been observed over the last 30 years. Most notably, the 1976 drought affected northern and western Europe, while the 1989 and 1991 droughts affected most of Europe (Bradford, 2000; Spinoni et al., 2015). In 2003, a prolonged drought affected most of Europe and since 1991, the yearly average economic impact of droughts in Europe has been €5.3 billion, with the economic damage of the 2003 drought in Europe amounting to at least €8.7 billion (Bradford, 2000; Mishra and Singh, 2011; Spinoni et al., 2015).

In Africa, since the late 1960s, the Sahel, a semi-arid region in West Africa between the Sahara Desert and the Guinea coast rainforest, has experienced droughts of unprecedented severity (Ward et al., 1999; Mishra and Singh, 2011). The drought has had a devastating impact on this ecologically vulnerable region and was a major impetus for the establishment of the United Nations Convention on Combating Desertification and Drought (Zeng, 2003). Although the frequency of droughts in the region is thought to have increased from the end of the 19th century, three long droughts have had dramatic environmental and societal effects on the Sahel nations. Famine followed severe droughts in the 1910s, 1940s, 1960s, 1970s, and 1980s although a partial recovery occurred from 1975 to 1980. Although at least one particularly severe drought has been confirmed in each century since the 1600s, the frequency and severity of the recent Sahelian drought stand out (Anteneh, 2012). Famine and dislocation on a massive scale—from 1968 to 1974 and again in the early and mid-1980s—were blamed on two spikes in the severity of the 1960–1980s drought periods (Hulme, 1992; Mishra and Singh, 2011; Nicholson, 2001).

5.2 Drought and human health

Regarding human health, the increase in drought frequency and severity across multiple geographic regions of the world, directly and indirectly affect human health (Salvador et al., 2020). This is through multiple pathways of disease transmission such as water-, food-, and vector-borne diseases, malnutrition, cardiovascular and respiratory illness, mental health disorders, renal diseases, and mortality (Salvador et al., 2023). This leads to short-term shocks or changes in temporal patterns of climate-sensitive diseases (Ebi et al., 2021). For instance, Burrows et al. (2024) systematically reviewed the effects of chronic slow-onset conditions such as drought and changing temperature on mental health and observed that these exposures resulted in depression and anxiety symptoms including suicide and other nonspecific psychological distress. Drought conditions and associated high temperatures have been shown to exacerbate symptoms of diseases such as Sickle Cell Anemia through dehydration-induced vaso-occlusive and increased blood viscosity (Obeagu and Obeagu, 2024). Similarly, drought affects water accessibility in drought-prone regions, compromising hydration and thereby worsening Sickle Cell Anemia. Water shortages associated with drought increase the risks of infectious diseases such as cholera, dysentery, typhoid, and rotavirus (Levy et al., 2018). Usually, water scarcity during drought conditions forces people to use unsafe water for domestic uses or for irrigating fields, thereby compromising hygiene practices and increasing the risk of waterborne and foodborne diseases, and skin and eye infections (Asmall et al., 2021). The COVID-19 pandemic which ravaged the globe in the last 5 years was worsened by climate extremes amplifying morbidity and mortality risks, thereby posing challenges in responding to the pandemic (Romanello et al., 2021). Furthermore, Lee et al. (2023) showed higher concentrations of ozone concentration during drought, implying potential excess health burdens from O_3 exposures as well as potential confounding effects of drought on the health burdens. Recently, Levy et al. (2018) notes that there is "very high confidence" that increased risks of food- and water-borne diseases can be expected "if climate change continues as projected across the representative concentration pathway (RCP) scenarios until mid-century" (Levy et al., 2018). Already, studies have shown that the incidence and range of vectorborne, waterborne, and foodborne infectious diseases, such as malaria and diarrheal diseases, have shifted over past decades as a result of changes in climate extremes and climate variability (Romanello et al., 2021). Drought reduces food security and increases the risk of malnutrition through reducing crop yield and livestock production (Levy et al., 2018). Studies have shown differential impacts of droughts on human health with the young, elderly, chronic patients, and the poor being disproportionately affected (UNDRR, 2021). It can be observed that droughts have profound implications on human health and the impacts are likely to increase in severity with projected increases in extreme climate events, especially in tropical and subtropical regions. Therefore, there is a need to develop strategies aimed at reducing drought-related health impacts.

5.3 Drought and terrestrial ecosystems

The increase in climate-related studies over the past decades has shown that the Earth's climate system is changing at an accelerated rate than previously envisaged. These changes are resulting in the occurrence of extreme events such as droughts, which, in turn, have profound repercussions on ecosystem services, including the loss of sequestered forest carbon and associated water and carbon fluxes across the globe (Zhao and Running, 2010; Vicente-Serrano et al., 2021). At the plant scale, drought induces moisture stress resulting in poor growth, low turgidity, reduced leaf water content in the short term, and impairment of critical plant functions such as protein synthesis (Saneoka et al., 2004). Droughts are also key determinants of the geographical distribution and the composition of plant communities (Knapp et al., 2002) and the dependent animal species. Although drought has affected virtually all ecosystems, the impacts have been remarkable in subtropical ecosystems where droughts are a recurrent phenomenon. These droughts have not only affected the structure of ecosystems but their function (i.e., the ability to provide goods and services) as well. Different plants and animals respond differently to the prolonged shortages in the water supply that are often associated with drought. In general, plants exposed to drought cease to function normally and alter their morphological, physiological, and molecular makeup (Gaspar et al., 2002; Zia et al., 2021). For instance, xylem hydraulic failure has been linked to drought-induced tree mortality in many ecosystems and influences water balance and carbon metabolism (Martinez-Vilalta et al., 2019). As plant productivity forms the base of many food webs in ecosystems, any phenomenon that threatens this productivity has cascading effects on the rest of the ecosystem. Table 1 summarizes the effect of drought on plants, which in turn affect ecosystem productivity and attendant processes.

In a review of global studies, Allen et al. (2010) observed that while episodic mortality occurs in the absence of climate change, evidence from these studies suggests accelerated mortality in some forested ecosystems. This is especially concerning given the high natural tree mortality rates and die-offs, which are likely to be exacerbated by projected intense, more frequent, and prolonged droughts under a future climate. Moreover, studies have also shown that the propensity of the impacts of drought to persist well after the drought has subsided (Frank et al., 2015) in what is commonly referred to as "drought legacies" (Vilonen et al., 2022). According to a review by Müller and Bahn (2022), these drought legacies have had a significant impact on a number of ecosystem processes including altered nutrient cycling (carbon and nitrogen), phenology, species composition, herbivory, and increased plant mortality. Using satellite imagery, Zhao and Running (2010) demonstrated that large-scale droughts reduced net primary productivity, especially in the Southern Hemisphere, thereby weakening the terrestrial carbon sink and competing claims between

TABLE 1 Effects of drought on key plant processes.

Plant process affected by drought	Characteristics of the effect
Seed germination and early vegetative growth	Drought reduces the rate of germination, growth, and plant population, including seedling vigor. Growth parameters negatively affected by drought include plant height, root length, shoot and root fresh and dry weight, number of leaves, tillers, plant leaf, and root area
Stomatal conductance and photosynthesis	During a drought, plant hormone, abscisic acid, is transferred to leaves resulting in stomata closure. Persistent stomatal closure reduces water loss resulting in reduced photosynthesis. However, the effect is dependent on the growth stage and type of plant
Plant water relations	Drought-induced stress reduces relative water content and cell membrane stability leading to damaged cell membranes
Reactive oxygen species (ROS)	Water deficit is often associated with an imbalance in carbon resource use and enhanced generation of ROS. ROS causes oxidative damage and ultimately causes cellular damage and death
Plant yield	Drought negatively affects yield directly through its influence on key plant processes such as germination, growth, reproduction, and flowering. Although every growth stage of a plant is negatively affected by drought, the reproductive stage is the most affected
Seed quality	Seed quality is significantly affected by drought stress through its effect on every phonological stage of plant growth. Poor seed quality results in poor plant germination, low plant vigor, diseased plants, and in the worst case extinction due to failed reproduction

Adapted from Zia, R., Nawaz, M.S., Siddique, M.J., Hakim, S., Imran, A., 2021. Plant survival under drought stress: implications, adaptive responses, and integrated rhizosphere management strategy for stress mitigation. Microbiol. Res., 242, 126626.

food and biofuel production. Plants respond in multiple ways to drought at the cellular and organism levels (Levitt, 1972; Zia et al., 2021). These strategies fall into three categories, namely, (i) drought escape (fast phonological development), (ii) drought avoidance (e.g., closing stoma, modification of root structure), and drought tolerance (developing mechanisms to withstand drought through adjusting genetic make-up and metabolic pathways).

5.4 Drought and agriculture

Agriculture is the mainstay of human livelihoods through providing food, fuel, fibers, and raw materials. These requirements need to be provided within the context of an ever-increasing global population, a changing climate characterized by extreme events such as drought, as well as the need to ensure environmental sustainability. It is therefore critical to have an improved understanding of the impacts of these drivers (especially drought) on the agricultural system. Although droughts have traditionally been regarded as slow, creeping phenomena that can last for an extended period, recent evidence shows an increase in abrupt and intense droughts that are commonly referred to as flash droughts (Osman et al., 2022). The intense drying over a relatively short period has serious implications for hydrological and agricultural systems despite the short duration of the drought event (Black, 2024). Given the complex nature of droughts in terms of their occurrence, propagation, and effects on the agriculture subsystem, there is a need to improve our understanding of their impacts. This is critical for making the sector resilient to climate change through the development of comprehensive mitigation and adaptive strategies. Fig. 2 provides a summary of the effects of climate variability-induced drought on agriculture.

It is observed that climate variability and change indices change climate variables such as precipitation and temperature. These, in turn, directly affect the cropping and livestock system and indirectly via their influence on hydrology. Some of the effects of climate variability on the cropping system include changes in bioclimatically suitable areas for crop production, altered length of the growing season, and disease susceptibility while the livestock system is affected by changes in rangeland productivity as well as mating season. Indirectly, a reduction in precipitation would result in reduced water levels in rivers and underground, thereby affecting the amount of water available for agriculture. Thus, drought is regarded as a secondary stressor that exacerbates the

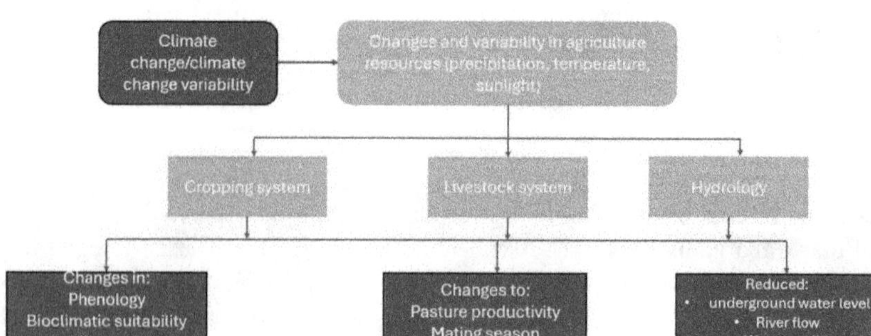

FIG. 2 Impacts of climate variability on the agricultural system. *(Adapted from Kim, C.-G., et al., 2009. Impacts and Countermeasures of Climate Change in Korean Agriculture (in Korean). Research Report No. 593. Korea Rural Economic Institute.*

effects of other stresses through pathogen attack and heat stress, ultimately reducing agricultural productivity as well as food security (Fadiji et al., 2022; Sári et al., 2024). With the projected increase in drought frequency and intensity, food insecurity is likely to worsen in sub-Saharan Africa due to overreliance on rain-fed agriculture and limited resources for mitigating the effects of repeated droughts (Ngwenya and Simatele, 2024). A detailed review of the climate variability impacts on the agriculture sector is provided in other sections of this book (see Chapter 4).

5.5 Drought mitigation and adaptation measures

A monitoring system is required to minimize the impacts of drought timely. Effective monitoring of droughts helps in developing a drought early warning system (Anteneh, 2012). To objectively evaluate droughts, several indices as stated previously have been developed to evaluate the water supply deficit in relation to the time duration of precipitation shortage (Bordi and Sutera, 2007). As explained previously, drought indices are based on the amount of precipitation and they measure the deviation of actual precipitation from a historically established average. However, some of the drought indices take into account other additional climatological indices such as temperature, evapotranspiration, or soil moisture (Anteneh, 2012). Drought indices developed over short-term time scales are useful in monitoring meteorological drought and agricultural drought, whereas indices developed over long-term time scales are useful for monitoring hydrologic and socioeconomic drought (Cacciamani et al., 2007). An important aspect of drought monitoring and the development of an early warning system is the ability to effectively forecast future drought events. Forecasting future dry events is very important for finding sustainable solutions to water management and risk assessment of drought occurrences (Bordi and Sutera, 2007). Drought forecasting techniques vary from the use of stochastic methods to the use of satellite imagery. Recently, neural networks have become increasingly popular tools for drought forecasting (Bordi and Sutera, 2007; Barua, 2010; Anteneh, 2012; Araneda-Cabrera et al., 2021).

6 Outlook on climate variability and adaptation

The combination of high climate variability and limited livelihood options due to high levels of poverty affecting a large proportion of the population together contribute to vulnerability to climate change (Kusangaya et al., 2014, 2021). High climate variability that results in frequent exposure to floods and droughts contributes significant vulnerability to current and future climate shocks to a population with limited resilience (Kusangaya et al., 2021). Large proportions of people in rural and urban areas already do not have adequate access to water and sanitation, which is exacerbated by extreme weather events such as floods and droughts; climate change will worsen these problems (Kunedzimwe et al.,

2021). Although efforts are being made to improve water resource management in southern Africa, existing systems such as dams developed for domestic, industrial, and municipal use are not able to satisfy the demand in most parts of southern Africa (Kusangaya et al., 2021). This is due to inadequate development of infrastructure and human capacity for managing these systems, another challenge compounded by climate change. A large proportion of the southern African population has very limited livelihood options due to poverty (Yanda et al., 2010). When floods and droughts occur, they have no cushion from these shocks: if existing poverty levels are not eliminated, climate change will exacerbate these problems. Environmental degradation in the form of soil erosion, deforestation, and water pollution adversely affect available water resources—also worsened by climate change, increasing the vulnerability of affected communities. Southern African communities are often caught unaware when floods and droughts are imminent. This is partly due to poorly developed early warning systems, which increases exposure to these shocks. A contributory factor is inadequate hydro-meteorological monitoring systems (Yanda et al., 2010; Bhattacharjee et al., 2011; Kusangaya et al., 2021) in large parts of southern Africa, which limits location-specific knowledge about the nature and effects of climate variability. Lack of locally relevant data constrains the development of adaptation measures appropriate for different environmental settings and communities. This often leads to "one-size-fits-all" recommendations, which can result in investment of limited financial resources into ineffective adaptation strategies.

7 Opportunities, progress, and prospects

Early drought detection is fundamental to proactive decision-making and disaster preparedness. Previous studies indicate that precipitation-based indicators (e.g., SPI) are better for drought detection compared with other indicators (e.g., SSI) (AghaKouchak and Nakhjiri, 2012; AghaKouchak et al., 2015). A number of satellite products (e.g., AIRS) provide near-surface air relative humidity information that is yet to be used for drought monitoring. Since near-surface air relative humidity directly influences evaporation and as such is connected to precipitation, it is reasonable to expect that it could provide valuable drought information and improve early drought onset detection.

Kusangaya et al. (2021) stated that the availability of water resources in the future and the implications of these findings are as follows: (1) Water availability is likely to be compromised, since, for example, rainfall for a season may now be limited to a few days, and altered rainfall characteristics (intensity, duration, etc.) will in turn eventually impact on flow regimes, and hence available spatial and temporal water resources; (2) Water management measures are likely to become strict and this makes the management of water resources, for example, implementation of integrated water resources management systems urgent; (3) Most of the population in southern Africa depends on rain-fed agriculture and lives in rural areas, where mean annual rainfall is below approximately 600 mm and therefore the impacts of climate variability are likely to be most severe. These conditions make the completion of overdue dams, construction of other water infrastructure, and establishment of irrigation infrastructure to support agriculture urgent to support smallholder irrigation schemes to ensure food security; (4) Similarly, due to rainfall variability and increased temperatures, available water for electricity generation is likely to be reduced. This will be particularly severe for dry season flows, since for hydroelectric power generation, water levels must meet certain thresholds for sustainable power generation. This is exemplified by the current case in Zimbabwe (2019) when the Zimbabwe Electricity Transmission and Distribution Company issued a statement acknowledging high electricity demand and warning citizens of potential power cuts. A power cut schedule has since been developed and strictly respected. To mitigate the impacts of future such extremes, alternative power sources such as solar must be seriously considered; and (5) The implications of these changes on agriculture and food security must be emphasized. Planting dates vary due, for example, to a shift in rainfall season onset. Changes in climatic patterns are also a basis for changes in growing patterns: moving a particular crop from a region that has become unsuitable to another may be more so. When fundamental processes for agriculture such as this are subject to change, food security is heavily compromised and food shortage is probable. This dynamic underscores the need for new genetically modified crop development.

On the contrary, the use of satellite remote sensing in studying droughts and their impacts on ecosystems has gathered momentum. Through satellite remote sensing observations, the biophysics of the drought phenomenon and the development of new drought indicators for research and practical applications have also gathered momentum (AghaKouchak et al., 2015; West et al., 2019; Marumbwa et al., 2021). In addition, there are opportunities to develop new indicators that quantify the effects of drought on carbon and nitrogen cycling (AghaKouchak et al., 2015). Several recently launched satellite missions have provided atmospheric CO_2 information and offer an opportunity to develop indicators for assessing the relationships between water deficits and nutrient cycling. Such indicators would be fundamental to assessing drought and climate variability impacts on the ecosystem and biogeochemical processes (AghaKouchak et al., 2015). AghaKouchak et al. (2015) pointed out that advances in remote sensing of microwave land emissivity show that there are opportunities to quantitatively assess surface soil moisture and vegetation water content since microwave-based vegetation monitoring offers unique ways for assessing drought impacts on vegetation. Furthermore, there is still a need to improve early drought detection by integrating satellite observations not currently used for drought monitoring (e.g., satellite near-surface air relative humidity and water vapor data), which can be used for drought early warning. However, there is a need for more research on the uncertainty quantification of satellite drought characterization, which will substantially improve the usability of satellite-based drought information (AghaKouchak et al., 2015).

References

Adou, M., et al., 2024. Drought analysis using normalized difference vegetation index and land surface temperature over Niamey region, the southwestern of the Niger between 2013 and 2019. J. Hydrol. Reg. Stud. 52 (February), 101689. https://doi.org/10.1016/j.ejrh.2024.101689.

AghaKouchak, A., Nakhjiri, N., 2012. A near real-time satellite-based global drought climate data record. Environ. Res. Lett. 7 (4), 44037. https://doi.org/10.1088/1748-9326/7/4/044037.

AghaKouchak, A., et al., 2015. Remote sensing of drought: progress, challenges and opportunities. Rev. Geophys. 53 (2), 452–480. https://doi.org/10.1002/2014RG000456.

Ahmad, M.M., Yaseen, M., Saqib, S.E., 2022. Climate change impacts of drought on the livelihood of dryland smallholders: implications of adaptation challenges. Int. J. Disaster Risk Reduct. 80 (May), 103210. https://doi.org/10.1016/j.ijdrr.2022.103210.

Ahmed, K., Shahid, S., Nawaz, N., 2018. Impacts of climate variability and change on seasonal drought characteristics of Pakistan. Atmos. Res. 214 (July), 364–374. https://doi.org/10.1016/j.atmosres.2018.08.020.

Akhtari, R., Mahdian, M.H., Morid, S., 2007. Assessment of spatial analysis of SPI and EDI drought indices in Tehran province. Iran-Water Resour. Res. 2 (3), 27–38.

Alemaw, B.F., Matondo, J.I., 2020. Impacts of possible climate change and variability on the water resources of southern African: a regional modelling approach. In: Sustainable Development Goals Series. Springer International Publishing, pp. 57–70, https://doi.org/10.1007/978-3-030-31543-6_6.

Ali, S., et al., 2023. Spatio-temporal variations in trends of vegetation and drought changes in relation to climate variability from 1982 to 2019 based on remote sensing data from East Asia. J. Integr. Agric. 22 (10), 3193–3208. https://doi.org/10.1016/j.jia.2023.04.028.

Allen, C.D., Macalady, A.K., Chenchouni, H., Bachelet, D., McDowell, N., Vennetier, M., Kitzberger, T., Rigling, A., Breshears, D.D., Hogg, E.T., Gonzalez, P., 2010. A global overview of drought and heat-induced tree mortality reveals emerging climate change risks for forests. Forest Ecol. Manag. 259 (4), 660–684.

Anteneh, B.A., 2012. Short-Term and Long-Term SPI Drought Forecasts Using Wavelet Neural Networks and Wavelet Support Vector Regression in the Awash River Basin of Ethiopia. MSc thesis, McGill University.

Anyamba, A., et al., 2003. Seasonal to interannual variability of vegetation and fires at SAFARI 2000 sites inferred from advanced very high resolution radiometer time series data. J. Geophys. Res. Atmos. 108 (13). https://doi.org/10.1029/2002jd002464.

Araneda-Cabrera, R.J., Bermudez, M., Puertas, J., 2021. Revealing the spatio-temporal characteristics of drought in Mozambique and their relationship with large-scale climate variability. J. Hydrol. Reg. Stud. 38 (September). https://doi.org/10.1016/j.ejrh.2021.100938.

Asmall, T., Abrams, A., Röösli, M., Cisse, G., Carden, K., Dalvie, M.A., 2021. The adverse health effects associated with drought in Africa. Sci. Total Environ. 793, 148500.

Bagheri, F., 2016. Mapping drought hazard using SPI index and GIS (a case study: Fars province, Iran). Int. J. Environ. Geoinform. 3 (1), 22–28. https://doi.org/10.30897/ijegeo.304419.

Barrows, C.W., Rotenberry, J.T., Allen, M.F., 2010. Assessing sensitivity to climate change and drought variability of a sand dune endemic lizard. Biol. Conserv. 143 (3), 731–736. https://doi.org/10.1016/j.biocon.2009.12.013.

Barua, S., 2010. Drought Assessment and Forecasting Using a Nonlinear Aggregated Drought Index. PhD thesis, Victoria University, p. 178.

Battersby, J., Watson, V., 2018. Urban Food Systems Governance and Poverty in African Cities. Routledge, https://doi.org/10.4324/9781315191195.

Belayneh, A., 2012. Short-term and long-term SPI drought forecasts using wavelet neural networks and wavelet support vector regression in the Awash River Basin of Ethiopia. Master of Science Thesis. McGill University.

Bhaga, T.D., et al., 2020. Impacts of climate variability and drought on surface water resources in sub-Saharan Africa using remote sensing: a review. Remote Sens. 12 (24), 1–34. https://doi.org/10.3390/rs12244184.

Bhattacharjee, R., et al., 2011. Regional impacts of climate change: Africa. In: Crop Adaptation to Climate Change. Wiley, pp. 66–77, https://doi.org/10.1002/9780470960929.ch6.

Bhavani, P., et al., 2017. Long-term agricultural performance and climate variability for drought assessment: a regional study from Telangana and Andhra Pradesh states, India. Geomat. Nat. Hazards Risk 8 (2), 822–840. https://doi.org/10.1080/19475705.2016.1271831.

Black, E., 2024. Global change in agricultural flash drought over the 21st century. Adv. Atmos. Sci. 41 (2), 209–220.

Bogachev, M.I., et al., 2023. Reversal in the drought stress response of the Scots pine forest ecosystem: local soil water regime as a key to improving climate change resilience. Heliyon 9 (11), e21574. https://doi.org/10.1016/j.heliyon.2023.e21574.

Bonaccorso, B., et al., 2003. Spatial variability of drought: an analysis of the SPI in Sicily. Water Resour. Manag. 17, 273–296.

Bordi, I., Sutera, A., 2007. Drought monitoring and forecasting at large scale. In: Methods and Tools for Drought Analysis and Management, pp. 3–27, https://doi.org/10.1007/978-1-4020-5924-7_1.

Bradford, R.B., 2000. Drought events in Europe. In: Drought and Drought Mitigation in Europe. Springer, pp. 7–20, https://doi.org/10.1007/978-94-015-9472-1_2.

Bunyavanich, S., et al., 2003. The impact of climate change on child health. Ambul. Pediatr. 3 (1), 44–52. https://doi.org/10.1367/1539-4409(2003)003<0044:tiocco>2.0.co;2.

Burrows, K., Denckla, C.A., Hahn, J., Schiff, J.E., Okuzono, S.S., Randriamady, H., Mita, C., Kubzansky, L.D., Koenen, K.C., Lowe, S.R., 2024. A systematic review of the effects of chronic, slow-onset climate change on mental health. Nat. Mental Health, 1–16.

Cacciamani, C., et al., 2007. Monitoring and forecasting drought on a regional scale: Emilia-Romagna region. In: Methods and Tools for Drought Analysis and Management, pp. 29–48, https://doi.org/10.1007/978-1-4020-5924-7_2.

Capra, A., Consoli, S., Scicolone, B., 2013. Long-term climatic variability in Calabria and effects on drought and agrometeorological parameters. Water Resour. Manag. 27 (2), 601–617. https://doi.org/10.1007/s11269-012-0204-0.

Castano, A., 2013. Monitoring Drought at River Basin and Regional Scale: Application in Sicily. PhD thesis, University of Catania.

Cheval, S., 2015. The standardized precipitation index—an overview. Rom. J. Meteorol. 12 (1–2), 17–64.

Collins, J.M., 2011. Temperature variability over Africa. J. Clim. 24 (14), 3649–3666. https://doi.org/10.1175/2011JCLI3753.1.

Desanker, P.V., Justice, C.O., 2001. Africa and global climate change: critical issues and suggestions for further research and integrated assessment modeling. Clim. Res. 17 (2 Special 8), 93–103. https://doi.org/10.3354/cr017093.

Desanker, P.V., et al., 2005. Requirements for integrated assessment modelling at the regional and national levels in Africa to address climate change. In: Climate Change and Africa. Cambridge University Press, pp. 260–270, https://doi.org/10.1017/CBO9780511535864.034.

Dore, M.H.I., 2005. Climate change and changes in global precipitation patterns: what do we know? Environ. Int. 31 (8), 1167–1181. https://doi.org/10.1016/j.envint.2005.03.004.

Easterling, D.R., et al., 2016. Detection and attribution of climate extremes in the observed record. Weather Clim. Extrem. 11, 17–27. https://doi.org/10.1016/j.wace.2016.01.001.

Ebi, K.L., Vanos, J., Baldwin, J.W., Bell, J.E., Hondula, D.M., Errett, N.A., Hayes, K., Reid, C.E., Saha, S., Spector, J., Berry, P., 2021. Extreme weather and climate change: population health and health system implications. Ann. Rev. Public Health 42 (1), 293–315.

Fadiji, A.E., Santoyo, G., Yadav, A.N., Babalola, O.O., 2022. Efforts towards overcoming drought stress in crops: revisiting the mechanisms employed by plant growth-promoting bacteria. Front. Microbiol. 13, 962427.

Falkenmark, M., 1989. Water scarcity and food production in Africa. In: Food and Natural Resources. Elsevier, pp. 163–190, https://doi.org/10.1016/b978-0-12-556555-4.50011-9.

Feyisa, G.L., et al., 2014. Automated Water Extraction Index: a new technique for surface water mapping using Landsat imagery. Remote Sens. Environ. 140, 23–35. https://doi.org/10.1016/j.rse.2013.08.029.

Field, C.B., et al., 2012. Managing the Risks of Extreme Events and Disasters to Advance Climate Change Adaptation: Special Report of the Intergovernmental Panel on Climate Change. Cambridge University Press, https://doi.org/10.1017/CBO9781139177245.

Frank, D., Reichstein, M., Bahn, M., Thonicke, K., Frank, D., Mahecha, M.D., Smith, P., Van der Velde, M., Vicca, S., Babst, F., Beer, C., 2015. Effects of climate extremes on the terrestrial carbon cycle: concepts, processes and potential future impacts. Glob. Change Biol. 21 (8), 2861–2880.

Frischen, J., et al., 2020. Drought risk to agricultural systems in Zimbabwe: a spatial analysis of hazard, exposure, and vulnerability. Sustainability 12 (3), 752. https://doi.org/10.3390/su12030752.

Ganguly, A.R., et al., 2014. Toward enhanced understanding and projections of climate extremes using physics-guided data mining techniques. Nonlinear Process. Geophys. 21 (4), 777–795. https://doi.org/10.5194/npg-21-777-2014.

Gaspar, T., Franck, T., Bisbis, B., Kevers, C., Jouve, L., Hausman, J.F., Dommes, J., 2002. Concepts in plant stress physiology. Application to plant tissue cultures. Plant Growth Regul. 37, 263–285.

Gebrechorkos, S.H., Hülsmann, S., Bernhofer, C., 2020. Analysis of climate variability and droughts in East Africa using high-resolution climate data products. Glob. Planet. Change 186 (September), 103130. https://doi.org/10.1016/j.gloplacha.2020.103130.

Gibbs, W.J., Maher, J.V., 1967. Rainfall Deciles as Drought Indicators. Bureau of Meteorology, Melbourne Bulletin No. 48.

Gond, S., et al., 2023. Assessment of drought variability using SPEI under observed and projected climate scenarios over Uttar Pradesh, India. Phys. Chem. Earth 131 (January), 103440. https://doi.org/10.1016/j.pce.2023.103440.

Guttman, N.B., 1991. A sensitivity analysis of the Palmer Hydrologic Drought Index. J. Am. Water Resour. Assoc. 27 (5), 797–807. https://doi.org/10.1111/j.1752-1688.1991.tb01478.x.

Guttman, N.B., 1998. Comparing the Palmer drought index and the standardized precipitation index. J. Am. Water Resour. Assoc. 34 (1), 113–121. https://doi.org/10.1111/j.1752-1688.1998.tb05964.x.

Gutzler, D.S., Robbins, T.O., 2011. Climate variability and projected change in the western United States: regional downscaling and drought statistics. Clim. Dyn. 37 (5), 835–849. https://doi.org/10.1007/s00382-010-0838-7.

Gwatida, T., et al., 2023. Is climate really changing? Insights from analysis of 30-year daily CHIRPS and station rainfall data in Zimbabwe. Sci. Afr. 19, e01581. https://doi.org/10.1016/j.sciaf.2023.e01581.

Gxokwe, S., et al., 2022. Using cloud computing techniques to monitor long-term variations in eco-hydrological dynamics of small seasonally-flooded wetlands in semi-arid South Africa. J. Hydrol. 612, 128080. https://doi.org/10.1016/j.jhydrol.2022.128080.

Hailesilassie, W.T., Ayenew, T., Tekleab, S., 2023. A comparative study of drought characteristics using meteorological drought indices over the central main Ethiopian Rift. Hydrol. Res. 54 (3), 313–329. https://doi.org/10.2166/nh.2023.091.

Han, X., et al., 2023. The influence of anthropogenic climate change on meteorological drought in the Lancang-Mekong River basin. J. Hydrol. 626 (PB), 130334. https://doi.org/10.1016/j.jhydrol.2023.130334.

Harding, R., et al., 2011. WATCH: current knowledge of the terrestrial global water cycle. J. Hydrometeorol. 12 (6), 1149–1156. https://doi.org/10.1175/JHM-D-11-024.1.

Hayes, M., et al., 2011. Investigating the connections between climate change, drought and agricultural production. In: Handbook on Climate Change and Agriculture, pp. 73–86, https://doi.org/10.4337/9780857939869.00011.

Hayes, M., Svoboda, M., Le Comte, D., Redmond, K.T., Pasteris, P., 2005. Drought monitoring: new tools for the 21st century. In: Wilhite, D. (Ed.), Drought and Water Crises: Science, Technology, and Management Issues. CRC Press, pp. 53–69.

Henchiri, M., et al., 2021. Meteorological drought analysis and return periods over North and West Africa and linkage with El Niño–Southern Oscillation (ENSO). Remote Sens. 13 (23). https://doi.org/10.3390/rs13234730.

Hulme, M., 1992. A 1951–80 global land precipitation climatology for the evaluation of general circulation models. Clim. Dyn. 7, 57–72.

Hulme, M., et al., 2001. African climate change: 1900-2100. Clim. Res. 17 (2 Special 8), 145–168. https://doi.org/10.3354/cr017145.

Ifejika Speranza, C., Kiteme, B., Wiesmann, U., 2008. Droughts and famines: the underlying factors and the causal links among agro-pastoral households in semi-arid Makueni district, Kenya. Glob. Environ. Change 18 (1), 220–233. https://doi.org/10.1016/j.gloenvcha.2007.05.001.

Intergovernmental Panel on Climate Change (IPCC), et al., 2023. Climate Change 2022—Impacts, Adaptation and Vulnerability. Cambridge University Press, https://doi.org/10.1017/9781009325844.

Jacobi, J., et al., 2013. A tool for calculating the Palmer drought indices. Water Resour. Res. 49 (9), 6086–6089. https://doi.org/10.1002/wrcr.20342.

Kabat, P., et al., 2002. Coping With Impacts of Climate Variability and Climate Change in Water Management: A Scoping Paper, Hydro Delft. International Secretariat of the Dialogue on Water and Climate, Wageningen.

Kenabatho, P.K., Parida, B.P., Moalafhi, D.B., 2012. The value of large-scale climate variables in climate change assessment: the case of Botswana's rainfall. Phys. Chem. Earth 50–52, 64–71. https://doi.org/10.1016/j.pce.2012.08.006.

Keyantash, J., Dracup, J.A., 2002. The quantification of drought: an evaluation of drought indices. Bull. Am. Meteorol. Soc. 83 (8), 1167–1180. https://doi.org/10.1175/1520-0477(2002)083<1191:TQODAE>2.3.CO;2.

Knapp, A.K., Fay, P.A., Blair, J.M., Collins, S.L., Smith, M.D., Carlisle, J.D., Harper, C.W., Danner, B.T., Lett, M.S., McCarron, J.K., 2002. Rainfall variability, carbon cycling, and plant species diversity in a mesic grassland. Science 298 (5601), 2202–2205.

Knox, J.W., et al., 2010. A preliminary assessment of climate change impacts on sugarcane in Swaziland. Agric. Syst. 103 (2), 63–72. https://doi.org/10.1016/j.agsy.2009.09.002.

Kruger, A.C., 2006. Observed trends in daily precipitation indices in South Africa: 1910-2004. Int. J. Climatol. 26 (15), 2275–2285. https://doi.org/10.1002/joc.1368.

Kuhn, C., et al., 2019. Performance of Landsat-8 and Sentinel-2 surface reflectance products for river remote sensing retrievals of chlorophyll-a and turbidity. Remote Sens. Environ. 224 (February), 104–118. https://doi.org/10.1016/j.rse.2019.01.023.

Kundzewicz, Z.W., et al., 2008. The implications of projected climate change for freshwater resources and their management. Hydrol. Sci. J. 53 (1), 3–10. https://doi.org/10.1623/hysj.53.1.3.

Kunedzimwe, F., Kupika, O.L., Kusangaya, S., 2021. Climate change vulnerability assessment for riparian based livelihoods in semi arid parts of Zimbabwe (a geotechnological approach). Environ. Hazards 5 (4), 145.

Kuri, F., et al., 2020. Differential impact of remotely sensed dry dekads on maize yield in Zimbabwe. Geocarto Int. 35 (13), 1489–1510. https://doi.org/10.1080/10106049.2019.1583774.

Kusangaya, S., et al., 2014. Impacts of climate change on water resources in southern Africa: a review. Phys. Chem. Earth 67–69, 47–54. https://doi.org/10.1016/j.pce.2013.09.014.

Kusangaya, S., et al., 2021. Climate change impact on hydrological regimes and extreme events in southern Africa. In: Climate Change and Water Resources in Africa: Perspectives and Solutions Towards an Imminent Water Crisis. Springer International Publishing, pp. 87–129, https://doi.org/10.1007/978-3-030-61225-2_5.

Lee, H.J., Bell, M.L., Koutrakis, P., 2023. Drought and ozone air quality in California: identifying susceptible regions in the preparedness of future drought. Environ. Res. 216, 114461.

Levitt, J., 1972. Responses of Plants to Environmental Stresses. Academic Press, New York.

Levy, K., Smith, S.M., Carlton, E.J., 2018. Climate change impacts on waterborne diseases: moving toward designing interventions. Curr. Environ. Health Rep. 5, 272–282.

Magadza, C.H.D., 1996. Climate change: some likely multiple impacts in southern Africa. In: Climate Change and World Food Security. Springer, Berlin, Heidelberg, pp. 449–483, https://doi.org/10.1007/978-3-642-61086-8_17.

Malisawa, M.S., De Rautenbach, C.J.W., 2012. Evaluating water scarcity in the Southern African Development Community (SADC) region by using a climate moisture index (CMI) indicator. Water Sci. Technol. Water Supply 12 (1), 45–55. https://doi.org/10.2166/ws.2011.113.

Manatsa, D., Matarira, C.H., Mukwada, G., 2011. Relative impacts of ENSO and Indian Ocean dipole/zonal mode on east SADC rainfall. Int. J. Climatol. 31 (4), 558–577. https://doi.org/10.1002/joc.2086.

Maphosa, B., 1994. Lessons from the 1992 drought in Zimbabwe: the quest for alternative food policies. Nord. J. Afr. Stud. 3 (1), 53–58.

Martinez-Vilalta, J., Anderegg, W.R., Sapes, G., Sala, A., 2019. Greater focus on water pools may improve our ability to understand and anticipate drought-induced mortality in plants. New Phytol. 223 (1), 22–32.

Marumbwa, F.M., Cho, M.A., Chirwa, P.W., 2020. An assessment of remote sensing-based drought index over different land cover types in southern Africa. Int. J. Remote Sens. 41 (19), 1–15. https://doi.org/10.1080/01431161.2020.1757783.

Marumbwa, F.M., Cho, M.A., Chirwa, P.W., 2021. Geospatial analysis of meteorological drought impact on southern Africa biomes. Int. J. Remote Sens. 42 (6), 2155–2173. https://doi.org/10.1080/01431161.2020.1851799.

Masroor, M., et al., 2020. Exploring climate variability and its impact on drought occurrence: evidence from Godavari Middle sub-basin, India. Weather Clim. Extrem. 30 (August), 100277. https://doi.org/10.1016/j.wace.2020.100277.

Matondo, J.I., Peter, G., Msibi, K.M., 2005. Managing water under climate change for peace and prosperity in Swaziland. Phys. Chem. Earth 30 (11–16 Special issue), 943–949. https://doi.org/10.1016/j.pce.2005.08.041.

McBean, G., Ajibade, I., 2009. Climate change, related hazards and human settlements. Curr. Opin. Environ. Sustain. 1 (2), 179–186. https://doi.org/10.1016/j.cosust.2009.10.006.

McKee, T.B., Doesken, N.J., Kleist, J., 1993. The relationship of drought frequency and duration to time scales. In: Proceedings of the 8th Conference on Applied Climatology, California, pp. 179–183.

Measho, S., et al., 2019. Spatio-temporal analysis of vegetation dynamics as a response to climate variability and drought patterns in the semiarid region, Eritrea. Remote Sens. 11 (6). https://doi.org/10.3390/RS11060724.

Mirza, M.M.Q., 2003. Climate change and extreme weather events: can developing countries adapt? Clim. Policy 3 (3), 233–248. https://doi.org/10.1016/S1469-3062(03)00052-4.

Mishra, A.K., Singh, V.P., 2010. A review of drought concepts. J. Hydrol. 391 (1–2), 202–216. https://doi.org/10.1016/j.jhydrol.2010.07.012.

Mishra, A.K., Singh, V.P., 2011. Drought modeling—a review. J. Hydrol. 403 (1–2), 157–175.

Morid, S., Smakhtin, V., Moghaddasi, M., 2006. Comparison of seven meteorological indices for drought monitoring in Iran. Int. J. Climatol. 26 (7), 971–985. https://doi.org/10.1002/joc.1264.

Mukherjee, S., Mishra, A., Trenberth, K.E., 2018. Climate change and drought: a perspective on drought indices. Curr. Clim. Change Rep. 4, 145–163.

Müller, L.M., Bahn, M., 2022. Drought legacies and ecosystem responses to subsequent drought. Glob. Change Biol. 28 (17), 5086–5103.

Mutanga, O., et al., 2020. Environmental sustainability with remote sensing in Africa. ISPRS J. Photogramm. Remote Sens. 169, 212–213. https://doi.org/10.1016/j.isprsjprs.2020.09.007.

Nagarajan, R., 2009. Drought indices. In: Drought Assessment. vol. 3, pp. 160–204, https://doi.org/10.1007/978-90-481-2500-5_5.

Ndayiragije, J.M., Li, F., 2022. Effectiveness of drought indices in the assessment of different types of droughts, managing and mitigating their effects. Climate 10 (9), 125. https://doi.org/10.3390/cli10090125.

New, M., Hewitson, B., Stephenson, D.B., Tsiga, A., Kruger, A., Manhique, A., Gomez, B., Coelho, C.A., Masisi, D.N., Kululanga, E., Mbambalala, E., 2006. Evidence of trends in daily climate extremes over southern and west Africa. J. Geophys. Res.: Atmos. 111 (D14).

Ngwenya, M., Simatele, M.D., 2024. Modeling future (2021–2050) meteorological drought characteristics using CMIP6 climate scenarios in the Western Cape Province, South Africa. Model. Earth Syst. Environ. 10, 2957–2975. https://doi.org/10.1007/s40808-023-01937-6.

Nhemachena, C., Mano, R., 2007. Assessment of the economic impacts of climate change on agriculture in Zimbabwe: a ricardian approach. In: Policy Research Working Paper 4292. The World Bank, p. 47, https://doi.org/10.1596/1813-9450-4292.

Nicholson, S.E., 2001. Climatic and environmental change in Africa during the last two centuries. Clim. Res. 17 (2 Special 8), 123–144. https://doi.org/10.3354/cr017123.

Nikulin, G., et al., 2012. Precipitation climatology in an ensemble of CORDEX-Africa regional climate simulations. J. Clim. 25 (18), 6057–6078. https://doi.org/10.1175/JCLI-D-11-00375.1.

Núñez, J., et al., 2014. On the use of Standardized Drought Indices under decadal climate variability: critical assessment and drought policy implications. J. Hydrol. 517, 458–470. https://doi.org/10.1016/j.jhydrol.2014.05.038.

Obeagu, E.I., Obeagu, G.U., 2024. Implications of climatic change on sickle cell anemia: a review. Medicine 103 (6), e37127.

Ojara, M.A., et al., 2022. Evaluation of drought, wet events, and climate variability impacts on maize crop yields in East Africa during 1981–2017. Int. J. Plant Prod. 16 (1), 41–62. https://doi.org/10.1007/s42106-021-00178-w.

Orimoloye, I.R., et al., 2022. Drought: a common environmental disaster. Atmosphere 13 (1). https://doi.org/10.3390/atmos13010111.

Osman, M., Zaitchik, B.F., Badr, H.S., Otkin, J., Zhong, Y., Lorenz, D., Anderson, M., Keenan, T.F., Miller, D.L., Hain, C., Holmes, T., 2022. Diagnostic classification of flash drought events reveals distinct classes of forcings and impacts. J. Hydrometeorol. 23 (2), 275–289.

Özger, M., Mishra, A.K., Singh, V.P., 2009. Low frequency drought variability associated with climate indices. J. Hydrol. 364 (1–2), 152–162. https://doi.org/10.1016/j.jhydrol.2008.10.018.

Pacheco, J., et al., 2022. Influence of ENSO on droughts and vegetation in a high mountain equatorial climate basin. Atmosphere 13 (12). https://doi.org/10.3390/atmos13122123.

Palmer, W.C., 1965. Meteorological Drought. vol. 30 US Department of Commerce, Weather Bureau.

Palmer, W.C., 1968. Keeping track of crop moisture conditions, nationwide: the new crop moisture index. Weatherwise 21 (4), 156–161. https://doi.org/10.1080/00431672.1968.9932814.

Popova, Z., et al., 2014. Vulnerability of Bulgarian agriculture to drought and climate variability with focus on rainfed maize systems. Nat. Hazards 74 (2), 865–886. https://doi.org/10.1007/s11069-014-1215-3.

Reason, J.C., 2007. Tropical cyclone Dera, the unusual 2000/01 tropical cyclone season in the South West Indian Ocean and associated rainfall anomalies over Southern Africa. Meteorol. Atmos. Phys. 97 (1–4), 181–188. https://doi.org/10.1007/s00703-006-0251-2.

Reason, C.J.C., Keibel, A., 2004. Tropical cyclone Eline and its unusual penetration and impacts over the southern African mainland. Weather Forecast. 19 (5), 789–805. https://doi.org/10.1175/1520-0434(2004)019<0789:tceaiu>2.0.co;2.

Romanello, M., McGushin, A., Di Napoli, C., Drummond, P., Hughes, N., Jamart, L., Kennard, H., Lampard, P., Rodriguez, B.S., Arnell, N., Ayeb-Karlsson, S., 2021. The 2021 report of the Lancet Countdown on health and climate change: code red for a healthy future. Lancet 398 (10311), 1619–1662.

Rosenzweig, C., et al., 2001. Climate change and extreme weather events—implications for food production, plant diseases, and pests. Glob. Change Human Health 2 (2), 90–104. https://doi.org/10.1023/A:1015086831467.

Ross, T.F., Lott, J.N., 2004. A climatology of recent extreme weather and climate events. Bull. Am. Meteorol. Soc., 5079–5088.

Rossi, G., 2003. An integrated approach to drought mitigation in Mediterranean regions. In: Tools for Drought Mitigation in Mediterranean Regions, pp. 3–18, https://doi.org/10.1007/978-94-010-0129-8_1.

Rossi, G., et al., 1992. On regional drought estimation and analysis. Water Resour. Manag. 6 (4), 249–277. https://doi.org/10.1007/BF00872280.

Rouse, J.W., Space, G., 1978. Monitoring the vernal advancement of retrogradation of natural vegetation. In: Proceedings of the 3rd Earth Resource Technology Satellite (ERTS) Symposium.

Salehpour Jam, A., et al., 2018. Investigation of drought characteristics in north-west of Iran using Deciles Index. Watershed Eng. Manag. 10 (4), 552–563.

Salvador, C., et al., 2020. Effects of droughts on health: diagnosis, repercussion, and adaptation in vulnerable regions under climate change. Challenges for future research. Sci. Total Environ. 703, 134912.

Salvador, C., et al., 2023. Public health implications of drought in a climate change context: a critical review. Annu. Rev. Public Health 44, 213–232.

Saneoka, H., Moghaieb, R.E., Premachandra, G.S., Fujita, K., 2004. Nitrogen nutrition and water stress effects on cell membrane stability and leaf water relations in *Agrostis palustris* Huds. Environ. Exp. Bot. 52 (2), 131–138.

Sári, D., Ferroudj, A., Dávid, S., El-Ramady, H., Faizy, S.E.D., Ibrahim, S., Mansour, H., Brevik, E. C., Solberg, S.Q., Prokisch, J., 2024. Drought stress under a nano-farming approach: a review. Egypt. J. Soil Sci. 64 (1), 135–151.

Schulze, R., 2000. Transcending scales of space and time in impact studies of climate and climate change on agrohydrological responses. Agric. Ecosyst. Environ. 82 (1–3), 185–212. https://doi.org/10.1016/S0167-8809(00)00226-7.

Schulze, R.E., Bulcock, L.M., 2011. Climate change and surface water and deep percolation losses from irrigated areas under demand irrigation scheduling. In: A 2011 Perspective on Climate Change and the South African Water Sector, pp. 243–248.

Seaton, D., Dube, T., Mazvimavi, D., 2020. Use of multi-temporal satellite data for monitoring pool surface areas occurring in non-perennial rivers in semi-arid environments of the Western Cape, South Africa. ISPRS J. Photogramm. Remote Sens. 167, 375–384. https://doi.org/10.1016/j.isprsjprs.2020.07.018.

Shabelle, L., 2011. Drought-related food insecurity: a focus on the horn of Africa. In: Drought Emergency. FAO, pp. 1–7.

Sharara, A., et al., 2022. Fine-scale multi-temporal and spatial analysis of agricultural drought in agro-ecological regions of Zimbabwe. Geomat. Nat. Hazards Risk 13 (1), 1342–1365. https://doi.org/10.1080/19475705.2022.2072774.

Shekede, M.D., et al., 2023. Using Landsat satellite imagery to monitor the spatial and temporal dynamics of aquatic weed extent in Lakes Chivero and Manyame, located in an urban catchment of Zimbabwe. Water SA 49 (1), 46–55. https://doi.org/10.17159/wsa/2023.v49.i1.3950.

Simms, A., 2015. Prelims—Africa: Up in Smoke? Oxfam Publishing, https://doi.org/10.3362/9780855986520.000.

Sofia, G., Zaccone, C., Tarolli, P.T., 2023. Agricultural drought severity in NE Italy: variability, bias, and future scenarios. Int. Soil Water Conserv. Res. 12, 403–418. https://doi.org/10.1016/j.iswcr.2023.07.003.

Spinoni, J., et al., 2015. The biggest drought events in Europe from 1950 to 2012. J. Hydrol. Reg. Stud. 3, 509–524. https://doi.org/10.1016/j.ejrh.2015.01.001.

Sridhara, S., Chaithra, G.M., Gopakkali, P., 2021. Assessment and monitoring of drought in Chitradurga district of Karnataka using different drought indices. J. Agrometeorol. 23 (2), 221–227.

Suryabhagavan, K.V., 2017. GIS-based climate variability and drought characterization in Ethiopia over three decades. Weather Clim. Extrem. 15 (November), 11–23. https://doi.org/10.1016/j.wace.2016.11.005.

Svoboda, M., Hayes, M., Wood, D., 2012. Standardized Precipitation Index: User Guide.

Tadross, M., Jack, C., Hewitson, B., 2005. On RCM-based projections of change in southern African summer climate. Geophys. Res. Lett. 32 (23), 1–4. https://doi.org/10.1029/2005GL024460.

Tesfamariam, B.G., Gessesse, B., Melgani, F., 2019. Characterizing the spatiotemporal distribution of meteorological drought as a response to climate variability: the case of rift valley lakes basin of Ethiopia. Weather Clim. Extrem. 26 (November), 100237. https://doi.org/10.1016/j.wace.2019.100237.

Tesfamariam, B.G., Gessesse, B., Melgani, F., 2022. MaxEnt-based modeling of suitable habitat for rehabilitation of Podocarpus forest at landscape-scale. Environ. Syst. Res. 11 (1), 4. https://doi.org/10.1186/s40068-022-00248-6.

Thierfelder, C., Mwila, M., Rusinamhodzi, L., 2013. Conservation agriculture in eastern and southern provinces of Zambia: long-term effects on soil quality and maize productivity. Soil Tillage Res. 126, 246–258. https://doi.org/10.1016/j.still.2012.09.002.

Tsakiris, G., et al., 2007. Drought characterization. In: Drought Management Guidelines Technical Annex. 58, pp. 85–102.

UNDRR, 2021. Annual Report 2020. United Nations Office for Disaster Risk Reduction, 4.

Van Loon, A.F., 2015. Hydrological drought explained. Wiley Interdiscip. Rev.: Water 2 (4), 359–392.

Vicente-Serrano, S.M., Peña-Angulo, D., Murphy, C., López-Moreno, J.I., Tomas-Burguera, M., Dominguez-Castro, F., Tian, F., Eklundh, L., Cai, Z., Alvarez-Farizo, B., Noguera, I., 2021. The complex multi-sectoral impacts of drought: evidence from a mountainous basin in the Central Spanish Pyrenees. Sci. Total Environ. 769, 144702.

Vilonen, L., Ross, M., Smith, M.D., 2022. What happens after drought ends: synthesizing terms and definitions. New Phytol. 235 (2), 420–431.

Vörösmarty, C.J., et al., 2005. Geospatial indicators of emerging water stress: an application to Africa. Ambio 34 (3), 230–236. https://doi.org/10.1579/0044-7447-34.3.230.

Wang, Z., et al., 2018. Climate change enhances the severity and variability of drought in the Pearl River Basin in South China in the 21st century. Agric. For. Meteorol. 249 (September), 149–162. https://doi.org/10.1016/j.agrformet.2017.12.077.

Warburton, M.L., Schulze, R.E., Jewitt, G.P.W., 2012. Hydrological impacts of land use change in three diverse South African catchments. J. Hydrol. 414–415, 118–135. https://doi.org/10.1016/j.jhydrol.2011.10.028.

Ward, M.N., et al., 1999. Climate variability in northern Africa: understanding droughts in the Sahel and the Maghreb. In: Beyond El Niño, pp. 119–140, https://doi.org/10.1007/978-3-642-58369-8_6.

Wardlow, B., Anderson, T., Tadesse, C., Hain, W., Rodell, M., Thenkabail, P.S., 2016. Remote sensing of drought: emergence of a satellite-based monitoring toolkit for the United States. In: Remote Sensing of Water Resources, Disasters, and Urban Studies. CRC Press, pp. 367–398.

West, H., Quinn, N., Horswell, M., 2019. Remote sensing for drought monitoring & impact assessment: progress, past challenges and future opportunities. Remote Sens. Environ. 232, 1–83. https://doi.org/10.1016/j.rse.2019.111291.

White, D.A., 2010. Quantification of agricultural drought for effective drought mitigation and preparedness: key issues and challenges. In: Agricultural Drought Indices Proceedings of a WMO Expert Meeting Held in Murcia, Spain, June, p. 15.

Wilhite, D.A., 2005. Drought and Water Crises: Science, Technology, and Management Issues. CRC Press.

Wilhite, D.A., 2016. Managing drought risk in a changing climate. Clim. Res. 70 (2–3), 99–102. https://doi.org/10.3354/cr01430.

Wilhite, D.A., 2017. National drought management policy guidelines: a template for action. In: Drought and Water Crises: Integrating Science, Management, and Policy, second ed. vol. 3, pp. 55–94, https://doi.org/10.1201/b22009.

Wilhite, D.A., Glantz, M.H., 1985. Understanding the drought phenomenon: the role of definitions. Water Int. 10, 111–120.

Wilhite, D.A., Pulwarty, R.S., 2017a. Drought and water crises: lessons drawn, some lessons learned, and the road ahead. In: Drought and Water Crises: Integrating Science, Management, and Policy, second ed, pp. 515–528, https://doi.org/10.1201/b22009.

Wilhite, D.A., Pulwarty, R.S., 2017b. Drought as hazard: understanding the natural and social context. In: Drought and Water Crises: Integrating Science, Management, and Policy. vol. 3, pp. 3–20, https://doi.org/10.1201/b22009.

Wilhite, D.A., Svoboda, M.D., 2005. Monitoring drought in the United States: status and trends. In: Monitoring and Predicting Agricultural Drought., https://doi.org/10.1093/oso/9780195162349.003.0017.

Wilhite, D.A., Svoboda, M.D., Hayes, M.J., 2007. Understanding the complex impacts of drought: a key to enhancing drought mitigation and preparedness. Water Resour. Manag. 21 (5), 763–774. https://doi.org/10.1007/s11269-006-9076-5.

Wilhite, D.A., Sivakumar, M.V.K., Pulwarty, R., 2014. Managing drought risk in a changing climate: the role of national drought policy. Weather Clim. Extrem. 3, 4–13. https://doi.org/10.1016/j.wace.2014.01.002.

Wilk, J., Andersson, L., Warburton, M., 2013. Adaptation to climate change and other stressors among commercial and small-scale South African farmers. Reg. Environ. Change 13 (2), 273–286. https://doi.org/10.1007/s10113-012-0323-4.

Winkler, K., Gessner, U., Hochschild, V., 2017. Identifying droughts affecting agriculture in Africa based on remote sensing time series between 2000-2016: rainfall anomalies and vegetation condition in the context of ENSO. Remote Sens. 9 (8), 831. https://doi.org/10.3390/rs9080831.

Winkler, D.E., et al., 2018. The interaction of drought and habitat explain space–time patterns of establishment in saguaro (Carnegiea gigantea). Ecology 99 (3), 621–631. https://doi.org/10.1002/ecy.2124.

Wuebbles, D., et al., 2016. Our Globally Changing Climate.

Yamba, F.D., et al., 2011. Climate change/variability implications on hydroelectricity generation in the Zambezi River Basin. Mitig. Adapt. Strateg. Glob. Chang. 16 (6), 617–628. https://doi.org/10.1007/s11027-011-9283-0.

Yanda, P., et al., 2010. Climate change, adaptation and higher education: securing our future. SARUA leadership dialogue. In: SARUA Leadership Dialogue Series. Southern African Regional Universities Association (SARUA).

Yihdego, Y., Vaheddoost, B., Al-Weshah, R.A., 2019. Drought indices and indicators revisited. Arab. J. Geosci. 12 (3), 1–12. https://doi.org/10.1007/s12517-019-4237-z.

Zaniolo, M., Giuliani, M., Castelletti, A., 2019. Data-driven modeling and control of droughts. IFAC-PapersOnLine 52 (23), 54–60. https://doi.org/10.1016/j.ifacol.2019.11.009.

Zeng, N., 2003. Drought in the Sahel. Science 302 (5647), 999–1000. https://doi.org/10.1126/science.1090849.

Zhao, Y., et al., 2022. Drought monitoring and performance evaluation based on machine learning fusion of multi-source remote sensing drought factors. Remote Sens. 14 (24), 6398. https://doi.org/10.3390/rs14246398.

Zhao, M., Running, S.W., 2010. Drought-induced reduction in global terrestrial net primary production from 2000 through 2009. Science 329 (5994), 940–943.

Zhou, H., et al., 2023. Climatological spatial scales of meteorological droughts in China and associated climate variability. J. Hydrol. 617 (PB), 129056. https://doi.org/10.1016/j.jhydrol.2022.129056.

Zia, R., Nawaz, M.S., Siddique, M.J., Hakim, S., Imran, A., 2021. Plant survival under drought stress: implications, adaptive responses, and integrated rhizosphere management strategy for stress mitigation. Microbiol. Res. 242, 126626.

Ziervogel, G., et al., 2014. Climate change impacts and adaptation in South Africa. Wiley Interdiscip. Rev. Clim. Change 5 (5), 605–620.

Chapter 11

Climate variability and aquatic weed proliferation

Kgabo Humphrey Thamaga[a] and Mahlatse Kganyago[b]
[a]Department of GIS and Remote Sensing, University of Fort Hare, Alice, South Africa, [b]Department of Geography, Environmental Management and Energy Studies, University of Johannesburg, Johannesburg, South Africa

Chapter outline

1 Introduction

Global climate change has brought adverse meteorological conditions such as heatwaves and climate-related disasters such as floods and droughts, which have increased in intensity and duration. The Kyoto Protocol aims to halt the increase in temperatures by 2°C by 2030. However, the latest reports show that the 2°C increase has already been reached in various parts of the world due to anthropogenic climate change (Abbass et al., 2022). For example, Spain, France, and Europe recorded the highest temperature on record in 2023. Temperature plays an essential role in regulating the hydrological cycle, causing evaporation from open water surfaces and soils and transpiration from plants. In open water surfaces such as rivers and lakes, the effect of high temperatures and low precipitation (i.e., drought) can be seen by shrinkage in the water surface area, water depth, and reduction in the streamflow rate, while the prolonged transpiration from plant leaves causes reduced vigor and stress (Thamaga and Dube, 2019). Apart from these, the temperature regulates the physicochemical reactions and microbial activity in the water and, thus, the aquatic life.

Remote Sensing of Climate. https://doi.org/10.1016/B978-0-443-21731-9.00018-1
Copyright © 2024 Elsevier Inc. All rights are reserved, including those for text and data mining, AI training, and similar technologies.

Unfortunately, anthropogenic activities compound the already dire situation, causing high pollution of the world's most precious resource. The rapidly changing land use from a natural state to build-up areas, mines and quarries, and croplands have been linked to a decline in water quality (Thamaga and Dube, 2018a, 2019). Amid climate change and variability, governments, communities, and land managers must be equipped with accurate and relevant information to aid decisions and drive strategic policymaking.

One of the most concerning impacts of climatic variability and change is the permanent shrinkage of water surface area due to the introduction and proliferation of weeds. Sedimentation of the dams, lakes, and rivers, as well as the drought conditions, which temporarily shrink the water levels, provide a conducive environment for weeds to spread, usually along the edges of rivers, lakes, and dams. Alarmingly, floating weeds such as algal bloom and water hyacinth reduce the amount of light penetrating the water and thus alter the chemical reactions and composition of the water, for example, pH. Therefore, monitoring the response of aquatic environments to various climate-related phenomena is critical to ensure sustainable aquatic ecosystems and ecosystem services. Unfortunately, the accurate and reliable traditional monitoring using in situ instrumentation and field surveys has several limitations, such as small spatial and temporal coverage, high operational costs, and the risk of human errors. The adoption of an Earth observation (EO) approach is critical to addressing these limitations and information needs of water resources management agencies at various scales.

EO satellites are particularly suitable to address the information needs of multiple stakeholders, such as regional policy and decision-makers, national and provincial governments, and other interested parties, such as nongovernmental organizations and land managers. To date, various EO satellites exist, with the breakthrough achieved by the availability of high-resolution satellite sensors such as Sentinel-2 Multispectral Instrument (MSI), which provide essential spectral wavebands (such as three red-edge bands) to facilitate vegetation characterization using many techniques at our disposal. Moreover, the continuous availability of Landsat mission data at a consistent resolution of 30 m and thermal bands at 100–120 m facilitates the long-term analysis of the dynamics of aquatic weeds in response to climate stressors. In contrast, the availability of several narrow wavebands on Sentinel-3 Sea and Land Surface Temperature Radiometer (SLTR) provides opportunities to explore the changes in water physicochemical composition and analyze the effect of climate variability and change in lakes and oceans in the absence of operational hyperspectral satellites. Indeed, satellite sensors such as the Environmental Mapping and Analysis Program and Hyperspectral Infrared Imager enabled much research on the suitable wavebands for mapping aquatic weeds, but they are limited to specific geographic areas, and acquisitions are project-based due to large data volumes. The presence of archives from predecessors such as Moderate Resolution Imaging Spectroradiometer (MODIS) and Medium Resolution Imaging Spectrometer (MERIS) is also critical to analyzing aquatic environment trends.

This chapter is organized as follows: Section 2 addresses the impacts of pollution, climate variability, and aquatic weed proliferation on open-water ecosystems; Section 3 discusses remote sensing of climate variability and aquatic weed proliferation by providing an overview of satellite-based products for climate variability and aquatic weed proliferation and advanced algorithms for assessing the impacts of climate variability on aquatic weed proliferation. Section 4 focuses on the challenges of remote sensing sensors for aquatic weed proliferation; Section 5 provides an overview of future remote sensing satellite-based sensors; and Section 6 provides a summary and conclusion of the chapter.

2 Impacts of pollution, climate variability, and aquatic species proliferation on open-water ecosystems

Freshwater ecosystems are rapidly declining worldwide due to increased pressures (i.e., environmental and anthropogenic) resulting from poor sewage disposal, agriculture, industrial intensification, and climate change. These pressures are disposed directly into rivers, dams, and lakes. However, increased temperatures and the movement of agricultural materials, such as fertilizers, through runoff during rainy seasons affect the cycle of aquatic life, especially in polluted freshwater habitats. But these also encourage the growth of aquatic weeds, like Kariba weed (*Salvinia molesta*), parrot's feather (*Myriophyllum aquaticum*), water lettuce (*Pistia stratiotes*), and water hyacinth (*Eichhornia crassipes*), which are among the worst species globally and have detrimental effects on the environment and the economy on invaded freshwater ecosystems (Singh et al., 2020; Shekede et al., 2023). Estimates show that aquatic weeds can quadruple their biomass through asexual reproduction in 2 weeks under ideal growing conditions and environmental circumstances of waterbodies (Thamaga and Dube, 2019). For instance, for an aquatic weed species such as water hyacinth, within 8 months, 10 plants may spawn about 655,360 plants, which may clog approximately half a hectare of the surface area (Gunnarsson and Petersen, 2007). However, high nutrient content recharge and rising temperatures are the main factors influencing the establishment and reproduction of aquatic species in open water bodies. One of the most invasive aquatic plants globally has spread most to water ecosystems (Singh et al., 2020; Shekede et al., 2023). Aquatic weed infestation, such as water hyacinth, can chock freshwater bodies by obstructing light penetration. If persistent, this will pose a crisis to the already scarce freshwater resources. The momentousness of these invasive aquatic weeds is exhibited by the scale of invasion to a global extent. To date, multiple aquatic weeds have spread throughout the world, threatening freshwater ecosystems.

3 Remote sensing of climate variability and aquatic weed proliferation

Satellite-based remote sensing emerged to counter the shortcomings of conventional approaches used to monitor freshwater ecosystems in dams, rivers, or lakes. Multispectral sensors, for instance, offer effective, inexpensive, and short-term monitoring of aquatic weed proliferation in various climatic scenarios. In large water bodies, aquatic weed proliferation have been detected, identified, and monitored spatiotemporally owing to the critical role of remote sensing and their derivatives. The analysis of aquatic weed infestation and climate influence has been evaluated using sensor capabilities such as Sentinel-1 Synthetic Aperture Radar (SAR), -2 MSI, -3 SLTR, Landsat Operational Land Imager (OLI), and ERA5, which vary in terms of spatial resolution, spectral bands, and vegetation spectral indices (Nyamekye et al., 2021; Shekede et al., 2023). For instance, Dube et al. (2017) investigated the ability of Landsat 8 satellite data derived spectral and vegetation indices generated to detect and discriminate water hyacinth in freshwater ecosystems of Lake Manyame in Zimbabwe. Although Landsat performs well in the detection process, not all these remotely sensed variables produce plausible performance. Most of the indices used were relevant in enhancing classification accuracies (Vidhya et al., 2014). Despite the high degree of reflection in the near-infrared (NIR) region, the Normalized Difference Vegetation Index (NDVI) has been found to positively correlate with plant health or vigor, concentrated green pigments, or active photosynthetic rate.

Over the years, several remote sensing satellite datasets have been used to evaluate threats and the continuous spread of aquatic weeds and climate variability on the already scarce water resource. The use of remote sensing sensors with high and improved spectral resolution coupled with the development of advanced machine and deep learning algorithms could have potential to improve the detection and classification accuracies of aquatic weeds. These algorithms emerged as proxies to conventional parametric algorithms; however, they produce high classification accuracies, are more accurate, and are capable of processing high datasets. Also, they have been implemented in the classification of vegetation stress and medium to long-term vegetation changes, effects of climate variability, and impacts of land use and land cover influence of freshwater ecosystems. Spectral discrimination between invasive aquatic weed types in complex environments is a challenging task since diverse species types may have similar spectral signatures. Classification of aquatic weed infestation in narrow river channels and dams using multispectral resolution results in misclassification, which leads to poor accuracies.

4 Advanced algorithms for assessing the impacts of climate variability and aquatic weed proliferation

Parametric techniques such as spectral vegetation indices (SVIs) are well-established since the initial indices, that is, the Ratio Vegetation Index or Simple Ratio (Jordan, 1969), the Difference Vegetation Index (Tucker, 1979), and the NDVI (Rouse et al., 1974; Tucker, 1979), which are based on the red (630–690 nm) and NIR (760–900 nm) bands, several SVIs have emerged, driven mainly by sensor technology developments and exploration for better indices based on previously identified limitations (Kganyago et al., 2024). For example, the saturation, atmospheric sensitivity, and soil influence affecting the NDVI are well-known in the literature, evolving into variants such as the Soil-Adjusted Vegetation Index (Huete, 1988), which includes the soil-adjustment factor, *L*, and the Enhanced Vegetation Index (Liu and Huete, 1995), which incorporates the background adjustment and atmospheric noise compensation using the blue band (450–520 nm). Recently, the wide availability of the red-edge bands in sensors such as RapidEye, Worldview-2, and Sentinel-2 facilitated the development of narrow-band or red-edge indices (Clevers, 1989; Clevers and Verhoef, 1993; Frampton et al., 2013). For instance, the Normalized Difference Red-edge Index (Gitelson and Merzlyak, 1994), later known as NDI45, replaces the NIR and red bands in the NDVI formula with red-edge bands centered at 740 and 705 nm, respectively. Indeed, robust SVIs to atmospheric contamination, residual errors after atmospheric correction, soil background, and saturation are critical for increased separability of aquatic weeds from other vegetation, particularly along the river and lake banks, based on distinct occurrences of biophysical and biochemical parameters. Moreover, they can be input into advanced machine learning algorithms to characterize the spatiotemporal variations of aquatic weeds in response to climate change and variability.

Machine learning algorithms (i.e., nonparametric and nonlinear algorithms) have been prominent, with studies demonstrating their robustness to linearity, multimodality, small datasets, and dimensionality. Algorithms such as Classification and Regression Tree (CART), Random Forest (RF), and Support Vector Machine (SVM) have become standard in many applications, including for aquatic weeds mapping. For instance, Bayable et al. (2023) examined the capability of various machine learning algorithms to detect water hyacinth and to estimate its seasonal spatial coverage in Lake Tana using Landsat-8 and Sentinel-2 data. Their study found that the RF was superior to the SVM and CART, achieving accuracies between 97% and 99% with Sentinel-2 and 95%–98% with Landsat-8 data. Another study by Nyamekye et al. (2021) used the RF algorithm and Landsat Enhanced Thematic Mapper Plus (ETM+) and Operational Land Imager images to characterize the distribution of the various water weeds in 1975, 2003, and 2020 to a mean overall accuracy of 83.44% and mean User Accuracy of 79.24%. Ade et al. (2022) focused on genus-level and species-level discrimination of water hyacinth and water primrose (*Ludwigia* spp.) in Vietnam using RF and Sentinel-2 images, to an overall accuracy of 90% and class-wise accuracies greater than 79%. The study also identified submerged and emergent aquatic vegetation at the community level. Inarguably, the RF algorithm is the most popular, and as demonstrated above, many studies have consistently found better performance. However, other studies have also shown better performance of deep learning algorithms rooted in Artificial Neural Networks. The study by Abeysinghe et al. (2019) investigated the effectiveness of utilizing data acquired by a Remotely Piloted Aerial System with four machine learning algorithms, that is, Neural network (NN), SVM, and k-Nearest Neighbor (kNN), to identify the invasive plant species *Phragmites australis* in the Old Woman Creek estuary, USA. Their results showed that the accuracies were not too different from a parametric algorithm, that is, the Maximum Likelihood Classifier, but NN had a superior performance with an overall accuracy of 94.80%. Rodriguez-Garlito et al. (2023) demonstrated that the spatiotemporal distributions of water hyacinth in the Guadiana River, Spain, can be monitored weekly using multitemporal analysis of Sentinel-2 MSI imagery and Convolutional Neural Network deep learning models. Such capability is critical to observe the seasonal behavior and dynamics of the invasive plants over an extended period and their response to the variability in climate and meteorological variables.

5 Challenges of remote sensing sensors for aquatic weeds proliferation

Accurate mapping and frequent monitoring of the spread of aquatic weeds and the impact of climate variability on freshwater ecosystems continue to be key global challenges. A study by Thamaga and Dube (2018b) demonstrated that studies focused on aquatic weed mapping and modeling used several satellite sensors; however, there are still gaps in future exploration and documentation. The use of multispectral sensors, which are cheap, their low-spatial resolution yields low accuracies, leading to poor classification accuracy. As a result, when detecting and mapping aquatic species, the satellite remote sensing community must make intrinsic trade-offs between spatial resolution, spatial coverage, costs, optimal predictions, and high classification accuracies. High-spatial resolution sensors, such as LIDAR and WorldView-2, can precisely detect and map invasive aquatic weeds. However, these sensors have not yet been tested due to acquisition costs. Lack of training programs and comprehensive documentation complicates the use of open software applications. In the era of cloud computing, major challenges arise in accessing and using these readily available platforms. Most of these platforms require advanced skills such as scripting, and programming in languages, such as Python and JavaScript. Therefore, several users are excluded from the technological developments, making remotely sensed analyses reserved for academic researchers and experts. Indeed, the free availability of platforms, such as Google Earth Engine (GEE), accompanied by various training materials and other documents, and its efficiency has facilitated greater adoption by the research community. This is demonstrated by a previous study by Kumar and Mutanga (2018), indicating that up to 300 journal articles were published between 2011 and mid-year 2017. The intuitiveness of the GEE platform and related ecosystem of services provided by Google, such as developer tools (Earth Engine Apps), Google Drive (popular cloud storage), and Collab, Analytics and Google Cloud, have also facilitated the capability to develop and share EO-based products and thus birthed various operational web applications including popular such as Global Forest Watch (https://www.globalforestwatch.org/) and Climate Engine (https://app.climateengine.org/). Novel Python packages, such as Geemap by Qiusheng Wu (https://geemap.org/), provide those already using Python with access to - cloud-optimized time series data or data cubes. Other initiatives include Digital Earth Africa (https://www.digitalearthafrica.org/) and EO-Africa Dunia service (https://dunia.esa.int/). However, these initiatives are project-based or donor-dependent, making their sustainability questionable. The gap between the expert and user, as well as the availability of services that do not address user needs, can be addressed through greater collaboration between data scientists, spatial and nonspatial data analysts, remote sensing experts, ecologists, and water resource experts. This will also help combat the underutilization and undervaluation of satellite remote sensing data and products.

6 Future remote sensing satellite-based sensors

Satellite-based sensors for the detection, mapping, and monitoring of aquatic weed infestation in freshwater environments have evolved significantly over time. Even though satellite sensors are continuously evolving, there is limited research that maps and monitors the distribution of aquatic weed infestation and the influence of climate variability. However, recent advances in the scientific investigation of satellite remote sensing capabilities are currently advocating the establishment of innovative, reliable systems that might be launched to manipulate biochemical and biophysical spectral data for vegetation mapping. Researchers need to participate in long-term monitoring and future prediction of the aquatic weeds' infestation and the influence of changing climate at a catchment scale. The optimal variables and predictive models that can be used with multispectral datasets require more investigation (Dube et al., 2016). More attention must be paid to the spectral analysis since aquatic weed evaluations can now be completed with the spatial and spectral resolution of multispectral remote sensing systems. Acquired satellite data ought to be utilized as a problem-solving tool for future studies. However, this can be accomplished by having a detailed overview of the effects that climate, terrain, soil, and anthropogenic factors have on the proliferation and spread of invasive aquatic weeds and the rapidity and direction of spread (Thamaga and Dube, 2018a). Furthermore, water resource managers and remote sensing scientists should collaborate to determine variables essential for monitoring and mapping aquatic weed proliferation and climate variability using remotely sensed datasets, meteorological datasets, and their respective spatial scales of investigation. Therefore, to extract detailed aquatic weed information used for strategic management and the alleviation of these species, future research must employ high-resolution tools (i.e., unmanned aerial vehicle (UAV)) and advanced modeling techniques such as deep learning. Additionally, UAVs can be extremely helpful in validating results obtained from freely available satellite multispectral sensors.

7 Summary and conclusion

This chapter provide an overview of the impact of climate variability and aquatic weed proliferation using satellite remote sensing and advanced techniques. Most studies focused on the detection, mapping, and monitoring of aquatic weed infestation and climate variability in freshwater ecosystems, as well as spatial distribution, including water hyacinth and *S. molesta* emergence. Global climate change variability favors rapid growth and uncontrollable spread in polluted, etc. The dearth of suitable spatial data necessitates a greater reliance on remote sensing, which offers improved predictions for the detection and mapping of the spatiotemporal distribution of aquatic weeds, particularly in neglected water bodies, which play a pivotal role in neighboring communities. The arrival of freely and available satellite sensors, with a high revisit time, extensive coverage, and high radiometric, spatial, and spectral resolution, is imperative in eradicating barriers restraining the rapid adoption of remote sensing technologies for the management of aquatic weeds.

References

Abbass, K., Qasim, M.Z., Song, H., Murshed, M., Mahmood, H., 2022. A review of the global climate change impacts, adaptation, and sustainable mitigation measures. Environ. Sci. Pollut. Res. 29, 42539–42559. https://doi.org/10.1007/s11356-022-19718-6.

Abeysinghe, T., Milas, A.S., Arend, K., Hohman, B., Reil, P., Gregory, A., Vázquez-Ortega, A., 2019. Mapping invasive Phragmites australis in the Old Woman Creek estuary using UAV remote sensing and machine learning classifiers. Remote Sens. (Basel) 11.

Ade, C., Khanna, S., Lay, M., Ustin, S.L., Hestir, E.L., 2022. Genus-level mapping of invasive floating aquatic vegetation using Sentinel-2 satellite remote sensing. Remote Sens. (Basel) 14, 1–20.

Bayable, G., Cai, J., Mekonnen, M., Legesse, S.A., Ishikawa, K., Imamura, H., Kuwahara, V.S., 2023. Detection of water hyacinth (*Eichhornia crassipes*) in Lake Tana, Ethiopia, using machine learning algorithms. Water 15, 880. https://doi.org/10.3390/w15050880.

Clevers, J.G.P.W., 1989. Application of a weighted infrared-red vegetation index for estimating leaf Area Index by Correcting for Soil Moisture. Remote Sens. Environ. 29, 25–37.

Clevers, J.G.P.W., Verhoef, W., 1993. LAI estimation by means of the WDVI: a sensitivity analysis with combined PROSPECT-SAIL model. Remote Sens. Rev. 7, 43–64.

Dube, T., Mutanga, O., Ismail, R., 2016. Quantifying aboveground biomass in African environments: a review of the trade-offs between sensor estimation accuracy and costs. Trop. Ecol. 57 (3), 393–405.

Dube, T., Mutanga, O., Sibanda, M., Bangamwabo, V., Shoko, C., et al., 2017. Evaluating the performance of the newly-launched Landsat 8 sensor in detecting and mapping the spatial configuration of water hyacinth (*Eichhornia crassipes*) in inland lakes, Zimbabwe. Phys. Chem. Earth 100, 101–111. https://doi.org/10.1016/j.pce.2017.02.015.

Frampton, W.J., Dash, J., Watmough, G., Milton, E.J., 2013. Evaluating the capabilities of Sentinel-2 for quantitative estimation of biophysical variables in vegetation. ISPRS J. Photogramm. Remote Sens. 82, 83–92.

Gitelson, A., Merzlyak, M.N., 1994. Spectral reflectance changes associated with autumn senescence of Aesculus hippocastanum L. and Acer platanoides L. leaves. Spectral features and relation to chlorophyll estimation. J. Plant Physiol. 143.

Gunnarsson, C.C., Petersen, C.M., 2007. Water hyacinths as a resource in agriculture and energy production: a literature review. Waste Manag. 27, 117–129.

Huete, A.R., 1988. A soil-adjusted vegetation index (SAVI). Remote Sens. Environ. 25.

Jordan, C.F., 1969. Derivation of leaf-area index from quality of light on the forest floor. Ecology 50.

Kganyago, M., Adjorlolo, C., Mhangara, P., Tsoeleng, L., 2024. Optical remote sensing of crop biophysical and biochemical parameters: an overview of advances in sensor technologies and machine learning algorithms for precision agriculture. Comput. Electron. Agric. 218, 108730.

Kumar, L., Mutanga, O., 2018. Google Earth Engine applications since inception: usage, trends, and potential. Remote Sens. 10 (10), 1509. https://doi.org/10.3390/rs10101509.

Liu, H.Q., Huete, A., 1995. Feedback based modification of the NDVI to minimize canopy background and atmospheric noise. IEEE Trans. Geosci. Remote Sens. 33.

Nyamekye, C., Ofosu, S.A., Arthur, R., Osei, G., Appiah, L.B., Kwofie, S., Ghansah, B., Bryniok, D., 2021. Evaluating the spatial and temporal variations of aquatic weeds (Biomass) on Lower Volta River using multi-sensor Landsat Images and machine learning. Heliyon 7, e07080. https://doi.org/10.1016/j.heliyon.2021.e07080.

Rodriguez-Garlito, E.C., Paz-Gallardo, A., Plaza, A., 2023. Monitoring the spatiotemporal distribution of invasive aquatic plants in the Guadiana River, Spain. IEEE J. Sel. Top. Appl. Earth Obs. Remote Sens. 16, 228–241.

Rouse, J.W., Hass, R.H., Schell, J.A., Deering, D.W., Harlan, J.C., 1974. Monitoring the vernal advancement and retrogradation (green wave effect) of natural vegetation. In: Final Report, RSC 1978-4. Texas A & M University, College Station, TX.

Shekede, M.D., Gondo, T., Mavhenge, M.M., Mazhindu, A.N., 2023. Using Landsat satellite imagery to monitor the spatial and temporal dynamics of aquatic weed extent in Lakes Chivero and Manyame, located in an urban catchment of Zimbabwe. Water SA 49 (1), 46–55. https://doi.org/10.17159/wsa/2023.v49.i1.3950.

Singh, G., Reynolds, C., Byrne, M., Rosman, B., 2020. A remote sensing method to monitor water, aquatic vegetation, and invasive water hyacinth at national extents. Remote Sens. 12 (24), 4021. https://doi.org/10.3390/rs12244021.

Thamaga, K.H., Dube, T., 2018a. Remote sensing of invasive water hyacinth (Eichhornia crassipes): a review on applications and challenges. Remote Sens. Appl. 10, 36–46. https://doi.org/10.1016/j.rsase.2018.02.005.

Thamaga, K.H., Dube, T., 2018b. Testing two methods for mapping water hyacinth (Eichhornia crassipes) in the Greater Letaba river system, South Africa: discrimination and mapping potential of the polar-orbiting Sentinel-2 MSI and Landsat 8 OLI sensors. Int. J. Remote Sens. 39 (22), 8041–8059. https://doi.org/10.1080/01431161.2018.1479796.

Thamaga, K.H., Dube, T., 2019. Understanding the seasonal mapping of invasive water hyacinth (*Eichhornia crassipes*) in the Greater Letaba River system using Sentinel-2 satellite data. GIsci. Remote Sens. 56 (8), 1355–1377. https://doi.org/10.1080/15481603.2019.1646988.

Tucker, C.J., 1979. Red and photographic infrared linear combinations for monitoring vegetation. Remote Sens. Environ. 8.

Vidhya, R., Vijayasekaran, D., Ahamed, F.M., Jai, S., Rohini, M., Sinduja, A., et al., 2014. Improved classification of mangroves health status using hyperspectral remote sensing data. Int. Arch. Photogramm. Remote Sens. Spatial Inf. Sci. 40, 667–670. https://doi.org/10.5194/isprsarchives-XL-8-667-2014.

Chapter 12

The impact of climate variability on water quality

Terence Darlington Mushore[a,b], Godfrey Muroyiwa[a], and Teddious Mhizha[a]

[a]*Department of Space Science and Applied Physics, Faculty of Science, University of Zimbabwe, Harare, Zimbabwe,* [b]*Discipline of Geography, School of Agricultural, Earth and Environmental Sciences, University of KwaZulu-Natal, Pietermaritzburg, South Africa*

Chapter outline

1 Introduction

Climate variability is a significant and growing concern for the quality of the Earth's water resources. As the global climate continues to change at an unprecedented rate, the impacts of these variations on water quality have become increasingly apparent. From shifting precipitation patterns and rising temperatures to more frequent and severe extreme weather events, climate variability is altering the availability, distribution, and quality of freshwater sources around the world.

One of the primary ways in which climate variability influences water quality is through changes in precipitation patterns. As temperatures rise and weather patterns become more unpredictable, regions are experiencing alterations in the timing, intensity, and duration of rainfall. This can lead to both droughts and floods, with profound implications for water quality. During periods of drought, water levels in rivers, lakes, and groundwater sources may decrease, resulting in reduced dilution of pollutants and higher concentrations of contaminants such as heavy metals, nutrients, and pathogens (Vörösmarty et al., 2010). Conversely, intense rainfall events can cause runoff of sediment, pesticides, and other pollutants from agricultural lands and urban

Remote Sensing of Climate. https://doi.org/10.1016/B978-0-443-21731-9.00019-3
Copyright © 2024 Elsevier Inc. All rights are reserved, including those for text and data mining, AI training, and similar technologies.

areas, contributing to the degradation of water quality in downstream water bodies (Dudgeon et al., 2006).

Moreover, rising temperatures associated with climate variability can exacerbate the spread of harmful algal blooms in freshwater systems. Warmer temperatures, along with nutrient pollution from sources like agriculture and wastewater treatment plants, create ideal conditions for the rapid growth of algae species that can produce toxic substances harmful to aquatic life and human health (Havens et al., 2019). These algal blooms can deplete oxygen levels in the water, leading to the killing of fish and the accumulation of toxins that pose risks to human consumption of contaminated water or seafood.

Another aspect of climate variability that impacts water quality is the occurrence of extreme weather events, such as hurricanes, cyclones, and heavy storms. These events can result in the disruption of infrastructure, including wastewater treatment plants and sewer systems, leading to the release of untreated sewage and contaminants into water bodies (Ashbolt et al., 2013). In addition, storm surges and flooding can transport pollutants from industrial sites, landfills, and other sources into rivers, estuaries, and coastal waters, posing risks to public health and ecosystems.

Furthermore, interactions between climate variability and water quality have important implications for ecosystem health and biodiversity. Changes in water temperature, nutrient levels, and oxygen concentrations can alter the composition and functioning of aquatic habitats, affecting the survival, reproduction, and distribution of species that depend on these environments (Poff et al., 2002). For example, the acidification of freshwater due to increased carbon dioxide levels in the atmosphere can have detrimental effects on aquatic organisms that are sensitive to changes in pH, such as fish, amphibians, and invertebrates (Hall et al., 2014).

In this chapter, we will explore the multifaceted impacts of climate variability on water quality, examining how these changes are affecting ecosystems, human health, and the overall sustainability of water resources.

2 Climate

Climate refers to the long-term average weather patterns observed in a particular region over a significant period of time, typically spanning decades to centuries (National Aeronautics and Space Administration [NASA], 2021). It encompasses various atmospheric conditions such as temperature, precipitation, humidity, wind patterns, and atmospheric pressure (Intergovernmental Panel on Climate Change [IPCC], 2014). The climate is influenced by a range of factors, including the Earth's geographical location, solar radiation, ocean currents, topography, and the composition of the atmosphere (Houghton et al., 2001). Climate can be classified into different types, such as tropical, arid, Mediterranean, temperate, and polar, based on the prevailing weather conditions in a particular area (NASA, 2021). These classifications are determined by factors like temperature ranges, seasonal variations, and precipitation patterns (IPCC, 2014).

Climate change refers to the long-term alteration in Earth's climate patterns, resulting from natural processes as well as human activities (United Nations Framework Convention on Climate Change [UNFCCC], 2015). Human-induced climate change, primarily driven by the emission of greenhouse gases (such as carbon dioxide) from the burning of fossil fuels, is causing the planet to warm at an unprecedented rate (Houghton et al., 2001). This global warming leads to a range of impacts, including rising temperatures, changing precipitation patterns, sea-level rise, extreme weather events, and disruptions to ecosystems and biodiversity (NASA, 2021). Understanding and addressing climate change is crucial for the long-term well-being of humanity and the planet (UNFCCC, 2015). Efforts to mitigate climate change involve reducing greenhouse gas emissions, transitioning to renewable energy sources, promoting energy efficiency, adopting sustainable land-use practices, and implementing international agreements and policies to limit global warming (IPCC, 2014). Adaptation measures are also necessary to help societies and ecosystems cope with the changes that are already occurring and those expected in the future (NASA, 2021).

3 Climate variability

Climate variability refers to the natural fluctuations and changes in climate conditions over various timescales (Thornton et al., 2014). It encompasses both short-term fluctuations, such as daily or seasonal weather patterns, and long-term changes, such as decadal or centennial climate shifts (IPCC, 2014). Several factors contribute to climate variability, including natural processes within the Earth's system and external influences (IPCC, 2014).

Some of the key drivers of climate variability include:

(a) El Niño-Southern Oscillation (ENSO): ENSO is a natural climate pattern that occurs in the tropical Pacific Ocean. It involves the interaction between the ocean and the atmosphere, leading to periodic changes in sea surface temperatures and atmospheric pressure. ENSO events, which include El Niño (warmer phase) and La Niña (cooler phase), can have significant impacts on global weather patterns (Thornton et al., 2014).

(b) Solar Activity: Variations in solar radiation and activity can influence climate variability. The sun undergoes cycles of increased and decreased energy output, which can impact Earth's climate over long timescales (IPCC, 2014).

(c) Volcanic Activity: Major volcanic eruptions can inject large amounts of volcanic gases and aerosols into the atmosphere. These particles can reflect sunlight back into space, leading to a temporary cooling effect on the planet. The resulting climate response can last for a few years (IPCC, 2014).

(d) Atlantic Multidecadal Oscillation (AMO): The AMO is a natural climate oscillation that affects sea surface temperatures in the North Atlantic Ocean. It operates on a timescale of several decades and influences temperature and precipitation patterns in the surrounding regions (Thornton et al., 2014).

(e) Pacific Decadal Oscillation (PDO): The PDO is a long-lived pattern of Pacific climate variability that affects the North Pacific Ocean. It can influence temperature and precipitation patterns over the Pacific and North America, operating on a timescale of several decades (Thornton et al., 2014).

Climate variability has important implications for ecosystems, agriculture, water resources, and human societies (IPCC, 2014). Understanding and predicting climate variability is crucial for various sectors, including agriculture, energy, and disaster preparedness (Thornton et al., 2014). It is important to differentiate climate variability from climate change, which refers to long-term shifts in average climate conditions due to human activities, primarily the emission of greenhouse gases (IPCC, 2014).

4 Climate change

Climate change refers to long-term shifts in temperature and weather patterns on Earth (IPCC, 2014). It is primarily caused by the increasing concentration of greenhouse gases in the atmosphere, primarily carbon dioxide (CO_2), resulting from human activities such as burning fossil fuels, deforestation, and industrial processes (IPCC, 2014). The greenhouse effect is a natural process that allows the Earth to maintain a habitable temperature. However, human activities have significantly intensified this effect by releasing large amounts of greenhouse gases into the atmosphere. These gases trap heat from the sun, leading to an overall warming of the planet (IPCC, 2014).

This increase in global temperatures has numerous effects on the environment and ecosystems (IPCC, 2014). Some of the key impacts of climate change include:

(a) Rising Temperatures: Global average temperatures have been increasing over the past century, leading to heatwaves and higher average temperatures worldwide. This warming trend affects weather patterns and contributes to more frequent and intense heatwaves (IPCC, 2014).

(b) Extreme Weather Events: Climate change is associated with an increase in extreme weather events such as hurricanes, droughts, floods, and wildfires. These events can cause significant damage to ecosystems, infrastructure, and human lives (IPCC, 2014).

(c) Sea-Level Rise: As global temperatures rise, glaciers and ice caps melt, leading to an increase in sea levels. Rising sea levels pose a threat to coastal communities, infrastructure, and ecosystems (IPCC, 2014).

(d) Changes in Precipitation Patterns: Climate change can disrupt rainfall patterns, leading to changes in the frequency and intensity of precipitation. Some regions may experience more frequent and severe droughts, while others may face increased rainfall and flooding (IPCC, 2014).

(e) Ecosystem Disruption: Climate change affects ecosystems and biodiversity. Many species are experiencing changes in migration patterns, shifts in habitats, and alterations in their life cycles. Some species may struggle to adapt or face extinction (IPCC, 2014).

(f) Ocean Acidification: Increased absorption of carbon dioxide by the oceans leads to ocean acidification. This can harm marine life, particularly organisms with calcium carbonate shells or skeletons, such as coral reefs and shellfish (IPCC, 2014).

Addressing climate change requires global cooperation and concerted efforts to reduce greenhouse gas emissions and transition to cleaner and more sustainable energy sources. This involves implementing renewable energy technologies, improving energy efficiency, promoting sustainable land-use practices, and adopting policies to mitigate and adapt to climate change (IPCC, 2014). International agreements, such as the Paris Agreement, aim to limit global warming to well below 2°C above preindustrial levels and to pursue efforts to limit the temperature increase to 1.5°C. These agreements emphasize the importance of reducing greenhouse gas emissions, providing financial support to developing countries, and enhancing climate resilience (UNFCCC, 2015). Individual actions, such as reducing energy consumption, using public transportation, and supporting sustainable practices, also play a crucial role in combating climate change (IPCC, 2014).

5 Water quality

Water quality refers to the chemical, physical, biological, and radiological characteristics of water (Environmental Protection Agency [EPA], 2021). It is an important aspect of assessing the health and usability of water resources, whether they are natural bodies of water like lakes, rivers, and oceans, or water supplies for drinking, irrigation, and industrial use. Several factors contribute to water quality, including:

(a) Chemical Characteristics: This includes the presence and concentration of various substances such as nutrients (nitrogen, phosphorus), heavy metals (lead, mercury), organic compounds (pesticides, pharmaceuticals), and other pollutants (Kolpin et al., 2002). Elevated levels of certain chemicals can pose risks to human health and the environment.

(b) Physical Characteristics: Physical parameters include temperature, turbidity (the presence of suspended particles), color, odor, and taste (United States Geological Survey [USGS], 2019). These characteristics can affect the aesthetic appeal of water, its suitability for certain uses, and the habitats of aquatic organisms.

(c) Biological Characteristics: Biological indicators in water include microorganisms (bacteria, viruses, protozoa), algae, and aquatic plants (USGS, 2019). The presence of certain organisms can indicate the overall health of a water body. For example, the presence of harmful bacteria or excessive algal blooms can indicate poor water quality.

(d) Radiological Characteristics: Water can also contain naturally occurring radioactive elements, such as radon and uranium (USGS, 2020). Human activities, such as nuclear power generation or mining, can introduce additional radioactive substances into water sources. Monitoring radiological characteristics is important to ensure water safety.

Water quality is assessed through various methods, including laboratory testing and on-site measurements. Regulatory agencies, such as the EPA in the United States, set standards and guidelines for water quality to safeguard public health and protect the environment (EPA, 2021). These standards often vary depending on the intended use of the water, such as drinking water standards being more stringent than standards for recreational water bodies. Improving water quality involves implementing measures to reduce or eliminate pollution sources, such as wastewater treatment, agricultural best practices, and industrial pollution control. Additionally, public awareness and education play a vital role in promoting responsible water use and protecting water resources.

6 Impact of climate variability on water quality

Water quality is directly influenced by changes in climatic factors such as temperature, precipitation, evaporation, and wind. Climate variability impacts water quality in several ways. Higher temperatures can cause thermal pollution, leading to a decrease in dissolved oxygen levels and an increase in the growth of harmful algal blooms and pathogenic bacteria (EPA, 2021). Extreme precipitation events often result in increased nutrient and sediment loads in waterways and contamination from agricultural runoff or untreated sewage. Conversely, periods of drought can lead to decreased water flows and increased concentrations of pollutants in rivers and lakes (Thornton et al., 2014).

Rising sea levels can cause saltwater intrusion into freshwater sources, restricting access to safe drinking water and leading to the degradation of crops and ecosystems (EPA, 2021). Additionally, increased atmospheric carbon dioxide levels can lead to increased acidity in surface waters, which can harm aquatic life and lead to biodiversity loss (USGS, 2019). Stronger storms can cause excessive erosion, leading to an elevated sediment load in water bodies, resulting in increased turbidity and decreased water clarity (Thornton et al., 2014). These climatic factors interact with water quality parameters, affecting the health of aquatic ecosystems and the availability of clean water resources for various uses. Addressing these climate-related impacts on water quality requires comprehensive management strategies that consider both environmental conservation and climate change adaptation measures.

7 Mitigation measures

The following are some potential measures that can be taken to mitigate the impacts of climate variability on water quality:

(a) Implement effective land-use practices to control soil erosion and nutrient runoff.
(b) Develop and implement effective treatment technologies for contaminated water sources.
(c) Increase the use of green infrastructure (e.g., rain gardens, green roofs) to reduce runoff and promote infiltration.
(d) Enhance the resilience of aquatic ecosystems through restoration activities and protection of critical habitats.
(e) Enhance monitoring and early warning systems to detect changes in water quality promptly.

8 Adaptive management strategies

Adaptive management strategies can be implemented to mitigate the effects of climate variability on water quality. These strategies focus on proactive and flexible approaches that consider the dynamic nature of climate change and its impacts on water resources. Some adaptive management strategies include:

Watershed Management: Implementing watershed management practices can help protect and improve water quality by reducing pollution runoff, sedimentation, and nutrient loading into water bodies. This includes promoting sustainable land-use practices, such as riparian buffer zones, cover cropping, and wetland restoration, to minimize the impact of climate variability on water quality.

Green Infrastructure: Investing in green infrastructure, such as rain gardens, permeable pavements, and green roofs, can help manage stormwater and reduce the risk of flooding and pollution in urban areas. Green infrastructure promotes natural water filtration processes, replenishes groundwater resources, and improves overall water quality by capturing and treating runoff before it reaches water bodies.

Water Conservation and Efficiency: Promoting water conservation and efficiency measures, such as fixing leaks, using water-saving fixtures, and implementing water reuse systems, can help reduce the demand for freshwater resources and minimize the impacts of droughts and water scarcity on water quality. By conserving water, communities can maintain adequate water supplies for both human consumption and ecosystem health.

Monitoring and Early Warning Systems: Developing monitoring programs and early warning systems for water quality can help detect changes in water quality parameters, such as temperature, dissolved oxygen levels, and nutrient concentrations, in response to climate variability. By consistently monitoring water quality data, stakeholders can identify emerging threats and take timely actions to protect water resources and public health.

Collaborative Governance and Stakeholder Engagement: Engaging stakeholders, including government agencies, local communities, environmental organizations, and industry partners, in collaborative decision-making processes can help develop effective water quality management plans that consider the diverse perspectives and needs of different stakeholders. By fostering collaboration and dialogue, adaptive management strategies can address complex water quality issues and build resilience to climate variability.

Research and Innovation: Investing in research and innovation to develop new technologies, tools, and approaches for monitoring and managing water quality can help improve the effectiveness of adaptive management strategies. By staying informed about the latest scientific developments and technological advancements, decision-makers can make informed decisions and implement evidence-based practices to protect water quality in the face of climate variability.

By implementing these adaptive management strategies, stakeholders can effectively mitigate the effects of climate variability on water quality, promoting the sustainable use and protection of freshwater resources for current and future generations. Collaboration, innovation, and community engagement are key components of successful water quality management in the context of climate change.

9 Conclusion

In conclusion, the impacts of climate variability on water quality are diverse and interconnected, with far-reaching consequences for ecosystems, human health, and the sustainability of water resources. As the global climate continues to change, it is essential to understand and address these challenges through proactive management strategies, including watershed protection, sustainable land-use practices, and the implementation of effective water quality monitoring and pollution control measures. By taking action to mitigate the impacts of climate variability on water quality, we can ensure the availability of clean and safe water for present and future generations. Climate variability has significant impacts on water quality worldwide. Effective management strategies are crucial to minimize the negative effects of climate variability on water quality. Mitigation measures should be put in place to reduce the impact of climate change on water resources. Collaboration between governments, researchers, and communities is crucial to ensure sustainable water management practices. Adapting to these changes and implementing strategies for sustainable water management are crucial to ensure the availability of clean and safe water resources in the face of climate variability.

References

Ashbolt, N., Grabow, C., Snozzi, M., 2013. Indicators of microbial water quality. In: Water Quality: Guidelines, Standards and Health, pp. 289–316.

Dudgeon, D., Sharpe, R.W., Campbell, J.L., 2006. Freshwater biodiversity: importance, threats, status and conservation challenges. Biol. Rev. 81 (2), 163–182.

Environmental Protection Agency [EPA], 2021. Water Quality Standards. Retrieved from https://www.epa.gov/wqs-tech/water-quality-standards.

Hall, S.R., McMahon, S.E., Beier, C.R., 2014. When and how should measures of abiotic habitat conditions be estimated in species-habitat models? Ecol. Model. 273, 348–355.

Havens, K.E., Bullerjahn, G.S., Utterback, R.E., 2019. Climate change and harmful algal blooms: insights and perspective. Harmful Algae 91, 101593.

Houghton, J.T., Ding, Y., Griggs, D.J., et al., 2001. Climate Change 2001: The Scientific Basis. Contribution of Working Group I to the Third Assessment Report of the Intergovernmental Panel on Climate Change. Cambridge University Press.

Intergovernmental Panel on Climate Change [IPCC], 2014. Climate Change 2014: Impacts, Adaptation, and Vulnerability. Cambridge University Press.

Kolpin, D.W., Furlong, E.T., Meyer, M.T., Thurman, E.M., Zaugg, S.D., Barber, L.B., Buxton, H.T., 2002. Pharmaceuticals, hormones, and other organic wastewater contaminants in US streams, 1999-2000: a national reconnaissance. Environ. Sci. Technol. 36 (6), 1202–1211.

National Aeronautics and Space Administration [NASA], 2021. Global Climate Change: Vital Signs of the Planet. NASA. climate.nasa.gov.

Poff, N.L., Allan, J.D., Bain, M.B., Karr, J.R., Prestegaard, K.L., Richter, B.D., et al., 2002. The natural flow regime. Bioscience 52 (8), 769–784.

Thornton, P.K., Ericksen, P.J., Herrero, M., Challinor, A.J., 2014. Climate variability and vulnerability to climate change: a review. Glob. Change Biol. 20 (11), 3313–3328.

United Nations Framework Convention on Climate Change [UNFCCC], 2015. Paris Agreement. Retrieved from https://unfccc.int/process-and-meetings/the-paris-agreement/the-paris-agreement.

United States Geological Survey [USGS], 2019. Water-Quality Indicators: Physical and Biological Characteristics of Water. Retrieved from https://www.usgs.gov/special-topic/water-science-school/science/water-quality-indicators-physical-and-biological.

United States Geological Survey [USGS], 2020. Basics of Water Quality: Waterbody Types and Uses. Retrieved from https://www.usgs.gov/special-topic/water-science-school/science/basics-water-quality-waterbody-types-and-uses.

Vörösmarty, C.J., McIntyre, P.B., Gessner, M.O., Dudgeon, D., Prusevich, A., Green, P., et al., 2010. Global threats to human water security and river biodiversity. Nature 467 (7315), 555–561.

Chapter 13

Climate variability and aquatic ecosystems

Terence Darlington Mushore[a,b], Charity Gororo[a], Tapiwa Michelle Masawi[a], Teddious Mhizha[a], Gilberta Luba Thwala[c], Emmanuel Mashonjowa[a], Onisimo Mutanga[b], and John Odindi[b]

[a]*Department of Space Science and Applied Physics, Faculty of Science, University of Zimbabwe, Harare, Zimbabwe,* [b]*Discipline of Geography, School of Agricultural, Earth and Environmental Sciences, University of KwaZulu-Natal, Pietermaritzburg, South Africa,* [c]*Mining and Mineral processing Engineering, Faculty of Engineering, Manicaland State University of Applied Sciences, Mutare, Zimbabwe*

Chapter outline

Remote Sensing of Climate. https://doi.org/10.1016/B978-0-443-21731-9.00009-0
Copyright © 2024 Elsevier Inc. All rights are reserved, including those for text and data mining, AI training, and similar technologies.

1 Introduction to climate variability and aquatic ecosystems

Aquatic ecosystems are a vital component of the global environment although they are constantly threatened by natural and human induced changes. They do not only play a vital role in maintaining ecological diversity production but also offer a range of services to the human population, such as water for drinking and irrigation, leisure activities, and habitat for fisheries with significant economic value. Human civilizations need the products and services that well-functioning aquatic ecosystems provide because they are sustainable, continue to satisfy social demands and expectations, and retain their ecological structure and function through time (Meyer, 1997). Upholding biological variety and providing other essential goods and services, such as clean water and fish protein, depend on basic ecosystem functions like nutrient intake and cycling, primary and secondary production, decomposition, and food chain interactions. Fish for both recreational and commercial purposes are produced by food webs that are supported by primary production and inputs from the watershed. The biological uptake of nutrients cleans the water and safeguards downstream habitats. Water temperature, discharge range, and temporal patterns all have an impact on these critical processes' rates, and all of these factors may be affected by climatic variability.

Freshwaters have a diverse range of biological species; however, many of them are in danger of going extinct due to human activity and natural causes (Naiman et al., 1995). The expansion of foreign species, the increasing population fragmentation within species due to thermal limitations, and human adaptations to a varying climate (such as the increased need to build reservoirs for water delivery or flood control) are just a few ways that a changing climate may exacerbate these concerns. Natural aquatic ecosystems may be more sensitive to climate variability than controlled aquatic ecosystems. For instance, because control structures are present, regulated rivers may experience less flow variability due to climate variability than unregulated rivers. Still, there may be more competition for a finite amount of water as a result of indirect consequences of the climate, like a rise in the amount of electricity needed to run air conditioning. However, human activities have put a rising amount of direct and indirect harm on aquatic systems. Aquatic systems are likely to soon be subjected to additional stress brought on by climate variability, to add to the difficulties provided by changes in land use, environmental degradation, and diversion of water.

Climate variability influences water temperature, precipitation, and nutrient availability, which eventually affect the quantity and productivity of aquatic ecosystems. Coral reefs are extremely sensitive to temperature changes as high ocean temperatures such as those in dry years can lead to coral bleaching and mortality (Hoegh-Guldberg et al., 2017). On the other hand, variations in precipitation patterns affect river flow and distribution of freshwater compromising the ability of aquatic species to survive and reproduce (Poff et al., 2007).

Climate change can impact habitat and food availability for a number of marine ecosystems (Post et al., 2013). Climate variability's general impacts on oceanic environments are complicated and depend on an assortment of variables, including the particular environment, the seriousness and period of climatic dynamics, and the versatility of the species included.

2 Climate change and aquatic ecosystems: Impacts and vulnerabilities

Researchers have investigated the impacts of climate change on oceanic biological systems in incredible detail, which has highlighted the requirement for successful administration methodologies to protect imperiled species and their environments. For example, Arthington et al. (2018) found that changed supplement cycles, modified water temperature, stream, and quality, and an increment within the recurrence and seriousness of extraordinary weather events are among the results that climate change has on freshwater and marine environments. A study by Cheung et al. (2013) assessed the vulnerability of global marine fisheries to climate change, highlighting the disproportionate impacts on developing countries and the need for adaptive management strategies.

For the last few decades, climate change, food security, and their complex interaction have become a global issue (Plagányi, 2019). Rapid population growth has ruined the natural world and contaminated the surrounding waters. Every day, the amount of greenhouse gases in the atmosphere rises. As a result, we are threatened by both global warming and other weather-related phenomena like compromised air quality, floods, droughts, and cyclones. Changes in climate might happen throughout the planet or only in one particular area. However, it has an impact on all ecosystems, including aquatic ones. Due to the simultaneous changes in the average temperature of the air and water, aquatic creatures are particularly vulnerable to climate change. The primary causes of climate change in the aquatic system are rising sea levels and temperatures, altered monsoon patterns, extreme weather, and water stress, which have an impact on fish stocks and other aquatic organisms both directly and indirectly. It immediately affects an organism's physiological behavior and growth pattern, which reduces its ability to reproduce and ultimately results in death. The production, structure, function, and composition of aquatic ecosystems may all be affected indirectly.

Through the emission of greenhouse gases, humans are changing the climate on a worldwide scale (Wigley, 1999; NAST, 2000; Houghton et al., 2001). By 2100, it is predicted that the average worldwide surface temperature will rise by 1.5–5.8°C (Houghton et al., 2001), while rises may be larger in the United States (Wigley, 1999). Freshwater and coastal ecosystems, which are currently under stress, will experience more strain as a result of this predicted climate change. While temperature and hydrologic conditions differ significantly from year to year, aquatic systems are generally believed to be resilient and able to maintain a state of health and self-sufficiency. On the other hand, abrupt climate change might impose new environmental regimes that will strain aquatic ecosystems' capacity for resilience. Additionally, the degree of landscape fragmentation that surrounds aquatic ecosystems is unprecedented historically (Dale, 1997). Owing to the growing isolation and discontinuity of these systems, it will be extremely difficult for animals and plants to adapt to the rapid climate change. Thus, it is obvious that climate change poses a new, serious hazard to aquatic ecosystems that will interact in intricate ways with other pressures already brought on by humans (Carpenter et al., 1992; Firth and Fisher, 1992; Lake et al., 2000; Lodge, 2001; Poff et al., 2001).

The diversity of life on the earth is significantly impacted by human and climatic changes to ecosystems. Currently, natural variability is being compounded by a new type of human-caused climate change, which could accelerate the loss of biodiversity already endangered by previous human pressures. While water covers over 70% of the surface of the globe, aquatic ecosystem distribution and abundance are already altering because of climate change. The currents that move across the surface of the earth will vary if even a small variation in water temperature occurs. Ocean acidification, sea level rise, altered precipitation patterns, and variations in water temperature are only a handful of instances of how aquatic habitats are being impacted by a changing climate. On aquatic biodiversity, food webs, and ecosystem services like fisheries and water supply, these consequences may have a domino effect. Effective management and protection of aquatic environments depend on understanding and monitoring these effects.

2.1 Climate change within the ocean

Three essential variables are expected to alter as a result of the increase in greenhouse gases in atmosphere:

(i) Reduced total carbonate alkalinity

As CO_2 levels rise in atmosphere, seawater's overall carbonate alkalinity will decline (Gattuso et al., 1998; Kleypas et al., 1999). A number of studies have investigated the impact of climate change on total carbonate alkalinity. For instance, a study by Feely et al. (2004) found that the carbonate ion concentration in surface waters of the Pacific Ocean has decreased by around 30% since the Industrial Revolution, primarily attributed to the uptake of anthropogenic carbon dioxide. Similarly, a study by Cai et al. (2011) found that the total alkalinity of the East China Sea has decreased by around 12% since the 1980s, largely due to the absorption of carbon dioxide from the atmosphere. This specific variable is expected to have a major impact on the acidity and carbonate ion pool of the entire ocean. By 2050, doubling atmospheric carbon dioxide levels will result in a 30% reduction in the tropical aragonite saturation state.

(ii) Increased sea level

Numerous research works have confirmed the rise in sea level resulting from climate change. For instance, the Intergovernmental Panel on Climate Change (IPCC) reported in its latest assessment report (AR5) that global sea levels have risen by an average of 0.19 m (0.6 ft) since 1901, and that the rate of sea level rise has accelerated in recent decades (IPCC, 2013). Another study by Nerem et al. (2018) used satellite data to estimate the global mean sea level from 1993 to 2017. According to the study, during this time, sea levels have increased by 3.2 mm on average annually, with a notable acceleration of this pace of rise in recent years. The impact of sea level rise is already being felt around the world, with coastal flooding, erosion, and saltwater intrusion affecting many communities. A study by Hinkel et al. (2019) estimated that by 2100, up to 630 million people could be affected by coastal flooding caused by sea level rise, with many of these people living in low-lying areas in Asia.

(iii) Sea temperature increase

It has not been possible to distribute the notable rise in heat content evenly. Sea temperature has an impact on the marine ecosystem. Variations in global temperatures can have a direct impact on the rates and directions of ocean water migration because of their direct impacts on seawater density.

2.2 Deep sea biodiversity

Climate change can have a major coordinate and backhanded effect on deep-sea biodiversity. Temperature, circulation, and chemical varieties within the sea can influence species conveyance, plenitude, and environmental forms. The physiology and behavior of deep-sea creatures can be impacted by temperature changes within the sea, which can change their dispersion and plenitude. For illustration, hotter waters may cause run shifts or neighborhood termination for animals usual to cold temperatures. So also, changes in temperature can influence the development and metabolic rate of deep-sea animals, which in turn influences the species' capacity to outlive and replicate (Levin and Dayton, 2009; Morato et al., 2018). The ocean's expanding retention of carbon dioxide from the climate too changes the chemistry of the water by making it more acidic, which may have an effect on deep-sea life. Sea fermentation may moreover influence the survival and behavior of deep-sea angle, as they are helpless to changes in pH (Levin and Dayton, 2009; Morato et al., 2018). The quantity and quality of food for deep-sea species can also be impacted by climate variability. Benthic creature abundance and distribution can be impacted by changes in the quantity and properties of organic material submerged in the deep sea, for instance, which can be caused by variations in surface ocean production. Likewise, alterations in the range and quantity of pelagic creatures, such krill or tiny fish, might impact the food source for deep-sea predators (Danovaro et al., 2010).

2.3 Fish populations

Fish populations are being significantly impacted by climate change, which is changing their distribution, quantity, and productivity. These effects are a result of shifting ocean temperatures, currents, and chemistry. According to a 2013 study by Cheung et al., which examined data on fish catches worldwide, fish populations are already significantly declining as a result of climate change, especially in tropical areas. The study predicted that if greenhouse gas emissions stay the same in 2050, the world's fish catch might decrease by as much as 30%. Fish populations in the tropics were found to be especially susceptible to climate change (Pinsky et al., 2021). Changes in ocean temperatures are also affecting the distribution of fish populations. According to Poloczanska et al. (2013) many fish species are moving toward the poles as ocean temperatures warm, with some species shifting their ranges by as much as 400 km per decade.

2.4 Coral reef

Tropical ecosystems with a coral reef structure predominate in the intertidal and subtidal zones. Over the past 20 years, they have seen significant alterations, many of which are related to climate change and other pressures (Bryant et al., 1998). The coral reef is an important source of income for many developing countries (Parry, 2007). For about half of all marine species, coral serves as their habitat. However, the ecosystem's coral reefs are currently in grave danger. The coral reef is one of the most resistant ecosystems and too resilient to recover from weak chronic as well as acute stresses (Buddemeier et al., 2004). However, according to Hughes and coworkers (Hughes et al., 2003), the reef ecosystem is not able to sustain against chronic plus acute stress. Increasing acidity causes decreasing the pH of the ocean, which results in decreased aragonite saturation that can disrupt the calcification of coral (Orr et al., 2005). The world's ocean's increased acidity is a serious threat to coral reefs over the long run, but it is unclear how coral growth will affect the rising acidity. The saturation level of aragonite in deep cold water corals are 90–150 m (Feely et al., 2004). These deep-cold water corals clearly show the effects of acidification. Reef fish biodiversity suffers if corals are reduced as a result of negative effects from climate change. Coastal fisheries are badly affected by the warming of the climate and bleaching events. We can draw the conclusion that creatures that rely on coral reefs for their food and habitat suffer long-term effects when these reefs are destroyed.

2.5 Freshwater ecosystem

Freshwater ecosystems are particularly vulnerable to climate change. The lake's trophic status, depth, and size all affect how vulnerable this system is to climate change. According to Field et al. (2007), the negative impact was observed on the cold-water species and positive impact on the warm-water species. The freshwater lake system has altered in shape and distribution as a result of the severe consequences of climate change, and in many cases, they may have completely vanished. These are the attributes of the dynamics change in precipitation, evaporation, and runoff (Poff et al., 2002). Long-term increases in fish production are encouraged by climate change because it causes the production rate of invertebrate prey to grow logarithmically with temperature. According to Watson et al. (2001) for every 10°C increase in temperature, the rates increase by two to four times. However, the makeup of prey species will vary as a result of climate change. This change may cause antagonistic effects on the long-term enhancement of fish production (Watson et al., 2001). In short-time, climate change will cause a decrease in fish-production because of timing mismatch (Watson et al., 2001). The ability of the movement of the freshwater species is vital in determining the resistance of those species to withstand climate change (Poff et al., 2002).

3 Climate variability and water quality in aquatic ecosystems

The chemical and physical characteristics of water can be changed by variations in temperature, precipitation, and runoff. These changes can have a domino effect on aquatic biodiversity, food webs, and ecosystem services. Variations in the availability of nutrients are one way that climate variability can affect the quality of water. Changes in precipitation patterns can alter the timing and amount of nutrient inputs to aquatic ecosystems, which can impact the growth and productivity of algae and other primary producers (Breitburg et al., 2018). Changes in temperature can also impact the rates of nutrient cycling and decomposition, which can further affect nutrient availability in aquatic systems (Dodds and Oakes, 2004). Even if the watershed's fertilizer supply remains constant, rising temperatures alone may cause many lakes to become more eutrophic because warmer temperatures promote increased algae growth. Water quality will generally decrease as a result of climate change since noxious algae prefer to dominate warm, nutrient-rich waterways (see also Murdoch et al., 2000). Increased nutrient loading to lakes through storm-related erosion is predicted by climate change models along with an increase in severe precipitation episodes. Furthermore, if freshwater inputs during the summer are also reduced, dissolved nutrients will be retained in lakes longer, leading to an increase in output. These factors will affect the possibility of an increase in algal production, both independently and in combination.

Aquatic habitats' dissolved oxygen concentrations can be impacted by variations in the climate. In certain aquatic environments, warmer water can cause hypoxia (low oxygen levels) or anoxia (no oxygen levels) due to a decrease in the solubility of oxygen in the water (Rabalais et al., 2002). By affecting the amount of organic matter that enters aquatic environments and increasing the pace of decomposition and oxygen consumption, variations in precipitation patterns can also have an effect on dissolved oxygen levels (Diaz and Rosenberg, 2008). The intricate interplay between climate variability and water quality in aquatic ecosystems is contingent upon several aspects, such as the particular environment under consideration, the magnitude and duration of climatic changes, and the degree of adaptability exhibited by the species involved. For aquatic ecosystems to be effectively managed and conserved in the face of climate change, it is critical to comprehend these effects.

4 Impacts of climate variability on aquatic biodiversity

Climate variability can have a significant impact on the distribution, abundance, and productivity of aquatic species, which can then have an impact on aquatic biodiversity. A few of the main effects of climate variability on aquatic biodiversity include temperature changes associated with climate variability, which can have a significant impact on species distribution, abundance, and timing of life cycle events like reproduction, as well as an increase in the prevalence of diseases and parasites. Additionally, rising water temperatures brought on by climate change will cause changes in the spatial distribution of aquatic species and basic ecological processes. If species try to migrate to a suitable habitat in an effort to adapt, these effects might be lessened. However, human modification of possible migratory corridors may make it more difficult for certain species to migrate, which raises the risk of biodiversity loss and species extinction.

Additionally, the hydrology of aquatic ecosystems can be altered by variations in precipitation and runoff linked to climate variability, which can impact elements like water flow, sedimentation, and nutrient availability. Changes in the frequency, duration, and timing of significant precipitation events, such as floods and droughts, have a significant impact on the populations of aquatic organisms. Seasonal variations in the timing of snowmelt will impact stream flows and possibly the ability of many aquatic species to reproduce. These changes can have cascading effects on aquatic biodiversity, affecting the distribution, abundance, and productivity of species (Poff et al., 2007). Seasonal variations in precipitation and runoff will alter the species composition and ecological productivity of aquatic systems.

Another effect of climate variability on aquatic biodiversity is ocean acidification, which can have a major effect on aquatic biodiversity. The growth and survival of marine organisms can be impacted by acidification, especially those that depend on calcium carbonate skeletons or shells, such as certain planktonic species, mollusks, and corals (Doney et al., 2012). Furthermore, because of climatic variability, extreme weather events like storms, floods, and droughts may occur more frequently and be more severe, which might have a significant impact on aquatic biodiversity. Severe occurrences have the potential to damage ecosystems, change the quality of the water, and cause aquatic animals to die off or move (Hughes et al., 2018). The intricate effects of climate variability on aquatic biodiversity might differ based on a number of variables, such as the particular environment under consideration, the intensity and duration of climatic swings, and the ability of the species in question to adapt.

5 Climate variability and aquatic food web

Aquatic food webs are susceptible to notable effects from climate variability, which can influence the distribution, quantity, and efficiency of primary producers, consumers, and decomposers. Variations in temperature, precipitation, and runoff can alter the physical and chemical properties of aquatic habitats. The structure and functionality of food webs may be significantly impacted by these modifications. Climate variability can affect primary producers like phytoplankton and algae in aquatic food webs by altering their productivity. Climate variability-related variations in temperature, nutrition availability, and light levels can affect these species' growth and production, which can have a domino effect on higher trophic levels (Cayan et al., 2016).

Besides, the sum of natural matter that enters aquatic biological systems can alter due to varieties in precipitation and runoff caused by climate change. The generation and composition of the nourishment web may be affected by this. The accessibility of supplements and oxygen in aquatic environments may be affected by quicker rates of deterioration brought almost by expanded inputs of natural matter. Furthermore, they may result in hypoxic or anoxic conditions, which can be hurtful to oceanic life's well-being (Breitburg et al., 2018). Understanding the impacts of climatic change on aquatic biological systems is fundamental to creating compelling administration and preservation procedures for these frameworks in a changing and variable environment.

6 Climate variability and aquatic ecosystems services

The focal points that oceanic biological systems give to individuals are known as oceanic biological system administrations, and they can be essentially affected by climate change. A wide range of administrations are advertised by aquatic biological systems, such as provisioning nourishment and water, directing administrations like climate administration and water refinement, social administrations like aesthetics and diversion, and maintaining administrations like supplement cycling and essential generation. Climate change can lead to changes in temperature, precipitation, and runoff, which can influence the administrations given by aquatic biological systems. Changes in water temperature, for case, can influence the development and dispersal of angle and other oceanic life forms, which may influence human get to nourishment and recreational openings (Walther et al., 2002). The sum and quality of water that's accessible for human utilize, as well as the direction of water amount and quality, may be affected by varieties in precipitation and runoff (Poff et al., 2007).

Additionally, varieties within the climate can have an impact on the accessibility of social administrations like entertainment and aesthetics. Extraordinary climate occasions like surges and dry seasons can change the physical and tasteful characteristics of oceanic biological systems, which may have an effect on people's delight in and appreciation for these biological systems (Hughes et al., 2018). To summarize, the effect of climate change on aquatic environment administrations is complex and depends on a few variables, counting the particular biological system in address, the degree and length of temperature variances, and the degree of flexibility shown by the influenced human communities. It is basic to completely comprehend these impacts in arrange to oversee and preserve oceanic territories in a changing climate whereas adjusting human needs with the wellbeing of these biological systems.

7 Mitigation and adaptation strategies for aquatic ecosystems under climate variability

Adaptation and mitigation techniques are fundamental for the survival of oceanic environments. Mitigation measures point to decrease nursery gas emanations and constrain the variables that contribute to climate change, whereas adaptation procedures offer assistance toward biological systems and the communities that depend on them adjust to the impacts of climate change that are as of now being felt. By implementing initiatives like energy efficiency, renewable energy, and low-carbon transportation, communities can lessen their contribution to climate change and lessen the severity of its effects on aquatic ecosystems. Moreover, effective administration procedures like constraining deforestation, empowering economical horticulture, and defending wetlands and other characteristic biological systems can bolster or move forward carbon sequestration and capacity, which brings down the sum of carbon dioxide within the climate. Carbon offset program is another strategy that can help to incentivize greenhouse gas emissions reductions and support the implementation of mitigation projects that benefit aquatic ecosystems, such as wetland restoration or reforestation.

On the other hand, ecosystem-based adaptation strategies can be implemented such as restoring and protecting wetlands, mangroves, and coral reefs, which can help aquatic ecosystems build resilience to the effects of climate change, including rising sea levels, harsher weather, and acidification of the ocean. Water management practices can help to maintain or improve both the amount and the quality of water resources in aquatic ecosystems, lowering vulnerability to variations in precipitation and runoff. These practices include water conservation, reuse, and the use of natural infrastructure, such as rain gardens and green roofs. The identification and implementation of effective adaptation methods can be facilitated by integrated management techniques that take into account the impacts of climate change on aquatic ecosystems and the communities that rely on them, all while minimizing unintended consequences. More so, capacity building and education programs can help to build community resilience to climate change impacts by increasing awareness of the dangers and opportunities associated with a changing climate change and promoting the adoption of adaptive management strategies. Lastly the most effective strategies for managing aquatic ecosystems under climate variability will depend on the specific ecosystem and the local context and will require a collaborative and adaptive approach that engages stakeholders and incorporates best available science and management practices.

8 Strategies for managing aquatic ecosystems in a changing climate

Aquatic ecosystems can be managed in a changing climate using several strategies. Some of these strategies include regular monitoring and assessment of aquatic ecosystems as it is critical for understanding the effects of a climate's change and for identifying areas that may be particularly vulnerable to these impacts. Monitoring can help to identify changes in water quality, species distributions, and ecosystem function, providing important information for adaptive management. In addition, habitat restoration and protection can help to improve the management of aquatic ecosystems to climate change impacts. Strategies such as riparian buffer restoration, wetland creation, and habitat connectivity can help to maintain or enhance aquatic ecosystem function and the provision of ecosystem services.

Adding on the above, effective water management strategies can help to maintain or enhance both the quantity and quality of water resources in aquatic ecosystems. Strategies such as water conservation, water reuse, and the use of natural infrastructure such as green roofs and rain gardens can help to reduce water demand and improve water quality. Adaptive management is another strategy which can be used and is an iterative process of learning and adjustment that allows for management strategies to be modified in response to changing conditions. Identifying the best management practices to improve aquatic ecosystems' resistance to the effects of climate change can be aided by adaptive management. Finally, increasing public knowledge of the value of aquatic ecosystems and the effects of climate change on these ecosystems can be accomplished through outreach and education programs. Support for management approaches that encourage aquatic ecosystem resilience and the delivery of ecosystem services may increase as a result of this. In conclusion, the best management strategies for aquatic ecosystems in the face of climate change will depend on the specific ecosystem and the local context. These strategies will need to be collaborative and adaptive, involving stakeholders and incorporating the best available science and management practices.

9 The role of remote sensing in understanding the interactions between climate variability and aquatic ecosystems

9.1 How can remote sensing be used in climate variability and aquatic biodiversity?

The capacity to accumulate information over wide regions and on a customary premise makes remote sensing a profitable instrument for inquiring about oceanic biodiversity and climate variability. The temperature, precipitation, and vegetation cover variances are among the climate-related factors that can be utilized to screen. Experiences into the elements of the ocean–atmosphere framework can be picked up, for occurrence, by utilizing information to screen changes in ocean surface temperature and sea streams (Liu et al., 2019). Pettorelli et al. (2018) indicated that remote sensing can moreover be applied to track modifications in plant and land cover, subsequently giving experiences into how environments are adjusting to natural changes. The mapping and observing of oceanic biodiversity, counting the dispersion and plenitude of different species, is another application for remote sensing. For occasion, the degree and condition of coral reefs, which are significant environments for an assortment of marine species, can be mapped utilizing obsequious photography (Mishra and Gould, 2016). Wetlands and other environments are significant breeding and bolstering grounds for an assortment of species, and they may be mapped utilizing remote sensing (Kuenzer et al., 2014).

9.2 How remote sensing can be used in climate change and aquatic ecosystems?

Advances in remote sensing can be a valuable technique for following and assessing how changing climate is influencing oceanic biological systems. The wellbeing of oceanic environments and the impacts of climate change can be assessed by applying information on land cover changes, temperature, and quality that can be obtained by utilizing remote sensing. For occurrence, the escalation and degree of coral fading scenes, which can be brought on by warm water connected to climate change, are being observed by means of remote sensing. Changes within the reflectance of coral reefs can be recognized by satellite-based sensors, which can be utilized to outline the extent of dying occasions and track their seriousness (Hughes et al., 2017).

Comparably, changes in phytoplankton efficiency and dispersion can be checked utilizing satellite-based sensors that assess sea color, which can give data on chlorophyll concentrations and essential generation (Behrenfeld et al., 2017). Observing water quality parameters like chlorophyll-a, turbidity, and broken-up oxygen can offer assistance to determine how oceanic biological systems are reacting to climate change. Remote sensing can also be used to create adaptation and mitigation methodologies for oceanic frameworks on climate variability. Remote sensing can also be used to outline the defenselessness of coastal regions to storm surges and ocean level rise (Cucchiaro et al., 2020). The technique can be utilized to pinpoint districts that are powerless to disintegration or surges, which can offer assistance with decision-making with respect to the situation of the framework and the administration of coastal biological systems. Besides, the evaluation of living species in oceans, remote sensing can give experiences into their reactions to climate change (Pettorelli et al., 2014). Remote sensing can be a valuable instrument for checking and surveying the impacts of climate change on biological systems, but the translation of remotely detected information requires cautious thought of components such as sensor characteristics, barometrical conditions, and imagery analysis methods and ought to be coordinated with other sources of information and data.

9.3 How can remote sensing be used in climate variability and water quality in aquatic ecosystems?

Remote sensing can be used, among other purposes, to screen harmful algal sprouts (HABs) and the water quality in oceanic situations. HABs can have a major impact on human wellbeing, sea biodiversity, and water quality, which is why they are becoming a developing issue in numerous oceanic environments. By recognizing changes in the color of the water and the nearness of specific shades connected to harmful green growth, remote sensing can be utilized to recognize and track algal sprouts (HABs) (Gower et al., 2005). Water surface temperature can be derived from thermal infrared sensors, which can offer data on varieties in water temperature related to climatic change (Kaufman et al., 2002). Comparative to this, by recognizing varieties in water color connected to the nearness of oxygen-sensitive shades, remote sensing can be applied to track broken up oxygen levels in aquatic environments (Huang et al., 2008). Nevertheless, the interpretation of remotely sensed data requires the integration of several data and information sources and considers several factors, such as atmospheric conditions, sensor attributes, and image processing methods.

9.4 Limitations of remote sensing in monitoring water quality in aquatic ecosystems?

Remote sensing information especially from most space-borne sensors may not be sufficient to record capture changes in water quality due to limited temporal resolution, or it may not have a high enough spatial resolution to recognize small-scale contrasts in water quality. There is always a trade-off between spatial and temporal resolution, which makes it difficult to obtain desired attributes simultaneously in one sensor. Additionally, climatic obstructions from clouds and haze can adversely impact remote sensing information, constraining its precision and interpretability. The use of remote sensing to screen water quality in oceanic biological systems can moreover be prevented by sensor limits. Whereas diverse sensors have varying capacities and limitations, the choice of sensor might influence the exactness and quality of the information obtained by remote sensing. Specific strategies of water quality observation may be way better suited for specific sorts of sensors than for others. Overall, while remote sensing can provide a valuable tool for monitoring water quality in aquatic ecosystems, it should be used in conjunction with other sources of data and information, and its limitations should be carefully considered when interpreting the data. Validation can be another limitation in using remote sensing data to monitor water quality. To ensure the accuracy of the data and the validity of the interpretations, the data should be validated with ground-based water quality measurements.

10 Conclusion and policy implications

Climate variability and change are expected to have a major impact on aquatic ecosystems, affecting precipitation patterns, nitrogen cycle, water temperature, and biodiversity. The services that ecosystems give to human welfare, fisheries, water quality, recreation, and flood protection may all be significantly impacted by these consequences. The conservation of aquatic biodiversity maintains the diversity, quality, and value of these resources while ensuring their widespread use and steady supply. This means that there is a need to prevent extinction by prudent planning and management. Additionally, there is an urgent need to step up efforts to halt the loss of biodiversity and implement long-term preservation strategies. The conservation of biodiversity cannot be achieved solely by law enforcement but through the participation of all parties from the grassroots to the highest levels of decision-making. Reducing greenhouse gas emissions to lessen climate change, putting adaptive management strategies into place to address the effects of climate variability, and supporting ecosystem-based approaches to management that take ecosystem interactions into account are some of the policy implications for addressing these impacts. Overall, a mix of mitigation, adaptation, and ecosystem-based management techniques will be needed to address the effects of climate variability and change on aquatic ecosystems. These approaches will need to be customized to the unique features and difficulties of every ecosystem, necessitating cooperation and involvement from scientists, managers, and stakeholders. Remote sensing is a useful technique for spatial and temporal monitoring of water quality although it can be limited by sensor characteristics and environmental attributes such as clouds and haze.

References

Arthington, A.H., Kennen, J.G., Stein, E.D., Webb, J.A., 2018. Recent advances in environmental flows science and water management—innovation in the Anthropocene. Freshw. Biol. 63 (8), 1022–1034.

Behrenfeld, M.J., Hu, Y., O'Malley, R.T., Boss, E.S., Hostetler, C.A., Siegel, D.A., Sarmiento, J., Schulien, J., Hair, J., Lu, X., Rodier, S., Scarino, A.J., 2017. Annual boom–bust cycles of polar phytoplankton biomass revealed by space-based lidar. Nat. Geosci. 10 (2), 118–122.

Breitburg, D., et al., 2018. Declining oxygen in the global ocean and coastal waters. Science 359 (6371), eaam7240.

Bryant, D., Burke, L., McManus, J., Spaloling, M., 1998. Reefs at Risk: A Map-Based Indicator of Threats to the World's Coral Reefs. World Resources Institute, Washington, DC.

Buddemeier, R.W., Kleypas, J.A., Aronson, R.B., 2004. Potential Contributions of Climate Change to Stresses on Coral Reef Ecosystems. Coral Reefs and Global Climate Change. Pew Center on Global Climate Change, VA, USA.

Cai, W.J., et al., 2011. Decrease in the CO_2 uptake capacity in an ice-free Arctic Ocean basin. Science 329 (5991), 556–559.

Carpenter, S.R., Fisher, S.G., Grimm, N.B., Kitchell, J.F., 1992. Global change and freshwater ecosystems. Annu. Rev. Ecol. Syst. 23, 119–139.

Cayan, D.R., Dettinger, M.D., Pierce, D., Das, T., Knowles, N., Ralph, F.M., Sumargo, E., 2016. Natural variability, anthropogenic climate change, and impacts on water availability and flood extremes in the Western United States. In: Water Policy and Planning in a Variable and Changing Climate. Routledge, pp. 17–42.

Cheung, W.W., et al., 2013. Shrinking of fishes exacerbates impacts of global ocean changes on marine ecosystems. Nat. Clim. Change 3 (3), 254–258.

Cucchiaro, S., Fallu, D.J., Zhang, H., Walsh, K., Van Oost, K., Brown, A.G., Tarolli, P., 2020. Multiplatform-SfM and TLS data fusion for monitoring agricultural terraces in complex topographic and landcover conditions. Remote Sens. 12 (12), 1946.

Dale, V.H., 1997. The relationship between land-use change and climate change. Ecol. Process. 7, 753–769.

Danovaro, R., Company, D., Snelgrove, M., Dell'Anno, A., 2010. Can we predict the unpredictable? Deep-sea ecosystem dynamics in an era of climate change. Deep-Sea Res. II Top. Stud. Oceanogr. 57 (16), 1181–1188.

Diaz, R.J., Rosenberg, R., 2008. Spreading dead zones and consequences for marine ecosystems. Science 321 (5891), 926–929.

Dodds, W.K., Oakes, R.M., 2004. Headwater influences on downstream water quality. Environ. Manag. 34 (S1), S26–S35.

Doney, S.C., et al., 2012. Ocean acidification: the other CO_2 problem. Ann. Rev. Mar. Sci. 4, 469–492.

Feely, R.A., et al., 2004. Impact of anthropogenic CO_2 on the $CaCO_3$ system in the oceans. Science 305 (5682), 362–366.

Field, C.B., Mortsch, L.D., Brklasich, M., Forbes, D.L., Kovacs, P., Patz, J.A., et al., 2007. North America. In: Parry, M.L., Canziani, O.F., Palutikof, J.P., Hanson, P.J., Linden v d, C.E. (Eds.), Climate Change 2007: Impacts, Adaptation and Vulnerability. Cambridge University Press, Cambridge, pp. 617–652.

Firth, P., Fisher, S.G. (Eds.), 1992. Global Climate Change and Freshwater Ecosystems. Springer-Verlag, New York.

Gattuso, J.P., Frankignoulle, M., Bourge, I., Romaine, S., Buddemeier, R.W., 1998. Effect of calcium carbonate saturation of seawater on coral calcification. Global Planet. Change 18, 37–47.

Gower, J.F., et al., 2005. Detection of intense plankton blooms using the 709 nm band of the MERIS imaging spectrometer. Int. J. Remote Sens. 26 (10), 2005–2012.

Hinkel, J., et al., 2019. Coastal flood damage and adaptation costs under 21st century sea-level rise. Proc. Natl. Acad. Sci. U. S. A. 116 (23), 11195–11200.

Hoegh-Guldberg, O., et al., 2017. Coral reefs under rapid climate change and ocean acidification. Science 318 (5857), 1737–1742.

Houghton, J.T., Ding, Y., Griggs, D.J., Noguer, M., van der Linden, P.J., Xiaosu, V. (Eds.), 2001. Climate change 2001: the scientific basis. In: Intergovernmental Panel on Climate Change: Working Group I. Cambridge University Press, Cambridge, UK.

Huang, C., et al., 2008. Monitoring water quality in Lake Taihu, China, using EO-1 Hyperion. Remote Sens. Environ. 112 (4), 2254–2262.

Hughes, T.P., Baird, A.H., Bellwood, D.R., Card, M., Connolly, S.R., Folke, C., et al., 2003. Climate change, human impacts, and the resilience of coral reefs. Science 301 (5635), 929–933. https://doi.org/10.1126/science.1085046.

Hughes, T.P., Kerry, J.T., Álvarez-Noriega, M., Álvarez-Romero, J.G., Anderson, K.D., Baird, A. H., et al., 2017. Global warming and recurrent mass bleaching of corals. Nature 543 (7645), 373–377.

Hughes, B.B., et al., 2018. Ecological impacts of extreme weather events: processes, assessments, and challenges ahead. Front. Ecol. Environ. 16 (5), 273–281.

IPCC, 2013. Climate change 2013: the physical science basis. In: Contribution of Working Group I to the Fifth Assessment Report of the Intergovernmental Panel on Climate Change. Cambridge University Press.

Kaufman, Y.J., et al., 2002. Remote sensing of water clarity in turbid coastal waters using airborne multispectral data. J. Geophys. Res. Oceans 107 (C11), 3229.

Kleypas, J.A., Buddemeier, R.W., Archer, D., Gattuso, J.P., Langdon, C., Opdyke, B.N., 1999. Geochemical consequences of increased atmospheric CO_2 on coral reef. Science 284, 118–120.

Kuenzer, C., van Beijma, S., Gessner, U., Dech, S., 2014. Land surface dynamics and environmental challenges of the Niger Delta, Africa: remote sensing-based analyses spanning three decades (1986–2013). Appl. Geogr. 53, 354–368.

Lake, P.S., Palmer, M.A., Biro, P., Cole, J., Covich, A.P., Dahm, C., Gibert, J., Goedkoop, W., Martens, K., Verhoeven, J., 2000. Global change and the biodiversity of freshwater ecosystems: impacts on linkages between above sediment and sediment biota. BioScience 50, 1,099–1,107.

Levin, L.A., Dayton, P.K., 2009. Ecology of deep-ocean communities. Oceanography 22 (1), 65–79.

Liu, Y., Cai, W., Li, X., Zheng, F., 2019. Ocean remote sensing for climate research: advances and perspectives. Natl. Sci. Rev. 6 (3), 413–416. https://doi.org/10.1093/nsr/nwz034.

Lodge, D.M., 2001. Responses of lake biodiversity to global changes. In: Sala, O.E., Chapin, F.S., Huber-Sannwald, E. (Eds.), Future Scenarios of Global Biodiversity. Springer-Verlag, New York, pp. 277–313.

Meyer, J.L., 1997. Stream health: incorporating the human dimension to advance stream ecology. J. N. Am. Benthol. Soc. 16, 439–447.

Mishra, D.R., Gould Jr., R.W., 2016. Preface: remote sensing in coastal environments. Remote Sens. 8 (8), 665.

Morato, T., Durán Muñoz, E., Baco, A., Clark, M.R., 2018. Climate-induced changes in the distribution of deep-sea benthic communities in the eastern North Atlantic. Glob. Chang. Biol. 24 (3), e101–e115.

Murdoch, P.S., Baron, J.S., Miller, T.L., 2000. Potential effects of climate change on surface-water quality in North America 1. J. Am. Water Resour. Assoc. 36 (2), 347–366.

Naiman, R.J., Magnuson, J.J., McKnight, D.M., Stanford, J.A., 1995. The Freshwater Imperative. Island Press, Washington, DC. 165 pp.

NAST (National Assessment Synthesis Team), 2000. Climate Change Impacts on the United States: The Potential Consequences of Climate Variability and Change. U.S. Global Change Research Program, Washington, DC.

Nerem, R.S., et al., 2018. Climate-change–driven accelerated sea-level rise detected in the altimeter era. Proc. Natl. Acad. Sci. U. S. A. 115 (9), 2022–2025.

Orr, J.C., Fabry, V.J., Aumont, O., Bopp, L., Doney, S.C., Feely, R.A., et al., 2005. Anthropogenic Ocean acidification over the twenty-first century and its impact on calcifying organisms. Nature 437 (7059), 681–686. https://doi.org/10.1038/nature04095.

Parry, M.L. (Ed.), 2007. Climate Change 2007-Impacts, Adaptation and Vulnerability: Working Group II Contribution to the Fourth Assessment Report of the IPCC. vol. 4. Cambridge University Press.

Pettorelli, N., Safi, K., Turner, W., Satoh, F., 2014. Satellite remote sensing, biodiversity research and conservation of the future. Philos. Trans. R. Soc. B 369 (1643), 20130190. https://doi.org/10.1098/rstb.2013.0190.

Pettorelli, N., Schulte to Bühne, H., Tulloch, A., Dubois, G., Macinnis-Ng, C., Queirós, A.M., Keith, D.A., Wegmann, M., Schrodt, F., Stellmes, M., 2018. Satellite remote sensing of ecosystem functions: opportunities, challenges and way forward. Remote Sens. Ecol. Conserv. 4 (2), 71–93. https://doi.org/10.1002/rse2.59.

Pinsky, M.L., Fenichel, E., Fogarty, M., Levin, S., McCay, B., St. Martin, K., Young, T., 2021. Fish and fisheries in hot water: what is happening and how do we adapt? Popul. Ecol. 63 (1), 17–26.

Plagányi, É., 2019. Climate change impacts on fisheries. Science 363 (6430), 930–931. https://doi.org/10.1126/science.aaw5824.

Poff, N.L., Angermeier, P.L., Cooper, S.D., Lake, P.S., Fausch, K.D., Winemiller, K.O., Mertes, L. A.K., Oswood, M.W., Reynolds, J., Rahel, F.J., 2001. Fish diversity in streams and rivers. In: Chapin, F.S., Sala, O.E., Huber-Sannwald, E. (Eds.), Future Scenarios of Global Biodiversity. Springer-Verlag, New York, pp. 315–349.

Poff, N.L., Brinson, M.M., Day, J.W.J., 2002. Aquatic Ecosystems and Global Climate Change: Potential Impacts on Inland Freshwater and Coastal Wetland Ecosystems in the United States. Available from: www.pewclimate.org. (Retrieved: 26 June 2019).

Poff, N.L., et al., 2007. The natural flow regime: a paradigm for river conservation and restoration. BioScience 53 (7), 769–784.

Poloczanska, E.S., et al., 2013. Global imprint of climate change on marine life. Nat. Clim. Change 3 (10), 919–925.

Post, E., et al., 2013. Ecological dynamics across the Arctic associated with recent climate change. Science 341 (6141), 617–620.

Rabalais, N.N., et al., 2002. Hypoxia in the northern Gulf of Mexico: description, causes and change. Coast. Estuar. Stud. 58, 1–36.

Walther, G.R., et al., 2002. Ecological responses to recent climate change. Nature 416 (6879), 389–395.

Watson, R.T., Zinyowera, M.C., Moss, R.H., Dokken, D.J., 2001. IPCC Special Report: The Regional Impacts of Climate Change: An Assessment of Vulnerability. IPCC, United Nations.

Wigley, T.M.L., 1999. The Science of Climate Change: Global and U.S. Perspectives. Pew Center on Global Climate Change, Arlington, VA.

Chapter 14

Climate variability impacts on wetlands

Timothy Dube[a], Tatenda Musasa[a], Cletah Shoko[b], Munyaradzi Davis Shekede[c], and Terence Darlington Mushore[d]

[a]*Department of Earth Science, Institute for Water Studies, University of the Western Cape, Cape Town, South Africa,* [b]*Division of Geography, School of Geography, Archaeology and Environmental Studies, University of Witwatersrand, Johannesburg, South Africa,* [c]*Department of Geography Geospatial Sciences and Earth Observation, University of Zimbabwe, Harare, Zimbabwe,* [d]*Department of Space Science and Applied Physics, Faculty of Science, University of Zimbabwe, Harare, Zimbabwe*

Chapter outline

Remote Sensing of Climate. https://doi.org/10.1016/B978-0-443-21731-9.00004-1
Copyright © 2024 Elsevier Inc. All rights are reserved, including those for text and data mining, AI training, and similar technologies.

1 Introduction

Wetlands are among the most biologically diverse and productive ecosystems in the world which are essential for human livelihood. Global research estimates show that coastal and inland wetland areas occupy more than 12.1 million km^2 (Ramsar Convention Secretariat, 2018). This translates into areas that are permanently inundated (54%) and seasonally inundated (46%) (Ramsar Convention Secretariat, 2018). In Sub-Saharan Africa, these precious ecosystems cover about 7%, with freshwater marshes and floodplains being the most extensive types (Rebelo et al., 2010). Although wetlands occur within all the major river basins, the areal representation and the spatial distribution of the different wetland types vary greatly. Southern Africa is largely endowed with wetland types that range from floodplains, swamps, marshes, pans, dambos, and lakes with different names given to these depending on the location (Musasa and Marambanyika, 2020).

Wetlands provide essential goods and services which play an important role in people's livelihoods mainly through agriculture and fishing, especially in resource-constrained communities. This therefore aids toward the attainment of Sustainable Development Goals 1 and 2 on zero hunger and ending poverty, respectively (Junk et al., 2013; Pantshwa and Buschke, 2019; Rebelo et al., 2010). Despite their potential contribution, natural wetlands have been on the decline and facing degradation, which possibly would result in wetland loss hence putting pressure on the communities that depend on them (Junk et al., 2013; Ramsar Convention Secretariat, 2018). For example, research estimates show that between 1970 and 2015, inland and coastal wetlands were reduced by approximately 35% (Ramsar Convention Secretariat, 2018). It clearly shows that the provision of wetland ecosystem services is threatened by a combination of natural and human factors. While the human causes of wetland loss such as land reclamation and draining land for agriculture, pollution, proliferation of invasive species, sand and gravel harvesting, urban development, and mining (Rebelo et al., 2010; Junk et al., 2013; Ramsar Convention Secretariat, 2018) can be managed, natural factors such as climate change and drought are projected to have devastating effects on wetland ecosystems (Mitchell, 2013).

Since the 1950s, global climate systems have shown an unprecedented change. Literature shows that between 1880 and 2012, the land and ocean surface temperatures have increased by approximately 0.85°C (range between 0.65°C and 1.06°C) (Pachauri et al., 2014). Climate change has been identified as a major threat to wetlands. For instance, altered hydrology and rising temperature can result in changes in the biogeochemistry and function of a wetland to such an extent that some important services might be turned into disservices (Salimi et al., 2021). This means that they will, for example, no longer provide a water purification service and adversely they may start to decompose and release nutrients to the surface water causing problems such as eutrophication, acidification, and brownification in the water bodies. Wetlands play an important role in climate change, because of their capacity to modulate atmospheric concentrations of greenhouse gases such as methane, carbon dioxide, and nitrous oxide, which are dominant greenhouse gases contributing to about 60%, 20%, and 6% of the global warming potential, respectively (Erwin, 2009).

Climate change can affect wetlands directly and indirectly through the effects of rising temperature, changes in rainfall intensity and frequency, extreme climatic events such as drought, flooding, and the frequency of storms (Erwin, 2009). In the future, higher temperatures resulting in drought might shift the role of both constructed wetland and peatland from a sink to a source of carbon. Globally, the frequency of drought has been increasing in light of climate change (Marambanyika et al., 2021; Salimi et al., 2021), affecting the wetlands' ecological conditions to different extents. Southern Africa has not been spared, as evidenced by multidecadal rainfall variability associated with drought conditions. Drought, generally defined as a period of abnormally dry weather long enough to cause a serious hydrological imbalance, has implications on wetland existence and their potential to provide goods and services. This clearly shows that climatic variation occurs across multiple spatial and temporal scales, and understanding the impacts of both short- and long-term changes will provide valuable information to decision makers, hence contributing to wetland conservation and sound environmental monitoring.

Over the past decades, global climate change and land-use conversion have been cited as the major threats as they have increased the vulnerability of wetlands, leading to their disappearance (Kaplan et al., 2019). This has provoked wider thinking on the need for continuous monitoring of the wetlands and their behavior, which is crucial for their protection and sustainable management. To this end, remote sensing techniques have been the most viable option as they are often less costly and time-consuming for large geographic areas compared to conventional field mapping (Czajkowski et al., 2007) and have been a successful tool for monitoring wetlands in the past few decades. Over the past few decades, remote sensing technology has been the most effective tool to acquire both spatial and temporal wetlands data. Remote sensing data have been used in a number of wetland research areas such as mapping wetland changes (Schmidt and Skidmore, 2003), carbon cycle and climate warming in wetland environments (Guo et al., 2017), and hydrology dynamics in wetlands. Literature, e.g., Kaplan et al. (2019) shows that the number of wetland remote sensing publication has drastically risen since the 1990s in order to understand the impact of climate variability on wetlands. This work has demonstrated the contribution of medium spatial resolution data for wetland studies leveraging the need to advance this field so as to contribute to better improved decision-making. Both optical and microwave earth observation (EO) sensors can be used for monitoring wetlands; thus, optical satellites are mostly effective in vegetation monitoring as well as the wetlands' change mapping (Kaplan et al., 2019). However, data from optical satellites have been used for mapping surface wetland water using different techniques that applied range in complexity and general applicability.

2 Remote sensing principles and techniques

2.1 Overview of remote sensing

Remote sensing provides invaluable information to characterize and measure wetland condition and functioning (Kaplan et al., 2019). EO satellites can be used to delineate inundated areas, and thereby provide essential information on dynamic wetland extents (Guo et al., 2017). The rapid expansion of remote sensing provides recent and developed advances in monitoring wetlands. Since the 1960s, remote sensing observations serve as the most useful tool for gathering information such as in land-cover change or mapping features in wetland regions, climate warming in wetland ecosystems, species diversity and productivity, and hydrological processes in wetlands (Guo et al., 2017). Remotely sensed datasets and approaches provide frequent data with varying imaging characteristics, which are more practical and economical means to address issues of wetland identification, delineation, classification, vegetation or biomass, hydromorphic soil, hydrology and vegetation characteristics, productivity, and density. To this end, the availability of automated and near-real-time remotely sensed data is thus essential in gathering spatial information on the condition, distribution, and impact of climate variability on wetland ecosystems, from a local to a global scale.

The impacts of climate change tend to vary with the spatial distribution of wetlands and can be analyzed with the aid of multispectral and hyperspectral remote sensing satellite images such as Landsat, Moderate Resolution Imaging Spectroradiometer (MODIS), Satellite Pour l'Observation de la Terre (SPOT), and RapidEye (Thamaga and Dube, 2018; Kaplan et al., 2019). Some of these images have high spectral and spatial characteristics, which enable enhanced monitoring and mapping of wetland ecosystem characteristics. When compared to traditional labor-intensive field investigation, remote sensing information not only saves time but also enhances the prospect of characterizing wetland species through spectral and texture analytics. Recently, advances in remote sensing data have shown high potential to examine climate variability impacts threatening wetland ecosystem functioning and services. In addition, advances in sensor technologies have contributed toward the acquisition of freely available satellite imagery, such as Sentinel dataset. For example, Sentinel-2 is characterized by finer spatial (10 m) and higher spectral (13 spectral bands including red edge strategic bands) resolution, essential for extraction of wetland ecosystem characteristics, with varying geographical coverage (285 km) for the evaluation of wetland dynamics (Orimoloye et al., 2018). Remote sensing technology allows repetitive image acquisitions over the same area, which are required for the detection of temporal changes and patterns of wetland ecosystems. For instance, Sentinel-2 offers remotely sensed data at a high revisit frequency of between 5 and 19 days. In light of the advantages associated with the use of remotely sensed data, in Sub-Saharan Africa, researchers have used both passive, optical sensors, and active sensors to map and delineate the spatial distribution of wetlands in order to understand their status as well as associated impacts of climate variability under the changing environmental and anthropogenic pressure.

2.2 Sensors and platforms used in wetland monitoring

There have been some notable improvements in remote sensing platforms in terms of operational time, reliability, simplicity, and temporal resolution, which affects the spatial resolution. The period from 1960 to date has experienced some major changes in the field of remote sensing in a variety of ways, for example, the primary platform used to carry remotely sensed instruments shifted from airplanes to satellites. Both optical and microwave EO sensors have been used for monitoring wetlands at global and national scales. Land surface temperature (LST), as one of the most important variables in the physical processes of the Earth, is one of the unexplored parameters for studying wetland dynamics which is imperative. Increases in temperature therefore need to be largely monitored so as to inform decision makers (Kaplan et al., 2019). Apparently, there exist two main types of remote sensing systems that can be classified according to the source of signal they use to explore the object, that is active and passive. Active remote sensing instruments are that operate with their own source of emission or light, while passive ones rely on the reflected one. Radiation also differs by wavelengths that fall into short (visible, Near Infra Red, Middle Infra Red) and long (microwave). Passive sensors in remote sensing do not streamline energy of their own to the researched object or surface, unlike active ones (Kaplan et al., 2019). Passive remote sensing employs multispectral or hyperspectral sensors that measure the acquired quantity with multiple band combinations. These combinations differ by the number of channels (two wavelengths and more). The scope of bands includes spectra within and beyond human vision (visible, IR, NIR, TIR, microwave). Landsat Multi Spectral Sensor, Landsat Thematic Mapper (TM), SPOT, and Indian Remote Sensing Satellite (IRS-1B) Linear Imaging Self-scanning Sensor (LISS-II) are satellite remote sensing systems that have been used to study wetlands (Mandishona and Knight, 2019).

Aerial photography has been used for wetland mapping for many decades (United Nations, 2008). The recent technological advances have therefore meant that the image quality collected by aerial photography has been improving, from initially black and white (panchromatic) to true color (RGB), and then to color infrared (CIR). Aerial photographs are commonly collected by states and local governments. For example, the State of Massachusetts collected

1: 12,000 scale CIR aerial photographs to conduct a statewide inventory of potential vernal pool habitats. The statewide National Agriculture Imagery Program (NAIP) imagery can be freely downloaded from the USDA Geospatial Data Gateway. These high-resolution natural color and CIR aerial imagery have been used in numerous wetland studies (Vanderhoof et al., 2016; Wu and Lane, 2016). Similar to aerial photographs, multispectral satellite imagery is also collected by passive sensors. The most commonly used multispectral satellite sensors for wetland mapping include Landsat MSS/TM/ETM+/OLI, MODIS, AVHRR, SPOT-4/5/6/7, IKONOS, QuickBird, GeoEye-1, RapidEye, Sentinel-2, WorldView-1/2/3/4, among others.

Medium spatial resolution data which include Landsat TM/Enhanced Thematic Mapper Plus (ETM+), SPOT 1–4, Sentinel Multispectral Instrument (MSI), Landsat Multispectral Scanner, Landsat 8 Operational Land Imager (OLI), and thermal infrared sensor (TIRS) have also been applied in wetland monitoring. The Landsat program is the longest-serving Earth Observation Platform (EOP) that started in 1972 (Kaplan et al., 2019) and has been widely used in most wetland studies. In many cases, Landsat MSS images were used with TM which has a fairly better spatial resolution and strategically positioned spectral bands. Mandishona and Knight (2019) combined Landsat TM (30 m resolution) with MSS to detect wetland change in the Kafue Flats flood plain of Zambia. In this case, the high cost of acquisition for high-resolution datasets necessitated the use of MSS images. Landsat TM images can detect moisture content in wetlands, an attribute essential in the hydrological classification of wetlands in the transboundary basins of Southern Africa. As such, Landsat TM datasets have been used to monitor the rate of evapotranspiration in Seke, and Chihota wetlands in Zimbabwe.

Mandishona and Knight (2019) used ETM (August, 2000), Landsat TM, and SPOT images (August, 2000) to map spatial extent and temporal changes that occurred in wetlands in Harare, in particular Borrowdale and Mukuvisi wetlands. Fakarayi et al. (2015) used Landsat TM images to quantify land-cover/use changes in the Driefontein wetland, where grasslands were observed to be decreasing between 1995 and 2005. Sentinel MSI is an emerging new crop of the sensor with better sensing characteristics that can provide data for wetland change detection. Nevertheless, current literature reveals that, at the transboundary scale, very few research works have been done to monitor and assess environmental issues using sentinel data (Thamaga and Dube, 2018). Of the limited research, Thamaga and Dube (2018) used sentinel data to detect and map aquatic water invasive species in Letaba River systems in Tzaneen, South Africa.

2.3 Data acquisition and preprocessing

Any remote sensing data acquisition system consists of the selection of three major aspects: (i) platform (spaceborne, airborne), (ii) remote sensing system (optical, synthetic aperture radar (SAR)), and (iii) sensor (data from any specific satellite/airborne/Unmanned Aerial Vehicle (UAV) sensor). At present, satellite images offer satisfactory geometrical and semantic information for various applications regarding data of Earth's surface (Kaplan et al., 2019). To get geolocated data from satellite images, they need to be orthorectified. There is a need to provide orthorectification for different satellite images, such as WorldView, GeoEye, QuickBird, Ikonos, Spot, which are not acquired freely despite improved performance and accuracy of results (Thamaga and Dube, 2018).

The Landsat program is the longest EOP that spans from 1972, showing great strides in advancing the remote sensing field as the data are freely available. To date the mission boost of eight satellites with the most recent satellite, Landsat 9, which was launched in September 2021. However, its applicability is still yet to be fully explored as it is in its earlier stages. The Landsat 8 is the more commonly recently improved sensor that was launched in February 2013 (Thamaga and Dube, 2018). Over the years, the Landsat data as a valuable resource for global research have been used in different applications. Landsat 8 carries two instruments, OLI and TIRS. OLI collects data from nine, while TIRS collects data from two spectral bands and can be downloaded freely from the United States Geological Survey (USGS) at http://glovis.usgs.gov/. Taking into consideration the diverse properties of wetlands as they provide a transition between terrestrial and open-water aquatic ecosystems, wetlands and their properties are not easily detectable with optical satellite sensors (Kaplan et al., 2019). However, wetlands are more sensitive to some parts of the electromagnetic spectrum than others. To this end, the near-infrared which is sensitive to biomass content, in combination with the green band, has potential to provide valuable information for the water content, as well as for the soil wetness which is essential in wetland monitoring (Fakarayi et al., 2015; Guo et al., 2017). On the other hand, the short-wave infrared bands are more sensitive to the moisture content of both soil and vegetation, and they are particularly useful in separating wetlands from dry lands.

Literature also shows satellite data for instance, Kaplan et al. (2019) in their study acquired Landsat 5 (1990, 2000, and 2010) and Landsat TM8 (2020) data (30 m × 30 m), obtained from the geospatial data Cloud platform (https://www.gscloud.cn/). Remote sensing data were selected from mid-May to mid-September because images in this period have high quality, zero or very few clouds, and meet the requirements of reflecting surface information to the maximum extent. Subsequently, the images were reprojected and mosaicked later followed by atmospheric correction, using the dark object subtraction (DOS1) model in Quantum Geographical Information System version 2.1.8 software. Other data that could be included in this study mainly included 1:50,000 topographic maps, drainage maps, hydrogeological maps, some soil distribution maps, geomorphological maps, some field survey data, some land-use maps, and administrative division maps of state farms in the study area, which were obtained from local government databases (Kaplan et al., 2019).

Complete preprocessing of remotely sensed data can be conducted using the image processing software ENVI5.4 (Exelis Visual Information Solutions, Inc.), e.g., radiometric correction, geometric correction, image enhancement, mosaicking, and image cropping of remote sensing images in different periods (Kaplan et al., 2019). Based on the classification system and the actual land-use situation in an area, land-use types can be classified into cropland, forestland, grassland, water area, built-up land, and wetland. This can be aided with the help of field survey data, collected land-use data, and the land-use interpretation of the study area. In addition, the quantitative extraction of the normalized difference vegetation index (NDVI), difference vegetation index, and ratio vegetation index was achieved by ENVI 5.4 in a study conducted in China by Kaplan et al. (2019).

The latest SAR satellite developed by the European Space Agency, Sentinel-1, is an imaging radar satellite that has a constellation of two satellites, Sentinel-1A, launched on April 3, 2014, and Sentinel-1B, launched on April 22, 2016 (Thamaga and Dube, 2018; Kaplan et al., 2019). The instruments mainly cover applications that include monitoring sea ice zones and the Arctic environment; surveillance of marine environment; mentoring land surface motion risks; mapping of land surfaces like forest, water, and soil and agriculture; and mapping in support of humanitarian aid on crisis situation (Zhang et al., 2018). As a result of a number of environmental circumstances and system factors, satellite images can be often distorted in geometry and brightness, which requires preprocessing of the images before their use. SAR data are exposed to radiometric and geometric distortions that should be removed or minimized. The preprocessing of the Sentinel-1 SAR data can be easily done with the Sentinel-1 Toolbox integrated into Sentinel Application Platform (SNAP), and it contains a few steps: (i) data preparation, (ii) radiometric calibration, (iii) multilooking, (iv) spackle reduction, (v) terrain correction, and (vi) dB conversion. The data preparation consists of selecting the study area, selecting the data type needed for the study, selecting the date, and downloading the SAR product. Radiometric calibration corrects the SAR image so that the pixel values represent the radar backscatter of the reflected surface. SAR calibration provides imagery in which the pixel values can be directly related to the radar backscatter of the scene.

Additionally, too, satellite imagery datasets obtained from Sentinel 2 MSI can be downloaded in TIFF picture format from Copernicus access hub. Geometric and radiometric distortions that emanate from variations in sensor attributes and atmospheric conditions need to be corrected and for all images; the registration is usually done using the Universal Transverse Mercator conformal (coordinate system) projection employing ARC1950 as the Datum zone at 35°S. Radiometric corrections will also be done through image enhancement. The image analysis option in ArcMap 10.5 is often used to enhance brightness, contrast, and transparency of images, which can be confirmed using the swipe ban option. Metadata can later be used to normalize spectral signatures of the Earth's surface using the Raster calculator before the images can then converted to composite bands. The selected bands will then be imported into software such as ArcMap 10.5 before data cleaning is done so as to remove the no data value region.

2.4 Remote sensing data types relevant to climate variability impacts

Satellite sensors can provide multispectral imagery with finer spectral and better temporal resolutions, which are essential for classifying wetland vegetation types and analyzing wetland water dynamics. In addition to multispectral imagery, Light Detection and Ranging (LiDAR) data have increasingly been incorporated into the wetland mapping process. High-resolution digital elevation models (DEMs) can then be derived from LiDAR point clouds by using interpolation algorithms. Importantly, these LiDAR-based DEMs can be used to compute various topographic metrics, which serve as essential wetland indicators. Since most LiDAR sensors operate in the near-infrared spectrum, laser lights are strongly absorbed by water, resulting in very weak or no signal returns. As a result, water areas appear as dark features in the LiDAR intensity imagery. Therefore, LiDAR intensity data are particularly useful for mapping surface water and wetland inundation extent. A number of studies have reported improved accuracy of wetland inundation mapping by using LiDAR intensity data with simple thresholding techniques (Wu and Lane, 2016).

MODIS with high temporal resolution also showed its significant advantages in mapping wetland extent and change over time. Klein et al. (2015) mapped daily open water bodies using MODIS time series data and threshold technique; the technique depicts annual water changes. Frazier et al. (2003) used Landsat images to assess before and after flood occurrence to describe the relationship between flow regulation and inundation of floodplain wetlands. Results from their study highlighted that river regulation could reduce the duration and frequency of inundation. Other sensors, such as Sentinel, SPOT, have been employed to examine wetland hydrological regimes and mapping wetlands extent at various scales with satisfactory results (Muro et al., 2016; Slagter et al., 2020). Indices such as Land Surface Water Index (LSWI) and Normalized Difference Water Index (NDWI) have also been extensively used to improve accuracies. For instance, LSWI is known to be sensitive to the total amount of liquid water in vegetation and associated soil background. Using hydrological models such as the soil water assessment tool (SWAT) together with satellite data, soil and climatic information seem to be promising in assessing wetland hydrology as well as soil quality and quantity.

Remote sensing is an important component of climate system observations and has gradually become a leading research method in climate change studies. Satellite data also contribute significantly to the improvement of meteorological reanalysis products that are widely used for climate change research, for example, the National Center for Environmental Prediction reanalysis (Gibson, 2013; Jin et al., 2013). The Advanced Very High-Resolution Radiometer on board the

National Oceanic and Atmospheric Administration satellites allows us to monitor the Sea Surface Temperature (SST) worldwide. An increase in SST has been observed in all ocean basins since the 1970s, with an average estimated increase of 0.28°C from 1984 to 2006. This clearly shows that satellite data provide an independent way to investigate global temperature trends, particularly at the ocean surface and in the atmosphere.

LST is one of the most important variables in the physical processes of the Earth, and it is one of the unexplored parameters for studying wetland dynamics (Muro et al., 2016; Kaplan et al., 2019). LST is closely related to the surface energy balance and the water status of the land cover, and it depends on the radiative energy that the land absorbs (Mira et al., 2017). With the latest technological developments in remote sensing, many EO satellites like Landsat, Sentinel-3, MODIS, and ASTER operate in the thermal infrared region offering thermal bands for retrieving thermal maps of the Earth's surface (Kaplan et al., 2019). Landsat 8 is the latest satellite from the Landsat legacy, and it offers 100-m thermal data. Retrieving LST using Landsat data have been the subject in many studies resulting in several methods and algorithms (Wang et al., 2015; Mira et al., 2017). LST from satellite images has been used in different studies such as studies related to urban heat islands, earthquake monitoring, water extraction (Wang et al., 2015), climate changes, etc. One of the important parameters to understand the extensive range of existing processes in the wetland areas is the LST (Avdan and Jovanovska, 2016).

3 Climate variability and its impacts on wetlands

3.1 Understanding climate variability patterns

Fragile as they are, wetlands are highly vulnerable to sustainable exploitation as well as natural factors such as climate change due to the primary importance of the hydrological regime in controlling their ecological characteristics (Wang et al., 2015). Nevertheless, the ever changing climate that manifests through warming will alter precipitation and evapotranspiration rates and result in changes to runoff and groundwater levels. To this end, projected changes in climate are likely to substantially impact wetland hydrological conditions that will in turn have implications for wetland ecology; hence, it is more important to understand climate variability patterns (House et al., 2015). Analysis of information shows that temperature and precipitation are important variables related to climate that significantly alter wetland conditions, which therefore needs closely monitoring.

Climate change which is growing concern is expected to result in increased loss and degradation of most wetlands not only because of the complex relationships and feedbacks between climate and wetlands (Ramsar Convention Secretariat, 2018). This can also be attributed to the likely synergistic effects of the nonclimatic and climatic drivers, especially in coastal and low-lying areas. It is widely reported that global change has potential to cause abrupt changes in wetlands, which can be large in magnitude and difficult, expensive, or impossible to reverse (Millennium Ecosystem Assessment, 2005). Assessing ecohydrological impacts of climate change requires models that can accurately simulate water levels at the fine-scale resolution to which species and communities respond. Hydrological impacts of climate change differ through the wetland over short distances depending on the degree of groundwater/surface-water interaction. The effects of climate change on regional aquifers and catchment runoff may cause intricate and significantly detrimental impacts to wetlands underlain by permeable geology.

Climate change has been identified as a major threat to wetlands as it can influence a wetland ecosystem by increasing temperature and changing hydrological patterns, which in turn can alter the biogeochemistry of the ecosystem (Erwin, 2009). Since wetlands are often located in a transition zone between an aquatic and a terrestrial ecosystem, their hydrological fluctuation is inevitable. Although they have been known to be resilient to change in general, they may still be highly susceptible to hydrological changes, especially when this change is exacerbated by other sources of disturbance such as climate change, pollution, urbanization, and changes in land use. The impact of climate change on a wetland system can be predicted by using various approaches such as modeling, field survey, and experiments.

Changes in the hydrological cycle, for example, in the freshwater flow as well as a rise in sea level and increases in storm surges driven by global warming will result in a series of environmental problems. This will likely result in the enhanced erosion of shores and coastal habitats, salinization of groundwater and estuaries, altered tidal ranges in rivers and embayment, changes in sediment inputs and nutrient loadings, and increased coastal flooding and consequently decrease of freshwater availability for humans and ecosystems. Wetlands, such as mangroves, saltmarshes, and floodplains, can play a critical role in the physical buffering of climate change impacts, which if unchecked and remain unprotected will threaten wetland existence.

3.2 Key climate variables affecting wetland ecosystems

Wetlands are dynamic and complex ecosystems that are extremely vulnerable to changes in environmental conditions largely driven by land-use changes. Zedler and Kercher (2005) report that approximately almost half of wetland ecosystems have already been lost and degraded due to land-use changes, mainly through the conversion of wetlands to agricultural lands. Cognisant of the threats attributed to land-use changes, there has been growing international and regional recognition focusing on the impact of anthropogenic causes of climate change on the structure and functions of wetland ecosystems (Mitsch and Gosselink, 2015). Latifovic and Pouliot (2013) demonstrates that the Earth's climate is changing at an alarming rate with global temperatures rising, glaciers melting, rising sea levels, and exacerbated changes in weather and climate. To this effect, the current degree of change associated with the anthropogenic warming of the earth through sea level rise (SLR) and exacerbated variations in rainfall patterns is projected to result in a cascade of effects on human livelihoods and their well-being and natural ecosystems such as wetlands (Mitsch and Gosselink, 2015).

Climate variability has resulted in changes to inflows associated with increasing temperatures, more variable rainfall and more extremes are affecting rivers and wetlands nationally (Kaplan et al., 2019). This therefore mean there is reduced surface area for evaporation which has significant bearing on wetland ecosystem stability, which compromises their ability to provide goods and services in the near future. Additionally, high temperatures result in high evaporation rates, thus SLR and storm surge, which largely impact wetlands and the surrounding communities; hence, it remains a salient aspect. Climate change may cause more evapotranspiration and consequently a water level drawdown and also more flood inundation, which might critically affect the biogeochemistry, water quality (Waddington et al., 2015), and gas emission from wetland ecosystems. A warmer climate will also accelerate the loss of water through evapotranspiration. This leads to a water table drop, resulting in the exposure of top layers of the wetland to the oxygen where the organic matter can be oxidized readily. It has been predicted that exposure to warmer temperatures and drier conditions associated with climate change will shift the balance between ecosystem photosynthesis and respiration. Therefore, understanding wetland changes is important for the formulation of national policies and regulations, environmental protection, and ecological restoration.

Wetlands are globally recognized as the Earth's most valuable and productive ecosystems, providing a variety of ecosystem services to the biophysical environment and human livelihoods. Despite this alarming truth, it is evident that droughts and floods are a recurring phenomenon in many parts of the world southern Africa with no exception. Engelbrecht et al. (2015) report that as a result of increasing greenhouse gas emissions investigated under low-mitigation scenarios, the impact of climate change across southern Africa is projected to change annual and seasonal patterns in rainfall and temperature. These variations are known to have major consequences for human livelihoods (e.g., Ankrah, 2018) and natural ecosystems such as wetlands. Naturally, wetlands are complex and dynamic; their physical structure are shaped by movements of mass and their functional characteristics are influenced by earth systems processes (Tooth and McCarthy, 2007; Ellery et al., 2016). As a result, these ecosystems are extremely sensitive to the influence of droughts and floods associated with climatic oscillations such as ENSO and the 18-year oscillation. Understanding the different dimensions of climate variability, such as historical changes attributed to climate variability in the region, their impacts, and their occurrence, is also a crucial step toward developing effective models to predict and investigate different impacts of climate variables impacting wetlands. The prediction of climatic variables such as drought occurrence permits drought preparedness, which necessitates the development of specific contingency plans, for instance water rationing and the use of alternative water sources, thus ameliorating some of the foreseeable impacts.

3.3 Impacts of climate variability on wetland hydrology, vegetation, and biodiversity

Climate change is expected to intensify the loss and degradation of many wetlands. Climate change is likely causing parts of the water cycle to speed up as warming global temperatures increase the rate of evaporation worldwide. The influence of short-term climatic scenarios such as El Niño/La Niña and related events are largely associated with direct and indirect changes to the functioning of wetlands (Xie et al., 2019). It has been observed that during drought conditions rainfall is limited resulting in dramatic changes in wetland hydrology. For example, the decline in rainfall results in limited water inputs such as direct precipitation, overland flow, through flow, and groundwater recharge. This therefore alters surface waters in wetlands, often leading to temporary or permanent desiccation, thus causing shifts in wetland vegetation compositions and stress on fauna and avifauna. The hydrologic stress placed on groundwater has been widely documented (Niemi et al., 2004), with concerns related to the pressure placed on groundwater during dry periods, particularly by human activities for instance excessive abstraction of water from underground aquifer. Furthermore, flooding caused by excessive rainfall may replenish wetland waters and aquifers, but could possibly increase soil erosion and submerge flood-intolerant vegetation.

Climatic factors play a substantial role in the field of wetland changes. Temperature and precipitation were shown to be the main natural driving forces (Xi et al., 2021). The meteorological data adopted by Xi et al. (2021) showed that in the past two decades, the average temperature had increased by approximately by 0.8°C and precipitation had decreased by approximately 100 mm. Results indicated that compared with the annual average temperature and annual precipitation, the influence of warm-season temperature and precipitation on wetland changes was more critical. This may be attributed to the fact that in the warm season, some glaciers and snow melt, which can supply the water source of the wetland. The warm season is also the growing season of wetland vegetation, which clearly shows that temperature and precipitation affect the growth of vegetation, which is a key aspect in wetland ecologic conditions. However, climate variability makes vegetation more dependent on water conditions, and annual precipitation plays a key role in vegetation survival.

Climate variability has resulted in significant changes in wetland of various types. For instance, palustrine wetlands do not have a stable water supply and rely on replenishment from precipitation. This therefore means that they rely on more precipitation input which leads to the more hydrological recharge of palustrine wetlands, hence enhancing their ability to provide habitats for several key species found in these special types of wetlands (Kaplan et al., 2019). To this end, this type of wetlands is largely influenced by precipitation more than temperature changes, which also determines the goods and services derived from the particular ecosystem. In the warm season, glaciers and frozen soils are greatly affected by temperature. However, in different environments, the higher temperature causes the melting of glaciers and frozen soils, which benefits the hydrological conditions of riverine wetlands. The influence of precipitation on the riverine wetlands was not apparent, among them, the impact of warm-season precipitation was clearly not significant (Xi et al., 2021). This result may represent a sharp increase in warm-season temperature, which leads to increased evaporation and offsets the supplement of precipitation (Peng et al., 2023). Due to the significant temperature rise, the evapotranspiration consumption of riverine wetlands in the source area of the Yellow River was greater than the replenishment of glaciers and snow melting.

4 Remote sensing applications for wetland monitoring

4.1 Wetland mapping and classification using remote sensing

Wetlands are natural ecosystems that continue to be exposed to degradation as a result of anthropogenic and natural disturbances in recent decades. Despite the fact that some advancements have been made, effective methods that can produce large-scale wetland maps with detailed categories are still lacking due to the diversity and complexity of wetland ecosystems (Peng et al., 2023). To address this issue, scholars such as Peng et al. (2023) proposed a novel algorithm for detailed wetland type classification integrating k-fold random forest and hierarchical decision tree, and so named two-step classification algorithm. In this case, phenology-based features were composited based on time series Sentinel-1/2 images, and the k-fold random forest was used to extract five rough wetland types in Google Earth Engine (GEE) platform (Phinzi et al., 2021). Second, the hierarchical decision tree designed based on geometric features was used to separate the rough wetland types into 14 detailed types. Application of the two-step classification method in Northern, Central, and Southern Asia resulted in a continental-scale wetland map with an overall accuracy (OA) of $90.0 \pm 0.5\%$. Wetland types, including inland marsh, lake, river, coastal swamp, estuarine water, lagoon, shallow marine water, reservoir, canal/channel, and agricultural pond, had good accuracy with both User Accuracy and Producer Accuracy over 77%. The remaining wetland types had moderate accuracy, with both UA and PA over 58% (Peng et al., 2023). This therefore shows that remote sensing is an indispensable tool in as far as wetland mapping and classification is concerned.

Due to the low cost of data acquisition and high spatial-temporal resolution, remote sensing technology has been widely used in large-scale wetland mapping. In the past decades, several wetland datasets have been derived from satellite images, which aids in wetlands mapping and classification. These have been categorized as land-cover and land-use change (LCLUC) datasets, single-type wetland datasets, and multitype wetland datasets. The global LCLUC datasets, for instance, GlobeLand30 and GLC_FCS30, only contain limited wetland classes (i.e., water and wetland) in their classification scheme, which somehow limits their applicability as these datasets fail to separate the coarse classes into detailed wetland types. The single-type wetland datasets focus on one wetland type (i.e., water, mangroves, tidal flats) and can accurately delineate the extent of specific wetland types (Xi et al., 2021). Accurate and up-to-date wetland maps can offer useful information to improve understanding of ecosystem conditions and support wetland-related Sustainable Development Goal targets.

With the advancements of remote sensing, various classification methods have been developed to implement wetland classification for their designed classification scheme. Some of the commonly used methods include maximum likelihood classification (MLC)), random forest, support vector machine (Wang et al., 2015), and deep learning. These methods have been widely used in local or global wetland mapping producing accurate results, hence necessitating wetland mapping and classification. Cognizant of all this, supervised classification methods such as MLC adopted at pixel level may only be suitable to classify rough-type wetlands as some detailed wetland types are dynamic and share similar spectral characteristics, which may require advanced robust methods (Mao et al., 2020). Peng et al. (2023) reiterate that due to similar spectral characteristics, similar waterbody types (i.e., river, lake, and reservoir) cannot be well classified by the pixel-based supervised classification algorithm. However, these wetland types mentioned have different geometric characteristics, which can therefore be well separated by object-based decision tree algorithms at both the local or national scale (Mao et al., 2020).

The emergence of the GEE platform has meant that many different data sources, such as Landsat, Sentinel-2, MODIS, and SAR images, can now be easily accessed and processed for large-scale remote sensing classification (Xie et al., 2019). For example, the Sentinel-2 images are known to have potential for accurate wetland classification at a large scale due to the high spatial resolution and short repetition cycle. This mission offers 10-m resolution images with revisit intervals of 2–5 days, which has good potential in complex wetland monitoring, such as macrophytes mapping in the wetland (Carlson et al., 2022). Due to the sensitivity of certain wetlands and the capacity of penetrating clouds and vegetation, Sentinel-1 SAR data combined with optical images have often been used to improve wetland mapping accuracy. In addition, elevation-derived metrics can provide useful information and are also commonly used for wetland classification. Thus, the full use of multitemporal and multisource satellite images exhibits promising potential for large-scale classification with detailed wetland types.

4.2 Remote sensing-based assessment of wetland water dynamics

High spatial and temporal resolution remote sensing data are becoming readily available, which has made the use of remote sensing to monitor and quantify spatiotemporal changes in surface waters feasible and efficient (Alawadi et al., 2023). More recently, remote sensing methods that are based on spectral indices have been used to assess the changes in inundated areas and water storage in the Hawizeh marsh (south of Iraq) during the 2019 flood. In this case, the Normalized Difference Water Index (NDWI) and Normalized Difference Moisture Index (NDMI), were used to understand wetland water dynamics so as to inform decision-making. Water indices have been frequently utilized to detect water bodies because of their particular spectral properties in the visible and infrared spectrum (Kaplan et al., 2019). The accuracy assessment of the water areas extracted by the remote sensing indices showed an acceptable degree of reliability, which reflected positively on the water inflow calculations. This therefore shows that remote sensing has emerged as a powerful tool to monitor and map wetland dynamics, providing valuable data and insights into conservation efforts.

In order to characterize the spatial-temporal variation of surface water area accurate surface water dynamic maps can be used (Perennou et al., 2018). This implies that the use of optical remote sensing images to map surface water has been widely studied, using different methods and sensors. However, wetlands are highly dynamic and heterogenous, which thus calls for higher temporal and spatial resolutions data, with up to 10-m spatial resolution and 5 days' revisit time; the freely available Sentinel-2 data might be a good opportunity for mapping the surface water dynamics of complex wetlands. Slagter et al. (2020) investigated the effectiveness and limitations of Sentinel-2 time series data in combination with a novel Water Change Tracking algorithm for mapping the surface water dynamics of St. Lucia wetland in South Africa. The results showed that Sentinel-2 bands B2, B3, B4, B8A, B11, and B12 in combination with a fixed Minimum Normalized Water Score (MNWS) threshold of 3.0 was found suitable to binarize the MNWS image to water and nonwater pixels. The study revealed that the MNWS model was excellent at detecting surface water areas over thin clouds and those affected by surface sunlight.

Although significant advancements in mathematical models for surface water quality using reflectance values from satellite imagery are given, improving existing models using multiple linear regression to estimate water quality using Landsat 8 imagery remains an indispensable tool. Sharaf El Din and Zhang (2017) have proposed a regression-based technique to estimate surface water quality parameters using Landsat 8 OLI imagery. They propose a stepwise regression (SWR) to minimize the number of predictor variables and to maximize the precision of the water quality estimation. Highly accurate results were achieved when using the Landsat 8-based-SWR approach ($r^2 > 85\%$). This therefore shows that although applicability is still limited the same type of sensor has the potential to contribute to surface wetland water monitoring, thus contributing to sound environmental performance.

The Sentinel-2-based Wetland Use Intensity (WUI) captures spectral changes on the wetland surface over time, and the Sentinel-1-based Surface Water Occurrence reflects presence of surface water over a given period of time (Steinbach et al., 2023). The Sentinel-1 and -2 imagery and Landsat 7 and 8 imagery were acquired through the GEE cloud processing platform's data catalog. Sentinel-1 Ground Range Detected imagery in Interferometric Wide Swath mode from relative orbit number 74 was used for surface water detection and subsequent calculation of the Surface Water Dynamics layer for the same observation period. Landsat 7 and 8 surface reflectance (SR) imagery at 30-m spatial resolution and a 16-day revisit time (Schroeder et al., 2016) served for spectral comparison between the time of field data acquisition and the observation period. The results suggest that the adapted WUI indicator is informative, does not overlap with existing indicators, and is applicable to wetland management.

Many researchers used remote sensing data combined with field survey data to carry out many wetland studies (Haidary et al., 2013). Therefore, the combination of in situ data (ground truth) and Landsat would be beneficial in understanding land processes and in making management decisions about wetland management. The advantages of using remote-sensed data such as Landsat data in monitoring wetlands dynamics are: the images can be downloaded free of charge, records of the historic data are available on global scale, Landsat TM and Landsat ETM have multispectral bands, with good spatial and temporal resolution, and less image processing time is needed (Grundling et al., 2013). There are some of the few studies that were done by Kulawardhana et al. (2006) and Nhamo et al. (2017) that have shown that time series of Landsat data provide very useful information for mapping wetlands and LULC changes. However, there are several limitations such as cloud cover that usually limit the usable of the imagery and that usually affect the reliability of monitoring LULC and wetland.

4.3 Monitoring vegetation changes in wetland ecosystems

Wetland ecosystems play critical roles as water reservoirs for human consumption, habitats for plants and animals, and links between land and water ecosystems, supporting socioeconomic development and environmental protection (Mitsch and Gosselink, 2000). In this study, by using time series of Landsat data on GEE, a long-term cloudless remote sensing images dataset of Dongting Lake during the dry season was constructed (Peng et al., 2023). The multiyear wetland vegetation from 2003 to 2020 was mapped using the random forest model. Relationships between wetland vegetation dynamics and hydrological conditions (e.g., water level and inundation character) were examined using correlation and linear analysis (Hu et al., 2015).

Due to frequent cloudy and rainy conditions, frequent changes, and high heterogeneity of wetland landscapes, most previous studies that either used MODIS images with coarse spatial resolution or sparse Landsat images (Deng et al., 2014; Hu et al., 2015) have largely failed to accurately describe the annual changes of wetland vegetation. It is still necessary to accumulate Sentinel-2 data for long-term wetland vegetation monitoring. Although continuous remote sensing dataset provides valuable results, it is still a challenging task for the massive workload of data screening and image remediation with traditional personal computer platforms. With the help of remote sensing cloud platforms like GEE, we can obtain long-term data to perform multisource integration to map certain multiseason or annual wetland (Peng et al., 2023).

Atmospherically corrected Landsat 5 TM, Landsat 7 ETM+, and Landsat 8 OLI Level-2 SR products with a 30-m spatial and a 16-day temporal resolution were generated from the USGS EROS's Science Processing Architecture (ESPA) ordering system using an R access script (Kaplan et al., 2019). Results of the approach adopted have found statistically significant ($P < 0.05$) long-term changes in wetland ecosystems, as calculated from MNDWI and NDVI image series on both the MODIS and the Landsat scales (Kovács et al., 2022). Results showed higher users, producers, and OA when using a dense Landsat time series ($OA = 0.89 \pm 0.01$), outperforming the MOD09A1 time series ($OA = 0.37 \pm 0.03$).

4.4 Assessment of wetland biodiversity using remote sensing

Remote sensing can provide data over a large area in a consistent, objective manner and has been used to detect plant biodiversity in a range of ecosystems, typically based on relating spectral properties to the distribution of habitat, species, or functional groups. Wetlands are one of the most valuable natural resources on earth and play an important role in preserving biodiversity (Ramsar Convention Secretariat, 2018). However, many wetlands across the world have deteriorated and disappeared over the past several decades. To this end, the adoption of remote sensing images and point of interest (POI) data, and several frameworks, i.e., knowledge-based raster mapping (KBRM)-based framework have been implemented to assess wetland ecological conditions (Chen, 2023; Yang et al., 2021). Density maps of waterbodies, vegetation covers, imperviousness, roads, and POI values were derived and used as five ecological indicators that can represent the ecological conditions of wetlands (Chen, 2023). The KBRM approach was used to integrate these indicators into an overall rating and map wetland ecological conditions efficiently (Sun et al., 2023). Thus, spatial variations in wetland ecological conditions can be distinguished and represented in detail.

Multiband, multispectral, and high-resolution remote sensing technology and Geographic Information System (GIS) have been widely used in monitoring wetland dynamics and to research the patterns and causes of wetland degradation (Houborg et al., 2015; Gxokwe et al., 2020). Zhu et al. (2022) extracted the lake area, vegetation cover, and soil degradation from multitemporal remote sensing (TM, ETM, and OLI) and constructed a new wetland degradation index to assess the degree of lake wetland degradation in the Bashang Plateau. Yang et al. (2021) used remote sensing and POI data to assess the ecological status of a wetland in Suzhou, China. It has been found that 3S technology (GIS; Global Positioning System, GPS; and Remote Sensing, RS) can accurately reveal the spatiotemporal variation in wetland ecological degradation across scales (Zhu and Woodcock, 2014). In recent years, advances in technology and data processing have enabled the integration of remote sensing with machine learning algorithms. This fusion allows for automated wetland mapping and classification as well as biodiversity assessment, reducing the time and effort required to analyze vast amounts of data. Machine learning models can learn to distinguish between different wetland types and detect changes with a high degree of accuracy, making wetland monitoring and conservation efforts more efficient and effective (Chen, 2023). RS is therefore an important tool because of its ability to provide timely and accurate information on wetland extent vegetation hydrology and human disturbances, which enables policymakers to make informed decisions and implement effective conservation strategies.

5 Methodologies for analyzing climate variability impacts

5.1 Time series analysis of remote sensing data

In remote sensing, time series analysis is an essential element as it provides essential tools and methods that can be used to extract information about the landscape characterized by both spectral and temporal changes (Zhu and Woodcock, 2014). It is important to note that, unlike most time series analysis, forecasting future events can be based on past events, thus aiding the quality of assessment (Chen, 2023). In remote sensing, time series analysis tools are often applied to individual pixels independently (i.e., no interaction between pixels). The key is to apply the same method to every pixel, which means the same application over millions of pixels/locations, and this means many time series methods when applied over a spatial domain (e.g., a country) are computationally expensive—and may require a parallel computing environment (Kaplan et al., 2019; Peng et al., 2023).

Literature shows that data from high time frequency sensors (e.g., MODIS, AVHRR, SPOT Vegetation, Sentinel-3) are often regularly spaced in time and allow established time series methods to be applied (Kaplan et al., 2019). Data from (traditionally) time infrequent sensors (e.g., Landsat, Sentinel etc.) are irregularly observed and contain many noise (e.g., clouds) so traditional time series methods may not always apply. Time series data from Landsat-like sensors are often called "dense time stacks." Time series analysis methods in the preprocessing stage include gap fill, averaging (smoothing), and smart filtering (Peng et al., 2023). As with many other remotely sensed observations, the time series data tend to be noisy and must be "preprocessed" before any information can be extracted. The key to successful preprocessing is to find the key areas where neither too much information is lost nor too much noise is retained, which may have significant bearing (Chen, 2023).

Time series of remote sensing data can also be adapted to monitor changes in landscapes at both local and regional scales. Basically, land-cover change detection techniques are based on the comparison of sequential land-cover maps derived from remote sensing data or other sources for the same area. For example, for every sampling unit of the maps, the land-cover categories at the two dates are compared to detect subtle changes within broad land-cover classes. Thus, rather than detecting changes on the basis of land-cover categories, change detection is better performed on the basis of the continuous variables defining these categories, whether these are reflectance values measured by a satellite sensor or biophysical attributes derived by model inversion.

Times series satellite imagery is being utilized in a variety of ways to monitor land, water, and the atmosphere environments and to determine their interactions with human activities and climate change (Fakarayi et al., 2015; Thamaga and Dube, 2018). With the availability of a wide range of satellite datasets, the frequency and lengths of time series analysis are being expanded to investigate the progress and drivers of change. More applications are expected in the future toward the use of data fusion of multiple sensors acquired with various spatiotemporal details. Over the last few decades, satellite remote sensing has witnessed a paradigm shift in how multiple-sensor data are processed, archived, integrated, and applied in understanding changes in land, water, and air. The major reasons driving this explosive growth in the use of time series remotely sensed datasets include their public and free access, standardized products, and growth in computing power.

Mutisi and Nhamo (2015) applied aerial photographs to assess the rate and direction of wetland change in Belvedere, Zimbabwe as they could give a unique overview of a large area. Marambanyika and Beckedahl (2016) used 1985 and 1996 aerial photographs to detect land-use/cover changes in Dufuya, Madigane, Guruguru, Zungwi, and Chebvuterembatemwa wetlands in the Midlands Province. However, the use of aerial photos in wetland change detection at the transboundary level has become limited as government departments are no longer producing this type of dataset. Mandishona and Knight (2019) used ETM (August, 2000), Landsat Thematic Mapper, and SPOT images (August, 2000) to map spatial extent and temporal changes that occurred in wetlands in Harare, in particular Borrowdale and Mukuvisi wetlands. Fakarayi et al. (2015) used Landsat Thematic Mapper images to quantify land-use changes in the Driefontein wetland, where grasslands were observed to be decreasing between 1995 and 2005. Sentinel MSI is an emerging new crop of the sensors with better sensing characteristics that can provide data for wetland change detection. Nevertheless, current literature reveals that, at the transboundary scale, very few research works have been done to monitor and assess environmental issues using sentinel data (Thamaga and Dube, 2018). Of the limited research, Thamaga and Dube (2018) used sentinel data to detect and map aquatic water invasive species in Letaba River systems in Tzaneen, South Africa. Therefore, it is essential for us to provide deeper insights into this fast-developing direction.

The increase in the use of remote sensing data in mapping wetland ecosystems is linked to its ability to offer a variety of new applications that can quickly and synoptically monitor and manage large areas. Recent studies have indicated that the use of satellite imagery provides the most reliable primary data for the detection, monitoring, and mapping of wetland ecosystems. For example, Nhamo et al. (2017) mapped wetlands in Mpumalanga using Landsat 8 and MODIS-based NDVI and found that the wetland extent declined by 19%. Nineteen percent of degraded land has been mainly replaced by urban and agricultural development, which affected the ecohydrological processes and functions. In a different study, Orimoloye et al. (2018) assessed the potential of Landsat data to understand the status of Isimangaliso wetland in South Africa. Results obtained from the study showed that the extent of the wetland shrunk from 655.416 km^2 (1987) to 429.489 km^2 (2017) and achieved an overall classification accuracy of 97.55% and kappa coefficient of 0.941. This unequivocally underscores the important role of time series data in monitoring change and climate change impacts on wetlands.

5.2 Modeling approaches for assessing climate-wetland interactions

Hydrological connectivity between rivers and wetlands is considered one of the key critical factors for the integrity of floodplain landscapes. Abdel-Karim et al. (2020) simulated flood inundation using a two-dimensional hydrodynamic model and the connectivities between wetlands and rivers were calculated using geoprocessing tools in Aeronautical Reconnaissance Coverage Geographic Information System (ArcGIS) producing significant results. As expected, a large flood produces longer duration of connectivity relative to a small flood. Results also showed that a reduction in mean connectivity under a dryer climate (up to 1.8 days) is higher than the possibility of an increase under a wet climate (up to 1 day). The impacts of a water storage, in the headwater catchment, are highly pronounced in terms of inundation and wetland connectivity (e.g., mean connectivity reduced by 1.7 days). The relative change in connectivity is higher for a small flood compared to that of a large event (Abdel-Karim et al., 2020). These results demonstrate that there is a possibility of both an increase and a decrease in connectivity under future climate.

The SWAT is a continuous-time, semidistributed, and processed-based model developed and supported by the USDA Agricultural Research Service (Neitsch et al., 2011; Arnold et al., 2012). Digital Elevation Model, land-use-land-cover data, soil types, and different daily weather data include details of precipitation, maximum and minimum air temperatures, solar radiation, wind speed, and relative humidity. SWAT has received international acceptance as a robust interdisciplinary catchment-scale modeling tool that contribute to climate prediction steps. Major input data for SWAT include DEM, LULC, soil properties, and daily weather data (precipitation, maximum and minimum air temperature, relative humidity, wind speed, and solar radiation). Although hydrologic models play a critical role in the management of natural resources in arid areas, their application is challenged by the scarcity of data for calibration and validation. Mengistu and Belay (2019) attempted to configure, calibrate, and validate the SWAT model in a data-scarce catchment by using the regionalization with physical similarity approach. This approach adopted the dual calibration and validation procedure, i.e., one in the donor catchment by using SWAT-CUP (SWAT Calibration and Uncertainty Programs). Results of the study showed that the P-factor was 0.73 and 0.65 for calibration and validation, respectively, whereas the R-factor was 0.93 for calibration and 0.66 for validation periods (Mengistu and Belay, 2019). Generally, good model uncertainty is expressed by a higher value of the P-factor (toward 100%) and a lower value of the R-factor (toward 0).

5.3 Integration of remote sensing with climate models

Remote sensing is the most cost-effective approach to monitor vegetation cover and its changes over time, as it provides wide spatial coverage and repeated measurements over a short period, which are difficult to achieve otherwise (Paz-Kagan et al., 2019). The availability of multitemporal satellite data at different resolutions has immense importance as it enables the modeling of the distribution of invasive species and wetland ecosystem change detection (Buermann et al., 2008; Paz-Kagan et al., 2019).

Importantly, ongoing developments in remote sensing technologies lead to steadily improved sensors that have the potential to better support predictions of climate variability impacts (Buermann et al., 2008; Huang and Asner, 2009; Leitão and Santos, 2019). With the availability of freely available Copernicus satellites (in particular Sentinel-2), unprecedented detail is offered every 5 days at the equator, for the identification of invasive species (Rajah et al., 2019; Thamaga and Dube, 2018), as well as crop and vegetation mapping (Immitzer et al., 2019; Shoko and Mutanga, 2017). The tremendous additional potential is offered by a steadily growing fleet of commercial satellites, UAV, and hyperspectral sensors (Feilhauer et al., 2013; Piiroinen et al., 2018).

6 Case studies and examples

6.1 Case study 1: Remote sensing of climate impacts on coastal wetlands

As a distinct ecosystem, the coastal wetland links the ocean and land and plays an important role in the local natural environment and social economy (United Nations Environment Programme (UNEP), 2006; McGranahan et al., 2007). It has been reported that up to one-third of the world's population lives in limited coastal areas (~4% of the total land area [Commito et al., 2008] and more than 10% of the population lives in coastal areas with terrain elevations of less than 10 m [Costanza et al., 2008; Chen et al., 2010]). Hence, as an important component of coastal areas, coastal wetlands have become a significant natural resource for humans. Coastal wetland not only provides habitats for different types of wildlife but also affects sedimentation and controls storms and floods (Chen et al., 2010). However, a trend of reduced coastal wetland has been reported globally, decreasing at a rate of 0.949% yr^{-1} during the 20th and early 21st century (Costanza et al., 2008).

Literature shows that time series Landsat images and object-oriented classifications have been widely adopted to monitor the trends and status of coastal wetlands. Additionally, SLAMM (Sea Level Affecting Saltmarsh Model) was utilized to evaluate the influence of SLR on the wetlands from the seaward direction in 2100. The results showed that (1) more than 15,000 ha of the coastal wetlands were lost or degraded from 1989 to 2013 and (2) most of these were transformed into drained lands for agriculture, aquaculture, or industrial or residential use under the rapid expansion of the city. Moreover, 54% of the agricultural land was converted to built-up areas or other uses. (3) The loss of agricultural land and increase in built-up areas will continue in the short-term future (up to 2025), and the sprawl of the built-up areas will extend in every direction in the PNA.

Hidalgo-Corrotea et al. (2023) also explored the main drivers and impacts associated with the vulnerability of each coastal wetland as a result of climate change impacts by performing a Principal Component Analysis with Agglomerative Hierarchical Clustering. This allowed them to group coastal wetlands according to the drivers analyzed with results showing that 42.6 ± 9.2% of the coastal wetlands evaluated have high or very high vulnerability, with higher vulnerability at the 500-m scale (51.4%). Results further identified four groups of coastal wetlands: two located in central Chile, mainly affected by climate change-associated drivers (41.9 ± 2.1%), and one in central Chile which is affected by land-cover change (52.8 ± 6.2%); the latter has a lower vulnerability level (Hidalgo-Corrotea et al., 2023). This therefore shows that remote sensing is an important tool that enables exploration of climate change impacts on wetland, which affects their size and possibly result in degradation and their disappearance.

6.2 Case study 2: Monitoring climate-induced changes in peatland ecosystems

Global climate is changing faster than humankind has ever experienced. Model-based predictions of future climate are becoming more complex and precise, but they still lack crucial information about the reaction of some important ecosystems, such as peatlands. Peatlands belong to one of the largest carbon stores on the Earth. They are mostly distributed in high latitudes, where the temperature rises faster than in the other parts of the planet. Warmer climate and changes in precipitation patterns cause changes in the composition and phenology of peatland vegetation. Despite the fact that peatlands cover only around 3% of the terrestrial surface or 1% of the total Earth surface, they are the ecosystem with one of the highest concentrations of stored carbon. It is estimated that about 30% of all carbon stored in the world's soils is stored in peatlands (Gorham, 1991). Some areas of peatland occurrence are so remote that the only way of monitoring is remote sensing through satellites. Vegetation indices such as NDVI are a useful tool to obtain information about the beginning or the peak of a vegetation season. They may be used not only for satellite remote sensing but also for ground measurements in climate change experiments (May et al., 2020). Despite the wide availability of satellite products for NDVI calculation with fair spatial and temporal resolution (Cao et al., 2018), one must be careful when monitoring peatland vegetation.

Since climate is the most important determinant of the distribution and character of peatlands, a strong influence of any future changes of climate on these ecosystems is expected (Gignac and Vitt, 1994; Heijmans et al., 2008). A warmer and drier climate with higher rates of evapotranspiration will accelerate the decomposition of organic matter and the amount of nutrients in the ecosystem will increase. Consequently, higher levels of CO_2 and other greenhouse gases will be released into the atmosphere. The changes in hydrology followed by climate changes could affect the distribution and ecology of plant and animal species in the peatland ecosystem (Parish et al., 2008). In addition, the tree cover of bogs will increase as the result of lower water tables (Gignac and Vitt, 1994). On the other hand, peatlands affect the climate via a series of feedback effects, which include the sequestration of carbon dioxide, as well as exchanges of heat and moisture balance (Thompson et al., 2004).

6.3 Case study 3: Assessing wetland responses to extreme weather events

Among other Vegetation Indices, a recent study by Khanna et al. (2017) used a modified NDVI to evaluate the recovery of the salt marshes in Barataria Bay, LA, USA, after Hurricane Isaac in 2012 and the Deep Water Horizon oil spill in 2010 (Wang et al., 2010). In 2012, after Hurricane Isaac, there was a significant loss of green vegetation to water, soil, and nonphotosynthetic vegetation (NPV) in both oiled and oil-free sites. Also, vegetation in narrow stands of the Bay was considerably more stressed than vegetation in block stands. Another study developed an approach for identifying and estimating forest damage from Hurricane Katrina (Wang et al., 2010). The statistical analysis and comparison with the damage severity revealed that the Normalized Difference Infrared Index (NDII) was an optimal indicator for detecting hurricane-induced forest damage among the five commonly used VIs, including NDVI, EVI, NDII, LAI, and Fraction of Photosynthetically Active Radiation.

A number of studies used satellite image-derived VIs for drought monitoring other than CWs that paved the path to use the VIs for drought monitoring in CWs as well (Wang et al., 2010). For example, the combination of MODIS-derived NDVI as well as LST provided very useful information for agricultural drought monitoring and early warning system for farmers. Zhang et al. (2017) examined the impacts of the 2009/2010 drought in southwestern China on vegetation by calculating the standardized anomalies of NDVI, EVI, and NDWI LST using MODIS satellite images. The results implied that the NDVI, EVI, and NDWI declined, while LST increased in the 2009/2010 drought-stricken vegetated areas during the drought period. Ji and Peters (2003) established that while NDVI is a useful variable for monitoring vegetation conditions, the nature of the relationship between the NDVI and drought conditions varies based on the seasonal timing and variations in vegetation and soil type.

Jain et al. (2005) mapped the flood-affected areas in Koa catchment, Bihar in India using RS satellite data (IRS LISS III, 1999 and Landsat TM, 1995). They derived water indices using image processing techniques and indicated that a NDWI-based approach was able to identify flood inundation. NDWI derived from Landsat TM and ETM+ images were also used to map the inundation in New South Wales, NSW, Australia (Thomas et al., 2015). Another study generated a flood inundation map using a geomorphologic approach employing Shuttle Radar Topographic Mission (SRTM) DEM and satellite image data (ASTER and Landsat). They integrated both landform classification and spectral analysis for flooded area prediction by applying MNDWI and elevation range to assess flood inundation condition of an alluvial plain in central Vietnam (Ho et al., 2010). The study suggested that the extraction of moist soil by MNDWI can help to detect flooded sites; results were also compared with the landform classification map, SRTM DEM elevation ranges, and land-cover classification (Ho et al., 2010).

7 Challenges and future directions

7.1 Challenges and data quality, uncertainties issues

Despite its numerous advantages, remote sensing also has its limitations in wetland monitoring. Cloud cover can obstruct satellite imagery, leading to gaps in data collection (Jain et al., 2005). Additionally, the spatial resolution of some sensors may not be sufficient to capture small-scale changes in wetlands. Therefore, a combination of remote sensing data with ground-based observations and field surveys remains essential for comprehensive wetland monitoring and conservation (Ho et al., 2010; Thomas et al., 2015). Literature shows that quite significant progress has been made in the application of remote sensing techniques in wetland ecosystems research with plausible results being attained. Remote sensing methods play an important role in the detection, classification, and mapping of areas impacted by different forms of anthropogenic and natural activities including climate change. Therefore, the use of remote sensing to detect and map wetland ecosystems across Sub-Saharan Africa has gained attention in the last decade.

While several studies have successfully utilized remotely sensed data in wetland research, there are still challenges that still need to be addressed. Spatial studies of these ecosystems require versatile and robust computational methods to help deal with nonlinear relationships, high-order interactions, and missing data. Despite these difficulties, the methods used for mapping the distribution of wetlands should be clear to understand and easy to interpret. Wetland ecosystems are important to society and there is a need to establish digital efforts for wetland conservation. Furthermore, wetlands resources surveys, legislation, management, and research need to be revised since there is still much work to be done to protect future wetlands.

Guo et al. (2017) reiterates that there are also challenges related to data processing and analysis, sensor limitations, and integration with other data sources. Satellite observations have provided a wealth of data on global change research. Phinzi et al. (2021) observed that there are still some drawbacks, such as the lack of guarantees for the continuity of the data, and the data uncertainty as discussed in the preceding sections. To add on the fact that the network of multi-source remote sensing data has not been optimized also presents problems, that is the data assimilation process lacks data validation, and that EO information has not yet met the needs of model parameters (Guo et al., 2017).

7.2 Advancements and future prospects in remote sensing technology

Remote sensing is a powerful tool for monitoring and mapping wetland dynamics for biodiversity conservation. Its ability to provide timely and accurate information on wetland extent vegetation hydrology and human disturbances enables policymakers to enhance the quality of environmental monitoring. As remote sensing technology continues to advance, it will play an increasingly important role in climate change research (Guo et al., 2017; Phinzi et al., 2021). Future developments may include the deployment of new satellite missions, the integration of artificial intelligence and machine learning techniques for data analysis, and the increased use of drone-based remote sensing platforms for localized studies. Through a variety of methods such as multisatellite, multisensor, and long-term time series remote sensing data, it has been possible to derive climate sensitivity factors and aid in the study of the spatial variability of terrestrial ecosystems (e.g., glaciers, lakes, and vegetation) and in the understanding how they respond to global climate change (Klump et al., 2006; Guo et al., 2017; Abera et al., 2019). EO technology has provided highly useful information in global climate change research over the past few decades and greatly promoted its development, especially by providing biological, physical, and chemical parameters on a global scale (Guo et al., 2017). EO data have the

4V features (volume, variety, veracity, and velocity) of big data that are suitable for climate change research (Mauthner and Parry, 2013). It is clear that this large amount of data available from scientific satellites plays an important role as it aids in data availability, hence its fullness.

8 Conclusions

The chapter explored the role of remote sensing in assessing the impact of climate variability on wetland ecosystems. Wetlands are fragile ecosystems that provide essential goods and services although they are subjected to unsustainable exploitation. The study findings demonstrate that satellite data have brought a paradigm shift in understanding the impact of climate change on wetland ecological conditions. A wide range of both active and passive sensors have been adopted, which have proved to be an indispensable tool that aids the quality of assessment. The increase in the use of remote sensing data in mapping wetland ecosystems is linked to its ability to offer a variety of new applications that can quickly and synoptically monitor and manage large areas. However, its application has been masked by some uncertainties and limitations. For instance, there are issues to do with the resolution of the sensor applied, which results in a specific sensor failing to detect fine-scale variations, which therefore calls for the need for a multipronged approach that combines multiple datasets to complement each other's weaknesses. This has been evident in time series analysis classification maps, and the gaps for instance which were caused by the gap of Sentinel-1 images were filled using various methods such as gap filling. In future works, many satellite images with similar or higher resolutions, such as Landsat, Worldview-3, and GaoFen-2, could be used as potential data sources for accurate wetland mapping, which could extend this study for long-term wetland mapping. In addition, the sample migration algorithm is another future research point that could migrate the single-date samples to multiperiod samples. Migrating samples allows large-scale wetland mapping to be created over longer time series, which is important for wetland protection and sustainable management.

References

Abdel-Karim, R., Reda, Y., Abdel-Fattah, A., 2020. Nanostructured materials-based nanosensors. J. Electrochem. Soc. 167 (3), 037554.

Abera, W., Tamene, L., Abegaz, A., Solomon, D., 2019. Understanding climate and land surface changes impact on water resources using Budyko framework and remote sensing data in Ethiopia. J. Arid Environ. 167, 56–64.

Alawadi, W.A., Raheem, Z.A.H.A., Yaseen, D.A., 2023. Use of remote sensing techniques to assess water storage variations and flood-related inflows for the Hawizeh wetland. Environ. Monit. Assess. 195, 1246. https://doi.org/10.1007/s10661-023-11838-x.

Ankrah, J., 2018. Climate change impacts and coastal livelihoods; an analysis of fishers of coastal Winneba, Ghana. Ocean Coast. Manag. 161, 141–146.

Arnold, J.G., Moriasi, D.N., Gassman, P.W., Abbaspour, K.C., White, M.J., Srinivasan, R., Santhi, C., Harmel, R.D., Van Griensven, A., Van Liew, M.W., Kannan, N., 2012. SWAT: model use, calibration, and validation. Trans. ASABE 55 (4), 1491–1508.

Avdan, U., Jovanovska, G., 2016. Algorithm for automated mapping of land surface temperature using LANDSAT 8 satellite data. J. Sens. 2016.

Buermann, W., Saatchi, S., Smith, T.B., Zutta, B.R., Chaves, J.A., Milá, B., Graham, C.H., 2008. Predicting species distributions across the Amazonian and Andean regions using remote sensing data. J. Biogeogr. 35 (7), 1160–1176.

Cao, R., Chen, Y., Shen, M., Chen, J., Zhou, J., Wang, C., Yang, W., 2018. A simple method to improve the quality of NDVI time-series data by integrating spatiotemporal information with the Savitzky-Golay filter. Remote Sens. Environ. 217, 244–257.

Carlson, C.J., Albery, G.F., Merow, C., Trisos, C.H., Zipfel, C.M., Eskew, E.A., Olival, K.J., Ross, N., Bansal, S., 2022. Climate change increases cross-species viral transmission risk. Nature 607 (7919), 555–562.

Chen, J., 2023. Monitoring and mapping wetland dynamics for biodiversity conservation using remote sensing. J. Remote Sens. GIS 12, 309.

Chen, S., Geyer, W.R., Sherwood, C.R., Ralston, D.K., 2010. Sediment transport and deposition on a river-dominated tidal flat: an idealized model study. J. Geophys. Res. Earth Surf. 115, C10040.

Commito, J.A., Como, S., Grupe, B., Dow, W.E., 2008. Species diversity in the soft-bottom intertidal zone: biogenic structure, sediment, and macrofauna across mussel bed spatial scales. J. Exp. Mar. Biol. Ecol. 366, 70–81.

Costanza, R., Pérez-Maqueo, O., Martinez, M.L., Sutton, P., Anderson, S.J., Mulder, K., 2008. The value of coastal wetlands for hurricane protection. Ambio 37, 241–248.

Czajkowski, K., Torbick, N., Lawrence, P., 2007. Application and assessment of a GIScience model for jurisdictional wetlands identification in Northwestern Ohio. In: Wetland and Water Resource Modeling and Assessment: A Watershed Perspective. CRC Press, pp. 2–12.

Deng, F., Wang, X., Cai, X., Li, E., Jiang, L., Li, H., Yan, R., 2014. Analysis of the relationship between inundation frequency and wetland vegetation in Dongting Lake using remote sensing data. Ecohydrology 7 (2), 717–726.

Ellery, W.N., Grenfell, S.E., Grenfell, M.C., Powell, R., Kotze, D.C., Marren, P.M., Knight, J., 2016. Wetlands in southern Africa. Quaternary Environmental Change in Southern Africa. Cambridge University Press, Cambridge, pp. 188–202.

Engelbrecht, F., Adegoke, J., Bopape, M.J., Naidoo, M., Garland, R., Thatcher, M., McGregor, J., Katzfey, J., Werner, M., Ichoku, C., Gatebe, C., 2015. Projections of rapidly rising surface temperatures over Africa under low mitigation. Environ. Res. Lett. 10 (8), 085004.

Erwin, K.L., 2009. Wetlands and global climate change: the role of wetland restoration in a changing world. Wetl. Ecol. Manag. 17 (1), 71–84.

Fakarayi, T., Mashapa, C., Gandiwa, E., Kativu, S., 2015. Pattern of land-use and land cover changes in Driefontein Grassland Important Bird Area, Zimbabwe. Trop. Conserv. Sci. 8 (1), 274–283.

Feilhauer, H., Thonfeld, F., Faude, U., He, K.S., Rocchini, D., Schmidtlein, S., 2013. Assessing floristic composition with multispectral sensors—a comparison based on monotemporal and multiseasonal field spectra. Int. J. Appl. Earth Obs. Geoinf. 21, 218–229.

Frazier, P., Page, K., Louis, J., Briggs, S., Robertson, A.I., 2003. Relating wetland inundation to river flow using Landsat TM data. Int. J. Remote Sens. 24 (19), 3755–3770.

Gibson, L.A., 2013. The Application of the Surface Energy Balance System Model to Estimate Evapotranspiration in South Africa. PhD Thesis, University of Cape Town.

Gignac, L.D., Vitt, D.H., 1994. Responses of northern peatlands to climate change: effects on bryophytes. J. Hattori Bot. Lab. 75, 119–132.

Gorham, E., 1991. Northern peatlands: role in the carbon cycle and probable responses to climatic warming. Ecol. Appl. 1 (2), 182–195.

Grundling, A.T., Van den Berg, E.C., Price, J.S., 2013. Assessing the distribution of wetlands over wet and dry periods and land-use change on the Maputaland Coastal Plain, north-eastern KwaZulu-Natal, South Africa. S. Afr. J. Geomat. 2 (2), 120–138.

Guo, M., Li, J., Sheng, C., Xu, J., Wu, L., 2017. A review of wetland remote sensing. Sensors 17 (4), 777.

Gxokwe, S., Dube, T., Mazvimavi, D., 2020. Multispectral remote sensing of wetlands in semi-arid and arid areas: a review on applications, challenges and possible future research directions. Remote Sens. 12 (24), 4190.

Haidary, A., Amiri, B.J., Adamowski, J., Fohrer, N., Nakane, K., 2013. Assessing the impacts of four land use types on the water quality of wetlands in Japan. Water Resour. Manag. 27, 2217–2229.

Heijmans, M.M., Mauquoy, D., Van Geel, B., Berendse, F., 2008. Long-term effects of climate change on vegetation and carbon dynamics in peat bogs. J. Veg. Sci. 19 (3), 307–320.

Hidalgo-Corrotea, C., Alaniz, A.J., Vergara, P.M., Moreira-Arce, D., Carvajal, M.A., Pacheco-Cancino, P., Espinosa, A., 2023. High vulnerability of coastal wetlands in Chile at multiple scales derived from climate change, urbanization, and exotic forest plantations. Sci. Total Environ. 903, 166130.

Ho, L.T.K., Umitsu, M., Yamaguchi, Y., 2010. Flood hazard mapping by satellite images and SRTM DEM in the Vu Gia-Thu Bon alluvial plain, Central Vietnam. Int. Arch. Photogramm. Remote Sens. Spat. Inf. Sci. 38, 275–280.

Houborg, R., Fisher, J.B., Skidmore, A.K., 2015. Advances in remote sensing of vegetation function and traits. Int. J. Appl. Earth Obs. Geoinf. 43, 1–6.

House, A.R., Thompson, J.R., Acreman, M.C., 2015. Projecting impacts of climate change on hydrological conditions and biotic responses in a chalk valley riparian wetland. J. Hydrol. 534, 178–192.

Hu, Y.X., Huang, J.L., Du, Y., Han, P.P., Wang, J.L., Huang, W., 2015. Monitoring wetland vegetation pattern response to water-level change resulting from the Three Gorges Project in the two largest freshwater lakes of China. Ecol. Eng. 74, 274–285.

Huang, C.Y., Asner, G.P., 2009. Applications of remote sensing to alien invasive plant studies. Sensors 9 (6), 4869–4889.

Immitzer, M., Neuwirth, M., Böck, S., Brenner, H., Vuolo, F., Atzberger, C., 2019. Optimal input features for tree species classification in Central Europe based on multi-temporal Sentinel-2 data. Remote Sens. 11 (22), 2599.

Jain, S.K., Singh, R.D., Jain, M.K., Lohani, A.K., 2005. Delineation of flood-prone areas using remote sensing techniques. Water Resour. Manag. 19, 333–347.

Ji, L., Peters, A.J., 2003. Assessing vegetation response to drought in the northern Great Plains using vegetation and drought indices. Remote Sens. Environ. 87, 85–98.

Jin, X., Guo, R., Xia, W., 2013. Distribution of actual evapotranspiration over Qaidam Basin, an arid area in China. Remote Sens. 5 (12), 6976–6996.

Junk, W.J., An, S., Finlayson, C.M., Gopal, B., Květ, J., Mitchell, S.A., Mitsch, W.J., Robarts, R.D., 2013. Current state of knowledge regarding the world's wetlands and their future under global climate change: a synthesis. Aquat. Sci. 75, 151–167.

Kaplan, G., Avdan, Z.Y., Avdan, U., 2019. Mapping and monitoring wetland dynamics using thermal, optical, and SAR remote sensing data. In: Wetlands Management: Assessing Risk and Sustainable Solutions, p. 87.

Khanna, S., Santos, M.J., Koltunov, A., Shapiro, K.D., Lay, M., Ustin, S.L., 2017. Marsh loss due to cumulative impacts of Hurricane Isaac and the Deepwater Horizon oil spill in Louisiana. Remote Sens. 9 (2), 169.

Klein, I., Gessner, U., Dietz, A.J., Kuenzer, C., 2015. Global WaterPack – A 250 m resolution dataset revealing the daily dynamics of global inland water bodies. Remote Sens. Environ. 198, 345–362.

Klump, J., Bertelmann, R., Brase, J., Diepenbroek, M., Grobe, H., Höck, H., Lautenschlager, M., Schindler, U., Sens, I., Wächter, J., 2006. Data publication in the open access initiative. Data Sci. J. 5, 79–83.

Kovács, G.M., Horion, S., Fensholt, R., 2022. Characterizing ecosystem change in wetlands using dense earth observation time series. Remote Sens. Environ. 281, 113267.

Kulawardhana, W., Thenkabail, P.S., Masiyandima, M., Biradar, C.M., Vithanage, J., Finlayson, C. M., Gunasinghe, S., Alankara, R., 2006. Evaluation of Different Methods for Delineation of Wetlands in Limpopo River Basin Using Landsat ETM+ and SRTM Data. European Space Agency.

Latifovic, R., Pouliot, D., 2013. Approaches to IPCC land-use and land-use change reporting in agriculture areas with remote sensing. J. Chem. Inf. Model 53 (9), 1689–1699.

Leitão, P.J., Santos, M.J., 2019. Improving models of species ecological niches: a remote sensing overview. Front. Ecol. Evol. 7, 9.

Mandishona, E., Knight, J., 2019. Users' perceptions and understanding of two urban wetlands in Harare, Zimbabwe. S. Afr. Geogr. J. 101 (3), 326–348.

Mao, D., Wang, Z., Du, B., Li, L., Tian, Y., Jia, M., Zeng, Y., Song, K., Jiang, M., Wang, Y., 2020. National wetland mapping in China: a new product resulting from object-based and hierarchical classification of Landsat 8 OLI images. ISPRS J. Photogramm. Remote Sens. 164, 11–25.

Marambanyika, T., Beckedahl, H., 2016. Wetland utilisation patterns in semi-arid communal areas of Zimbabwe between 1985 and 2013 and the associated benefits to livelihoods of the surrounding communities. Trans. R. Soc. S. Afr. 71 (2), 175–186.

Marambanyika, T., Mupfiga, U.N., Musasa, T., Ngwenya, K., 2021. Local perceptions on the impact of drought on wetland ecosystem services and associated household livelihood benefits: the case of the Driefontein Ramsar site in Zimbabwe. Land 10, 587. https://doi.org/10.3390/land10060587.

Mauthner, N.S., Parry, O., 2013. Open access digital data sharing: principles, policies and practices. Soc. Epistemol. 27 (1), 47–67.

May, J.L., Hollister, R.D., Betway, K.R., Harris, J.A., Tweedie, C.E., Welker, J.M., Gould, W.A., Oberbauer, S.F., 2020. NDVI changes show warming increases the length of the green season at tundra communities in Northern Alaska: a fine-scale analysis. Front. Plant Sci. 11, 1174.

McGranahan, G., Balk, D., Anderson, B., 2007. The rising tide: assessing the risks of climate change and human settlements in low elevation coastal zones. Environ. Urban. 19, 17–37.

Mengistu, D.A., Belay, T., 2019. Land use and land cover dynamics and drivers in the Muga watershed, Upper Blue Nile basin, Ethiopia. Remote Sens. Appl. 15, 100249.

Millennium Ecosystem Assessment, 2005. Ecosystems and Human Well-Being. Island Press, Washington, DC.

Mira, M., Ninyerola, M., Batalla, M., Pesquer, L., Pons, X., 2017. Improving mean minimum and maximum month-to-month air temperature surfaces using satellite-derived land surface temperature. Remote Sens. 9 (12), 1313.

Mitchell, S.A., 2013. The status of wetlands, threats and the predicted effect of global climate change: the situation in sub-Saharan Africa. Aquat. Sci. 75 (1), 95–112.

Mitsch, W.J., Gosselink, J.G., 2000. The value of wetlands: importance of scale and landscape setting. Ecol. Econ. 35 (1), 25–33.

Mitsch, W.J., Gosselink, J.G., 2015. Wetlands. John Wiley & Sons.

Muro, J., Heinmann, S., Strauch, A., Menz, G. (Eds.), 2016. Land surface temperature retrieval in wetlands using normalized difference vegetation index-emissivity estimation and ASTER emissivity product. In: Living Planet Symposium.

Musasa, T., Marambanyika, T., 2020. Threats to sustainable utilization of wetland resources in Zimbabwe: a review. Wetl. Ecol. Manag. 28, 681–696.

Mutisi, L., Nhamo, G., 2015. Blue in the green economy: land use change and wetland shrinkage in Belvedere North and Epworth localities, Zimbabwe. J. Public Adm. 50 (1), 108–124.

Neitsch, S.L., Arnold, J.G., Kiniry, J.R., Williams, J.R., 2011. Soil and Water Assessment Tool Theoretical Documentation Version 2009. Texas Water Resources Institute.

Nhamo, L., Magidi, J., Dickens, C., 2017. Determining wetland spatial extent and seasonal variations of the inundated area using multispectral remote sensing. Water SA 43 (4), 543–552.

Niemi, G., Wardrop, D., Brooks, R., Anderson, S., Brady, V., Paerl, H., Rakocinski, C., Brouwer, M., Levinson, B., McDonald, M., 2004. Rationale for a new generation of indicators for coastal waters. Environ. Health Perspect. 112 (9), 979–986.

Orimoloye, I.R., Mazinyo, S.P., Nel, W., Kalumba, A.M., 2018. Spatiotemporal monitoring of land surface temperature and estimated radiation using remote sensing: human health implications for East London, South Africa. Environ. Earth Sci. 77, 1–10.

Pachauri, R.K., Allen, M.R., Barros, V.R., Broome, J., Cramer, W., Christ, R., Church, J.A., Clarke, L., Dahe, Q., Dasgupta, P., Dubash, N.K., 2014. Climate change 2014: synthesis report. In: Contribution of Working Groups I, II and III to the Fifth Assessment Report of the Intergovernmental Panel on Climate Change. IPCC, p. 151.

Pantshwa, A.O., Buschke, F.T., 2019. Ecosystem services and ecological degradation of communal wetlands in a South African biodiversity hotspot. R. Soc. Open Sci. 6 (6), 181770.

Parish, F., Sirin, A.A., Charman, D., Joosten, H., Minaeva, T.Y., Silvius, M., 2008. Assessment on Peatlands, Biodiversity and Climate Change. Online available at https://elibrary.ru/item.asp?id=22008727.

Paz-Kagan, T., Silver, M., Panov, N., Karnieli, A., 2019. Multispectral approach for identifying invasive plant species based on flowering phenology characteristics. Remote Sens. 11 (8), 953.

Peng, H., Xia, H., Shi, Q., Chen, H., Chu, N., Liang, J., Gao, Z., 2023. Monitoring spatial and temporal dynamics of wetland vegetation and their response to hydrological conditions in a large seasonal lake with time series Landsat data. Ecol. Indic. 142, 109283.

Perennou, C., Guelmami, A., Paganini, M., Philipson, P., Poulin, B., Strauch, A., Tottrup, C., Truckenbrodt, J., Geijzendorffer, I.R., 2018. Mapping Mediterranean wetlands with remote sensing: a good-looking map is not always a good map. Adv. Ecol. Res. 58, 243–277. Academic Press.

Phinzi, K., Ngetar, N.S., Ebhuoma, O., 2021. Soil erosion risk assessment in the Umzintlava catchment (T32E), Eastern Cape, South Africa, using RUSLE and random forest algorithm. S. Afr. Geogr. J. 103 (2), 139–162.

Piiroinen, R., Fassnacht, F.E., Heiskanen, J., Maeda, E., Mack, B., Pellikka, P., 2018. Invasive tree species detection in the Eastern Arc Mountains biodiversity hotspot using one class classification. Remote Sens. Environ. 218, 119–131.

Rajah, P., Odindi, J., Mutanga, O., Kiala, Z., 2019. The utility of Sentinel-2 vegetation indices (VIs) and Sentinel-1 synthetic aperture radar (SAR) for invasive alien species detection and mapping. Nat. Conserv. 35, 41–61.

Ramsar Convention Secretariat, 2018. Global Wetland Outlook: State of the World's Wetlands and Their Services to People. Ramsar Convention Secretariat, Gland.

Rebelo, L.M., McCartney, M.P., Finlayson, C.M., 2010. Wetlands of sub-Saharan Africa: distribution and contribution of agriculture to livelihoods. Wetl. Ecol. Manag. 18, 557–572.

Salimi, S., Almuktar, S.A., Scholz, M., 2021. Impact of climate change on wetland ecosystems: a critical review of experimental wetlands. J. Environ. Manag. 286, 112160.

Schmidt, K., Skidmore, A., 2003. Spectral discrimination of vegetation types in a coastal wetland. Remote Sens. Environ. 85 (1), 92–108.

Schroeder, T., Loveland, T., Wulder, M.A., Irons, J.R., 2016. Landsat Science Team: 2016 winter meeting summary. The Earth Observer, 19–23.

Sharaf El Din, E., Zhang, Y., 2017. Estimation of both optical and nonoptical surface water quality parameters using Landsat 8 OLI imagery and statistical techniques. J. Appl. Remote. Sens. 11 (4), 046008.

Shoko, C., Mutanga, O., 2017. Examining the strength of the newly-launched Sentinel 2 MSI sensor in detecting and discriminating subtle differences between C3 and C4 grass species. ISPRS J. Photogramm. Remote Sens. 129, 32–40.

Slagter, B., Tsendbazar, N.E., Vollrath, A., Reiche, J., 2020. Mapping wetland characteristics using temporally dense Sentinel-1 and Sentinel-2 data: a case study in the St. Lucia wetlands, South Africa. Int. J. Appl. Earth Obs. Geoinf. 86, 102009.

Steinbach, S., Hentschel, E., Hentze, K., Rienow, A., Umulisa, V., Zwart, S.J., Nelson, A., 2023. Automatization and evaluation of a remote sensing-based indicator for wetland health assessment in East Africa on national and local scales. Ecol. Inform. 75, 102032.

Sun, Y., Wu, G., Mao, M., Duan, X., Hu, J., Zhang, Y., Xie, Y., Qiu, X., Gong, W., Liu, T., Liu, T., 2023. Remote sensing and environmental assessment of wetland ecological degradation in the Small Sanjiang Plain, Northeast China. Front. Ecol. Evol. 11, 1125775. https://doi.org/10.3389/fevo.2023.1125775.

Thamaga, K.H., Dube, T., 2018. Testing two methods for mapping water hyacinth (Eichhornia crassipes) in the Greater Letaba river system, South Africa: discrimination and mapping potential of the polar-orbiting Sentinel-2 MSI and Landsat 8 OLI sensors. Int. J. Remote Sens. 39 (22), 8041–8059.

Thomas, R.F., Kingsford, R.T., Lu, Y., Cox, S.J., Sims, N.C., Hunter, S.J., 2015. Mapping inundation in the heterogeneous floodplain wetlands of the Macquarie Marshes, using Landsat Thematic Mapper. J. Hydrol. 524, 194–213.

Thompson, R.C., Olsen, Y., Mitchell, R.P., Davis, A., Rowland, S.J., John, A.W., McGonigle, D., Russell, A.E., 2004. Lost at sea: where is all the plastic? Science 304 (5672), 838.

Tooth, S., McCarthy, T.S., 2007. Wetlands in drylands: geomorphological and sedimentological characteristics, with emphasis on examples from southern Africa. Prog. Phys. Geogr. 31 (1), 3–41.

United Nations, 2008. Principles Relating to Remote Sensing of the Earth From Space. http://www.un.org/documents/ga/res/41/a41r065.htm. (Accessed 25 August 2023).

United Nations Environment Programme (UNEP), 2006. Marine and Coastal Ecosystems and Human Wellbeing: A Synthesis Report Based on the Findings of the Millennium Ecosystem Assessment. UNEP, Nairobi, Kenya. 76 pp.

Vanderhoof, M.K., Alexander, L.C., Todd, M.J., 2016. Temporal and spatial patterns of wetland extent influence variability of surface water connectivity in the Prairie Pothole Region, United States. Landsc. Ecol. 31, 805–824.

Waddington, J.M., Morris, P.J., Kettridge, N., Granath, G., Thompson, D.K., Moore, P.A., 2015. Hydrological feedbacks in northern peatlands. Ecohydrology 8 (1), 113–127.

Wang, W., Qu, J.J., Hao, X., Liu, Y., Stanturf, J.A., 2010. Post-hurricane forest damage assessment using satellite remote sensing. Agric. For. Meteorol. 150 (1), 122–132.

Wang, F., Qin, Z.H., Song, C.Y., Tu, L.L., Karnieli, A., Zhao, S.H., 2015. An improved mono-window algorithm for land surface temperature retrieval from Landsat 8 thermal infrared sensor data. Remote Sens. 7 (4), 4268–4289.

Wu, Q., Lane, C.R., 2016. Delineation and quantification of wetland depressions in the Prairie Pothole Region of North Dakota. Wetlands 36 (2), 215–227.

Xi, Y., Peng, S., Ciais, P., Chen, Y., 2021. Future impacts of climate change on inland Ramsar wetlands. Nat. Clim. Change 11 (1), 45–51.

Xie, J., He, N., Fang, L., Plaza, A., 2019. Scale-free convolutional neural network for remote sensing scene classification. IEEE Trans. Geosci. Remote Sens. 57 (9), 6916–6928.

Yang, Z., Bai, J., Zhang, W., 2021. Mapping and assessment of wetland conditions by using remote sensing images and POI data. Ecol. Indic. 127, 107485.

Zedler, J.B., Kercher, S., 2005. Wetland resources: status, trends, ecosystem services, and restorability. Annu. Rev. Environ. Resour. 30, 39–74.

Zhang, X., Yamaguchi, Y., Li, F., He, B., Chen, Y., 2017. Assessing the impacts of the 2009/2010 drought on vegetation indices, normalized difference water index, and land surface temperature in Southwestern China. Adv. Meteorol. 2017, 6837493.

Zhang, Y., Li, W., Sun, G., Miao, G., Noormets, A., Emanuel, R., King, J.S., 2018. Understanding coastal wetland hydrology with a new regional-scale, process-based hydrological model. Hydrol. Process. 32, 3158–3173. https://doi.org/10.1002/hyp.13247.

Zhu, Z., Woodcock, C.E., 2014. Continuous change detection and classification of land cover using all available Landsat data. Remote Sens. Environ. 144, 152–171.

Zhu, L., Ke, Y., Hong, J., Zhang, Y., Pan, Y., 2022. Assessing degradation of lake wetlands in Bashang Plateau, China based on long-term time series Landsat images using wetland degradation index. Ecol. Indic. 139, 108903.

Chapter 15

Climate variability impacts on aquatic macrophytes: The role of remote sensing technologies

Tatenda Dalu[a,b], Farai Dondofema[c], Kudzai S. Mpakairi[d,e], Chipo P. Mungenge[f], and Timothy Dube[d]

[a]*Aquatic Systems Research Group, School of Biology and Environmental Sciences, University of Mpumalanga, Nelspruit, South Africa,* [b]*South African Institute for Aquatic Biodiversity, Makhanda, South Africa,* [c]*Department of Geography and Environmental Sciences, University of Venda, Thohoyandou, South Africa,* [d]*Department of Earth Science, Institute for Water Studies, University of the Western Cape, Cape Town, South Africa,* [e]*School of Wildlife Conservation, African Leadership University, Kigali, Rwanda,* [f]*Department of Zoology and Entomology, Rhodes University, Makhanda, South Africa*

Chapter outline

Remote Sensing of Climate. https://doi.org/10.1016/B978-0-443-21731-9.00012-0
Copyright © 2024 Elsevier Inc. All rights are reserved, including those for text and data mining, AI training, and similar technologies.

1 Introduction

Anthropogenic activities impact the earth's climate dynamics, causing rapid global climate change and increasing climate variability (IPCC, 2007, 2012). Climate change is inevitably resulting in changes in climate variability and the frequency, intensity, spatial extent, duration, and timing of extreme weather and climate events (IPCC, 2012). The ensuing changes in the frequency and severity of these extreme climate events coupled with the variability of weather patterns have substantial implications for both human societies and natural ecosystems (Thornton et al., 2014). This climate variability has substantial impacts on biological systems and reverberates through the communities and countries reliant on them, with far-reaching effects extending to aquatic ecosystems.

Across diverse aquatic ecosystems, encompassing rivers, lakes, reservoirs, and wetlands, macrophytes flourish and exhibit remarkable adaptability. Macrophytes can be classified into distinct functional groups or morphotypes, as categorized by Sculthorpe (1967), namely submerged with floating leaves, rooted and submerged, free-floating, emergent, and free-floating and submerged (Madzivanzira et al., 2023a,b). Macrophytes have a vital role in the function and structure of many aquatic ecosystems. Their manifold role includes altering the water movement regimes (flow and wave impact conditions), serving as a food source, and altering water and sediment quality within these aquatic ecosystems (Chambers et al., 2008). Macrophytes offer a complex structural environment and serve as providers of shelter, refuge, and diverse niches for a wide range of organisms. Their intricate habitat supports various interactions between predators and prey (Dibble et al., 2006). Macrophytes are also involved in ecosystem processes such as biomineralization, transpiration, sedimentation, elemental cycling, material transformations, and the release of biogenic trace gases into the atmosphere (Cronin et al., 2006). The depth, density, diversity, and types of macrophytes present in a system are also indicators of waterbody health. The absence of macrophytes may indicate water quality problems such as excessive turbidity, herbicides, or salinization, which interfere with plant growth and development.

Both biotic and abiotic factors play significant roles in shaping the structural attributes of all macrophytes, including species composition, distribution, abundance, and diversity. These factors collectively drive the intricate characteristics of macrophyte communities. At the forefront of these factors are light, water temperature, water quality changes and nutrient enrichment, sediment composition, and fluctuations in water levels. The distribution of macrophytes in aquatic ecosystems is intrinsically linked to climate variability. Globally, as many countries continue to experience shifts in climate patterns, understanding the implications of these changes on macrophyte communities becomes imperative for informed conservation and management efforts. Preserving the delicate balance between macrophytes and their environment emerges as pivotal in sustaining the health and resilience of aquatic ecosystems in the face of an ever-changing climate.

The multifaceted functions of macrophytes in freshwater habitats outlined above are intricately tied to their distribution, which, in turn, is influenced by many factors. Despite the pivotal contributions of certain macrophytes to these aquatic systems, the southern African region is grappling with several invasive alien floating aquatic plant species, particularly in lentic systems notably water lettuce or Nile cabbage *Pistia stratiotes*, Kariba weed *Salvinia molesta*, water pennywort *Hydrocotyle* spp., and the water hyacinth *Pontederia crassipes*. These invasive macrophytes have been extensively studied and documented by various researchers (e.g., Chikwenhere, 1994; Chikwenhere, 2001; Brendonck et al., 2003; Dalu and Chauke, 2020; Madzivanzira et al., 2023a,b).

Understanding the spatiotemporal structure of macrophyte communities and functional groupings, as well as competition with other primary producers, necessitates the understanding of seasonal dynamics in real time (Villa et al., 2015; Bolpagni, 2020), in addition to the possible effect of invading species (Wolkovich and Cleland, 2011). Furthermore, to address serious challenges related to aquatic vegetation and how they are linked to global environmental change, natural resource managers and policymakers must understand phenological processes at increasingly large spatiotemporal dimensions (White and Nemani, 2003).

For efficient macrophyte monitoring, remote sensing (RS) provides near-optimal capabilities in terms of spatiotemporal resolution, synoptic view and coverage, sensitivity to structural and physiological vegetation characteristics, frequent sampling, and operational repeatability (Malthus, 2017). To monitor and quantify intra- and interannual changes in aquatic vegetation change, long-term, reliable satellite data are required (Villa et al., 2015). The 30-m spatial resolution of the Landsat constellation sensors was used in certain research for the phenological study of macrophytes (Hestir et al., 2015); however, the 16-day revisiting time may not provide complete capture of macrophytes growth temporal variability. As a result, various monitoring projects have concentrated on collecting spatial data on species composition, maximum depth colonization, density, percentage of horizontal area covered by macrophytes, biomass, and plant height (Johnson and Newman, 2011). Nowadays, airborne or satellite sensors enable several observations at enormous sizes at the same time, with relatively high spatiotemporal resolutions that are continually increasing. In fact, multispectral satellite data have been routinely employed to map macrophyte distribution across huge regions (Nelson et al., 2006; Dogan et al., 2009) and evaluate aquatic vegetation's spatiotemporal dynamics (Zhao et al., 2012).

Significant advances have recently been achieved in image processing enabling repeated and automated submerged macrophyte mapping across broad regions using punctual field data and diverse modeling methodologies. Using strong machine learning techniques, several researchers have convincingly proved the viability of submerged macrophyte mapping (Artificial Neural Networks, Random Forest, Support Vector Machines, or K-Nearest Neighbors) (Dogan et al., 2009; Kotta et al., 2013).

RS empowers the mapping and tracking of macrophyte distribution, biomass, phenology, and growth patterns through spectral data analysis (Dandois et al., 2020; Zhu et al., 2023). Spectral data, acquired from RS platforms like satellites, enable researchers to explore macrophyte distribution across extensive spatial scales. By deciphering the distinctive spectral signatures of diverse macrophyte species, researchers can differentiate and map their presence within aquatic systems. This insight illuminates spatial macrophyte patterns, identifying thriving and declining areas (Sardans et al., 2021). Furthermore, the integration of RS data with ancillary information, such as plant height and leaf area index, facilitates accurate biomass estimation. Quantifying biomass is pivotal for gauging macrophyte community productivity and ecological significance (Xiao et al., 2019). These supplementary parameters refine biomass estimations, deepening our comprehension of macrophyte dynamics. This chapter seeks to provide recommendations for consistent and spatialized data regarding how RS may be used to investigate important macrophyte phenological elements, such as the start and end of the growth season, to better understand the seasonal dynamics of aquatic vegetation (Ghirardi et al., 2019).

2 Factors affecting macrophyte distribution

2.1 Light

Photosynthesis and plant growth heavily rely on light, making it a fundamental aspect. In the case of submersed macrophytes, they require at least 1%–5% of solar radiation to be available at the surface for their survival (Piedade et al., 2022). Macrophytes have developed strategies to overcome light limitations in their environment. They achieve this by positioning their photosynthetic organs either near or above the water surface. These adaptations are observed in free-floating species, plants with floating leaves, and emergent species (Arthaud et al., 2021). The depth distribution of macrophytes in aquatic ecosystems is often limited by light. In limited light, macrophytes can exhibit shade tolerance and show light-related morphological variations, which in turn affects macrophyte community composition (Barko and Smart, 2006). The growth and biomass production of macrophytes heavily relies on the amount of light they receive underwater. High levels of illumination as well as low levels, to a certain extent, were observed to affect the overall development and productivity of three specific submerged macrophyte species, namely, *Hydrilla verticillata*, *Vallisneria natans*, and *Ceratophyllum demersum* (Jin et al., 2020). The decline in biomass production in these plants could be attributed mainly to a reduction in light intensity occurring with increasing depth within the lake environment. Since photosynthesis is considered one of the most crucial metabolic activities for submerged macrophytes' survival and growth, any decrease or fluctuation in available light significantly impacts their ability to thrive efficiently. Plants require light intensity as an essential environmental factor and a limiting factor for their growth (Collier et al., 2012). Within eutrophic systems, light availability may be reduced by abiotic turbidity from the high densities of phytoplankton and epiphytic algae. As a result, the distribution of macrophytes at different depths is often restricted due to limited access to sufficient light (Barko and Smart, 1981; Madzivanzira et al., 2023a).

2.2 Temperature

Temperature drives the metabolic processes in most plant species (Lind et al., 2022). Freshwater-submerged macrophytes appear to be eurythermic and can thrive within a wide range of temperatures. Temperature increases can influence the community composition, phenology, abundance, productivity, distribution, and population growth of macrophytes (Li et al., 2018). However, it is important to note that these effects may vary among species. For example, an increase in temperature can stimulate the growth of certain species (Pip, 1989), whereas others may respond differently (Lind et al., 2022). Various physiological reactions are impacted by temperature, including dormancy and turion formation, seed germination development, photosynthetic rate, and oxygen consumption. Undoubtedly, temperature plays a crucial role in the growth of macrophytes.

2.3 Precipitation

Precipitation is essential for macrophytes as it provides a reliable water source that they need to grow and survive. Additionally, precipitation helps transport nutrients from the catchment into aquatic ecosystems, thereby enriching the water with nutrients (Iteba et al., 2021) leading to macrophyte growth and productivity. Moreover, when it rains, temporary or seasonal bodies of water may form. These offer favorable conditions for macrophytes to reproduce and spread their seeds or propagules effectively (Williams, 2006). This improves distribution across different areas and enhances genetic diversity within these plant populations. Precipitation plays an important role in maintaining stable water levels in their habitats, preventing drastic changes that could harm these macrophyte communities. Most reservoirs in tropical Africa are generally shallow and polymictic. These reservoirs fill up during the rainy season but significantly decrease in volume throughout the dry season due to various factors such as evaporation, sedimentation, water withdrawal, and municipal usage (Mustapha, 2009). The poor ability to spread laterally is common in other tropical macrophyte communities (Thomas et al., 2000; Madzivanzira et al., 2023b). Drought conditions that are synonymous with low precipitation could therefore negatively impact macrophyte community assemblages within tropic aquatic ecosystems.

2.4 Nutrient availability

Nutrient concentrations in lentic systems define the trophic status of aquatic ecosystems. The systems are classified as either oligotrophic (system with low nutrient concentration), mesotrophic (system with moderate nutrient concentration), or eutrophic (system with high nutrient concentration) (Kalff, 2000). Most rooted macrophytes can obtain nutrients from the sediment even in eutrophic systems as well as from the water (Barko and Smart, 1981; Prajapati et al., 2017; Marufu et al., 2018) and may thereby be quite effective in satisfying their nutritional requirements (Barko and Smart, 2006). Root uptake is the primary pathway for nitrogen, phosphorus, and micronutrients, whereas foliage uptake is the primary pathway for calcium, magnesium, sodium, potassium, and sulfate (Barko et al., 1991). Volume, water residence time, and nutrient input from the surrounding watershed drive the natural trophic state of aquatic ecosystems (Dalu et al., 2013; Kalff, 2000). Nonetheless, anthropogenic activities such as enrichment and climate change can also modify the nutrient status of aquatic systems. Except for free-floating invasive plants, invasive macrophytes are hypothesized to be limited by the availability of nutrients in the sediment. Sediments containing higher nutrient levels increase the likelihood of colonization and successful invasion by rooted macrophytes (Thomaz et al., 2015; Dalu et al., 2021). For instance, a study conducted at Lake Nasser (Egypt) indicated a correlation between water chemistry, calcium content, phosphorus levels in sediments, and the expansion of macrophyte invasive species (Ali and Soltan, 2006). In nutrient-limited environments, models by Van Gerven et al. (2015) predict that with increasing nutrient levels, submerged plants are outcompeted by floating plants. Macrophytes such as *Pontederia crassipes*, *Salvinia auriculata*, and *Hydrocotyle umbellata* are common in tropical regions. According to Fernández et al. (1990), these macrophytes have a rapid growth rate and reproductive capacity, allowing them to quickly spread throughout various water habitats. This is particularly true in environments with high nutrient content. The extensive water surface coverage by these macrophytes can negatively impact aquatic ecosystems, disrupt ecological balance, affect water flow, and obstruct dams and navigation in waterways (Heneidy et al., 2019). This was observed in Lake Chivero (Zimbabwe) where massive growth in *Pontederia crassipes*, due to high nutrient concentrations, led to increased growth and blockage of waterways in the Manyame River (Fig. 1).

2.5 Hydrological regimes

Hydrological regimes, including seasonal variations in rainfall and water flow, greatly influence water availability in African freshwater systems (Dalu et al., 2012, 2021). Hydrological fluctuations can alter the physical characteristics of aquatic habitats, such as sediment composition, light penetration, and nutrient availability. These changes can directly affect the suitability of habitats for different macrophyte species, leading to shifts in community composition based on

FIG. 1 Dense mats of *Pontedoria crassipes* at the spillway *(top)* and at the mouth of Manyame River *(bottom)*, Zimbabwe. *(Photo by Chipo P. Mungenge.)*

their tolerance to varying habitat conditions (Deegan et al., 2012). The timing and intensity of hydrological events, such as flood pulses, can influence the germination and establishment of macrophyte species. Floods, for instance, can create suitable conditions for germination by depositing nutrient-rich sediments and reducing competition from other plant species (Catian et al., 2018). Hydrological variations can also impact the competitive interactions among macrophyte species. Rapid changes in water levels may favor opportunistic, fast-growing species that can quickly colonize newly exposed areas. Over time, these species might outcompete slower-growing species, leading to shifts in the macrophyte community's composition and structure (Wagner and Falter, 2002). Hydrological regimes can influence the dispersal of macrophyte propagules (seeds, spores, fragments) across aquatic systems. Fluctuations in water flow can aid in the dispersal of propagules downstream, promoting genetic exchange among different populations and influencing the spatial distribution of species (Riis et al., 2004). Overall, changes in hydrological regimes, especially those driven by climate change or human activities, can, therefore, disrupt the balance of macrophyte communities, potentially leading to cascading effects on other ecosystem components, such as fish populations, water quality, and overall biodiversity.

3 Climate variability and macrophyte distribution patterns

Evidence indicates that Africa's warming rate surpasses the global average, with expectations for this trend to persist (Hulme et al., 2001). However, it is important to note that the climatic consequences will vary significantly depending on specific regions within the continent rather than a unified impact across Africa as a whole. Consequently, certain areas may experience increased aridity while others witness amplified precipitation. Nonetheless, the overarching consensus is that most locales will face unfavorable effects from these changes (Collier et al., 2008). The current rates of climate change globally are unparalleled, and biological responses to these climatic changes have also been rapid at the different levels of ecosystems, communities, and species (Heino et al., 2009). The effects these changes have on different species can vary, as shifts in climate may alter various environmental factors such as temperature and precipitation patterns. Among the ecosystems susceptible to global climate change, freshwater ecosystems stand out as particularly vulnerable environments in the face of global climate change (Bates et al., 2008), with macrophytes facing heightened susceptibility within these ecosystems.

The potential implications of global warming on African aquatic ecosystems, as projected by climate change models, remain uncertain. Nevertheless, there have been reports indicating some likely outcomes. Regarding lentic systems such as lakes and ponds, it is anticipated that climate change will likely result in a greater abundance and wider distribution of emergent and floating species while reducing the presence and spread of submerged macrophytes (Fig. 2). Changes in water level variations in water bodies because of climate change will

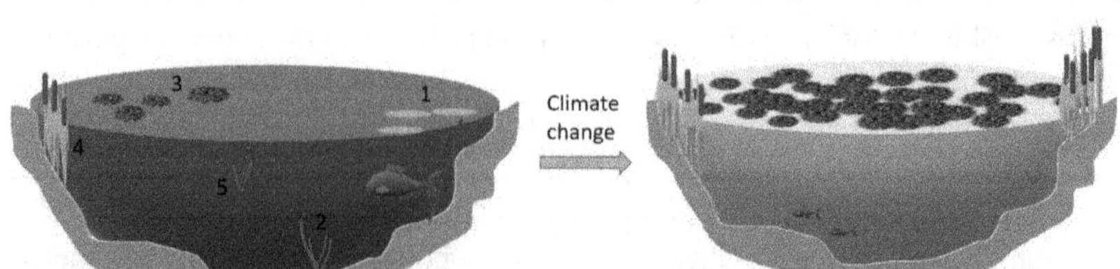

FIG. 2 Within lentic systems, climate change will lead to an increase in temperature, precipitation, and carbon dioxide levels, which will directly lead to a decline in submerged species populations while promoting the proliferation of emergent and floating invasive macrophytes. Additionally, it will result in an overall increment in phytoplankton abundance but with correspondingly reduced light intensity. *(Source: Lind, L., Eckstein, R.L., Relyea, R.A., 2022. Direct and indirect effects of climate change on distribution and community composition of macrophytes in lentic systems. Biol. Rev. 97(4), 1677–1690.)*

likely lead to shoreline erosion and loss of macrophytes (Dalu et al., 2012). For much of tropical arid and semiarid Africa, potentially harmful and problematic algal species may likely proliferate (Dalu and Wasserman, 2018). If temperatures continue to rise due to global warming, deep lakes like Kariba, Malawi, Tanganyika, and Victoria could experience reduced depths for light transparency (Ndebele-Murisa et al., 2010). Extreme changes in water levels and flow patterns will also impact the distribution and growth of macrophyte communities, as some species are adapted to specific water level ranges and may struggle to survive during droughts or flooding (Carmignani and Roy, 2017).

Researchers are also continuously studying how the impact of climate change on aquatic invasive species can affect how invasive species establish themselves and spread in both their native habitats and new environments. According to predictions from Thomas et al. (2000), these climate shifts could potentially give invasive aquatic species an advantage over others, leading to a reshaping of their ecological influence (Heneidy et al., 2019).

4 Introduction to remote sensing of macrophytes

The monitoring of emergent macrophytes (Alahuhta et al., 2011), floating plants (Madzivanzira et al., 2023a), and their interaction with submerged macrophytes (Li et al., 2017) has been mainly done using in situ approaches. These methods facilitate meticulous, in situ assessments of macrophyte health and community composition. However, these methods are costly, time consuming, and laborious for extensive water resources in need of monitoring. To overcome some of the limitations of in situ approaches, RS is increasingly being employed in combination with traditional techniques to monitor macrophytes effectively (Visser et al., 2013; Duffy et al., 2019; Free et al., 2020; Costa et al., 2021). For a comprehensive knowledge of macrophytes, researchers at the local, regional, and global levels are increasingly relying on RS for timely and cost-effective results (Ashraf et al., 2010; Free et al., 2020).

As RS is continuously growing, people who are inexperienced in the field may be unaware of the applications, technical options, and products that are accessible to them. Because data are obtained from a distance, the use of RS in macrophyte research helps to overcome accessibility concerns, allowing the survey of fragile or dangerous locations while the operator stays at a single point of safety (Kalacska et al., 2017; Rhee et al., 2018). RS data on macrophytes may be used in quantitative research and compared across spatiotemporal scales when gathered and processed appropriately (Jia et al., 2017; Klemas, 2016). The exact precision of quantitative macrophyte data collected by RS may also be used to evaluate gradual, progressive changes on RS platforms (Fig. 3). It is becoming evident that RS is slowly becoming one of the leading effective tools for monitoring macrophytes in remote and impoverished locations (Duffy et al., 2019).

Aerospace technology advancements have enabled the creation of airborne sensors and satellite platforms, each of which has increased the spatial scale of

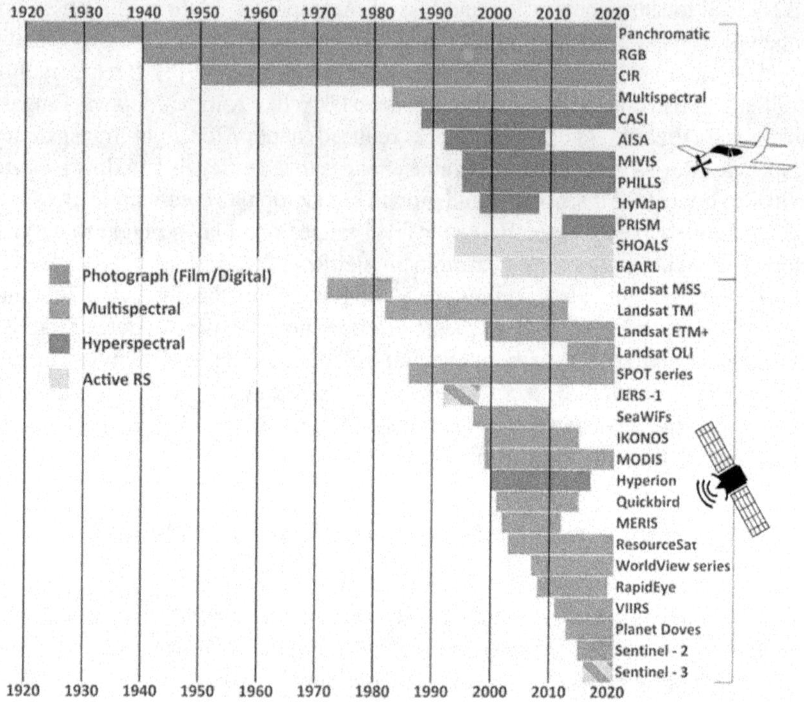

FIG. 3 Operation chronology of a selection of remote sensors used in macrophyte research by platform and sensor type.

RS data accessible for macrophyte study. As indicated by the large number of studies that use this data source and the wide spectrum of active sensors, satellite RS is gradually becoming the major instrument for macrophyte research (Palmer et al., 2015). Technological advancements, notably, in the field of unmanned aerial vehicles (UAVs), are constantly opening new doors. UAVs are becoming more affordable due to their recent commercialization and advancements providing on-demand data for macrophyte research (Anderson and Gaston, 2013; Linchant et al., 2015; Rhee et al., 2018).

It is important to highlight that no RS approach can be used effectively in all settings and for all research issues (Dörnhöfer and Oppelt, 2016; Stocks et al., 2019), As a result, it is critical to be aware of the various solutions, their advantages, and disadvantages, and how they might be combined to better understand the spatiotemporal distributions of macrophytes. While RS experts are critical for technological innovation and method advancement, it is users from other professions who use RS procedures in their own work that propel RS forward through new and imaginative applications (Farster et al., 2021).

5 Detecting stress-induced changes in macrophyte communities

The responses of macrophyte communities to stressors induced by extreme climatic events hold paramount ecological significance and serve as barometers of ecosystem health and resilience. Present methodologies for monitoring macrophytes predominantly encompass underwater surveys and ecological modeling (Scott et al., 2002). Underwater surveys, whether conducted via scuba diving or remotely operated vehicles (ROVs), facilitate meticulous, in situ assessments of macrophyte health and community composition (Wingfield et al., 2005; Olden et al., 2022). Simultaneously, ecological modeling, encompassing habitat suitability models and population dynamics simulations, contributes to predictive insights into macrophyte responses amid evolving environmental conditions (Riis et al., 2017; Guan et al., 2020; Saha et al., 2020). These complementary approaches collectively enhance our comprehension of macrophyte dynamics and their pivotal role in bolstering ecosystem resilience, thereby empowering informed conservation and management strategies in an era of escalating climate uncertainty. Additionally, advancements in satellite and drone technologies offer a nonintrusive and cost-effective means to monitor macrophyte communities over large spatial scales and at frequent intervals. This provides valuable data on macrophyte distribution, biomass, and response to stressors.

5.1 Remote sensing and monitoring macrophytes

Stressors from drought, floods, or heatwaves cause distinctive biochemical and physiological alterations in macrophytes (Woodward et al., 2016; Reitsema et al., 2018; Huang et al., 2021). These changes manifest as subtle shifts in the reflectance patterns across the electromagnetic spectrum (Wolf et al., 2013; Malthus, 2017; Rowan and Kalacska, 2021). RS instruments, calibrated to capture specific wavelength ranges, can decipher these spectral shifts, unveiling stress-induced responses even before visible signs of distress emerge (Fritz et al., 2017; Rowan and Kalacska, 2021). The use of RS, characterized by its noninvasive nature and its ability to capture data across diverse spatial and temporal scales, can significantly enhance our ability to manage and conserve aquatic ecosystems effectively (Villa et al., 2017; Huber et al., 2022).

RS techniques emerge as transformative toolsets, offering the capacity to discern stress-induced variations in macrophyte communities with unprecedented precision and scope (Villa et al., 2017). For example, vegetation indices, calculated from RS data, are potent indicators of macrophyte health and stress (Jakubauskas et al., 2000). The normalized difference vegetation index (NDVI), for instance, quantifies the differential reflectance between visible and near-infrared spectral regions (e.g., in Proctor and He, 2013). These spectral regions are essential in monitoring plant response to environmental stimuli based on the changes and shifts of macrophyte response curves (Jakubauskas et al., 2000; Khanna et al., 2011; Malthus, 2017). For example, decreased NDVI values may signify reduced chlorophyll content and compromised photosynthetic activity in response to stressors (Proctor and He, 2013). Complementary indices, such as the enhanced vegetation index (EVI) and the photochemical reflectance index (PRI), also offer nuanced insights into stress-induced changes in pigment concentration and photoprotective mechanisms of macrophytes (Jakubauskas et al., 2000; Stratoulias, 2015). In addition, classification approaches and close-range UAVs offer a wealth of information for analyzing macrophyte communities (Jakubauskas et al., 2000; Silva et al., 2010; Khanna et al., 2011; Malthus, 2017). These sensors enable researchers to gather data at fine spatial resolutions over specific areas of interest and can be used to monitor floating macrophytes (Jakubauskas et al., 2000; Silva et al., 2010). This integration of RS technologies not only enhances our capacity to scrutinize the dynamic responses of macrophyte communities but also furnishes a robust foundation for informed decision-making in the conservation and management of these vital aquatic ecosystems.

The wall-to-wall monitoring provided by RS accommodates investigations that span microhabitat responses to broader landscape-level changes (Nelson et al., 2006; Albright and Ode, 2011). This is imperative since macrophyte communities operate across multiple spatial scales, from individual plants to entire ecosystems, and field methods might be time-consuming and costly to monitor macrophytes at these scales (Nelson et al., 2006; Albright and Ode, 2011). Furthermore, this capability is particularly pertinent when studying stressors with varying magnitudes of impact (Albright and Ode, 2011). By capturing stress-induced changes at diverse scales, RS enriches our understanding of the ecological implications of stressors on macrophyte communities.

5.2 Remote sensing and climate models: An overview

Macrophytes play a pivotal role in shaping the structure and functioning of aquatic ecosystems (Jones et al., 2020). However, several studies underscore that their dynamics are profoundly influenced by climate variability (Johnson et al., 2021). The integration of RS data with climate models has emerged as a promising approach to enhance our comprehension and prognostications of macrophyte dynamics under evolving climate conditions (Brown et al., 2022). Remote Sensing techniques offer a valuable means to gather spatially explicit data on macrophytes, enabling tasks such as mapping, biomass estimation, phenological monitoring, and growth pattern analysis (Peterson et al., 2020). The spectral data captured by RS sensors assume a critical role in macrophyte mapping and species identification. Unique spectral signatures distinguish different macrophyte species, allowing their differentiation and mapping based on distinct reflectance patterns across various wavelengths

(Hansen et al., 2021). Through spectral data analysis, researchers have demonstrated the capability to delineate the distribution and abundance of macrophyte species within aquatic ecosystems (Wu et al., 2023).

Satellite imagery, aerial photography, and UAVs constitute the primary sources of RS data. These sources furnish diverse types of data, encompassing spectral, thermal, and light detection and ranging (LiDAR) data (Smith et al., 2023). Spectral data capture the distinctive spectral signatures of macrophyte species, facilitating species identification and mapping. Thermal data yield insights into macrophyte stress and growth rates. Derived from RS platforms, thermal data offer valuable information regarding macrophyte stress and growth rates (Chen et al., 2021). Temperature has proven to be a pivotal environmental factor influencing macrophyte physiological processes, and thermal data can depict thermal signatures reflecting the thermal state of macrophyte communities. Changes in these signatures can serve as indicators of stress responses to factors like temperature anomalies, water level fluctuations, or nutrient imbalances. Monitoring thermal patterns can yield early warning signals of macrophyte health and aid in identifying vulnerable areas. On the other hand, LiDAR data deliver detailed insights into macrophyte structure and height, offering a comprehensive understanding of their habitat. Concurrently, climate models have been harnessed to simulate and forecast climate variables that significantly impact macrophyte dynamics (Smith et al., 2023). The fusion of RS data and climate models permits a holistic analysis and prediction of macrophyte responses to climate variability.

5.3 Remote sensing macrophyte dynamics under climate variability

The temporal dimension is equally important in macrophyte studies, and RS aptly assists in monitoring phenology. Tracking spectral changes over time enables researchers to grasp macrophyte life cycle stages, elucidating flowering, seed production, and senescence timings (Ma et al., 2021). Understanding phenological shifts illuminates how climate fluctuations influence macrophyte reproductive success and overall fitness. Macrophyte growth patterns are equally amenable to RS scrutiny. Changes in structure and derived vegetation indices serve as indicators of growth dynamics (Tong et al., 2023). These patterns unveil expansions, contractions, and spatial heterogeneities within macrophyte populations. Consequently, our grasp of macrophyte responses to environmental shifts deepens, informing conservation strategies.

Integration of RS data with climate models elevates our predictive capabilities, foreseeing macrophyte responses under varied climate scenarios (Peng et al., 2022). By coupling climate variables like temperature, precipitation, and water availability, researchers have improved our understanding of how climate change impacts macrophyte dynamics (Madzivanzira et al., 2023a). These models guide our understanding of future macrophyte distributions, phenology shifts, and growth patterns, furnishing crucial insights for effective ecosystem management and conservation strategies.

5.4 Remote sensing—Climate model integration

The integration of RS data with climate models has significantly advanced our understanding of macrophyte dynamics within the context of climate variability. The interactions between aquatic plants, or macrophytes, and their surrounding environment are complex and strongly influenced by changing climate conditions. RS technology, which encompasses satellite imagery, aerial photography, and UAVs, offers a robust means to capture spatial and temporal information on macrophyte distribution, growth patterns, and phenology. The spectral data acquired from RS platforms enable researchers to analyze macrophyte distribution across large spatial scales (Dandois et al., 2020; Wang et al., 2022). The distinct spectral signatures of various macrophyte species allow for differentiation and mapping, contributing to the identification of areas where macrophytes thrive or face decline (Sardans et al., 2021). The integration of ancillary data, such as plant height and leaf area index, enhances biomass estimation accuracy, thereby offering a more comprehensive understanding of macrophyte dynamics and productivity (Xiao et al., 2019). Additionally, RS facilitates the monitoring of macrophyte phenology by tracking spectral changes over time, thereby shedding light on the timing and duration of life cycle stages (Ma et al., 2021). This insight into phenological shifts aids in assessing how climate fluctuations impact macrophyte reproductive success and overall ecological fitness.

Climate models proved to provide a complementary tool in unravelling the intricate relationship between macrophyte dynamics and climate variability. These models simulate climate variables like temperature, precipitation, and water availability, enabling researchers to predict how changing climatic conditions will impact macrophyte responses (Peng et al., 2022; Madzivanzira et al., 2023a). By integrating RS data with climate models, researchers can project macrophyte behavior under different climate scenarios. This integration enhances our capacity to anticipate macrophyte distributions, phenology shifts, and growth patterns, thus contributing to effective ecosystem management and conservation strategies (Brown et al., 2022). Ultimately, this combined approach provides a holistic understanding of how macrophytes respond to varying climate conditions, allowing for the formulation of strategies that promote their resilience in the face of climate variability.

6 Challenges and opportunities in remote sensing—Climate model integration for macrophyte modeling

Integrating RS data with climate models presents both challenges and promising opportunities in understanding macrophyte dynamics within the context of climate variability. Despite the potential benefits, this integration encounters several challenges that warrant careful consideration. Data availability and quality are critical concerns. Access to high-quality and consistent RS data can be restricted, impacting the reliability of integrated analyses (Pettorelli et al., 2018). Variability in data collection methods, sensors, and calibration procedures introduces uncertainties and affects data reliability (Wang and Menenti, 2021; Smith et al., 2023).

Moreover, spatial and temporal resolution constraints of RS data can undermine the accuracy and applicability of modeling macrophyte dynamics. Fine-scale variations and rapid changes in macrophyte distribution and phenology may not be adequately captured, limiting the effectiveness of the integration. Scaling issues further complicate the integration process. Integrating data across different spatial and temporal scales requires careful consideration. Variations in data resolution, particularly when integrating RS data with climate models, can introduce uncertainties and affect the accuracy of integrated analyses. Synchronization between different temporal scales presents challenges in capturing rapid changes (Sun et al., 2023; Wang et al., 2024).

Furthermore, model validation and calibration are paramount for accurate predictions, yet these aspects present significant challenges. Reliable ground truth data for macrophyte dynamics are often limited, making it difficult to validate and calibrate the integrated models effectively. Obtaining comprehensive and representative ground truth data across large spatial scales is logistically challenging and resource-intensive, adding complexities to validation and calibration efforts (Chen et al., 2022).

However, the integration of RS data with climate models also offers promising opportunities that can potentially overcome these challenges. Advanced RS platforms, such as high-resolution satellite sensors and UAVs, offer improved spatial and temporal coverage, enabling finer-scale observations and more frequent data collection. This enhanced monitoring capacity contributes to a deeper understanding of macrophyte responses to climate variability. Multisensor data integration, by combining optical, radar, and hyperspectral sensors, provides a multidimensional view of macrophyte dynamics, which enhances accuracy and understanding (Li et al., 2021).

Machine learning techniques have emerged as powerful tools that revolutionize the processing and analysis of RS datasets. These algorithms automate classification, feature extraction, and pattern recognition, accelerating data processing and enhancing accuracy in macrophyte mapping and monitoring. Additionally, data assimilation techniques, which integrate RS measurements with numerical models, offer real-time updates and enhance the accuracy of model predictions. These techniques can provide more reliable projections of macrophyte responses under varying climate scenarios (Wu et al., 2019; Kim et al., 2022).

Furthermore, the integration of RS with climate models fosters interdisciplinary collaboration among specialists, ecologists, climatologists, and modelers. These collaborative efforts facilitate innovative approaches, refine methodologies, and address complex research questions related to macrophyte dynamics and ecosystem responses to climate variability. This cross-disciplinary exchange advances the understanding of the interactions between climate, RS, and macrophyte dynamics, fostering growth in both RS and ecological fields. Thus, while RS is a powerful tool for unlocking the intricate dynamics of macrophyte responses to extreme climatic events, it is full of challenges and limitations. An astute recognition of these constraints is vital to ensure the precise interpretation of results and the refinement of methodologies.

6.1 Spectral complexity and mixed pixels

With their intricate interplay of water, submerged macrophytes, sediments, and dissolved constituents, aquatic environments manifest a spectral intricacy that challenges RS observations (Silva et al., 2010; Malthus, 2017). In such complex landscapes, mixed pixels often arise, where the spectral signature of multiple constituents is blended within a single pixel (Silva et al., 2008). The disentanglement of macrophyte-specific information from this spectral amalgam represents a formidable challenge. However, advanced spectral unmixing algorithms and data acquired at satisfactory spatial resolutions can be crucial tools in unravelling this complexity (Giardino et al., 2014).

6.2 Scale mismatch

An enduring challenge in RS lies in the inherent scale mismatch between the observational scale of RS instruments and the localized ecological processes occurring within macrophyte communities (Hofstra et al., 2020). The responses of macrophytes to stressors are frequently nuanced and occur at fine spatial scales, often shaped by microhabitat conditions. In contrast, RS data typically encompass larger spatial extents, which can dilute the capacity to capture localized changes with the requisite precision (Giardino et al., 2014; Hofstra et al., 2020). Researchers should increasingly integrate ground-truth data from comprehensive field surveys to bridge this scale disparity, facilitating a more nuanced interpretation of RS observations.

6.3 Temporal resolution

Extreme climatic events, characterized by their rapid onset and transient nature, demand the utmost temporal resolution to accurately capture the swift shifts in macrophyte responses (Thamaga and Dube, 2018). However, many RS platforms, particularly satellites, may need more temporal revisit rates to observe these changes (e.g., Landsat-8 operational land imager (OLI)). Consequently, RS may inadvertently overlook ephemeral stress events or short-term phenological changes (Giardino et al., 2014). To surmount this limitation, a mosaic approach is often adopted, harnessing the complementary strengths of multiple data sources, including satellites, drones, and unmanned aerial systems, to enhance both temporal coverage and resolution.

6.4 Atmospheric interference

The aquatic environment is notably susceptible to atmospheric interference, which introduces complexities capable of attenuating, scattering, or absorbing electromagnetic radiation (Duffy et al., 2019). These atmospheric effects can veil the accurate retrieval of aquatic features and potentially compromise the reliability of stress detection (Duffy et al., 2019; Topp et al., 2020). To address this, the stringent application of atmospheric correction techniques to RS data is indispensable, rectifying these distortions and safeguarding the integrity of ecological interpretations.

6.5 Sensor limitations and spectral bands

RS sensors are confined to specific spectral ranges, which may not always encompass the most informative wavelengths for detecting stress-induced changes in macrophytes (Crowder and Painter, 1991). Stress responses in macrophytes might manifest subtly in spectral regions not covered by available sensors (Crowder and Painter, 1991; Topp et al., 2020). This limitation prompts ongoing innovation in sensor design, with the emergence of hyperspectral sensors promising to extend the boundaries of our capacity to discern nuanced stress signals.

6.6 Complex underwater environments

The subsurface aquatic realm introduces a stratum of complexity for RS. The penetration depth of electromagnetic radiation in water is constrained by the optical properties of the water itself, thus dictating the depth at which accurate observations of macrophyte dynamics can be achieved (Duffy et al., 2019; Haroon, 2022). This can be particularly problematic during flooding events. Furthermore, factors such as water turbidity and suspended particles can introduce additional obscurities, potentially hindering the precise assessment of macrophyte health (Haroon, 2022). Advanced techniques such as LiDAR and sonar have been incorporated to complement passive RS data, providing multidimensional insights into underwater landscapes.

7 Way forward

Despite significant advancements, the utilization of RS for monitoring aquatic environments remains limited. This limitation primarily arises from the interference of the water column, which attenuates the intensity of benthic reflectance signals and introduces variability within a scene (Nelson et al., 2006). In the early applications of RS to assess macrophyte cover and distribution, researchers exclusively relied on qualitative visual analysis of aerial photos (Ackleson and Klemas, 1987; Nelson et al., 2006). Due to the small size of macrophytes, acquiring very high spatial resolution data is essential, but it can be costly and often beyond the reach of studies with limited resources (Nelson et al., 2006). Furthermore, the variability in the calibration and correction processes used to account for air and water column effects poses challenges since these differ for most environments (i.e., Case-1 and Case-2 waters) and sensors (Malthus, 2017).

The cogency of RS technology in unravelling the intricate ecological interactions of macrophytes to extreme climatic events cannot be overstated. It facilitates the documentation of ecological perturbations and the elucidation of underlying mechanisms and responses, thereby bolstering our capacity to fortify macrophyte-dominated ecosystems against the tumultuous backdrop of global environmental change (Nelson et al., 2006; Malthus, 2017). As a sentinel of our aquatic ecosystems, RS stands as a bastion of knowledge, a tool indispensable in pursuing ecological resilience and safeguarding these vital aquatic domains (Thamaga and Dube, 2018; Topp et al., 2020; Rowan and Kalacska, 2021). RS tools can uncover and decipher the responses of macrophyte communities to climatic extremes over different spatial and temporal scales (Wolf et al., 2013; Krishnaveni and Vishnu, 2015). This renders RS in macrophyte monitoring not only practical but also cost-efficient (Nelson et al., 2006).

By harnessing RS's capacity to capture ecological dynamics across expansive spatial and temporal dimensions, we advance our potential to manage and conserve aquatic ecosystems judiciously and sustainably (Krishnaveni and Vishnu, 2015). This innovative avenue not only empowers our scientific toolkit but also bestows a vantage point for comprehending the broader ramifications of climate perturbations on natural domains (Nelson et al., 2006; Krishnaveni and Vishnu, 2015).

References

Ackleson, S.G., Klemas, V., 1987. Remote sensing of submerged aquatic vegetation in lower chesapeake bay: a comparison of Landsat MSS to TM imagery. Remote Sens. Environ. 22, 235–248.

Alahuhta, J., Heino, J., Luoto, M., 2011. Climate change and the future distributions of aquatic macrophytes across boreal catchments. J. Biogeogr. 38, 383–393.

Albright, T.P., Ode, D.J., 2011. Monitoring the dynamics of an invasive emergent macrophyte community using operational remote sensing data. Hydrobiologia 661, 469–474.

Ali, M.M., Soltan, M.A., 2006. Expansion of *Myriophyllum spicatum* (Eurasian water milfoil) into Lake Nasser, Egypt: invasive capacity and habitat stability. Aquat. Bot. 84 (3), 239–244.

Anderson, K., Gaston, K.J., 2013. Lightweight unmmaned aerial vehicles will revolutionize spatial ecology. Front. Ecol. Environ. 11, 138–146.

Arthaud, F., Toury, J., Romestaing, C., Bornette, G., 2021. Photosynthetic and morphological traits control aquatic plant distribution according to light stress. Evol. Ecol. 35, 739–760.

Ashraf, S., Brabyn, L., Hicks, B.J., Collier, K., 2010. Satellite remote sensing for mapping vegetation in New Zealand freshwater environments: a review. N. Z. Geogr. 66, 33–43.

Barko, J.W., Smart, R.M., 1981. Comparative influences of light and temperature on the growth and metabolism of selected submersed freshwater macrophytes. Ecol. Monogr. 51 (2), 219–236.

Barko, J.W., Smart, R.M., 2006. Mobilization of sediment phosphorus by submersed freshwater macrophytes. Freshw. Biol. 10 (3), 229–238.

Barko, J.W., Gunnison, D., Carpenter, S.R., 1991. Sediment interactions with submersed macrophyte growth and community dynamics. Aquat. Bot. 41 (1–3), 41–65.

Bates, B.C., Kundzewicz, Z.W., Wu, S., Palutikof, J.P., 2008. Climate Change and Water. Technical Paper of the Intergovernmental Panel on Climate Change. IPCC Secretariat, Geneva.

Bolpagni, R., 2020. Linking vegetation patterns, wetlands conservation, and ecosystem services provision: from publication to application. Aquat. Conserv.: Mar. Freshw. Ecosyst. 30 (9), 1734–1740.

Brendonck, L., Maes, J., Rommens, W., Dekeza, N., Nhiwatiwa, T., Barson, M., Callebaut, V., Phiri, C., Moreau, K., Gratwicke, B., Stevens, M., Allyn, N., Holsters, E., Ollevier, F., Marshall, B., 2003. The impact of water hyacinth (*Eichhornia crassipes*) in a eutrophic subtropical impoundment (Lake Chivero, Zimbabwe). II. Species diversity. Arch. Hydrobiol. 158 (3), 389–405.

Brown, A., et al., 2022. Integrating remote sensing and climate models to understand macrophyte dynamics. Environ. Sci. Technol. 56 (2), 789–798.

Carmignani, J.R., Roy, A.H., 2017. Ecological impacts of winter water level drawdowns on lake littoral zones: a review. Aquat. Sci. 79, 803–824.

Catian, G., Da Silva, D.M., Suarez, Y.R., Scremin-Dias, E., 2018. Effects of flood pulse dynamics on functional diversity of macrophyte communities in the Pantanal Wetland. Wetlands 38, 975–991.

Chambers, P.A., Lacoul, P., Murphy, K.J., Thomaz, S.M., 2008. Global diversity of aquatic macrophytes in freshwater. In: Freshwater Animal Diversity Assessment, pp. 9–26, https://doi.org/10.1007/s10750-007-9154-6.

Chen, Y., et al., 2021. Thermal remote sensing for macrophyte stress assessment. Remote Sens. Rev. 42 (3), 215–228.

Chen, J., Chen, S., Fu, R., Li, D., Jiang, H., Wang, C., Peng, Y., Jia, K., Hicks, B.J., 2022. Remote sensing big data for water environment monitoring: current status, challenges, and future prospects. Earth's Future 10 (2), e2021EF002289.

Chikwenhere, G.P., 1994. Biological control of water lettuce in various impoundments of Zimbabwe. J. Aquat. Plant Manage. 32, 27–29.

Chikwenhere, G.P., 2001. Current strategies for the management of water hyacinth on the Manyame River system in Zimbabwe. In: Biological and Integrated Control of Water Hyacinth, *Eichhornia crassipes*. Proc. Second Global Working Group Meeting for the Biological and Integrated Control of Water Hyacinth. ACIAR Proceedings (No. 102), pp. 105–108.

Collier, P., Conway, G., Venables, T., 2008. Climate change and Africa. Oxford Rev. Econ. Pol. 24 (2), 337–353.

Collier, C.J., Waycott, M., McKenzie, L.J., 2012. Light thresholds derived from seagrass loss in the coastal zone of the northern Great Barrier Reef, Australia. Ecol. Indic. 23, 211–219.

Costa, V., Serôdio, J., Lillebø, A.I., Sousa, A.I., 2021. Use of hyperspectral reflectance to non-destructively estimate seagrass *Zostera noltei* biomass. Ecol. Indic. 121, 107018.

Cronin, G., Lewis Jr., W.M., Schiehser, M.A., 2006. Influence of freshwater macrophytes on the littoral ecosystem structure and function of a young Colorado reservoir. Aquat. Bot. 85 (1), 37–43.

Crowder, A., Painter, D.S., 1991. Submerged macrophytes in Lake Ontario: current knowledge, importance, threats to stability, and needed studies. Can. J. Fish. Aquat. Sci. 48 (8), 1539–1545.

Dalu, T., Chauke, R., 2020. Assessing macroinvertebrate communities in relation to environmental variables: the case of Sambandou wetlands, Vhembe Biosphere Reserve. Appl. Water Sci. 10 (1), 1–11.

Dalu, T., Wasserman, R.J., 2018. Cyanobacteria dynamics in a small tropical reservoir: understanding spatio-temporal variability and influence of environmental variables. Sci. Total Environ. 643, 835–841.

Dalu, T., Nhiwatiwa, T., Clegg, B., 2012. Aquatic macrophytes in a tropical African reservoir: diversity, communities and the impact of reservoir-level fluctuations. Trans. R. Soc. S. Afr. 67 (3), 117–125.

Dalu, T., Nhiwatiwa, T., Clegg, B., 2013. Temporal variation of the plankton communities in a small tropical reservoir (Malilangwe, Zimbabwe). Trans. R. Soc. S. Afr. 68 (2), 85–96.

Dalu, T., Dlamini, P., Wasserman, R.J., Mokgoebo, M.J., Mutshekwa, T., Dondofema, F., Cuthbert, R.N., 2021. Effects of environmental variables on littoral macroinvertebrate community assemblages in subtropical reservoirs. Chem. Ecol. 37 (5), 419–436.

Dandois, J.P., Olano, M., Ellis, E.C., 2020. Remote sensing of vegetation structure using computer vision. Remote Sens. Environ. 244, 111764.

Deegan, B.M., White, S.D., Ganf, G.G., 2012. Nutrients and water level fluctuations: a study of three aquatic plants. River Res. Appl. 28 (3), 359–368.

Dibble, E.D., Thomaz, S.M., Padial, A.A., 2006. Spatial complexity measured at a multi-scale in three aquatic plant species. J. Freshwater Ecol. 21 (2), 239–247.

Dogan, M., Saygideger, S.D., Colak, U., 2009. Effect of lead toxicity on aquatic macrophyte *Elodea canadensis* Michx. Bull. Environ. Contam. Toxicol. 83, 249–254.

Dörnhöfer, K., Oppelt, N., 2016. Remote sensing for lake research and monitoring—recent advances. Ecol. Indic. 64, 105–122.

Duffy, J.E., Benedetti-Cecchi, L., Trinanes, J., Muller-Karger, F.E., Ambo-Rappe, R., Boström, C., Buschmann, A.H., Byrnes, J., Coles, R.G., Creed, J., Cullen-Unsworth, L.C., Diaz-Pulido, G., Duarte, C.M., Edgar, C.J., Fortes, M., Goni, G., Hu, C., Huang, X., Hurd, C.L., et al., 2019. Toward a coordinated global observing system for seagrasses and marine macroalgae. Front. Mar. Sci. 6, 317.

Farster, B., Huynh, Q., Messinger, D., Middleton, C., Resmini, R.G., 2021. Panel: how to meet the need for spectral expertise. In: Spectral Sessions. Available online: https://www.l3harrisgeospatial.com/Company/Events/Tradeshows/Spectral-Sessions. (27 January 2021).

Fernández, O.A., Sutton, D.L., Lallana, V.H., Sabbatini, M.R., Irigoyen, J.H., 1990. Aquatic Weed Problems and Management in South and Central America. Oxford University Press, Oxford (UK).

Free, G., Bresciani, M., Trodd, W., Tierney, D., O'Boyle, S., Plant, C., Deakin, J., 2020. Estimation of lake ecological quality from Sentinel-2 remote sensing imagery. Hydrobiologia 847, 1423–1438.

Fritz, C., Dörnhöfer, K., Schneider, T., Geist, J., Oppelt, N., 2017. Mapping submerged aquatic vegetation using Rapid Eye satellite data: the example of Lake Kummerow (Germany). Water 9 (7), 510.

Ghirardi, N., Bolpagni, R., Bresciani, M., Valerio, G., Pilotti, M., Giardino, C., 2019. Spatiotemporal dynamics of submerged aquatic vegetation in a deep lake from Sentinel-2 data. Water 11 (3), 563.

Giardino, C., Bresciani, M., Stroppiana, D., Oggioni, A., Morabito, G., 2014. Optical remote sensing of lakes: an overview on Lake Maggiore. J. Limnol. 73, S1.

Guan, C., Saha, M., Weinberger, F., 2020. Simulated heatwaves lead to upregulated chemical defense of a marine foundation macrophyte against microbial colonizers. Front. Mar. Sci. 7, 463.

Hansen, R., et al., 2021. Spectral signatures for macrophyte species identification. Int. J. Appl. Earth Obs. Geoinf. 84, 110–119.

Haroon, A.M., 2022. Review on aquatic macrophytes in Lake Manzala, Egypt. Egypt. J. Aquat. Res. 48 (1), 1–12.

Heino, J., Virkkala, R., Toivonen, H., 2009. Climate change and freshwater biodiversity: detected patterns, future trends and adaptations in northern regions. Biol. Rev. 84 (1), 39–54.

Heneidy, S.Z., Halmy, M.W.A., Fakhry, A.M., El-Makawy, A. M., 2019. The status and potential distribution of *Hydrocotyle umbellata* L. and *Salvinia auriculata* Aubl. under climate change scenarios. Aquat. Ecol. 53, 509–528.

Hestir, E.L., Brando, V.E., Bresciani, M., Giardino, C., Matta, E., Villa, P., Dekker, A.G., 2015. Measuring freshwater aquatic ecosystems: the need for a hyperspectral global mapping satellite mission. Remote Sens. Environ. 167, 181–195.

Hofstra, D., Schoelynck, J., Ferrell, J., Coetzee, J., de Winton, M., Bickel, T.O., Champion, P., Madsen, J., Bakker, E.S., Hilt, S., Matheson, F., Netherland, M., Gross, E.M., 2020. On the move: new insights on the ecology and management of native and alien macrophytes. Aquat. Bot. 162, 103190.

Huang, Y., Chen, X.S., Li, F., Hou, Z.Y., Li, X., Zeng, J., Deng, M., Zou, Y.A., Xie, Y.H., 2021. Community trait responses of three dominant macrophytes to variations in flooding during 2011–2019 in a Yangtze River-connected floodplain wetland (Dongting Lake, China). Front. Plant Sci. 12, 604677.

Huber, S., Hansen, L.B., Nielsen, L.T., Rasmussen, M.L., Sølvsteen, J., Berglund, J., Paz von Friesen, C., Danbolt, M., Envall, M., Infantes, E., 2022. Novel approach to large-scale monitoring of submerged aquatic vegetation: a nationwide example from Sweden. Integr. Environ. Assess. Manage. 18 (4), 909–920.

Hulme, M., Doherty, R., Ngara, T., New, M., Lister, D., 2001. African climate change: 1900–2100. Climate Res. 17 (2), 145–168.

IPCC, 2007. Climate change 2007: impacts, adaptation and vulnerability. In: Parry, M.L., Canziani, O.F., Palutikof, J.P., van der Linden, P.J., Hanson, C.E. (Eds.), Contribution of Working Group II to the Fourth Assessment Report of the Intergovernmental Panel on Climate Change. Cambridge University Press, Cambridge, UK.

IPCC, 2012. Managing the risks of extreme events and disasters to advance climate change adaptation. In: Field, C.B., Barros, V., Stocker, T.F., et al. (Eds.), A Special Report of Working Groups I and II of the Intergovernmental Panel on Climate Change. Cambridge University Press, Cambridge, NY, USA.

Iteba, J.O., Hein, T., Singer, G.A., Masese, F.O., 2021. Livestock as vectors of organic matter and nutrient loading in aquatic ecosystems in African savannas. PLoS One 16 (9), e0257076.

Jakubauskas, M., Kindscher, K., Fraser, A., Debinski, D., Price, K.P., 2000. Close-range remote sensing of aquatic macrophyte vegetation cover. Int. J. Remote Sens. 21 (18), 3533–3538.

Jia, Q., Cao, L., Yésou, H., Huber, C., Fox, A.D., 2017. Combating aggressive macrophyte encroachment on a typical Yangtze River lake: lessons from a long-term remote sensing study of vegetation. Aquat. Ecol. 51, 177–189.

Jin, S., Ibrahim, M., Muhammad, S., Khan, S., Li, G., 2020. Light intensity effects on the growth and biomass production of submerged macrophytes in different water strata. Arab. J. Geosci. 13, 1–7.

Johnson, J.A., Newman, R.M., 2011. A comparison of two methods for sampling biomass of aquatic plants. J. Aquat. Plant Manage. 49 (1), 1–8.

Johnson, L., et al., 2021. Climate variability impacts on aquatic macrophyte dynamics. Aquat. Ecol. 55 (3), 389–402.

Jones, M., et al., 2020. The role of aquatic macrophytes in ecosystem functioning. Front. Ecol. Evol. 8, 145.

Kalacska, M., Chmura, G.L., Lucanus, O., Bérubé, D., Arroyo-Mora, J.P., 2017. Structure from motion will revolutionize analyses of tidal wetland landscapes. Remote Sens. Environ. 199, 14–24.

Kalff, J., 2000. Limnology. Prentice Hall, Upper Saddle River, New Jersey.

Khanna, S., Santos, M.J., Ustin, S.L., Haverkamp, P.J., 2011. An integrated approach to a biophysiologically based classification of floating aquatic macrophytes. Int. J. Remote Sens. 32 (4), 1067–1094.

Kim, J., Ryu, Y., Dechant, B., 2022. Development of a filter-based near-surface remote sensing system to retrieve far-red sun-induced chlorophyll fluorescence. Remote Sens. Environ. 283, 113311.

Klemas, V., 2016. Remote sensing of submerged aquatic vegetation. In: Finkl, C.W., Makowski, C. (Eds.), Seafloor Mapping Along Continental Shelves: Research and Techniques for Visualizing Benthic Environments. vol. 13. Springer International Publishing, Cham, Switzerland, pp. 125–140.

Kotta, J., Kutser, T., Teeveer, K., Vahtmäe, E., Pärnoja, M., 2013. Predicting species cover of marine macrophyte and invertebrate species combining hyperspectral remote sensing, machine learning and regression techniques. PLoS One 8 (6), e63946.

Krishnaveni, R.Y., Vishnu, B., 2015. Temporal variation of aquatic invasive plants in Kuttanad wetland ecosystem using remote sensing techniques. Trends Biosci. 8 (21), 5845–5853.

Li, Z., He, L., Zhang, H., Urrutia-Cordero, P., Ekvall, M.K., Hollander, J., Hansson, L.A., 2017. Climate warming and heat waves affect reproductive strategies and interactions between submerged macrophytes. Glob. Chang. Biol. 23 (1), 108–116.

Li, C., Wang, T., Zhang, M., Xu, J., 2018. Maternal environment effect of warming and eutrophication on the emergence of curled pondweed, *Potamogeton crispus* L. Water 10 (9), 1285.

Li, Z., Xu, Z., Yang, Y., Stewart, R.I., Urrutia-Cordero, P., He, L., et al., 2021. Heat waves alter macrophyte-derived detrital nutrients release under future climate warming scenarios. Environ. Sci. Technol. 55 (8), 5272–5281.

Linchant, J., Lisein, J., Semeki, J., Lejeune, P., Vermeulen, C., 2015. Are unmanned aircraft systems (UASs) the future of wildlife monitoring? A review of accomplishments and challenges. Mammal Rev. 45 (4), 239–252.

Lind, L., Eckstein, R.L., Relyea, R.A., 2022. Direct and indirect effects of climate change on distribution and community composition of macrophytes in lentic systems. Biol. Rev. 97 (4), 1677–1690.

Ma, Q., Zhang, H., Bai, Z., Zhao, H., 2021. Spectral characteristics and phenological characteristics of the aquatic vegetation in different regions of Poyang Lake from 2000 to 2020. Remote Sens. (Basel) 13 (10), 1906.

Madzivanzira, T.C., Coetzee, J.A., Dalu, T., 2023a. Factors structuring aquatic macrophytes. In: Kumar, S., et al. (Eds.), Aquatic Macrophytes: Ecology, Functions and Services. Elsevier/Springer Nature, Cambridge/Singapore.

Madzivanzira, T.C., Mungenge, C.P., Dube, T., Dalu, T., 2023b. From benthic to floating: phytoplankton dynamics in African freshwater lakes and reservoirs. In: El-Sheekh, M., Elsaied, H. (Eds.), Lakes of Africa: Microbial Diversity and Sustainability. Elsevier, Cambridge.

Malthus, T.J., 2017. Bio-optical modeling and remote sensing of aquatic macrophytes. In: Mishra, D.R., Ogashawara, I., Gitelson, A.A. (Eds.), Bio-optical Modeling and Remote Sensing of Inland Waters. Elsevier, Amsterdam, The Netherlands.

Marufu, L.T., Dalu, T., Crispen, P., Barson, M., Simango, R., Utete, B., Nhiwatiwa, T., 2018. The diet of an invasive crayfish, *Cherax quadricarinatus* (Von Martens, 1868), in Lake Kariba, inferred using stomach content and stable isotope analyses. Bioinvasions Rec. 7 (2), 121–132.

Mustapha, M.K., 2009. Problems, challenges and management of small, shallow tropical African reservoirs: a case study of Oyun Reservoir, Offa, Nigeria. Int. J. Lakes Rivers 2 (1), 163–174.

Ndebele-Murisa, M.R., Musil, C.F., Raitt, L., 2010. A review of phytoplankton dynamics in tropical African lakes. S. Afr. J. Sci. 106 (1), 13–18.

Nelson, S.A., Cheruvelil, K.S., Soranno, P.A., 2006. Satellite remote sensing of freshwater macrophytes and the influence of water clarity. Aquat. Bot. 85 (4), 289–298.

Olden, J.D., Miler, O., Bijaye, A., 2022. Lake-wide mapping of littoral habitat using underwater videography. Knowl. Manage. Aquat. Ecosyst. 423, 18.

Palmer, S.C., Kutser, T., Hunter, P.D., 2015. Remote sensing of inland waters: challenges, progress and future directions. Remote Sens. Environ. 157, 1–8.

Peng, D., Song, X., Fan, X., 2022. Effects of climate change on the growth of submerged macrophytes in Lake Taihu, China. Water 14 (7), 2032.

Peterson, K.T., Sagan, V., Sloan, J.J., 2020. Deep learning-based water quality estimation and anomaly detection using Landsat-8/Sentinel-2 virtual constellation and cloud computing. GISci. Remote Sens. 57 (4), 510–525.

Pettorelli, N., Schulte to Bühne, H., Tulloch, A., Dubois, G., Macinnis-Ng, C., Queirós, A.M., Keith, D.A., Wegmann, M., Schrodt, F., Stellmes, M., Sonnenschein, R., 2018. Satellite remote sensing of ecosystem functions: opportunities, challenges and way forward. Remote Sens. Ecol. Conserv. 4 (2), 71–93.

Piedade, M.T.F., Parolin, P., Junk, W., Schöngart, J., Wittmann, F., Demarchi, L.O., et al., 2022. Vegetation. In: Dalu, T., Wasserman, R.J. (Eds.), Fundamentals of Tropical Freshwater Wetlands: From Ecology to Conservation Management. Elsevier, Cambridge.

Pip, E., 1989. Water temperature and freshwater macrophyte distribution. Aquat. Bot. 34 (4), 367–373.

Prajapati, M., van Bruggen, J.J., Dalu, T., Malla, R., 2017. Assessing the effectiveness of pollutant removal by macrophytes in a floating wetland for wastewater treatment. Appl. Water Sci. 7 (8), 4801–4809.

Proctor, C., He, Y., 2013. Estimation of foliar pigment concentration in floating macrophytes using hyperspectral vegetation indices. Int. J. Remote Sens. 34 (22), 8011–8027.

Reitsema, R.E., Meire, P., Schoelynck, J., 2018. The future of freshwater macrophytes in a changing world: dissolved organic carbon quantity and quality and its interactions with macrophytes. Front. Plant Sci. 9, 629.

Rhee, D.S., Kim, Y.D., Kang, B., Kim, D., 2018. Applications of unmanned aerial vehicles in fluvial remote sensing: an overview of recent achievements. KSCE J. Civ. Eng. 22, 588–602.

Riis, T., Biggs, B.J., Flanagan, M., 2004. Colonisation and temporal dynamics of macrophytes in artificial stream channels with contrasting flow regimes. Arch. Hydrobiol. 159 (1), 77–96.

Riis, T., Levi, P.S., Baattrup-Pedersen, A., Jeppesen, K.G., Rosenhøj Leth, S., 2017. Experimental drought changes ecosystem structure and function in a macrophyte-rich stream. Aquat. Sci. 79, 841–853.

Rowan, G.S., Kalacska, M., 2021. A review of remote sensing of submerged aquatic vegetation for non-specialists. Remote Sens. (Basel) 13 (4), 623.

Saha, M., Barboza, F.R., Somerfield, P.J., Al-Janabi, B., Beck, M., Brakel, J., Ito, M., Pansch, C., Nascimento-Schulze, J.C., Jakobsson Thor, S., Weinberger, F., Sawall, Y., 2020. Response of foundation macrophytes to near-natural simulated marine heatwaves. Glob. Chang. Biol. 26 (2), 417–430.

Sardans, J., Bartrons, M., Margalef, O., Gargallo-Garriga, A., Janssens, I.A., Peñuelas, J., 2021. Nonlinear trade-offs between temperature, phosphorus and nitrogen availability in natural ecosystems. Glob. Ecol. Biogeogr. 30 (2), 442–458.

Scott, W.A., Adamson, J.K., Rollinson, J., Parr, T.W., 2002. Monitoring of aquatic macrophytes for detection of long-term change in river systems. Environ. Monit. Assess. 73, 131–153.

Sculthorpe, C.D., 1967. The Biology of Aquatic Vascular Plants. Edward Arnold Ltd., UK, London.

Silva, T.S., Costa, M.P., Melack, J.M., Novo, E.M., 2008. Remote sensing of aquatic vegetation: theory and applications. Environ. Monit. Assess. 140, 131–145.

Silva, T.S.F., Costa, M.P., Melack, J.M., 2010. Spatial and temporal variability of macrophyte cover and productivity in the eastern Amazon floodplain: a remote sensing approach. Remote Sens. Environ. 114 (9), 1998–2010.

Smith, H.D., Dubeux, J.C., Zare, A., Wilson, C.H., 2023. Assessing transferability of remote sensing pasture estimates using multiple machine learning algorithms and evaluation structures. Remote Sens. 15 (11), 2940.

Stocks, J.R., Rodgers, M.P., Pera, J.B., Gilligan, D.M., 2019. Monitoring aquatic plants: an evaluation of hydroacoustic, on-site digitising and airborne remote sensing techniques. Knowl. Manage. Aquat. Ecosyst. 420, 27.

Stratoulias, D., 2015. Assessment of Physiological Status and Spatial Distribution of Emergent Macrophytes Based on Imaging Spectroscopy (Doctoral dissertation). University of Leicester.

Sun, Y., Wen, J., Gu, L., Joiner, J., Chang, C.Y., van Der Tol, C., Porcar-Castell, A., Magney, T., Wang, L., Hu, L., Rascher, U., 2023. From remotely-sensed solar-induced chlorophyll fluorescence to ecosystem structure, function, and service: Part II—Harnessing data. Glob. Change Biol. 29 (11), 2893–2925.

Thamaga, K.H., Dube, T., 2018. Remote sensing of invasive water hyacinth (*Eichhornia crassipes*): a review on applications and challenges. Remote Sens. Appl.: Soc. Environ. 10, 36–46.

Thomas, S., Cecchi, P., Corbin, D., Lemoalle, J., 2000. The different primary producers in a small African tropical reservoir during a drought: temporal changes and interactions. Freshw. Biol. 45 (1), 43–56.

Thomaz, S.M., Mormul, R.P., Michelan, T.S., 2015. Propagule pressure, invasibility of freshwater ecosystems by macrophytes and their ecological impacts: a review of tropical freshwater ecosystems. Hydrobiologia 746, 39–59.

Thornton, P.K., Ericksen, P.J., Herrero, M., Challinor, A.J., 2014. Climate variability and vulnerability to climate change: a review. Glob. Chang. Biol. 20 (11), 3313–3328.

Tong, X., Duan, L., Liu, T., Yang, Z., Wang, Y., Singh, V.P., 2023. Estimation of grassland aboveground biomass combining optimal derivative and raw reflectance vegetation indices at peak productive growth stage. Geocarto Int. 38 (1), 2186497.

Topp, S.N., Pavelsky, T.M., Jensen, D., Simard, M., Ross, M.R., 2020. Research trends in the use of remote sensing for inland water quality science: moving towards multidisciplinary applications. Water 12 (1), 169.

Van Gerven, L.P., de Klein, J.J., Gerla, D.J., Kooi, B.W., Kuiper, J.J., Mooij, W.M., 2015. Competition for light and nutrients in layered communities of aquatic plants. Am. Nat. 186 (1), 72–83.

Villa, P., Bresciani, M., Bolpagni, R., Pinardi, M., Giardino, C., 2015. A rule-based approach for mapping macrophyte communities using multi-temporal aquatic vegetation indices. Remote Sens. Environ. 171, 218–233.

Villa, P., Pinardi, M., Toth, V., Hunter, P., Bolpagni, R., Bresciani, M., 2017. Remote sensing of macrophyte morphological traits: implications for the management of shallow lakes. Limnology 76 (S1), 109–126.

Visser, F., Wallis, C., Sinnott, A.M., 2013. Optical remote sensing of submerged aquatic vegetation: opportunities for shallow clearwater streams. Limnologica 43, 388–398.

Wagner, T., Falter, C.M., 2002. Response of an aquatic macrophyte community to fluctuating water levels in an oligotrophic lake. Lake Reserv. Manage. 18 (1), 52–65.

Wang, X., Ding, Y., Jiang, Y., 2022. Remote sensing-based assessment of macrophyte variability in a large eutrophic lake: implications for lake management. Water 14 (6), 1564.

Wang, Z., Menenti, M., 2021. Challenges and opportunities in Lidar remote sensing. Front. Remote Sens. 2, 641723.

Wang, F., Zhu, X., Cheng, X., Zhang, Y., Li, Y., 2024. Mmkdgat: multi-modal knowledge graph-aware deep graph attention network for remote sensing image recommendation. Expert Syst. Appl. 235, 121278.

White, M.A., Nemani, R.R., 2003. Canopy duration has little influence on annual carbon storage in the deciduous broad leaf forest. Glob. Chang. Biol. 9 (7), 967–972.

Williams, D.D., 2006. The Biology of Temporary Waters. Oxford University Press, USA.

Wingfield, R., Murphy, K., Gaywood, M., 2005. Lake habitat suitability for the rare European macrophyte *Najas flexilis* (Willd.). Rostk. and Schmidt. Aquat. Conserv.: Mar. Freshw. Ecosyst. 15 (3), 227–241.

Wolf, P., Rößler, S., Schneider, T., Melzer, A., 2013. Collecting in situ remote sensing reflectances of submersed macrophytes to build up a spectral library for lake monitoring. Eur. J. Remote Sens. 46 (1), 401–416.

Wolkovich, E.M., Cleland, E.E., 2011. The phenology of plant invasions: a community ecology perspective. Front. Ecol. Environ. 9 (5), 287–294.

Woodward, G., Bonada, N., Brown, L.E., Death, R.G., Durance, I., Gray, C., Hladyz, S., Ledger, M. E., Milner, A.M., Pawar, S., 2016. The effects of climatic fluctuations and extreme events on running water ecosystems. Philos. Trans. R. Soc. B: Biol. Sci. 371 (1694), 20150274.

Wu, X., Xiao, Q., Wen, J., You, D., Hueni, A., 2019. Advances in quantitative remote sensing product validation: overview and current status. Earth-Sci. Rev. 196, 102875.

Wu, Q., Xie, M., Jin, L., Dong, H., Yuan, D., Yang, T., Pan, Y., 2023. Water exchange unevenness alters the species dominance and community composition of submerged macrophytes in Erhai Lake and the potential mechanisms revealed by laboratory experiment. Sci. Total Environ. 882, 163624.

Xiao, H., Wu, J., Li, Q., Xiang, X., 2019. Remote sensing-based biomass estimation of aquatic vegetation: influencing factors and accuracy assessment. Sci. Total Environ. 651, 2969–2979.

Zhao, D., Jiang, H., Yang, T., Cai, Y., Xu, D., An, S., 2012. Remote sensing of aquatic vegetation distribution in Taihu Lake using an improved classification tree with modified thresholds. J. Environ. Manage. 95 (1), 98–107.

Zhu, X., Zhang, M., Zhao, S., Liu, J., 2023. Lake water quality change monitoring and assessment of eco-environmental response to land use and climate change in the Lake Taihu Basin, China. Environ. Manage. 71 (2), 348–362.

Chapter 16

Climate variability and urban environments

Terence Darlington Mushore[a,b], Tapiwa Michelle Masawi[a], Charity Gororo[a], Teddious Mhizha[a], Gilberta Luba Thwala[c], Emmanuel Mashonjowa[a], Onisimo Mutanga[b], and John Odindi[b]

[a]*Department of Space Science and Applied Physics, Faculty of Science, University of Zimbabwe, Harare, Zimbabwe,* [b]*Discipline of Geography, School of Agricultural, Earth and Environmental Sciences, University of KwaZulu-Natal, Pietermaritzburg, South Africa,* [c]*Mining and Mineral Processing Engineering, Faculty of Engineering, Manicaland State University of Applied Sciences, Mutare, Zimbabwe*

Chapter outline

1 Introduction

Climate variability is the term used to depict the recurrent varieties within the Earth's climate that happen all through time and are caused by a variety of factors, such as varieties in sun-oriented radiation, volcanic movement, and Earth's circle. In recent decades, human activities such as the burning of fossil fuels and changes in land utilization have altogether expanded the barometrical concentrations of nursery gasses, rising worldwide temperatures and animating the rate of climate variability (Allen et al., 2018). This variability incorporates events such as El Nino/La Nina, dry seasons, and varieties in temperature and precipitation patterns over time. It too ranges between different time and spatial

Remote Sensing of Climate. https://doi.org/10.1016/B978-0-443-21731-9.00013-2
Copyright © 2024 Elsevier Inc. All rights are reserved, including those for text and data mining, AI training, and similar technologies.

scales. Urban ranges, moreover, impact climate variability due to several reasons, such as changes in land use, expanded vitality use, and emanations from transportation.

Due to the removal of vegetation and other penetrable surfaces in favor of impermeable substances like concrete and blacktop, which assimilate and hold more heat, urbanization regularly comes about in an increment in surface temperature. The impacts of climate variability can be exacerbated by the urban heat island (UHI) effect, coming about in frequent and strong heat waves that can be especially perilous for populations that are at risk, such as the elderly and individuals with preexisting restorative issues. Due to their high population densities, the concentration of buildings, and dependence on complex socioecological frameworks, urban regions are especially vulnerable to the impacts of climatic variability. The well-being and economic activities of urban individuals can be altogether affected by changes in temperature, precipitation, and extraordinary climate occasions (Seto et al., 2014). A valuable approach for analyzing how climate inconstancy influences urban zones is through the application of remote sensing. Analysts can identify changes in temperature, precipitation, and other climatic factors over time in urban locales and assess the impacts of these changes on territorial biological systems and social frameworks by assessing using data from space-, air-, and ground-based sensors (Weng, 2009).

It is important to comprehend how climatic variability and urban settings are connected to formulate strategies that will successfully assist in reducing the impacts of climate change and preparing for projected changes. Remote sensing and other procedures for observing and surveying the implications of urbanization on temperature flow can give imperative experiences into these intuitive, empowering urban planning and administration procedures to create more flexible cities and protect vulnerable populations from possible adverse extremes.

2 Remote sensing of urbanization and temperature dynamics

The dynamics of temperature in metropolitan regions, where the UHI effect is a common occurrence, have been widely studied using remote sensing. The UHI effect is the term used to describe the elevated temperatures seen in urban regions when compared to the nearby rural areas. This is because buildings and other urban infrastructure absorb and re-emit heat while vegetation, water, and natural environments mostly keep heat away. Due to the removal of vegetation and other penetrable surfaces in favor of impermeable substances like concrete and black-top, which assimilate and hold warmer, urbanization frequently comes about in an increment in surface temperature. With the assistance of remote sensing, it is conceivable to track changes in temperature patterns, land use, and land cover over time. Thermal infrared and optical imageries are broadly applied remote sensing data for inquiring about urbanization and temperature changes. Applying remote sensing technologies, it is possible to determine how much heat radiation the Earth's surface transmits, and how that relates to surface temperature. Changes in urban temperature patterns can be quantified by studying thermal infrared imageries over time.

Analysts can recognize patterns in land cover and land use, such as the development of urban ranges and the vanishing of vegetation, by looking at multispectral imageries. For instance, researchers can apply thermal infrared imageries to find hotspots, while using multispectral optical imageries to find the sorts of land cover that are responsible for those temperatures, like impermeable surfaces or green locales. Progresses in thermal infrared and multispectral remote sensing and analysis techniques have made it possible to identify both hot and cold surfaces; applying diverse resolutions of surface temperature conveyances, temperature maps have ended up more broadly accessible. Generally, remote sensing offers a powerful device for exploring the complicated connections between urbanization and temperature patterns.

In places with rapid urbanization and sparse monitoring networks, important insights into the spatial pattern of thermal environmental modifications can be obtained from a variety of remote sensing data. Examples include measurements made with the Moderate Resolution Imaging Spectroradiometer (MODIS), which provide observations of the land surface temperature with a spatial resolution of 1000m, or Landsat data with more accurate 120m/100m resolutions, with increasing datasets becoming available with the launch of new satellites and improved sensors (Li et al., 2022). With intricate interactions among biophysical factors (such as soil qualities, vapor pressure, humidity percentage, and breeze circulation), urban thermal environment dynamics and consequences can vary dramatically. In addition, urban environments change albedo and nocturnal radiation, and urban mobility increases the amount of greenhouse gases released into the atmosphere, which is likely to raise the local temperature (Li et al., 2022). In several studies, remote sensing has been applied to examine the geographical and temporal dynamics of UHI and its effects on urban settings. According to a study that employed satellite data (Imhoff et al., 2010), the UHI effect varied depending on the size and morphology of the urban region and was the greatest in the summer in the United States. Another study (Zhang et al., 2015) applied remote sensing to illustrate how the UHI impact in Shanghai was affected by components like land use, population density, and surface materials. Investigations on the impacts of UHI on human well-being and vitality have, moreover, used remote sensing. In specific, the impact of UHI assessed the impact of UHI on health in Phoenix, Arizona, using remote sensing and ground-based information, and found that the UHI impact on health increased by up to 5% between 2007 and 2017 (Ge et al., 2017). The UHI was also associated with an 8.7% increase in heat-related mortality in the city of Wuhan, China (Liu et al., 2019). Evidently, historical records of remotely sensed data are useful in understanding the interaction between urban surface properties and temperature dynamics at a variety of spatial and temporal scales.

Remote sensing offers a strong apparatus for exploring the perplexing connections between temperature elements and urbanization, which can help direct urban planning and administration approaches to reduce the impacts of urbanization on nearby climatic conditions. Remote sensing offers a variety of data for examining urban buildings, land cover, and biophysical aspects in addition to meteorological measurements. Understanding the urban thermal environment requires a thorough grasp of these urban characteristics, particularly, the connections between land use and land cover, vegetation indices, and temperatures (Li et al., 2022). Observed patterns underline the need for proactive urban planning and management methods to reduce the effects of UHI on human health and energy consumption. Remote sensing can research the dynamics of temperature in urban settings.

3 Remote sensing of wind patterns in urban areas

Urban wind patterns are crucial for the dispersion of air pollutants and the distribution of heat and moisture. Researchers can track changes in wind patterns in urban areas over time and investigate the causes of these changes, such as variations in land use and urban shape, by analyzing satellite- and ground-based data. This is significant because wind has a significant influence on temperature, air quality, and other environmental parameters in urban microclimates. Building height and direction, urban shape, and the existence of wind forecast verification of green spaces are only a few of the variables that affect wind patterns. LIDAR (Light Detection and Ranging) is remote sensing that can be used to evaluate wind patterns in urban areas. LIDAR creates accurate 3D representations of urban areas by using laser pulses to measure the location and distance of objects (Filioglou et al., 2022). Due to their ability to offer the high-precision horizontal region spatial variability and the vertical contour of the horizontal breeze remote sensing systems like Doppler lidars (Light Detection and Ranging) present a particularly advantageous opportunity for wind forecast verification. By contrasting the modeled winds with Doppler lidar observations under various atmospheric conditions and from season to season, in the coastal environment of Helsinki, Finland, the performance of numerical weather prediction models, analysis systems, and large-eddy simulation models were evaluated (Filioglou et al., 2022). Researchers can locate regions with strong wind and turbulence, as well as regions with decreased wind due to the presence of structures or other obstructions, by examining LIDAR data. In meteorological applications, the spatial distribution of clouds is used to identify pressure systems and consequently wind directions and speeds. High pressure areas are associated with descending winds and suppressed cloud development while low pressure and strong winds can coincide with areas of vertical cloud development all observable on satellite images.

Chen et al. (2015) used satellite data to demonstrate that the effects of buildings and other urban infrastructure caused wind speeds in urban regions to be lower than those in nearby rural areas. Yang et al. (2018) applied remote sensing to illustrate how changes in Singapore's land utilization and urban morphology have influenced wind patterns, which had repercussions for the scattering of discussed contamination. The impacts of wind designs on urban settings, such as the scattering of pollutants and the dispersion of heat and moisture, have also been examined applying remote sensing. One research, for instance, used remote sensing to foresee the dispersion of heat and moisture in Phoenix, Arizona's urban locales and found that wind patterns essentially affected these conveyances (Li et al., 2018). Another study by Jain et al. (2019) found that wind patterns influenced the spatial dissemination of contaminants and applied remote sensing to evaluate the scattering of discuss poisons in Delhi, India. Remote sensing of barometrical movement vectors is another strategy for looking at wind patterns in metropolitan situations. To decide the wind speed and course within the lower climate, this strategy tracks the development of clouds through time using satellite pictures. This data can, at that point, be utilized to advise urban settlement and administration methods. Despite the reality that wind speeds tend to rise with elevation from the Earth's surface, they can, in some cases, drop at more prominent levels because of winds being redirected over tall structures and retarded by urban surfaces.

In general, remote sensing offers a strong approach for exploring the perplexing connections between wind patterns and urban settings, which can direct administration and arranging activities for more feasible and bearable cities. Analysts can get a more careful understanding of the impacts of wind patterns on urban microclimates by combining remote sensing methods with other sorts of information such as meteorological information and urban climate models. They can, at that point, create procedures to moderate the impacts of wind-related dangers and improve the, in general, quality of the urban environment. These understandings highlight the need for proactive urban arrangement and administration strategies to diminish the impacts of wind patterns on quality and human well-being, as well as the potential of remote sensing for examining wind patterns in urban settings.

4 Remote sensing of the influence of urban areas on the distribution and structure of the clouds

Remote sensing can be used to track changes in cloud properties in urban districts over time and explore the causes of these changes, such as modifications in land utilization and urban shape. The dispersion and structure of clouds can be essentially affected by urban centers, which can at that point affect territorial climate characteristics and precipitation. Satellites with optical and infrared ray sensors can offer an around-the-world viewpoint of cloud properties and their relationships to cities. Remote sensing analysis can reveal locales with overwhelming cloud cover and screen changes in cloud properties over time. Moreover, they can use obsequious information to explore how cities influence cloud arrangement, particularly, how urban warm islands influence territorial air steadiness. LIDAR is another remote sensing technology applied to investigate how urban zones influence the dispersion and structure of clouds. This helps to decide where clouds are most likely to develop and what effect cities have in affecting cloud structure by looking at LIDAR information. A more comprehensive picture of how metropolitan zones influence cloud shape and scattering can be obtained by combining these remote sensing strategies with ground-based air perceptions.

Wang et al. (2018) applied remote sensing to illustrate how changes in land use and urban morphology in Houston, Texas have influenced changes in cloud qualities, counting changes in cloud recurrence and length. Clouds were found to have a cooling impact on metropolitan areas in the city of Shanghai (Zhang et al., 2017). In Delhi, India, remote sensing analysis revealed that clouds contributed to lower the concentration of toxins in the atmosphere (Saha et al., 2020).

5 Remote sensing of interaction between urbanization and rainfall patterns

Urbanization can impact the hydrological cycle, which can influence how much water is accessible and how frequently it downpours. Water administration, agribusiness, and other businesses that depend on tried and true water sources may be altogether affected by this. Radar, optical, and thermal imageries can be used to inquire about how urbanization and precipitation patterns are associated. Changes in precipitation patterns in metropolitan districts can be tracked through time to explore the causes of these changes. Estimations of precipitation's intensity and scattering as well as changes in patterns over time can be done using radars. Optical remote sensing is useful to investigate how urbanization and precipitation patterns are associated.

Various studies have illustrated that urbanization can essentially change nearby precipitation patterns, affecting the intensity, recurrence, and duration of precipitation. For instance, urban districts within the US saw more frequent and intense rainstorm occasions than the adjacent provincial zones (Shepherd et al., 2002). Li et al. (2017) illustrated how urbanization in China diminished precipitation in metropolitan zones and expanded the chance of dry spells using remote sensing. Regions where urbanization has changed neighborhood hydrological patterns, such as surface runoff, penetration, and evapotranspiration, can be located by analyzing remote sensing information. These variations may have a critical effect on the recurrence and duration of precipitation in a certain locale, which has repercussions for water administration and other businesses.

The linkage between urbanization and precipitation patterns can be better captured by combining remote sensing methods with hydrological models and other datasets. This may offer assistance in direct urban settlement and administration activities to reduce how urbanization influences territorial hydrological properties, improve water administration, and energize more naturally inviting urban development. By and large, remote sensing offers a strong apparatus for exploring the complex connections between urbanization and precipitation patterns. Kumar et al. (2017) applied remote sensing to illustrate how urbanization in India resulted in a reduction in surface water accessibility and an increment in reliance on groundwater assets.

6 Projections of implications of urban growth on climate variability and change

The average temperature of the entire planet has risen by around 1°C (1.7°F) since 1880. By 2050, it is expected that the world's temperature will have increased by around 1.5°C and 2–4°C by 2100. Given that cities can modify local climatic patterns through changes in land use, energy consumption, and transportation systems, urban expansion has substantial consequences for climate variability and change. Chen et al. (2017) projected the effects of urbanization on China's temperature and precipitation patterns using a global climate model. They found that urban areas would experience greater warming and more intense precipitation events than rural areas. The UHI effect in the United States would rise by up to 4°C by the end of the 21st century, with the southeastern and western regions experiencing the most increases (Shahmohamadi et al., 2010). By 2030, it is anticipated that roughly 50% of Africans will live in urban areas (Cartwright, 2015) and this will affect temperatures, winds, and precipitation. Generally, predictions indicate that, in the future decades, urbanization will have major effects on climate variability and change, highlighting the need for proactive urban planning and management techniques to lessen these effects and encourage more resilient and sustainable urban environments.

7 Challenges to the application of remote sensing to detect climate variability in urban areas

Although remote sensing can recognize climate variability in urban settings, its application comes with several troubles. These challenges comprise innovative and methodological limitations, as well as issues with information accessibility and quality. A few of these challenges include the potential spatial resolution limitations of remote sensing information, which can make it challenging to capture the fine-scale variability of climatic factors in metropolitan situations. Urban districts, where microclimates can vary significantly over small distances, require high spatial resolution datasets for analysis. High spatial and temporal resolution do not coexist on a given sensor, which limits spatiotemporal analysis, especially for rapidly changing phenomena such as wind patterns. Due to this, it may be challenging to determine detailed climatic variations over time in metropolitan settings. Moreover, cloud cover increases the gap between available data, which causes additional limitations to the value of numerous remote sensing datasets, especially in surface analysis. Furthermore, certain information (especially high resolution) might not be freely accessible. Even though remote sensing information can abdicate plenty of data, it can be difficult to comprehend the discoveries, particularly for nonexperts. To be able to supply more information on the climate framework, remote sensing information may have to be combined with other forms of information, such as ground-based observations since it is not a direct measurement approach.

The determination of remote sensing information might not be sufficient to identify microclimates created by particular structures such as road canyons and building sheds in urban locales. The elucidation of remote sensing information may moreover give challenges, especially when looking to distinguish between normal and artificial causes affecting climate variability. However, the application of remote sensing to understand climatic variability in urban situations has progressed altogether despite these impediments.

8 Conclusion

Climatic variability creates significant issues for urban ranges, which are domestic to sizable and extending populaces that are subjected to extraordinary climate events and other impacts of climate change, with potential impacts on human well-being and financial movement. Urbanization itself contributes to climate variability through an assortment of factors including changes in land use and land cover as well as anthropogenic pollution such as industrial and vehicular emissions. In this manner, it is pivotal to comprehend the complex connections between climate variability and urban settings to make effective plans for lessening and adapting to associated climate change impacts. In an arrangement to understand the impacts of urbanization on climate change, remote sensing offers powerful data for detailed multitemporal and multiscale analysis. This can aid urban planning and administration techniques to make climate-informed and resilient cities and protect vulnerable populations. Analysts can pinpoint locales where climate inconstancy is most likely to have an impact and focus on mediations to reduce those impacts and empower more versatile urban arrangements and management by examining remote sensing information. Given climatic changes and impacts, it is important to apply a variety of strategies to illuminate the issues posed by climatic variability and urbanization, including cutting greenhouse gas emissions and improving energy use efficiency in industries and transportation frameworks.

References

Allen, M., Dube, O.P., Solecki, W., Aragón-Durand, F., Cramer, W., Humphreys, S., et al., 2018. Global warming of 1.5°C. An IPCC Special Report on the impacts of global warming of 1.5°C above pre-industrial levels and related global greenhouse gas emission pathways, in the context of strengthening the global response to the threat of climate change, sustainable development, and efforts to eradicate poverty. IPCC.

Cartwright, A, 2015. Better growth, better cities: Rethinking and redirecting urbanisation in Africa. The New Climate Economy Working Paper.

Chen, F., et al., 2015. Distinct urban and rural CO_2, CH_4 and sensible heat fluxes in northern China. Sci. Rep. 5, 14601.

Chen, L., et al., 2017. Urbanization-induced surface warming in eastern China. Proc. Natl. Acad. Sci. U. S. A. 114 (19), 9597–9602.

Filioglou, M., Preissler, J., Troiville, A., Thobois, L., Vakkari, V., Auvinen, M., Fortelius, C., Gregow, E., Hämäläinen, K., Hellsten, A., Järvi, L., O'Connor, E., Schönach, D., Hirsikko, A., 2022. Evaluating modelled winds over an urban area using ground-based Doppler lidar observations. Meteorol. Appl. 29 (2), e2052. https://doi.org/10.1002/met.2052.

Ge, C., et al., 2017. Estimating the impact of the urban heat island effect on energy consumption in Phoenix, Arizona using remote sensing and demographic data. Appl. Energy 185, 1569–1578.

Imhoff, M.L., et al., 2010. Remote sensing of the urban heat island effect across biomes in the continental USA. Remote Sens. Environ. 114 (3), 504–513.

Jain, A., et al., 2019. Remote sensing of particulate matter (PM2.5) in Delhi using MODIS data. Remote Sens. Appl.: Soc. Environ. 13, 149–159.

Kumar, S., et al., 2017. Urbanization and the groundwater regime: a remote sensing-based assessment in India. J. Hydrol. 550, 468–478.

Li, X., et al., 2017. Quantifying the effects of urbanization on changes in surface runoff in Beijing using remote sensing and GIS. J. Hydrol. 546, 1–10.

Li, X., et al., 2018. Remote sensing of land surface temperature and evapotranspiration in urban areas of Phoenix, Arizona: impacts of urban morphology and land use. J. Geophys. Res.: Atmos. 123 (15), 7989–8003.

Li, X., Stringer, L.C., Dallimer, M., 2022. The impacts of urbanisation and climate change on the urban thermal environment in Africa. Climate 10 (11), 164. https://doi.org/10.3390/cli10110164.

Liu, X., et al., 2019. Estimating the mortality burden attributable to the urban heat island effect during extreme heat events: a case study of Wuhan, China. Environ. Health Perspect. 127 (5), 57006.

Saha, A., et al., 2020. Role of clouds in modulating air quality in an urban environment: a remote sensing and modeling study of Delhi, India. Atmos. Environ. 224, 117316.

Seto, K.C., et al., 2014. Human settlements, infrastructure and spatial planning. In: Climate Change 2014: Impacts, Adaptation, and Vulnerability. Part A: Global and Sectoral Aspects. Contribution of Working Group II to the Fifth Assessment Report of the Intergovernmental Panel on Climate Change, Cambridge University Press, pp. 923–975.

Shahmohamadi, P., Che-Ani, A.I., Ramly, A., Maulud, K.N.A., Mohd-Nor, M.F.I., 2010. Reducing urban heat island effects: a systematic review to achieve energy consumption balance. Int. J. Phys. Sci. 5 (6), 626–636.

Shepherd, J.M., et al., 2002. Detection of rainfall trends and variability in downtown Atlanta, Georgia. Int. J. Climatol.: J. R. Meteorol. Soc. 22 (13), 1643–1662.

Wang, D., et al., 2018. Changes in cloud properties over Houston, Texas, as a function of land use and urban morphology. J. Geophys. Res.: Atmos. 123 (18), 10,605–10,620.

Weng, Q., 2009. Remote sensing of impervious surfaces in the urban areas: requirements, methods, and trends. Remote Sens. Environ. 113 (Suppl. 1), S56–S70.

Yang, W., et al., 2018. Urbanization and its impacts on local wind patterns in Singapore. J. Geophys. Res.: Atmos. 123 (5), 2717–2733.

Zhang, X., et al., 2015. Spatial and temporal variations of the urban heat island in Shanghai, China since the 1990s. Int. J. Climatol. 35 (6), 1319–1333.

Zhang, X., et al., 2017. The impact of clouds on surface radiation and temperature in Shanghai, China. J. Geophys. Res.: Atmos. 122 (3), 1761–1778.

Chapter 17

Climate management and policy development: An earth observation perspective

Timothy Dube[a] and Oshneck Mupepi[b]

[a]*Department of Earth Science, Institute for Water Studies, University of the Western Cape, Cape Town, South Africa,* [b]*Department of Geography, Environmental Sustainability and Resilience Building, Midlands State University, Gweru, Zimbabwe*

Chapter outline

Remote Sensing of Climate. https://doi.org/10.1016/B978-0-443-21731-9.00016-8
Copyright © 2024 Elsevier Inc. All rights are reserved, including those for text and data mining, AI training, and similar technologies.

1 Introduction to remote sensing in climate and policy development

1.1 Background and context

Remote sensing has become an important science in policy and decision making in all aspects of management across the globe. Its capacity to provide a synoptic view of the global panorama made it an option when decision making is being done especially for management and monitoring of the impacts of climate change on physical aspects of the earth over large geographical areas. Remote sensing has revolutionized the understanding of our dynamic environment. When aerial photography, which had been available since the early 20th century, was supplemented by digital satellite-borne imagery in the 1970s, the range of remotely sensed earth observation (EO) techniques grew (De Leeuw et al., 2010). Since many environmental remote sensing systems are publicly sponsored and the number of nations investing in EO infrastructure is rising, the number of EO systems designed to observe and monitor environmental processes is increasing, and governments are strongly promoting this science. Support for EO among the general population is largely fueled by environmental concerns. The reverse is true as well, since EO stimulates environmental awareness and concern.

1.2 Scope and objectives of the chapter

The chapter discusses the applications of remote sensing in climate analysis and policy development. It gives a background to remote sensing techniques and how they are used when assessing climatic variables for easy management. Remote sensing data used in climate management and policy development are considered, and the sources of this data and its acquisition are discussed to inform policy development and climate analysis from an EO perspective. Since climate management involves an understanding of natural resources that act as climate control factors, the application of remote sensing in monitoring these resources is outlined in this chapter. Applications of remote sensing in disaster management and monitoring of climate policies are also covered in this chapter to provide an understanding of how disasters emanating from climate change are managed using remote sensing data and how management policies are monitored. Challenges associated with the use of remote sensing data are also analyzed whilst possible solutions are discussed to ensure improved exploitation of remote sensing, data utilization opportunities in climate management and policy development, as well as overcoming the shortcomings involved. Recommendations for future research and integration of remote sensing data in climate management and policy development are also outlined and discussed for improvements in this field with time. Therefore, the aims of this chapter are to (1) examine the application of remote sensing in climate analysis; (2) analyze the application of remote sensing data in policy development and monitoring; (3) outline challenges and opportunities associated with the application of remote sensing data in climate analysis and policy development; and (4) assessing future directions and emerging trends in the application of remote sensing in climate management and policy development.

1.3 Overview of remote sensing technology

Remote sensing is generally referred to as the science of collecting data using noncontact techniques (Lillisand et al., 2015; Mirzakarimova, 2023). This term used to be used to refer to satellite platforms acquiring data by optical and radar sensors (Campbell, 2002, Schowengerdt, 2007). The term "remote sensing" has most recently been used to refer to any technique for acquiring images and spatial data, including airborne surveying and photogrammetry (Mikhail et al., 2001), though the terms "satellite-based" or "airborne" remote sensing, as well as "mobile mapping," for terrestrial platforms, are still occasionally used. In addition to the main classification based on platforms, remote sensing technology is distinguished by the imaging sensor employed, such as a large-format aerial camera and by the application domain, such as close-range photogrammetry (Fraser et al., 2005; Martnez et al., 2013; Mirzakarimova, 2023). The objective of any remote sensing technology is to provide observation of some physical parameters in specific areas at a given period. The physical space is broadly defined and, from the shape of the Earth, includes man-made objects, vegetation, atmospheric parameters, and anything in between (Toth and Jóź ków, 2016). Remote sensing science identifies the earth's surface and atmospheric phenomena using electromagnetic radiation as a medium of interaction. Since there is a need to specify geographical precise space and precise time during which specific phenomena occur using propagated energy, sensors are characterized by spatial, temporal, and spectral resolutions. Spectral resolution is typically defined as the number of portions of the electromagnetic spectrum that are sensed by the remote sensing device. These portions are referred to as “bands.” A second factor that is important in spectral resolution is the width of the bands, which determines the degree to which phenomena can be distinguished. Spatial resolution is defined by the pixel size of the imagery. A pixel or picture element is the smallest two-dimensional area sensed by the remote sensing device. This defines the smallest objects that can be detected by a specific sensor, whilst temporal resolution is defined by how often a particular remote sensing device can image a particular area of interest (Congalton, 2010).

Remote Sensing (RS) refers to the science of identification of earth surface features and estimation of their geo-biophysical properties using electromagnetic radiation as a medium of interaction (Congalton, 2010; Campbell and Wynne, 2011; Di and Yu, 2023; Mirzakarimova, 2023). Spectral, spatial, temporal, and polarization signatures are major characteristics of the sensor/target, which facilitate target discrimination. Earth surface data, as seen by the sensors in different wavelengths (reflected, scattered, and/or emitted), is radiometrically and geometrically corrected before extraction of spectral information. Remote sensing data, with its ability for a synoptic view, and repetitive coverage with calibrated sensors to detect changes and observations at different resolutions, provides a better alternative for natural resources management as compared to traditional methods. Today a lot of active (e.g., RADAR) and passive (e.g., optical like Landsat, Aster, Sentinel 2) remote sensors are available to provide data about the earth (e.g., Landsat, SPOT, Aster, MODIS, and Sentinel), atmosphere, and oceans (e.g., NOAA).

1.4 Importance of remote sensing in climate and policy development

Remote sensing can monitor dynamic resources and parameters that can be used for climate prediction and analysis. Besides climate studies, decision makers require data about the earth’s resources that are distributed over large geographical areas. Traditional methods of monitoring resources using field surveys are tiresome and sometimes ineffective due to the dynamic nature of climate and earth resources. This made remote sensing an important tool to aid decision making especially in the field of climate management (Yang et al., 2013). Climate depends on earth’s resources and can also influence the behavior of these resources, which makes it necessary to use resources like sea surface temperature, sea level, carbon sinks in the form of forests, and changes in inland water resources as proxies when studying climate changes as well as managing its changes. When making decisions about the climate, they need to be transboundary, as climate impacts and drivers of change have no boundaries. This calls for the need to employ remote sensing with a synoptic view of large geographical areas at good temporal resolutions to monitor and manage climate change drivers like deforestation and greenhouse gas emissions. This creates baseline databases that can be used to monitor the effectiveness of international management efforts like the United Nations Convention on Climate Change and the United Nations Convention to Combat Desertification. These agreements provide for domestication of provisions agreed upon to manage global issues, hence remote sensing can be the best to collect required data at regional and global levels from time to time.

Gore (2006) described how a single image of the earth, taken in December 1968 by a crew member of the Apollo 8 mission, "exploded into the consciousness of mankind." During the same period of this picture being taken, the modern environmental movement was born. The Clean Air Act, the Clean Water Act, the Natural Environmental Policy Act, and the first Earth Day all came about within a few years of this picture being seen for the first time" (Gore, 2006). This indicates how remote sensing-based data availability enhanced awareness and guided policy development and environmental management legislation. Satellite-borne remote sensing and modern environmental policy thus came of age at the same time. The launch of the first Landsat satellite in early 1972 coincided with the development of the first environmental policies, and the subsequently established environmental protection agencies created the initial demand for environmental EO products (Johnston and Cordes, 2003). The demand for environmental EO, however, depends considerably on policy and market opportunities, and should thus be considered in the context of societal demand and the policies the EO serves. The countries represented in the Group on Earth Observation (GEO) recognized this dependency on societal demand while agreeing in February 2005 to develop a Global Earth Observation System of Systems (GEOSS) to develop societal benefits of EO (Christian, 2005; Lautenbacher, 2006). This increased the application of remote sensing in environmental management policies.

2 Remote sensing techniques for climate analysis

2.1 Remote sensing data sources and platforms

Remote sensing data is acquired by sensors onboard various satellite platforms like Landsat, MODIS, Sentinel, KOMPASAT, and RADARSAT among others. However, drones are also being used to collect high-quality earth surface data at very local scales (Simic Milas et al., 2018). The digital raster data received at ground stations from various satellites is processed and formatted to remove random and systematic distortions and to calibrate radiometrically and rectify geometrically (Jia, 2016). This is done by users or data suppliers before they are suitable for input into a Geographic Information System (GIS). Once acquired, these data are stored in data and information centers, which are available in several nations around the world. Both national (public sector) and commercial remotely sensed data can typically be accessed and acquired through these data centers directly or with the aid of an intermediary, such as a commercial vendor, a governmental agency, or an academic institution working in the area. Remote sensing data may be accessed in a variety of formats and a variety of forms, ranging from full scene to quarter scene to specific area coverage (Sajjad and Kumar, 2019). The data may also be acquired as raw system-corrected data or in a variety of processed formats, including geometrically and atmospherically corrected, and georeferenced data, registered to a given geospatial coordinate system. This data is accessible from online portals where it can be freely downloaded or paid for. Some of the main sources that provide remote sensing data include the United States Geological Survey (USGS) (https://earthexplorer.usgs.gov/), which provides a wide range of remote sensing products, including MODIS, Landsat, NOAH AVRR, ASTER, Radar data, digital elevation models (DEM), Sentinel among other data, and some commercial satellite imagery including IKONOS and OrbView3 and many other datasets. The European Space Agency's Copernicus Open Access Hub (https://scihub.copernicus.eu/) provides Sentinel data products, whilst the Earth Observation for Sustainable Development's Land Viewer (https://eos.com/products/landviewer/) provides options to download processed or raw data from a wide range of satellite sensors, including Landsat, Sentinel, MODIS, KOMPASAT, and SPOT among others. Other sources include NASA Earth Observation, NOAA Class, and Glovis among others.

2.2 Satellite sensors and products for climate analysis

Satellite sensors for climate analysis have been increasing and improving over time to provide reliable data about climate parameters like precipitation, temperature, wind, water vapor, and humidity, among others. The launching of passive microwave sensors in the 1980s made it easy to collect important information from within the clouds unlike only on the cloud top, as was the case before (Levizzani and Cattani, 2019). The launch of the Tropical Rainfall Measuring Mission (TRMM) (Kummerow et al., 1998) with the first Precipitation Radar (PR). That active sensor introduced more important information on precipitation formation within clouds. This PR demonstrated that cloud horizontal and vertical structure sensing is necessary to improve our ability to measure precipitation at the local and global levels through the use of passive and active microwave (MW) sensors. TRMM was followed by CloudSat in 2006 (Stephens et al., 2002) and by the Global Precipitation Measurement (GPM) mission in 2014, with the new Dual-frequency Precipitation Radar (DPR) (Hou et al., 2014). Most datasets are nowadays gauge-corrected to ensure the necessary calibration; the most used for climate studies is the Global Precipitation Climatology Project (GPCP) dataset (Adler et al., 2018), which is the longest satellite precipitation record (40 years). Global infrared (IR)-based datasets cover reasonably long time spans, such as the Climate Hazards Centre's Infrared Precipitation with Stations (CHIRPS) (Funk et al., 2015) and the Climate Prediction Centre (CPC) Merged Analysis of Precipitation (CMAP) (Xie and Arkin, 1997).

Among the merged products, it is worth noting the Precipitation Estimation from Remotely Sensed Information using Artificial Neural Networks (PERSIANN) (Nguyen et al., 2019) the CPC MORPHing technique (CMORPH) (Xie et al., 2017), and the Global Satellite Mapping of Precipitation (GSMaP) (Mega et al., 2018). Climate data records can be obtained from the National Oceanic and Atmospheric Administration's (NOAA) Climate Data Record Program (CDRP) producing and preserving climate records from the National Oceanic and Atmospheric Administration's (NOAA) satellite sensor (Bates et al., 2016). This Climate Data Record (CDR) provides Fundamental Climate Data Records (FCDRs) for numerous satellite records and Thematic Climate Data Records (TCDRs) for atmospheric, oceanic, and terrestrial variables, some of which include brightness, temperature, precipitation, aerosol optical thickness, cloud properties, sea surface temperature, and sea ice concentration among other variables accessible from the CDRP website (http://www.ncdc.noaa.gov/cdr), which are important in climate studies (Yang et al., 2016). The European Space Agency's (ESA's) climate change initiative is making use of remote sensing data to provide essential remote sensing data for variables that include greenhouse-gas concentrations, sea-ice extent and thickness, and sea-surface temperature and salinity among others (Ehret et al., 2008).

Scanning Imaging Absorption spectrometer for Atmospheric Cartography (SCIAMACHY) on board Environmental Satellite (ENVISAT) (2002–2012) and Thermal and Near infrared Sensor for Carbon Observation-Fourier Transform Spectrometer (TANSO-FTS) on board Greenhouse gases Observing Satellite (GOSAT) (2009-ongoing) are providing greenhouse gas data being deployed ESA, whilst ESA-CCI Sea Ice Concentration (SIC) dataset (http://esa-cci.nersc.no/) provide sea ice concentration data.

European Organization for the Exploitation of Meteorological Satellite's (EUMETSAT's) Satellite Application Facility (SAF) on Climate Monitoring (CM) (CMSAF (http://www.cmsaf.eu/) develops, generates, archives, and distributes high-quality satellite-derived products of the global energy and water cycle and related sustained services in support to understand climate (Bates et al., 2016). One of the major highlights of CMSAF is the long-term commitment to provide funding until 203 (Yang et al., 2016). The main data sets from CMSAF include SSM/I FCDR for brightness, temperature, Free Tropospheric Humidity (FTH), and the Hamburg Ocean Atmosphere Parameters and Fluxes from Satellite (HOAPS) data set (http://www.hoaps.zmaw.de/). Others include CM SAF Clouds, Albedo and Radiation Advanced Very High Resolution Radiometer (AVHRR) data edition 1 (CLARA A1) surface radiation properties, cloud properties, and surface albedo datasets (https://climatedataguide.ucar.edu/climate-data/clara-a1-cloud-propertiessurface-albedo-and-surface-radiation-products-based-avhrr), which is a global dataset derived from measurements of the series of AVHRR on the NOAA satellite series, including METOP-A satellite (Yang et al., 2016).

2.3 Data acquisition and preprocessing techniques

Conventionally, remote sensing data used to be acquired through downloading raw images from online sources like Glovis (https://glovis.usgs.gov/), USGS (https://earthexplorer.usgs.gov/), and Sentinel Open Access Hub (https://scihub.copernicus.eu/) among others for processing using various GIS software (Guo et al., 2021). This is still a route that most researchers take. This includes downloading images and archiving in computers from importation into GIS software, like ArcGIS, QGIS, and ILWIS, among others. Preliminary procedures include preprocessing followed by processing and accuracy assessment. Image quality is important to ensure accurate interpretation of remotely sensed images. Nevertheless, noise often exists inevitably in images that make target phenomena blurred to certain degrees (Mullah et al., 2020; Yao et al., 2021). This makes it difficult for human observers to discriminate finer details like edges of features. There are many kinds of noises interfering with the imaging system, principally from the photoelectric detectors, for example, photon shot noise, dark current noise, thermal noise, and so on (Zhang et al., 2018). This makes image preprocessing important to ensure high-quality images. Thus, the design of the preprocessing algorithm is particularly critical. Preprocessing of remotely sensing data can be done using GIS software or plugins like the Semi-automatic Classification Plugin in QGIS or the Sentinel Application toolbox. These are used to perform atmospheric correction, orthorectification, resampling, and terrain correction among other preprocessing procedures to remove geometric and radiometric errors. Recently, cloud computing has improved the methods of acquiring and preprocessing remote sensing data by ensuring the integration of multisource data for easy analysis. Platforms like Google Earth Engine provide options to perform all required preprocessing procedures through the use of machine languages like JavaScript. Therefore image acquisition through this method does not require downloading and storing individual images for preprocessing in computers but in online databases where this information can be retrieved or modified at any time.

2.4 Climate variables derived from remote sensing data

Remote sensing can collect climatic variable data for spatial and temporal analysis. Variables like precipitation, water vapor, temperature, humidity, and wind can be remotely observed and subjected to further analysis in GIS environments. Microwave and infrared remote sensing sensors are onboard various satellite platforms (for instance the TRMM) (Igri et al., 2022) and the Climate Hazards Centre's Infrared Precipitation with Stations (CHIRPS) (Anjum et al., 2022) to collect global precipitation data at specific periods, which allows easy analysis of trends in spatial and temporal distribution. The same applies to other variables like land surface temperature and water vapor from NOAA and MODIS sensors as well as humidity and wind data from AMSR-2/AMSR-E,

ASCAT, Fengyun-3D Hyperspectral Infrared Radiation Atmospheric Sounding (HIRAS), and SMAP sensors (Zhu et al., 2020; Sim et al., 2022). These datasets can be processed to provide high-quality up-to-date data for climate analysis and prediction. Some variables important for analyzing future climate trends, for instance, pollution concentration, are remotely sensed to understand their influence on climate. Remote sensors are calibrated to observe various aspects of the atmosphere and the earth's surface to allow for data analysis and decision making. Climate analysis is currently being done based on variables like radiation, precipitation, aerosols, clouds, water vapor, and wind speed among others. Therefore, sensors are developed technically to sense the response of this phenomenon, especially Radar and Lidar sensors. Moderate Resolution Imaging Spectroradiometer (MODIS) is one of the first passive satellite spectroradiometers meant to observe aerosols from as high as 700 km. The MODIS AOD available from Giovanni (http://disc.sci.gsfc.nasa.gov/giovanni) is used to analyze spatial variations of aerosol loading, which assists in climate impact analysis. Mapping of pollution in the atmosphere is being done using various sensors like the MOPITT instrument onboard the Terra using nadir sound. This is one of the techniques used for measuring atmospheric carbon concentration. These parameters are used to understand the state of the atmosphere and potential changes in the future.

2.5 Case studies on remote sensing applications in climate analysis

Remote sensing has been and is still being applied to analyze the climate and its dynamics based on key climatic variables. In 2007, Working Group I of the IPCC, based on remotely sensed climate variables, released the Fourth Assessment Report, which describes estimations of projected future climate change (IPCC, 2007). As is well known, since it was set up by WMO and UNEP, with the help of other bodies, in 1988, the IPCC has been very successful in terms of raising the profile of carbon dioxide emissions and the enhanced greenhouse effect in scientific circles and making the public and politicians aware of the hazards of global warming. This was acknowledged by the award of the Nobel Peace Prize in December 2007 to the IPCC for its relentless pursuit of scientific evidence, digesting it and presenting it (Cracknell and Varotsos, 2011).

Plisnier et al. (2000) assessed the impacts of the ElNino Southern Oscillation based on remotely sensed climatic data. Their objective was to improve understanding of the impact of ENSO on East African ecosystems by measuring teleconnections between an ENSO index and several ecosystem variables in a spatially explicit way and for different time lags. They analyzed the spatial patterns of teleconnections in the region by combining time series of climate variables measured for meteorological stations with a time series of vegetation index and surface temperature data measured by remote sensing. Based on this they confirmed the ENSO impact on the climatic and ecological variability in East Africa, and that its impact is highly differentiated in space, as the direction, magnitude, and timing of this impact are controlled by the local climate system, the presence of large lakes, and proximity to the coast and, possibly, local topography and land cover. This is one of the cases that indicates application of remote sensing in climate analysis.

Kumar et al. (2020) applied remote sensing observations (passive microwave satellite imagery and model reanalysis) to quantify the magnitude of Arctic sea-ice loss in the boreal summer (July-September), especially in September at different timescales (daily, monthly, annual, and decadal). They indicated that Arctic sea-ice declined rapidly in the boreal summer during 1979–2018, while the highest decline in sea-ice extent (SIE) was observed in September. They attributed the loss of SIE and sea-ice concentration to the impacts of land-ocean warming and the northward heat advection into the Arctic Ocean, which has resulted in considerable thinning of sea-ice thickness (SIT) and reduction in the multiyear ice (MYI) for summer 2018. Global and Arctic land-ocean temperatures have increased by ~0.78°C and ~3.1°C, respectively, over the past 40 years (1979–2018), while substantial warming rates have been identified in the Arctic Ocean (~3.5°C in the last 40 years) relative to the Arctic land (~2.8°C in the last 40 years). This study demonstrated the linkages of sea-ice dynamics to ice drifting and accelerated melting due to persistent low pressure and high air-ocean temperatures, supplemented by the coupled ocean-atmospheric forcing.

These case studies indicate how remote sensing has become important in climate change analysis, especially due to its ability to provide long-term data for time series analysis at a global scale, which is a great contribution to policy development at the global level.

3 Remote sensing applications in policy development

3.1 Role of remote sensing in environmental policy

Good governance requires effective and efficient policies. Policies, in turn, require information to assess the need and urgency to develop policy to mobilize instruments for implementation, to control and enforce proper execution, and to monitor and evaluate policy effects. Lack of information, which previously may have hampered the efficiency and effectiveness of policies, is rapidly overcome, while emerging information techniques change the practice of environmental policy (Esty, 2004). Hence, given this changing technical arena and our ambition to better understand the relationship between EO and policy, it is critical to consider the contribution of remote sensing from a functional perspective or the contribution of remote sensing toward enhancing the effectiveness and efficiency of environmental policies.

It is acknowledged that policy remote sensing is fundamental in assessing the necessity of developing policies (Revenga, 2005). There is, however, a paucity of publicly available literature, indicating cases of how remote sensing influenced actual policy development. The detection of the ozone hole is one of the best-documented examples of remote sensing triggering policy development (Christie, 2004). Concern about the detrimental effect of chlorofluorocarbons (CFCs) on the ozone layer (Molina and Rowland, 1974) stimulated the US Congress to commission NASA to develop the Total Ozone Mapping Spectrometer (TOMS) sensor to monitor the state of the ozone layer. The sensor, launched in 1978 on board the NIMBUS satellite, did not report any anomalies until 1986 when NASA confirmed the deepening and enlargement of the size of the ozone hole since the late 1970s (Stolarski et al., 1986). This reporting after continued analysis of TOMS data resulted in updates that woke up the world by indicating the decline of spring ozone concentration over one British Antarctic Survey research station (Farman et al., 1985). In response to these confirmations, the 1987 Montreal Protocol (Velders et al., 2007) prescribed a 50% reduction and, 4 years later, a complete ban on the use of hard CFCs. This information indicates the great role of remote sensing in identifying environmental problems and informing policy development for management.

However, in some cases, recognition of environmental problems is frequently confirmed but rarely initiated by EO. Other sources of information generally precede quantification by EO. This can be noted during remotely sensed detection of increased water turbidity due to dredging in Northern Poyang Lake, China (Wu et al., 2007). This was followed by a ban on sand mining, but, in the first place, this was preceded by reports expressing concern about the impacts of dredging (Zhong and Chen, 2005; Fok and Pang, 2006). This shows that remote sensing is an important tool for confirming existing problems and informing policymakers on how to effectively tackle problems.

Other policy-relevant information from remote sensing includes the recent assessments of the rate of polar sea ice cover reduction (Nghiem et al., 2007) and the melting of the Arctic (Chen et al., 2006) and Antarctic ice caps and the Third and Fourth Assessment Reports of the Intergovernmental Panel on Climate Change (IPCC, 2001, 2007). Through these channels, remote sensing contributed without doubt to the development of policies on climate change adaptation and mitigation, deforestation, and trade in forest products. It is clear that EO infrastructure and products, such as satellites, imaging devices, and imagery, are required to enable the support of international policies, and there have been various initiatives to assess the needs and existing capacity. For example, the second report on the adequacy of the global observing systems for the climate in support of the United Nations Framework Convention on Climate Change (UNFCCC) (Mason et al., 2003), which identified that climate observation systems were not adequate, led to a stronger commitment of governments to address these inadequacies and to produce a truly global climate observing infrastructure. As in this case, remote sensing provides information for examining the necessity to address environmental issues. Remote sensing has, however, unrivaled potential in verifying objectively whether problems exist, and to assess their extent and magnitude synoptically across a range of scales. Additionally, with the gradual construction of the remotely sensed archive, remote sensing can make assessments over longer periods than most other data collection techniques could.

3.2 Remote sensing for natural resource management

Remote sensing has become an important science for monitoring and assessment of natural resources (Verbyla, 1995; Johannsen et al., 2003; Ustin, 2004; Li et al., 2020; Al-Djazouli et al., 2021; Teck et al., 2023), especially with the launching of earth surface monitoring satellite sensors like Landsat series in the early 1970s. With growing interest to understand the changes in available resources due to climate change and anthropogenic pressures, it became difficult to have data about the spatial and temporal distribution of these resources without mapping sciences. Given the opportunity to have a synoptic view of earth resources over large areas from time to time using satellite remote sensing, it became easy to understand how dynamic resources like water, vegetation, and land in general change with time (Sharma et al., 2021; Verbyla, 2022). This gave birth to time series analysis of environmental phenomena with enhanced accuracy across the globe. Remote sensing contributes to high-level influential policy studies. The Food and Agricultural Organization's (FAO's) global forest resource assessments (Marklund and Schoene, 2006; Potapov et al., 2011), which despite being based on reports from national forest surveys, have a separate remote sensing survey since the 1990s, as a complement to the main assessment report. This indicates how important remote sensing is in informing natural resource management based on the constant provisioning of spatial and temporal data for natural resource observation.

3.3 Remote sensing in disaster risk reduction and resilience planning

Given the spatial and frequently dynamic nature of all aspects of disaster risk and also of the commonly considered phases of the disaster cycle (mitigation, preparedness, early warning, response, and recovery), remote sensing is superbly suited to study and monitor disaster risk (Kerle et al., 2008). This, however, does not happen in a policy context that is always equally well established. Hazard and risk assessment are increasingly part of urban and regional planning policies, with spatial analysis, frequently supported by remote sensing data, serving to assess the probability of loss in a given area and for given periods due to environmental hazards. In particular, policies designed to limit losses due to floods or wildfires are employed and aided by RS Data (Busenberg, 2004; Fairbrother and Turnley, 2005). After the devastation caused by Hurricanes Irma and Maria in Puerto Rico, remote sensing data and analyses were a useful source of information to understand the extent and severity of damage and to inform potential strategies for recovery actions (Marlier et al., 2022). Remote sensing has also become a critical tool for hazard types that are more difficult to assess, such as seismic hazards (Dhar et al., 2017). By providing detailed information on the hazard's spatiotemporal characteristics, remote sensing not only supports the implementation of policies aimed at loss reduction, such as requirements for adequate construction standards or no-build zones, but also for such policies to be defined in detail in the first place. While remote sensing-based response to natural disasters is naturally ad hoc and case-driven, a policy framework for international support to disaster response based on satellite imagery has been developed (Cabrera et al., 2013; Vanderhorst et al., 2021). The International Charter "Space and Major Disasters", established in 1999, provides for rapid image acquisition and damage map production and is widely seen as a successful example of international cooperation. However, while the Charter operates within the framework of Space Law, i.e., the legally binding Outer Space Treaty, and the United Nations' nonbinding Remote Sensing Principles, it employs a best-effort approach with no legal binding, and no liability is assumed for the resulting mapping products (Ito, 2005).

3.4 Remote sensing for monitoring climate change policies

Data from both optical and SAR sensors are available to support the requirements of the Kyoto Protocol on a long-term basis (Rosenqvist et al., 2000, 2003; Patenaude et al., 2005). As time progresses, archives of these data are augmented, allowing long-term continuity of observation and monitoring of land cover change. As examples, data from the NOAA, Landsat, and SPOT sensors are available from the 1960s, early 1970s, and mid-1980s, respectively, and these archives are continually given the availability of NOAA AVHRR, MODIS, Landsat 8–9, and Sentinel successors. New sensors are often designed such that data are intercomparable with their predecessors, and successful deployment and provision of data (e.g., Terra-1 ASTER and MODIS) is likely to render these as key complementary or supplementary land observing systems in future years. The second report on the adequacy of the global observing systems for climate in support of the United Nations Framework Convention on Climate Change (UNFCCC) (Mason et al., 2003) identified the inadequacy of climate observation systems, which led to enhanced commitment of governments to address these inadequacies and to produce a true global climate observing infrastructure. This indicates the readiness to employ remote sensing in climate change management and policy development (Mason et al., 2003). Today, various EO systems are in place to monitor the drivers of climate change and mitigation measures in place. Remote sensing plays a role in monitoring climate change policies and compliance with international climate change treaties across the globe. Greenhouse gas (GHG) concentration-monitoring satellites (ENVISAT; GOSAT) (Ehret et al., 2008) are in space to ensure monitoring trends in the concentration of GHGs against implemented measures like afforestation and depletion of carbon sinks. Observation of carbon sinks and storage across the globe was eased through this remote sensing technology, and this improves assessment of the effectiveness of carbon sink conservation strategies and carbon emission reduction. This shows how important remote sensing is when monitoring climate change policy, as noted in IPCC climate change reports (IPCC, 2014).

3.5 Case studies on remote sensing in policy development

The International Society for Photogrammetry and Remote Sensing (ISPRS) together with the University of Michigan met in 1999 to discuss how remote sensing technology could be used to collect information required for the implementation of the Kyoto Protocol and monitor compliance with its terms (Rosenqvist et al., 2003). This included a collection of data on carbon emission sources and concentrations of carbon dioxide and other greenhouse gases that were contributing to climate change. Based on discussions the agreement was reached to reduce atmospheric carbon concentrations to 1990 levels or below, which made it necessary to employ remote sensing in monitoring the concentration of Green House Gases as well as available carbon sinks to ensure progress toward reducing climate change-inducing gasses. This can be observed when remote sensing is applied to monitor carbon emissions and carbon storage as well as sequestration measures (Chen et al., 2021; Gao et al., 2023). One of the US National Aeronautics Space Administration's (NASA) remote sensing missions is the Orbiting Carbon Observatory-2, whose principal science objective is to estimate the global geographic distribution of CO2 sources and sinks at Earth's surface, through time (Cressie, 2018). This data is being used to monitor carbon concentration in the atmosphere to provide data on whether carbon emissions are reducing as per the objectives under the United Nations Framework on Climate Change (Cressie, 2018).

4 Challenges and opportunities in remote sensing for climate and policy development

4.1 Data availability and accessibility

Despite the adoption of remote sensing for climate mapping and related decision making, some climate datasets are not easily accessible for decision makers. This is due to challenges related to computational requirements, data storage, and software needed to work with large amounts of data when working with large geographical areas like regional and global levels (Huntington et al., 2017). A good example can be that of deriving drought indices or evapotranspiration data from Landsat images for the whole region, which requires downloading and storage of so many images, each with a size between 150 and 1000 megabytes. This is a lot of data with a lot of time required to derive data at fine resolution like 30m with Landsat series or 10m with Sentinel products. However, some course resolution datasets exist for instance the Moderate Resolution

Imaging Spectroradiometer (MODIS), but they provide course information compared to Landsat series and Sentinel products. This forces researchers to adopt processed products at course resolution like evaporation or temperature data from the World Climate database at I km spatial resolution (Oyler et al., 2015). This becomes a drawback in the context of climate mapping using remote sensing data. However, cloud computing is becoming a better method to overcome the challenges of storage and processing capacities of machines when dealing with large volumes of data for climate mapping. Cloud computing platforms like Google Earth Engine (GEE) (Tamiminia et al., 2020; Zhao et al., 2021; Velastegui-Montoya et al., 2023) on which people can integrate a lot of datasets from various sources with ease and process them in very short space of time, most decision makers lack skills of using these platforms, given the technicality involved in using Java and Python machine languages. This leaves decision makers with less knowledge of cloud computing machine languages at the edge of remote sensing applications. Those who adopt image downloading and processing using various GIS software will have challenges in handling large volumes of data.

4.2 Data integration and fusion techniques

Remote sensing data fusion, as one of the most commonly used techniques, aims to integrate the information acquired with different spatial and spectral resolutions from sensors mounted on satellites, aircraft, and ground platforms to produce fused data that contains more detailed information than each of the sources (Zhang, 2010; Long et al., 2020). With the increase in data of different spectral, spatial, and radiometric qualities including textual data from other sources data integration and fusion using conventional methods (use of GIS software) is becoming challenging (Zhang, 2010). However, opportunities for data integration despite increasing volumes, variability, and sources of advanced technologies like cloud computing are shaping up the future of big data analysis in the context of remote sensing. Pixel-level fusion is usually used for optical images, but methods of fusing optical panchromatic (PAN) and multispectral (MS) images exist. High-level fusion includes feature-level and decision-level fusion of multisource data, such as synthetic aperture radar (SAR), optical images, LiDAR, GIS data, and ground data (Solberg et al., 1994; Gawlikowski et al., 2020). These techniques are not without their challenges in the field of remote sensing. Developing effective methods for automatic fusion and interpretation of the multisource, multitemporal sensor data still is a challenging activity. It remains necessary due to the fast development of various sensor technologies, such as the VHR optical sensor, SAR, and LiDAR sensors onboard airborne, spaceborne, and terrestrial platforms (Montgomery et al., 2019). The discontinuous and heterogeneous spectral and spatial characteristics of the multisource very high-resolution panchromatic and multispectral resolution images require precise registration and new pan-sharpening techniques. These can simultaneously improve the spatial resolution and retain the spectral fidelity of original multispectral data during pixel-level fusion. Although several methods have been proposed, problems still exist in computation efficiency and effectiveness.

4.3 Spatial and temporal scale challenges

Some challenges are related to spatial and temporal scales of various remote sensing data required to analyze climate data and policy development. For climate analysis, variables need to be assessed over 30 years or more. However, some recent sensors with good spatial revolution like sentinel have not yet provided data for longer time frames, which leaves climate change studies using these sensors impossible until the coming decades. Some sensors that have long operational records are no longer operational for instance the TRMM, which operated between 1997 and 2015. Data from this platform is good enough for climate analysis; but when it comes to climate change, one cannot adopt it alone. This results in the fusion of this data with some other satellite data, which may present accuracy challenges in the course as different platforms use different techniques to collect data, for example, infrared versus microwave sensing. Despite that high-quality remote sensing demands both high spatial and temporal resolutions, most freely available sensors fail to satisfy both temporal and spatial resolutions as they tend to satisfy one of these (Sajjad and Kumar, 2019). In this case, MODIS had a very high temporal resolution of 1 day but coarse spatial resolution of 250 m. This makes it good to provide temporal detail but with limited spatial detail, which is fundamental for understanding the existence of environmental phenomena. Sensors like Landsat and Sentinel have higher spatial resolutions (30 and 10 m respectively), but they cannot provide daily updates on environmental phenomena dynamics. This is a challenge that forces people to adopt coarse spatial resolution data (MODIS; NOAA) to assess some climatic variables like temperature and water vapor among others. Hyperspectral remote sensing data with high spectral resolution will hold a promising key to the solution to the future challenges of remote sensing technology (Sajjad and Kumar, 2019). This will reduce the time of image processing for various spectral bands. However, hyperspectral remote sensing data will also need to be specifically enhanced for spatial resolution to provide better results.

4.4 Accuracy assessment and validation of remote sensing products

Validation of remote sensing products is important to determine the reliability of remotely sensed data for specific uses. Remote sensing models for analyzing environmental phenomena require to be confirmed for agreement with ground truth data. There are large numbers of models based on remote sensing data applied by researchers in various fields (Sajjad and Kumar, 2019). All mathematical models do not have compatibility with all sensors, and these also do not work in all geographical locations (Sajjad and Kumar, 2019). However, the on-ground data to be used for model improvement tend to be too sparse, which increases problems when comparing remotely sensed data like precipitation and temperature to ground station data. In areas where ground stations are evenly distributed and in significant numbers, accuracy assessment is easy, and this improves the accuracy of results carried out by remote sensing data. The introduction of cloud and PRs in space has opened the way to correcting errors deriving from the misidentification of cloudy and precipitating scenes on the one hand, and from the sensitivity of the radiative transfer model to cloud parameters on the other (Stephens and Kummerow, 2007). Validation studies conducted over several decades have demonstrated that problems still arise in comparing satellite-derived precipitation products against those of ground radars when assessing their ability to represent the spatial variability of rainfall and precipitation intensity (Stephens and Kummerow, 2007). Differences in precipitation schemes and observation geometry are the basic reasons for this, and the advent of cloud radars and PRs in space is solving this issue.

4.5 Ethical considerations in remote sensing for climate policy development

With the advances in remote sensing technology in terms of the fine detail that can now be collected using high-resolution data, privacy and security are at stake, though this information is crucial for decision making. The possibility of surveillance and data gathering without consent is one of the main issues with remote sensing (Ma et al., 2015). Although some people gather this data for policy development, some people may not even be aware that their activities are being monitored in the context of global remote sensing, which causes anxiety and a perceived loss of privacy. This is especially important in the context of national security and law enforcement, where remote sensing technologies have been applied to track down people and gather information (Zipper et al., 2019). Some nations have put rules in place to control the use of remote sensing data in response to these worries. For instance, the General Data Protection Regulation (GDPR) of the European Union contains provisions requiring organizations to seek consent before collecting personal data and to make sure that the data is only used for the purposes for which it was obtained (Ma et al., 2015). However, due to the global nature of remote sensing, it may be challenging to enforce these laws, especially when data is gathered and processed across international borders. As a result, it is important to be honest, respect people's privacy, and keep some data private while still using it to make decisions. Furthermore, there is a risk that remote sensing technology would amplify global disparities already in place. When it comes to making decisions and allocating resources, having access to high-quality data can be quite helpful, potentially expanding the gap between those who can afford to invest in these technologies and those who cannot. Finding a balance between the clear advantages of these technologies and the potential hazards to privacy and equality is crucial for remote sensing ethics (Zipper et al., 2019). To create clear norms and best practices for the appropriate use of remote sensing data, continual discussion among stakeholders, including governments, industry, and civil society, will be necessary. It is possible to ensure that the potential of remote sensing is used for the greater good while upholding the rights and dignity of people and states by encouraging openness and accountability.

5 Future directions and emerging trends

5.1 Advancements in remote sensing technology

Remote sensing applications are increasingly advancing in climate science. Progress in the development of new suitable sensors and techniques for analyzing this data is making it ideal for climate management and policy development. Real-time and long-term monitoring, forecast initialization, model verification, and statistical analyses are some of the aspects in which remote sensing applications are advancing. Remote sensing records now span multiple decades and provide information on multiple processes in the climate system. The availability of long-term records of this climate data makes it possible to verify climate change over longer periods (30 years or more). The ever-changing Earth-observing satellite constellation and the development and deployment of new remote sensing capabilities, including NASA's Earth System Observatory and ESA's Sentinel missions, present a timely opportunity to make advances in these areas. This means the development of large amounts of remote sensing data that can be used to understand environmental dynamics including climate. This is motivating the development of new techniques to analyze and assimilate large volumes of data with high spatial and temporal resolution. The launch of passive microwave sensors in the 1980s increased the derivation of information about clouds and parameters within clouds, which made climate studies easy. The same applies to the launch of active radar sensors for global precipitation monitoring, which is advancing the monitoring of precipitation trends in response to climate change. Though big data introduces more technological challenges with regards to handling large amounts of data from various sensors, including data from newly developed sensors like sentinel products with a higher spatial resolution (10 m) than Landsat (30) and MODIS (250 m), advanced data handling platforms are being developed. An example of advanced data handling and processing platforms is cloud computing, which includes the use of Google Earth Engine making use of Python and Java machine languages. These are advances in the field of remote sensing, which requires more skills for those dealing with recent technical information about climate change for policy development. One of the main challenges of big data analytics in remote sensing and satellite imagery is the high cost and complexity of storing and processing massive amounts of data. Cloud computing and storage are currently offering a solution by providing scalable, flexible, and cost-effective resources for data management and analysis. Cloud platforms, such as Google Earth Engine, Amazon Web Services, and Microsoft Azure, enable users to access, share, and process remote sensing and satellite data without installing or maintaining any hardware or software. Cloud computing and storage also open avenues for the integration of different data sources and formats, and the development of new applications and services.

5.2 Integration of remote sensing with other technologies (e.g., GIS, big data)

Due to the increasing production of data every day and the generation of new data of different quality in terms of temporal, spectral, radiometric, and spatial resolution, the need for large storage systems and the capacity to fuse this heterogeneous data for detailed understanding of phenomena is growing. When remote sensing is integrated with other technologies like big data and GISs, advanced analyses of climate change can be done based on the fusion of multisource and a variety of data, which helps to produce more detailed analytics of climate and its dynamics over periods. A combination of GIS and remote sensing produces a lot of climate information—its past, present, and future dynamics prediction. Climate prediction models as well as advanced methods of analyzing complex attributes of the climate including modeling the relationships between specific variables of climate and anthropogenic activities. This informs decision making through quantifying human contribution to climate change, which informs possible policy development and its monitoring. Big data involves a variety of data from other domains like demographics and other textual data, which need to be integrated with remote sensing data in its variety to produce details information. This in conjunction with the GIS now constitutes robust hybrid techniques of modeling climate and its impacts as well as any environmental phenomena. GIS can overlay data from various sources to decide multicriteria perspectives like the use of fuzzy logic and analytic hierarchy process. This is important when incorporating big data and remote sensing to produce improved quality. Despite increasing opportunities of having large quantities of data, hardware capacity is being prioritized, which will result in further developments in the field of remote sensing and GIS.

5.3 Role of artificial intelligence and machine learning

Another key trend in big data analytics in remote sensing and satellite imagery is the application of artificial intelligence (AI) and machine learning (ML) techniques to automate and enhance data analysis and interpretation. AI and ML can help extract meaningful information from complex and noisy data, such as identifying objects, patterns, and anomalies, classifying land cover and use, estimating biophysical parameters, and predicting changes and trends. AI and ML can also improve the accuracy, efficiency, and scalability of data processing and analysis, and enable new capabilities such as real-time monitoring, anomaly detection, and decision support. AI has already proven to be a truly transformational and value-enhancing, disruptive technology in a variety of applications (Sejnowski, 2018). Already, various forms of AI, including ML, have been applied with varying levels of success to many satellite remote sensing and Numerical Weather Prediction (NWP) problems (Krasnopolsky et al., 2013), ranging from remote sensing (Ball et al., 2017) to severe weather prediction (McGovern et al., 2017). In operational applications to date, AI has played a significant role, and the successes have been hard-fought (Sirmacek and Vinuesa, 2022). However, AI demonstration projects have shown much promise for a wide range of geophysical problems from identifying critical situations to aid human interpretation, to discover new relationships in large datasets (Schlef et al., 2019), and correcting model forecasts (Campos et al., 2019). It is now clear that AI approaches, including recent advances in ML technology, such as Transfer Learning and Long and Short Term Memory Networks (LSTMs) (Schmidhuber and Hochreiter, 1997), Deep and Extreme Learning (Bengio et al., 2017; Schmidhuber, 2015; Odena et al., 2019), and Computer Vision, have the potential to meet increasing requirements for and by nowcast and forecast products, including numerical and statistical weather forecasts and climate projections.

5.4 Potential applications and benefits in climate and policy development

Advances in technologies for remote sensing data analysis mean big opportunities in leveraging big data and understanding the environmental aspects better compared to decades before. Employment of ML languages and AI opens up ways for the development of advanced models of understanding the climate and storage of high-quality information, which can be used to inform policy and decision making from time to time. Increased opportunities for the storage of big data and its retrieval show the great role remote sensing alongside ML, and cloud computing will play in the policy information and monitoring domain. If these techniques are effectively employed, decisions will be made proactively owing to the capacity of AI, MLL, and Cloud computing to ease the interpretation of complex historical remote sensing and big data, analyzing it, and using it to determine future trends in climate and environmental phenomena. This will result in increased awareness and preparedness regarding future climate dynamics.

5.5 Challenges to address for future progress

Despite a promising improvement in the quality of data and information derived from advanced remote sensing sensors, the challenges will still arise as technology improves with time. These challenges include the storage of data, which is increasing every day in terms of quantities, diversity, and quality. Through observation of previous trends in data production and handling challenges, proactive measures for big data storage, analysis, and integration need to be put in place. Some future remote sensing sensors need to be at par in terms of spatial, temporal, and spectral resolution with their predecessors to ensure continuity and reduction in data variety, which has been and is posing challenges in current big data analysis and integration.

6 Conclusion and recommendations

6.1 Implications of remote sensing for climate and policy decision-making

Advanced remote sensing technology is a great opportunity for information production about understanding the climate, its dynamics, implications, and mitigation measures. The increasing capacity to harness remotely sensed data using advanced data handling platforms like cloud computing techniques means quick analysis of data from various sources and development of informative analytics, which aid the implementation of various management strategies. Remote sensing is being used to model earth resources like vegetation, water surfaces, and human activities, which constitute indicators of changes in climatic conditions. This makes policy development easy owing to a deep understanding of the responses of environmental phenomena to climate as well as how these phenomena influence climate. Of importance is the mapping of atmospheric pollutants, including greenhouse gases using remote sensing, whilst reasoning around the effectiveness of the implemented carbon reduction and deforestation mitigation strategies. This implies strong monitoring of climate dynamics and the development of well-informed response measures. Making global decisions could be difficult without a synoptic view and understanding of the world. Many thanks to remote sensing technology, which brought the whole globe and its phenomena digital. As data increases, big data analytics need to be invested in a lot, and cloud computing platforms need to be capitalized and improved to ensure robust management of climate and development of robust, effective, and efficient policies that will make management of resources easy and holistic.

6.2 Recommendations for future research, data collection, and policy integration

As new technologies for data acquisition, processing, and analysis are increasing and data requiring integration increases, robust methods of fusing and handling this diverse information need to be developed. Therefore, future research should be channeled toward innovating current methods of data handling to keep up with voluminous data being developed daily as well as new climatic challenges requiring diverse data for robust response are arising. Research nowadays has to focus on understanding weaknesses and challenges being faced with current systems of data handling including the quality of data. Current challenges are handling big data and the lack of high-resolution data with long climatic records, especially temporal resolution. Therefore, sensor technology-related studies are required to inform the development of new sensors tailor-made to collect data about climatic variables daily and at high spatial resolution. The lifespan of these satellite sensors needs to be enhanced to ensure long-term climatic data records in the future. Alternatively, there is a need for continuity in data records with uniform quality to solve challenges related to variety in remote sensing data during data fusion. In this regard, policies need to be revised so that targets for reducing carbon emissions are reached in less than a decade. This ensures effective implementation and monitoring of pollution and related activities and quick correctional measures. However, short-term monitoring of polluting activities can be achieved if high temporal and spatial resolution data is made available and easy to interpret.

References

Adler, R.F., Sapiano, M.R., Huffman, G.J., Wang, J.J., Gu, G., Bolvin, D., Shin, D.B., 2018. The global precipitation climatology project (GPCP) monthly analysis (new version 2.3) and a review of 2017 global precipitation. Atmosphere 9 (4), 138.

Al-Djazouli, M.O., Elmorabiti, K., Rahimi, A., Amellah, O., Fadil, O.A.M., 2021. Delineating of groundwater potential zones based on remote sensing, GIS and analytical hierarchical process: a case of Waddai, eastern Chad. GeoJournal 86, 1881–1894.

Anjum, M.N., Irfan, M., Waseem, M., Leta, M.K., Niazi, U.M., ur Rahman, S., Nadeem, M.U., 2022. Assessment of PERSIANN-CCS, PERSIANN-CDR, SM2RAIN-ASCAT, and CHIRPS-2.0 rainfall products over a semi-arid subtropical climatic region. Water 14 (2), 147.

Ball, J.E., Anderson, D.T., Chan, C.S., 2017. Comprehensive survey of deep learning in remote sensing: theories, tools, and challenges for the community. J. Appl. Remote. Sens. 11 (4), 042609.

Bates, J.J., Privette, J.L., Kearns, E.J., Glance, W., Zhao, X., 2016. Sustained production of multi-decadal climate records: lessons from the NOAA climate data record program. Bull. Am. Meteorol. Soc. 97 (9), 1573–1581.

Bengio, Y., Goodfellow, I., Courville, A., 2017. Deep Learning. vol. 1 MIT Press, Cambridge, MA, USA.

Busenberg, G.J., 2004. Adaptive policy design for the management of wildfire hazards. Am. Behav. Sci. 48 (3), 314–326.

Cabrera, S., Langston, S.M., Antoniou, N., Urquijo, E., 2013. The progressive use of satellite technology for disaster management relief: challenges to a legal and policy framework. In: 64th International Astronautical Congress.

Campbell, J.B., 2002. Introduction to Remote Sensing. CRC Press.

Campbell, J.B., Wynne, R.H., 2011. Introduction to Remote Sensing. Guilford Press.

Campos, R.M., Krasnopolsky, V., Alves, J.H.G., Penny, S.G., 2019. Nonlinear wave ensemble averaging in the Gulf of Mexico using neural networks. J. Atmos. Ocean. Technol. 36 (1), 113–127.

Chen, J., Gao, M., Huang, S., Hou, W., 2021. Application of remote sensing satellite data for carbon emissions reduction. J. Chin. Econ. Bus. Stud. 19 (2), 109–117.

Chen, J.L., Wilson, C.R., Tapley, B.D., 2006. Satellite gravity measurements confirm accelerated melting of Greenland ice sheet. Science 313 (5795), 1958–1960.

Christian, E., 2005. Planning for the global earth observation system of systems (GEOSS). Space Policy 21 (2), 105–109.

Christie, M., 2004. Data collection and the ozone hole. Hist. Meteorol. 1, 99–105.

Congalton, R.G., 2010. Remote sensing: an overview. GIsci. Remote Sens. 47 (4), 443–459.

Cracknell, A.P., Varotsos, C.A., 2011. New aspects of global climate-dynamics research and remote sensing. Int. J. Remote Sens. 32 (3), 579–600.

Cressie, N., 2018. Mission CO2ntrol: a statistical scientist's role in remote sensing of atmospheric carbon dioxide. J. Am. Stat. Assoc. 113 (521), 152–168.

De Leeuw, J., Georgiadou, Y., Kerle, N., De Gier, A., Inoue, Y., Ferwerda, J., Narantuya, D., 2010. The function of remote sensing in support of environmental policy. Remote Sens. 2 (7), 1731–1750.

Dhar, S., Rai, A.K., Nayak, P., 2017. Estimation of seismic hazard in Odisha by remote sensing and GIS techniques. Nat. Hazards 86, 695–709.

Di, L., Yu, E., 2023. Remote sensing. In: Remote Sensing Big Data. Springer International Publishing, Cham, pp. 17–43.

Ehret, G., Kiemle, C., Wirth, M., Amediek, A., Fix, A., Houweling, S., 2008. Space-borne remote sensing of CO_2, CH_4, and N_2O by integrated path differential absorption lidar: a sensitivity analysis. Appl. Phys. B 90, 593–608.

Esty, D.C., 2004. Environmental protection in the information age. NYUL Rev. 79, 115.

Fairbrother, A., Turnley, J.G., 2005. Predicting risks of uncharacteristic wildfires: application of the risk assessment process. For. Ecol. Manag. 211 (1–2), 28–35.

Farman, J.C., Gardiner, B.G., Shanklin, J.D., 1985. Large losses of total ozone in Antarctica reveal seasonal ClO x/NO x interaction. Nature 315 (6016), 207–210.

Fok, M., Pang, T., 2006. Finless porpoises in Wuhan, China. Univ. Hong Kong Newslett. Dep. Ecol. Biodivers. 34, 18–20.

Fraser, C.S., Woods, A., Brizzi, D., 2005. Hyper redundancy for accuracy enhancement in automated close range photogrammetry. Photogramm. Rec. 20 (111), 205–217.

Funk, C., Peterson, P., Landsfeld, M., Pedreros, D., Verdin, J., Shukla, S., Michaelsen, J., 2015. The climate hazards infrared precipitation with stations—a new environmental record for monitoring extremes. Sci. Data 2 (1), 1–21.

Gao, F., Wu, J., Xiao, J., Li, X., Liao, S., Chen, W., 2023. Spatially explicit carbon emissions by remote sensing and social sensing. Environ. Res. 221, 115257.

Gawlikowski, J., Schmitt, M., Kruspe, A., Zhu, X.X., 2020. On the fusion strategies of Sentinel-1 and Sentinel-2 data for local climate zone classification. In: IGARSS 2020–2020 IEEE International Geoscience and Remote Sensing Symposium. IEEE, pp. 2081–2084.

Gore, A., 2006. An Inconvenient Truth: The Planetary Emergency of Global Warming and What We Can Do About It. Rodale.

Guo, W., Carroll, M.E., Singh, A., Swetnam, T.L., Merchant, N., Sarkar, S., Singh, A.K., Ganapathysubramanian, B., 2021. UAS-based plant phenotyping for research and breeding applications. Plant Phenomics. https://doi.org/10.34133/2021/9840192.

Hou, A.Y., Kakar, R.K., Neeck, S., Azarbarzin, A.A., Kummerow, C.D., Kojima, M., Iguchi, T., 2014. The global precipitation measurement mission. Bull. Am. Meteorol. Soc. 95, 701–722.

Huntington, J.L., Hegewisch, K.C., Daudert, B., Morton, C.G., Abatzoglou, J.T., McEvoy, D.J., Erickson, T., 2017. Climate engine: cloud computing and visualization of climate and remote sensing data for advanced natural resource monitoring and process understanding. Bull. Am. Meteorol. Soc. 98 (11), 2397–2410.

Igri, P.M., Tanessong, R.S., Vondou, D.A., Mba, W.P., Jores, T.K., Kaïssassou, S., Yepdo-Djomou, Z., 2022. Evaluation of the tropical rainfall measuring mission (TRMM) 3B42 and 3B43 products relative to synoptic weather station observations over Cameroon. In: Tshimanga, R.M.,

Moukandi N'kaya, G.D., Alsdorf, D. (Eds.), Congo Basin Hydrology, Climate, and Biogeochemistry: A Foundation for the Future, pp. 97–119.

IPCC, 2001. Climate Change 2001: The Scientific Basis. Contribution of Working Group I to the Third Assessment Report of the Intergovernmental Panel on Climate Change. Cambridge University Press.

IPCC, 2007. The Physical Science Basis. Contribution of Working Group I to the Fourth Assessment Report of the Intergovernmental Panel on Climate Change. vol. 996 Cambridge University Press, Cambridge, United Kingdom/New York, NY, USA, pp. 113–119.

IPCC, 2014. Fifth Assessment Report of the Intergovernmental Panel on Climate Change (Core Writing Team, R. K. Pachauri and L. A. Meyer (Eds.)). IPCC, Geneva, Switzerland (152,153,165).

Ito, A., 2005. Issues in the implementation of the international charter on space and major disasters. Space Policy 21 (2), 141–149.

Jia, Z.H.U., 2016. Analysis of Landsat8 satellite remote sensing data preprocessing. Remote Sens. Nat. Resour. 28 (2), 21–27.

Johannsen, C.J., Petersen, G.W., Carter, P.G., Morgan, M.T., 2003. Remote sensing changing natural resource management. J. Soil Water Conserv. 58 (2), 42A–45A.

Johnston, S., Cordes, J., 2003. Public good or commercial opportunity? Case studies in remote sensing commercialization. Space Policy 19 (1), 23–31.

Kerle, N., Heuel, S., Pfeifer, N., 2008. Real-Time Data Collection and Information Generation Using Airborne Sensors. Leiden, The Netherlands, Taylor & Francis/Balkema, pp. 43–74.

Krasnopolsky, V.M., Fox-Rabinovitz, M.S., Belochitski, A.A., 2013. Using ensemble of neural networks to learn stochastic convection parameterizations for climate and numerical weather prediction models from data simulated by a cloud resolving model. Adv. Artif. Neural Syst. 2013, 5.

Kumar, A., Yadav, J., Mohan, R., 2020. Global warming leading to alarming recession of the Arctic sea-ice cover: insights from remote sensing observations and model reanalysis. Heliyon 6 (7).

Kummerow, C., Barnes, W., Kozu, T., Shiue, J., Simpson, J., 1998. The tropical rainfall measuring mission (TRMM) sensor package. J. Atmos. Ocean. Technol. 15 (3), 809–817.

Lautenbacher, C.C., 2006. The global earth observation system of systems: science serving society. Space Policy 22 (1), 8–11.

Levizzani, V., Cattani, E., 2019. Satellite remote sensing of precipitation and the terrestrial water cycle in a changing climate. Remote Sens. 11 (19), 2301.

Li, J., Pei, Y., Zhao, S., Xiao, R., Sang, X., Zhang, C., 2020. A review of remote sensing for environmental monitoring in China. Remote Sens. 12 (7), 1130.

Lillesand, T., Kiefer, R.W., Chipman, J., 2015. Remote Sensing and Image Interpretation. John Wiley & Sons.

Long, D., Yan, L., Bai, L., Zhang, C., Li, X., Lei, H., Shi, C., 2020. Generation of MODIS-like land surface temperatures under all-weather conditions based on a data fusion approach. Remote Sens. Environ. 246, 111863.

Ma, Y., Wu, H., Wang, L., Huang, B., Ranjan, R., Zomaya, A., Jie, W., 2015. Remote sensing big data computing: challenges and opportunities. Future Gener. Comput. Syst. 51, 47–60.

Marklund, L.G., Schoene, D., 2006. Global assessment of growing stock, biomass and carbon stock. In: Forest Resources Assessment Programme Working paper, 106.

Marlier, M.E., Resetar, S.A., Lachman, B.E., Anania, K., Adams, K., 2022. Remote sensing for natural disaster recovery: lessons learned from hurricanes Irma and Maria in Puerto Rico. Environ. Sci. Pol. 132, 153–159.

Martínez, S., Ortiz, J., Gil, M.L., Rego, M.T., 2013. Recording complex structures using close range photogrammetry: the cathedral of Santiago De Compostela. Photogramm. Rec. 28 (144), 375–395.

Mason, P.J., Manton, M., Harrison, D.E., Belward, A., Thomas, A.R., Dawson, D.K., 2003. The second report on the adequacy of the global observing systems for climate in support of the UNFCCC. GCOS Rep. 82, 74.

McGovern, A., Elmore, K.L., Gagne, D.J., Haupt, S.E., Karstens, C.D., Lagerquist, R., Williams, J. K., 2017. Using artificial intelligence to improve real-time decision-making for high-impact weather. Bull. Am. Meteorol. Soc. 98 (10), 2073–2090.

Mega, T., Ushio, T., Takahiro, M., Kubota, T., Kachi, M., Oki, R., 2018. Gauge-adjusted global satellite mapping of precipitation. IEEE Trans. Geosci. Remote Sens. 57 (4), 1928–1935.

Mikhail, E.M., Bethel, J.S., McGlone, J.C., 2001. Introduction to Modern Photogrammetry. John Wiley Sons Inc.

Mirzakarimova, G.M., 2023. Remote sensing data: international experiences and applications. In: Actual Problems of Science and Education in the Face of Modern Challenges. vol. 14 No. 1.

Molina, M.J., Rowland, F.S., 1974. Stratospheric sink for chlorofluoromethanes: chlorine atom-catalysed destruction of ozone. Nature 249 (5460), 810–812.

Montgomery, J., Brisco, B., Chasmer, L., Devito, K., Cobbaert, D., Hopkinson, C., 2019. SAR and LiDAR temporal data fusion approaches to boreal wetland ecosystem monitoring. Remote Sens. 11 (2), 161.

Mullah, H.U., Deka, B., Prasad, A.V.V., 2020. Fast multi-spectral image super-resolution via sparse representation. IET Image Process. 14 (12), 2833–2844.

Nghiem, S.V., Rigor, I.G., Perovich, D.K., Clemente-Colón, P., Weatherly, J.W., Neumann, G., 2007. Rapid reduction of Arctic perennial sea ice. Geophys. Res. Lett. 34 (19).

Nguyen, V.C., Jeong, S., Ko, J., Ng, C.T., Yeom, J., 2019. Mathematical integration of remotely-sensed information into a crop modelling process for mapping crop productivity. Remote Sens. 11 (18), 2131.

Odena, A., Olsson, C., Andersen, D., Goodfellow, I., 2019. Tensorfuzz: debugging neural networks with coverage-guided fuzzing. In: International Conference on Machine Learning. PMLR, pp. 4901–4911.

Oyler, J.W., Ballantyne, A., Jencso, K., Sweet, M., Running, S.W., 2015. Creating a topoclimatic daily air temperature dataset for the conterminous United States using homogenized station data and remotely sensed land skin temperature. Int. J. Climatol. 35 (9), 2258–2279.

Patenaude, G., Milne, R., Dawson, T.P., 2005. Synthesis of remote sensing approaches for forest carbon estimation: reporting to the Kyoto protocol. Environ. Sci. Pol. 8 (2), 161–178.

Plisnier, P.D., Serneels, S., Lambin, E.F., 2000. Impact of ENSO on East African ecosystems: a multivariate analysis based on climate and remote sensing data. Glob. Ecol. Biogeogr. 9 (6), 481–497.

Potapov, P., Hansen, M.C., Gerrand, A.M., Lindquist, E.J., Pittman, K., Turubanova, S., Løyche Wilkie, M., 2011. The global Landsat imagery database for the FAO FRA remote sensing survey. Int. J. Digit. Earth 4 (1), 2–21.

Revenga, C., 2005. Developing indicators of ecosystem condition using geographic information systems and remote sensing. Reg. Environ. Chang. 5, 205–214.

Rosenqvist, A., Imhoff, M., Milne, A., Dobson, C., 2000. Remote sensing and the Kyoto protocol: a workshop summary. In: Congress of the International Society of Photogrammetry and Remote Sensing.

Rosenqvist, Å., Milne, A., Lucas, R., Imhoff, M., Dobson, C., 2003. A review of remote sensing technology in support of the Kyoto protocol. Environ. Sci. Pol. 6 (5), 441–455.

Sajjad, H., Kumar, P., 2019. Future challenges and perspective of remote sensing technology. In: Applications and Challenges of Geospatial Technology: Potential and Future Trends, pp. 275–277.

Schlef, K.E., Moradkhani, H., Lall, U., 2019. Atmospheric circulation patterns associated with extreme United States floods identified via machine learning. Sci. Rep. 9 (1), 7171.

Schmidhuber, J., 2015. Deep learning in neural networks: an overview. Neural Netw. 61, 85–117.

Schmidhuber, J., Hochreiter, S., 1997. Long short-term memory. Neural Comput. 9 (8), 1735–1780.

Schowengerdt, R.A., 2007. Remote Sensing: Models and Methods for Image Processing, third ed. Academic Press, p. 2 (ISBN: 978-0-12-369407-2).

Sejnowski, T.J., 2018. The Deep Learning Revolution. MIT Press.

Sharma, L.K., Gupta, R., Pandey, P.C., 2021. Future aspects and potential of the remote sensing technology to meet the natural resource needs. In: Advances in Remote Sensing for Natural Resource Monitoring. Wiley, pp. 445–464.

Sim, S., Lee, E., Seo, M., Seong, N.H., Jeong, D., Woo, J., Han, K.S., 2022. Deep neural network-based spatial gap-filling of MODIS ice surface temperatures over the Arctic using satellite and reanalysis data. Remote Sens. Lett. 13 (12), 1213–1221.

Simic Milas, A., Cracknell, A.P., Warner, T.A., 2018. Drones–the third generation source of remote sensing data. Int. J. Remote Sens. 39 (21), 7125–7137.

Sirmacek, B., Vinuesa, R., 2022. Remote sensing and AI for building climate adaptation applications. Results Eng. 15, 100524.

Solberg, A.H.S., Jain, A.K., Taxt, T., 1994. Multisource classification of remotely sensed data: fusion of Landsat TM and SAR images. IEEE Trans. Geosci. Remote Sens. 32 (4), 768–778.

Stephens, G.L., Kummerow, C.D., 2007. The remote sensing of clouds and precipitation from space: a review. J. Atmos. Sci. 64 (11), 3742–3765.

Stephens, G.L., Vane, D.G., Boain, R.J., Mace, G.G., Sassen, K., Wang, Z., Mitrescu, C., 2002. The CloudSat mission and the A-train: a new dimension of space-based observations of clouds and precipitation. Bull. Am. Meteorol. Soc. 83 (12), 1771–1790.

Stolarski, R.S., Krueger, A.J., Schoeberl, M.R., McPeters, R.D., Newman, P.A., Alpert, J.C., 1986. Nimbus 7 satellite measurements of the springtime Antarctic ozone decrease. Nature 322 (6082), 808–811.

Tamiminia, H., Salehi, B., Mahdianpari, M., Quackenbush, L., Adeli, S., Brisco, B., 2020. Google earth engine for geo-big data applications: a meta-analysis and systematic review. ISPRS J. Photogramm. Remote Sens. 164, 152–170.

Teck, V., Poortinga, A., Riano, C., Dahal, K., Legaspi, R.M.B., Ann, V., Chea, R., 2023. Land use and land cover change implications on agriculture and natural resource management of Koah Nheaek, Mondulkiri province, Cambodia. Remote Sens. Appl. 29, 100895.

Toth, C., Jóźków, G., 2016. Remote sensing platforms and sensors: a survey. ISPRS J. Photogramm. Remote Sens. 115, 22–36.

Ustin, S.L. (Ed.), 2004. Manual of Remote Sensing, Remote Sensing for Natural Resource Management and Environmental Monitoring. vol. 4. John Wiley & Sons.

Vanderhorst, H.R., Suresh, S., Renukappa, S., Heesom, D., 2021. Strategic framework of unmanned aerial systems integration in the disaster management public organisations of the Dominican Republic. Int. J. Disaster Risk Reduct. 56, 102088.

Velastegui-Montoya, A., Montalván-Burbano, N., Carrión-Mero, P., Rivera-Torres, H., Sadeck, L., Adami, M., 2023. Google earth engine: a global analysis and future trends. Remote Sens. 15 (14), 3675.

Velders, G.J., Andersen, S.O., Daniel, J.S., Fahey, D.W., McFarland, M., 2007. The importance of the Montreal protocol in protecting climate. Proc. Natl. Acad. Sci. 104 (12), 4814–4819.

Verbyla, D.L., 1995. Satellite Remote Sensing of Natural Resources. Vol. 4 CRC Press.

Verbyla, D.L., 2022. Satellite Remote Sensing of Natural Resources. CRC Press.

Wu, G., de Leeuw, J., Skidmore, A.K., Prins, H.H., Liu, Y., 2007. Concurrent monitoring of vessels and water turbidity enhances the strength of evidence in remotely sensed dredging impact assessment. Water Res. 41 (15), 3271–3280.

Xie, P., Arkin, P.A., 1997. Global precipitation: a 17-year monthly analysis based on gauge observations, satellite estimates, and numerical model outputs. Bull. Am. Meteorol. Soc. 78 (11), 2539–2558.

Xie, P., Joyce, R., Wu, S., Yoo, S.-H., Yarosh, Y., Sun, F., Lin, R., 2017. Reprocessed, bias-corrected CMORPH global high-resolution precipitation estimates. J. Hydrometeor. 18, 1617–1641.

Yang, J., Gong, P., Fu, R., Zhang, M., Chen, J., Liang, S., Xu, B., Shi, J., Dickinson, R., 2013. The role of satellite remote sensing in climate change studies. Nat. Clim. Change 3 (10), 875–883.

Yang, W., John, V.O., Zhao, X., Lu, H., Knapp, K.R., 2016. Satellite climate data records: development, applications, and societal benefits. Remote Sens. 8 (4), 331.

Yao, G., Yilmaz, A., Zhang, L., Meng, F., Ai, H., Jin, F., 2021. Matching large baseline oblique stereo images using an end-to-end convolutional neural network. Remote Sens. 13 (2), 2833–2844.

Zhang, J., 2010. Multi-source remote sensing data fusion: status and trends. Int. J. Image Data Fusion 1 (1), 5–24.

Zhang, M., Xu, G., Chen, K., Yan, M., Sun, X., 2018. Triplet-based semantic relation learning for aerial remote sensing image change detection. IEEE Geosci. Remote Sens. Lett. 16 (2), 266–270.

Zhao, Q., Yu, L., Li, X., Peng, D., Zhang, Y., Gong, P., 2021. Progress and trends in the application of Google earth and Google earth engine. Remote Sens. 13 (18), 3778.

Zhong, Y., Chen, S., 2005. Impact of dredging on fish in Poyang Lake. Jiangxi Fish. Sci. Technol. 1, 15–18.

Zhu, L., Bao, Y., Petropoulos, G.P., Zhang, P., Lu, F., Lu, Q., Xu, D., 2020. Temperature and humidity profiles retrieval in a plain area from fengyun-3D/HIRAS sensor using a 1D-VAR assimilation scheme. Remote Sens. 12 (3), 435.

Zipper, S.C., Stack Whitney, K., Deines, J.M., Befus, K.M., Bhatia, U., Albers, S.J., Beecher, J., Brelsford, C., Garcia, M., Gleeson, T., O'Donnell, F., 2019. Balancing open science and data privacy in the water sciences. Water Resour. Res. 55 (7), 5202–5211.

Chapter 18

National to global perspectives and legislation on remote sensing and climate variability

Timothy Dube and Tatenda Musasa
Department of Earth Science, Institute for Water Studies, University of the Western Cape, Cape Town, South Africa

Chapter outline

1 Introduction

The concept of climate change is not a new phenomenon, as it has become a global pressing problem. Climate change refers to a situation when the characteristics and distribution of climatic variables like humidity, temperature, rainfall, and wind deviate from the expected normal resulting in widely varying impacts (Giorgi, 2006; Solomon, 2007; Govender et al., 2022). Closely linked to this is the concept of climate variability, which is the way aspects of climate

Remote Sensing of Climate. https://doi.org/10.1016/B978-0-443-21731-9.00014-4
Copyright © 2024 Elsevier Inc. All rights are reserved, including those for text and data mining, AI training, and similar technologies.

(i.e., temperature and rainfall) differ from the average expected. In addition, the effects of climate variability have been felt in many different parts of the world, causing increases in global temperature, melting of permafrost, rising sea-level, alteration of the agricultural calendar, changes of vegetation zones, and increases in heat waves (Rosenzweig et al., 2008). These impacts are likely to be a long-standing problem in both developed and developing countries in the near future, because there are many levels that still remain uncertain (Solomon, 2007; Schowengerdt et al., 2007; Govender et al., 2022).

Diverse programmes and initiatives have been put in place in Europe and worldwide to drive the use and sharing of remote data to support decision makers (European Space Agency, 2012). For example, Copernicus processes these data and provides users with information through a set of services that address thematic areas including land, marine, atmosphere, climate change, and disaster management (Schowengerdt et al., 2007). The Global Earth Observation System of Systems (GEOSS) is a set of coordinated, independent Earth observation, information and processing systems that provide access to information for public and private sectors. Satellite remote sensing is therefore the most viable option as it can collect information about an object without physical contact through the electromagnetic spectrum using a space platform (Lillesand et al., 1994; Schott, 2007; Schowengerdt et al., 2007). For climate change analysis, RS is a required tool for up-to-date environmental data acquisition both at local and synoptic levels.

Remote sensing has been applied to monitor surface temperature changes across the world. For example, evapotranspiration (ET) is an important part of the planet's hydrological cycle, and it is likely to change in a warming world, as it is highly increased by high temperatures (Michon, 2008). Accurate ET estimation enables improvements in weather and climate forecasts, flood and drought forecasts, predictions of agricultural productivity, and assessment of climate change impacts (Sun et al., 2012). Hecker and Gieske (2004) posited that satellite data can be used for the determination of surface emissivity and temperature, rock emissivity mapping, and thermal hotspot detection such as forest fires, or underground coal fires, or to volcanic activity. Bradley et al. (2002) modeled spatial and temporal road thermal climatology in rural and urban areas of west midlands UK using satellite land cover classification, field analysis of urban canyon characteristics, physical variables of albedo, emissivity, and surface roughness within a GIS environment.

The use of remotely sensed data to estimate ET started in the 1980s and has since seen the development of various approaches and models that are guided by several international and national laws and policies. However, there is no consensus on which approach or model is the best because each of these approaches/models has its advantages and limitations (Kostenko, 2020). Despite the aforementioned contributions of Satellite Remote Sensing (SRS), there are concerns about the suitability of satellite data for monitoring and understanding climate change. Climate change studies require observations to be calibrated/validated and consistent, and to provide adequate temporal and spatial sampling over a long period. However, satellite data often contain uncertainties caused by biases in sensors and retrieval algorithms, as well as inconsistencies between continuing satellite missions. The use of satellite observations in climate change studies requires a clear identification of such limitations. This, therefore, motivates the present chapter to explore the evolving field of remote sensing and its relationship to climate variability from national to global perspectives.

1.1 Legislative perspectives on remote sensing

1.1.1 Overview of key national policies and regulations governing remote sensing activities

It is important to note that remote sensing is a space activity important for data acquisition, and as such it is governed by broad legal principles of international space law (Dempsey, 2016). Literature shows that the Outer Space Treaty requires that major space activities be carried out by international law, including the United Nations Charter (Sterling, 2007; See, 2017). National legislation is an important aspect since it safeguards the interests of maintaining international peace and security and understanding for better-informed decision (See, 2017). Remote sensing activities are subject to international law, including the UN Charter and the Space Treaty, 1967, and the relevant instruments of International Telecommunications Union (Guo et al., 2015; Dempsey, 2016). The UN General Assembly Resolution 41/65 of 1986 seeks to fulfil this void in international law by establishing a nonbinding international framework for the regulation of remote sensing activities (Guo et al., 2015).

Initially, only a few countries had access to remote sensing data; however, with the development of outer space technology and research, more states are acquiring the capacity to explore the use of remote sensing data. Likewise, the availability of private remote sensing data for commercial purposes has increased. As a result, several countries are increasingly turning to Intellectual Property (IP) Rights, particularly copyright, to protect and monopolize remote sensing data a position also confirmed by Sterling (2007). However, the existing IP laws on remote sensing data vary greatly in the domestic and international law arena, depending on the region and scale of operation. Surprisingly, the international laws on outer space do not adequately address the problems that afflict the copyright protection of remote sensing data. To this end, there has been the emergence of new legal issues concerning remote sensing, which as a result several countries have fashioned their country-specific laws on remote sensing.

The United States and Canada follow an approach similar to the UN Principles to govern its space activities. Under the US Land Remote Sensing Policy, 1992, remote sensing is defined as a collection of data that can be processed into imagery of surface features of Earth from a classified or unclassified satellite (Land Remote Sensing Policy Act, 1992). The US law makes a clear distinction between data and information depending on the processing applied. Canadian law similarly makes a distinction between raw data and remote sensing products based on processing (Remote Sensing Space Systems Act, 2005, Section 2). Laws in Europe do not generally follow the raw data and processed data distinction as within the UN Principles, the US, and Canada. For example, the German Satellite Data Security Law explicitly negates the importance of the distinction between raw and processed data (Satellite Data Security, 2007; Kostenko, 2020). The legislated laws of the States regulate remote sensing data generation and enhancement by private entities. The laws are wholly diverse among different States as, for example, US's remote sensing policy particularly focuses on remote sensing satellite technology and makes a distinction between private and public, while France and EU's focus appears to be on data-driven remote sensing, and there is no difference between public and private (European Space Agency, 2012). The differences in approaches to defining and differentiating remote sensing data may have very specific implications on the availability and the type of protection available to them.

1.1.2 Case studies highlighting effective national approaches

Several countries have adopted several laws with varying degrees of success. For instance, Algeria has the following: Presidential Decree No. 02-49 "Creation, organization and functioning of the Algerian Space Agency (ASAL)" of 16 January 2002; Presidential Decree No. 06-225 "Ratifying the Convention for Damage Caused by Space Objects" of 24 June 2006; and Presidential Decree No. 06-468 (European Space Agency, 2012). To this end, space activities in Algeria are regulated through the Algerian Space Agency (ASAL), which is charged with promoting the exploration and peaceful uses of outer space; reinforcing national capacities; ensuring the safety and well-being of the national community, and environmental protection (Kostenko, 2020). This clearly shows that the ASAL is playing a formidable role as far as regulation of space activities is concerned hence the advancement of RS technology.

The space activities in the Republic of Azerbaijan is conducted and regulated by the Space Agency of the Republic of Azerbaijan in accordance with Decree 1326 of the President of the Republic of Azerbaijan dated April 27, 2021 and Decision 54 dated of the Cabinet of Ministers of the Republic of Azerbaijan dated February 26, 2022. Satellites and other space objects are exempted from property taxes (Article 199.4.2 of the Tax Code of the Republic of Azerbaijan). Draft law "on Space activities" has been prepared based on UN treaties on outer space and other relevant international legislation, and currently is under review by the government of the Republic of Azerbaijan. The stated draft act regulates the space activities carried out by governmental bodies, citizens and legal entities in Earth orbit and outer space (European Space Agency, 2012; Dempsey, 2016). The supervision over the space activities including uplink of signals to satellites, satellite navigation, satellite earth observation is carried out by the Ministry of Digital Development and Transport of the Republic of Azerbaijan. This therefore shows that Azerbaijan has taken great strides to regulate RS activities in a bid show commitment to the international space organizations.

Colombia has Decree 2442, of 18 July 2006 on the Creation of the Colombian Commission of Space (CCE) to guide RS activities in the country (Marboe, 2019). The CCE is established to deal with consultation, coordination, orientation and planning with the purpose of guiding the execution of a national policy for the development and application of space technologies and to coordinate plans, programs and projects in the field. Decree No. 2442 of 18 July 2006, created the Colombian Space Commission as an intersectional agency with consultancy, coordination, guidance and planning functions, responsible for guiding the implementation of national policy for the development and application of space technologies and for coordinating the design of related plans, programmes and projects (Kostenko, 2020). This therefore shows that the CCE plays a pivotal role in as far as regulation of RS activities are concerned as it aids in planning and coordinating space activities.

Literature also shows that the Government of Japan exercises continuous supervision over activities of a space agency which conducts space-related activities (Dempsey, 2016). The Government also maintains a register of space objects and is involved in the planning, review and authorization of space activities. Japan Aerospace Exploration Agency (JAXA) is obliged to seek a compulsory insurance and special liability arrangements to prepare for possible third party damages caused by launching activities (Marboe, 2019). By establishing a system for such permission and a system for compensation for damage caused by the fall of artificial satellites, various treaties on the development and utilization of space are implemented accurately and public safety will be ensured. Analysis of information clearly shows that the JAXA is an important aspect whose purpose is to protect the victims of damage and thereby to contribute to the improvement of people's lives and the development of the economy.

The Government of Nigeria conducts supervision and authorization of space activities through the National Space Research and Development Agency (NASRDA) and the National Space Council which are charged with, among other things, capacity building in space science technology management and development, developing satellite technology, building and launching of space objects, licensing, and maintaining a register of space objects (Table 1). Elsewhere, South African National Space Agency (SANSA) is designated to promote the peaceful use of space, support the creation of an environment conductive to industrial development in space technology, foster research in space science, advance scientific competencies, and to foster international cooperation in space-related activities. The SANSA seeks; to promote the peaceful uses of outer space; to support the creation of an environment

TABLE 1 Effective national policies, their scope of application, and registration.

State and legislation	Scope of application	Authorization	Registration	Liability	Safety
Argentina – National Decree No. 995/91, Creation of the National Commission on Space Activities (28 May 1991); – Decree PEN No. 125/1995: Creates the "National Registry of Objects Launched into Outer Space" under the Authority of the National Commission of Space Activities (CONAE).	Persons carrying on various space activities or launches of high-power rockets in Argentina must have the activities or launches approved.	The operation of a launch facility requires a launch facility license.	A public national registry is established under the direct authority of CONAE (Arts 1, 8 Decree No. 125/95). Operators and owners shall effect registration of their space objects in the national registry (Art. 2 National Decree No. 125/95).	Information on the insurance arrangement shall be provided for inclusion in the national registry (Art. 5 Decree No. 125/95).	Information on precautions taken with regard to non-polluting of outer space, including celestial bodies (in particular pertaining to mechanisms for placement in a transfer orbit at the end of the useful life of the space object) and the anticipated date of disintegration, recovery or loss of contact with the space object shall be provided for inclusion in the national registry (Art. 5 Decree No. 125/95).

Australia – Space Activities Act 1998 (No. 123, 1998); – Statutory Rules No. 186 Space Activities Regulations 2001; – Space (Launches and Returns) Act 2018.	Australian nationals carrying on various space activities outside Australia must have the activities approved. The Act extends to Australian external Territories (Space (Launches and Returns) Act).	The operation of a launch facility requires a launch facility license.	There is a Register of Space Objects that is publicly available on the Department's website. The register includes details for certain space objects launched under an Australian launch permit.	The responsible party for the launch or return of a space object is liable to pay compensation for any damage the object causes to a third party (Section 63 Space (Launches and Returns) Act 2018).	The Minister must, by writing, appoint a Launch Safety Officer for the following: (i) a launch of a space object covered by an Australian launch permit; (ii) a return to a place or area in Australia of a space object covered by an Australian launch permit or a return authorization (Section 50 Space (Launches and Returns) Act 2018).
Nigeria – Nationals Space Research and Development Agency Act (NASRDA) Act 2010.	To establish the National Space Research and Development Agency (NASRDA) for, among other things, the encouragement of capacity building in space science and technology development and management.	A license condition may consist in the requirement to permit inspection and testing of its facilities and equipment.	NASRDA shall develop satellite technology for various applications and operationalize indigenous space systems for providing space services.	NASRDA shall maintain a register of space objects (Section 10 NASRDA Act).	A license condition is required to insure against liability incurred in respect of damages or loss suffered by third parties in the Federal Republic of Nigeria or elsewhere as a result of the activities authorized by the license (Section 9 NASRDA Act).

Continued

TABLE 1 Effective national policies, their scope of application, and registration—cont'd

State and legislation	Scope of application	Authorization	Registration	Liability	Safety
South Africa – Statutes of the Republic of South Africa—Trade and Industry No. 84 of 1993, Space Affairs Act, 1993; – Space Affairs Amendment Act, 1995.	Material scope: launching, operation of a launch facility, participation in space activities entailing obligations to the State in terms of international conventions, treaties or agreements ratified by South Africa or affecting national interest and any other space activity.	No activity falling into the scope of the Space Affairs Act shall be carried on without a license (Section 11 Space Affairs Act) which are issued by the South African Council for Space Affairs (Section 5, Space Affairs Act).		A license may contain conditions relating to the liability of the license for damages, security to be given by the license for such damages and the manner in which such security is to be given, and liability of the licensee resulting from international conventions.	The issuance of licenses may be subject to conditions, inter alia, taking into account minimum safety standards as determined by the Council (Section 11 Space Affairs Act).
Belgium – Law on the activities of launching, flight operations or guidance of space objects of 17 September 2005; – Royal Decree implementing certain provisions of the Law of 17 September 2005 on the activities of launching, flight operations and guidance of space objects.	Territorial and Personal jurisdiction: activities of launching, flight operations and guidance of space objects carried out by natural or legal persons in the zones placed under the jurisdiction or control of the Belgian State.	Authorization of the Minister is required (Art. 4 Law on the activities of launching).	A National Register of Space Objects shall be created and all space objects for which Belgium is the launching State shall be entered, except when the registration is made by another State or an international organization.	When the Belgian State is liable, pursuant to Article VII of the Outer Space Treaty, the provisions of the Liability Convention ion will guide operations.	The operator must inform immediately the crisis centre designated by the King of any maneuver, any malfunctioning or any anomaly of the space object, likely to result in a danger for persons on the ground, aircraft in flight or other space objects, or to cause any damage (Art. 16 Law on the activities of launching).

conductive to industrial development of space technology; to foster research in space science, advance scientific competencies; and to foster international cooperation in space-related activities. This therefore shows that several countries have been making great strides in a bid to regulate space activities so that their potential is fully explored.

1.2 International frameworks and agreements

1.2.1 Analysis of international treaties and agreements relevant to remote sensing

The Committee on the Peaceful Uses of Outer Space is the forum for the development of international space law (Sterling, 2007). The Outer Space Treaty was Adopted by the General Assembly in its resolution 2222 (XXI), opened for signature on 27 January 1967, entered into force on 10 October 1967 (Kostenko, 2020). The treaty reiterates that, States Parties to the Treaty shall be guided by the principle of cooperation and mutual assistance and shall conduct all their activities in outer space, with due regard to the corresponding interests of all other States Parties to the Treaty. The exploration and use of outer space bodies, shall be carried out for the benefit and in the interests of all countries, irrespective of their degree of economic or scientific development, and shall be the province of all mankind (Dempsey, 2016). Outer space, shall be free for exploration and use by all States without discrimination of any kind, on a basis of equality and in accordance with international law. Analysis of information shows that this treaty calls for international cooperation taking into cognisance the interests of all parties wherever exploration would take place for space activities.

Additionally, there is the "Rescue Agreement" which is an Agreement on the Rescue of Astronauts, the Return of Astronauts and the Return of Objects Launched into Outer Space. This was adopted by the General Assembly in its resolution 2345 (XXII), opened for signature on 22 April 1968, entered into force on 3 December 1968 (Dempsey, 2016; Kostenko, 2020). It reiterates that each Contracting Party which receives information or discovers that the personnel of a spacecraft have suffered accident or are experiencing conditions of distress not under the jurisdiction of any State shall immediately: (a) Notify the launching authority or, if it cannot identify and immediately communicate with the launching authority. It also states that if, owing to accident, distress, emergency or unintended landing, the personnel of a spacecraft land in territory under the jurisdiction of a Contracting Party, it shall immediately take all possible steps to rescue them and render them all necessary assistance (Marboe, 2019). If assistance by the launching authority would help to effect a prompt rescue or would contribute substantially to the effectiveness of search and rescue operations, the launching authority shall cooperate with the Contracting Party with a view to the effective conduct of search and rescue operations.

The "Liability Convention" is also another important convention that guides remote sensing activities. In this case, the Convention on International Liability for Damage Caused by Space Objects was adopted by the General Assembly in its resolution 2777 (XXVI), opened for signature on 29 March 1972, entered into force on 1 September 1972 (Marboe, 2019; Froehlich et al., 2020). The convention calls for States Parties to this Convention, to recognize the common interest of all mankind in furthering the exploration and use of outer space for peaceful purposes. Article II of this convention states that the Launching State shall be absolutely liable to pay compensation for damage caused by its space object on the surface of the Earth or to aircraft in flight. Article III reiterates that in the event of damage being caused elsewhere than on the surface of the Earth to a space object of one launching State or to persons or property on board such a space object by a space object of another launching State, the latter shall be liable only if the damage is due to its fault or the fault of persons for whom it is responsible (Dempsey, 2016). No exoneration whatever shall be granted in cases where the damage has resulted from activities conducted by a launching State. This clearly shows that the convention provides a safety net for all states carrying space activities to ensure that there are no liabilities and fosters for compensation of damage caused. The "Registration Convention" calls for the registration of all objects carrying out space activities. The Registration Convention also provides for registration of information regarding space objects, such as their purpose, location and parameters with the UN Secretary-General (Marboe, 2019). As is well known, the purpose of this convention is primarily to make the identification of the launching state in case of causing of damage due to launching of space objects. It also affirms that States shall bear international responsibility for their national activities in outer space. For instance, when a space object is launched into Earth orbit or beyond, the launching State shall register the space object by means of an entry in an appropriate registry which it shall maintain (Kostenko, 2020). Each launching State shall inform the Secretary-General of the United Nations of the establishment of such a registry. A mandatory system of registering objects launched into outer space would, in particular, assist in their identification and would contribute to the application and development of international law governing the exploration and use of outer space.

Literature shows that The "Moon Agreement" is an agreement Governing the Activities of States on the Moon and Other Celestial Bodies (Marboe, 2019). This agreement raises the important need to define and develop the provisions of these international instruments in relation to the Moon and other celestial bodies, having regard to further progress in the exploration and use of outer space (United Nations, 2008). It seeks to maintain the moon as an area of no conflicts attributed to space activities as it reiterates the need to use such exclusively for peaceful purposes. To this end, any threat or use of force or any other hostile act or threat of hostile act on the Moon is prohibited. It is likewise prohibited to use the Moon in order to commit any such act or to engage in any such threat in relation to the Earth (Dempsey, 2016). All activities on the Moon, including its exploration and use, shall be carried out in accordance with international law, in particular the Charter of the United Nations.

1.2.2 Importance of international cooperation and collaboration

International cooperation in the exploration and use of outer space for peaceful purposes is intended for the benefit and in the interest of all States. The importance of international cooperation has been clearly described in various legal instruments including those adopted under the framework of the United Nations (United Nations, 2008). It is widely recognized that tremendous success in the exploration and use of outer space for peaceful purposes has been accomplished as a result of international cooperation which has been an important principle from the beginning of the space age. SUPARCO has a strong element of international cooperation in its space development programme. It enjoys the permanent membership of several international organizations, institutes, scientific committees, United Nations bodies for international cooperation in the scientific and technical information, provision of data, joint projects, KHTT (Know-How Technology Transfer), training, collaborative studies. SUPARCO has also entered into many bilateral and multilateral cooperation agreements for collaboration in space-related activities. Pakistan is also party to all the five United Nations treaties on outer space (United Nations, 2008). At SUPARCO, great importance is attached to the role of facilitating international cooperation in exploration and peaceful uses of outer space. Therefore, SUPARCO strives for capacity building by organizing international/national and regional trainings/conferences/seminars in the realm of space science and technology.

Capabilities in space technology are turning into a corner stone of emergence of developing countries such as India and China. In a string of agreements for cooperation in outer space, India has developed several cooperation mechanisms with international partners to harness the benefits of space technology for mutual benefits. India has recently signed a multitude of cooperative agreements in outer space and these instruments of cooperation have advanced the space based capabilities and have mutually benefitted the partners. In one such recent engagement, India has signed a bilateral cooperation Memorandum of Understanding with Brazil in the implementation of an arrangement establishing cooperation in the augmentation of a Brazilian Earth station for receiving and processing the data from Indian Remote Sensing satellite (Murthi and Madhusudan, 2008). The agreement has been based on the interest of Brazil to receive the RS data for natural resources management, but not limited to it. In a foreign policy perspective, the Indo-Brazil agreement has been timed at a stage when the Sino-Brazil cooperation is faced with a challenge of recent failure and is set to provide India a push into a better soothe South cooperation (United Nations, 2008). The agreement has provided India an added location to downlink its satellite data and may well lay a foundation for future cooperation with other Latin American States and for further global strengthening of ties in forums.

Despite the significant progress made in fostering international cooperation in recent years, there is still much work to be done to fully realize the potential of remote sensing as a tool for global problem-solving (Guo et al., 2015). Key obstacles include ensuring that remote sensing data is accessible and usable by decision-makers in developing countries, as well as addressing concerns related to data privacy and security (Murthi and Madhusudan, 2008; United Nations, 2008). However, through continued international collaboration and data sharing, there is no doubt that remote sensing will play an increasingly vital role in shaping our planet's future. It is clear that, the global impact of remote sensing has been transformative, offering unprecedented insights into our planet's resources, environmental changes, and natural disasters. As the technology continues to advance, international collaboration and data sharing will be essential in ensuring that the benefits of remote sensing are realized by all, ultimately contributing to a more sustainable and resilient world.

2 Scientific principles of remote sensing

2.1 Overview of sensors and platforms used in remote sensing

Remote sensing platforms are continuously improving in terms of operational time, reliability, simplicity, and temporal resolution which affects the spatial resolution (Guo et al., 2015; Meftah et al., 2021). With diverse options available with specific characteristics, choosing a suitable platform depends on the nature of task that is to be conducted and scale of operation. The period from 1960 to today has experienced some major changes in the field of remote sensing in a variety of ways for example, the primary platform used to carry remotely sensed instruments shifted from airplanes to satellites (Meftah et al., 2021). Satellites can cover much more land space than planes, and can monitor areas on a regular basis. To add on, during this time, imagery became digital in format rather than analog. The digital format made it possible to display and analyze imagery using computers, a technology that was also undergoing rapid change during this period. In addition, digital sensors permitted the transmission of imagery from the platform itself to the ground station without a problematic film return system as was used on CORONA and other satellites. The sensors typically deployed on these platforms digital cameras, light-detection and ranging (LiDAR) systems, synthetic aperture radar (SAR) systems, and multispectral and hyperspectral scanners.

The turbulent social movements of the 1960s and 1970s awakened a new and continuing concern about the changes in the Earth's physical environment. Societal support was and generally continues to remain strong for remote sensing technology. As noted above, modern aerial sensors collect data using film, digital, LiDAR, SAR and other methods (Toth and Grzegorz, 2016). Although satellites are considered primary platforms (collecting much of the data used by governmental and research organizations), other platforms such as ground-based and airborne platforms (including both manned and unmanned aircraft) are quite prevalent. Unmanned airborne platforms (such as drones) have given more people the ability to remotely sense data at a relatively cost effective fashion. Different platforms can carry different types of sensors dependent upon the needs of the user and each platform offers advantages and disadvantages in terms of range, cost, stability, frequency, scale, and accuracy.

Ground-based platforms include hand held cameras (film or digital), cranes, ground vehicles which typically provide up to 50 m elevated remote sensing data (Toth and Grzegorz, 2016). These types of platforms are relatively inexpensive, stable, and due to their low altitude, provide high-resolution data. Airborne platforms include airplanes, helicopters, unmanned aerial systems (UAS), high-altitude aircraft, and free-floating balloons. Airborne platforms can provide up to 50 km of elevated remote sensing data, and are useful for acquiring atmospheric data, can pin-point locations, and can be used for specific tasks such as precision agriculture, building maintenance, emergency response, mapping, and environmental monitoring. The advantages of airborne platforms include: 1) the user can control variables such as altitude; 2) the time of coverage can be controlled; 3) it's easy to mobilize (Toth and Grzegorz, 2016). For example, clouds often appear and dissipate over a target over a period of several hours during a given day. Aircraft on site can respond with a moment's notice to take advantage of clear conditions, while satellites are locked into a schedule dictated by orbital parameters. The disadvantages of airborne platforms (especially relative to spaceborne platforms) are that they can be expensive (primarily the cost of an aircraft rental) and aerial platforms can drift off course. This shows that each platform has its own share of weakness which can however be overcome by its strengths to perform a specified task.

2.2 Applications of remote sensing in climate variability studies

2.2.1 Monitoring temperature and precipitation patterns

Precipitation plays a primary role in the global water and energy cycle, and its variations are closely linked to climate change hence the need for continuous monitoring (Yang et al., 2013). This has provoked wider thinking to better monitor the variability with time using RS technology. Sundara et al. (2012) reiterates that the spatial and temporal variability of precipitation on the global scale can be retrieved from observations made by infrared sensors on board geostationary satellites, passive microwave sensors carried by the polar-orbiting satellites and recently active radars on board the TRMM satellite and its successors (Yang et al., 2013). In their study, Sun et al. (2012) estimated ET from MODIS and Advanced Microwave Scanning Radiometer-Earth Observing System (AMSR-E) land surface temperature and moisture over of the Southern Great Plains. Spatial patterns of modeled ET and of key model parameters indicate that radiation and moisture availability are the controlling factors on ET in the study area (Wang, 2023). The spatial distributions of the estimated parameters appear reasonable: they are smooth and not overly patchy or random; and they are consistent with the land cover and soil moisture distribution through the area. Anderson et al. (2008) derived land surface temperature (LST) from remote sensing data in the thermal infrared (TIR) band (8–14 μm) to diagnose of biosphere stress resulting from soil moisture deficiencies. The study concluded that TIR imagery can serve as an effective substitute for precipitation data, providing much needed water information in data-poor regions of the world (Yang et al., 2013).

Several studies reviewed different remote sensing techniques used to estimate ET which is an important element of climate variability (Elhag et al., 2011; Gibson, 2013; Gibson et al., 2013; Jin et al., 2013; Nouri et al., 2013; Ferreira et al., 2016; Gokool et al., 2016; Zhang et al., 2016). Empirical methods that have been adopted involve the use of statistically-derived relationships between ET and vegetation indices such as the normalized difference vegetation index (NDVI) or the enhanced vegetation index (EVI), residual methods of surface energy balance which include the Surface Energy Balance Algorithm over Land (SEBAL), Surface Energy Balance System (SEBS) (Ferreira et al., 2016; Gokool et al., 2016; Zhang et al., 2016). Gibson et al. (2013) and Jin et al. (2013) also mention physically-based methods that involve the application of the combination of Penman-Monteith and Priestley-Taylor types of equations, and data assimilation methods adjoined to the heat diffusion equation and through the radiometric surface temperature sequences.

SRS is an important component of climate system observations and has gradually become a leading research method in climate change studies (Yang et al., 2013). Satellite data also contribute significantly to the improvement of meteorological reanalysis products that are widely used for climate change research, for example, the National Center for Environmental Prediction (NCEP) reanalysis (Gibson et al., 2013; Jin et al., 2013). The Advanced Very High Resolution Radiometer on board the National Oceanic and Atmospheric Administration (NOAA) satellites allows us to monitor the SST worldwide (Yang et al., 2013). An increase in SST has been observed in all ocean basins since the 1970s, with an average estimated increase of 0.28 °C from 1984 to 2006 (Yang et al., 2013). This clearly shows that satellite data provides an independent way to investigate global temperature trends, particularly at the ocean surface and in the atmosphere.

Lu et al. (2021) explored possible causes for the interdecadal shift in the interannual variability in summertime precipitation (IVSP) over South China (SC) following the mid-2000s. Zhou et al. (2021) investigate glacier velocity changes and glacier flow patterns in the Himalayas over the period from 1999 to 2018 using 220 scenes of Landsat-7 panchromatic images taken between 1999 and 2000 and Sentinel-2 panchromatic images taken between 2017 and 2018.

2.2.2 *Assessing land cover changes and their impact on climate*

Li et al. (2008) assessed land-use/land-cover change pattern in Lake Qinghai watershed between 1977 and 2004 by combining Landsat MSS, TM and

ETM data. Shrinkage in lake level and grassland degradation were discovered in the study area, with the first being attributed to climate change and the latter to anthropogenic disturbance. Determining the impact of LULC change and climate variability, on hydrological responses, using remote sensing has also been extensively studied (Ferreira et al., 2016; Govender et al., 2022). However, understanding the impact of these variables on ET remains limited when compared to surface run-off and streamflow. LULC change and climate variability both contribute to serious deterioration of the natural environment, hydrologic cycle and water resource availability. LULC change has been monitored and mapped for decades using remote sensing and GIS (Ferreira et al., 2016; Govender et al., 2022). Earlier methods consisted of integrating historical topographical map data, field observations and aerial photography in order to produce LULC maps (Ferreira et al., 2016). Overtime, there has been advancements in LULC classification methods, which aim to continuously improve accuracy in LULC mapping, in conjunction with advancements in satellite imagery. Hence, it is essential to ensure that advancements and progress in remote sensing are studied extensively to enhance future monitoring.

Climate variability has also contributed to severe effects on water resources in water scarce regions and has been studied as a stand-alone impact on hydrology and it has been studied in conjunction with LULC change (Ferreira et al., 2016; Govender et al., 2022). As a result of human population expansion, urban heat islands have contributed towards climate variability. Other negative effects include increasing sea surface temperatures that have caused variation in temperature and precipitation (Ferreira et al., 2016). In addition, climate variability has shown to worsen droughts and flooding. Therefore, there is a need to understand and quantify the effects of climate variability on hydrology. Although the effect of LULC and climate variability has been applied on hydrological responses, there is still a limited number of studies that focus on their effect on ET (Forkuor et al., 2018; Govender et al., 2022). Various hydrological models (HMs) have applied remote sensing to determine the effect of LULC change and climate variability on run-off and streamflow.

In a study conducted by de Hipt et al. (2019) in West Africa, climate change has a larger impact on discharge and sediment yield and showed results of an increase of 3.3% in ET due to climate variability. Besides, the role of LULC change is more dominant than that of climate variability in the annual surface runoff and ET responses as a result of LULC conversion and temperature variation across the watersheds during the study period (Govender et al., 2022). Studies show that climate variability proves to have a significant impact on run-off, ET and drought possibility in Sub-Saharan Africa. Accurate determination of regional ET and estimation of the evolution of the climate, water resources planning and management, agricultural water saving crop production simulation and environmental issues have important practical significance (Ferreira et al., 2016).

3 Integrating remote sensing data into climate models

3.1 Challenges and limitations of incorporating remote sensing data

3.1.1 Data quality and uncertainty issues

The issue of data quality (DQ) is of growing importance in RS, due to the widespread use of digital services that exploit remote sensing data (National Academies of Sciences, Engineering, and Medicine., 2007; Batini et al., 2017). A key challenge in the effective use of remotely sensed data is getting the appropriate data products into the hands of the appropriate decision makers, in time frames reasonable for action (Bielecka and Burek, 2019). International agencies (such as the United Nations Food and Agriculture Organization), and NGOs (such as conservation NGOs including the World Wildlife Fund and Conservation International) face setbacks in developing the required technical knowledge to take advantage of remote sensing technologies (Niu et al., 2016). The challenge is to develop the capacity for these groups to communicate so that the remote sensing community understands the needs of policy makers and policy makers understand the capabilities of remote sensing.

Atmospheric conditions particularly influence the usability of optical remote sensing data. As a consequence, a major quality dimension of RS data usability is cloud cover (Lowell, 2004; Bielecka and Burek, 2019). DQ dimension related to cloud cover can be described as data completeness, radiometric completeness. In addition to this, the distribution of cloud cover highly influences the usability of the images in spatial terms (Yang et al., 2013; Niu et al., 2016). Therefore, cloud cover distribution on images can be defined in terms of spatial completeness since it compromises the quality of results. Radiometric accuracy of image data is not only influenced by the effects of the atmosphere and the path length but topographic effects of shadows (Bielecka and Burek, 2019). Regarding optical systems shadows account for radiometric distortions in data. In contrast, topographic effects on active microwave data accounts for a higher geometric than radiometric distortion due to its oblique nature.

For the purposes of this chapter, two types of uncertainty can be recognized; ambiguity and vagueness. Ambiguity is, essentially, the problem of making a choice between two or more alternatives and is probably the more commonly studied form of uncertainty (Lowell, 2004; Bielecka and Burek, 2019). Vagueness is associated with making sharp or precise distinctions. Traditionally, one major source of uncertainty has been a lack of data or knowledge (Bielecka and Burek, 2019). The last four or five decades have seen tremendous technological advances in remote sensing and geographical information systems such that many topics of research have moved from being handicapped by a lack of data to now having, and often being unable to cope with, large data sets (Lowell, 2004; Yang et al., 2013). In some instances, the sizes of data sets are now so overwhelmingly large that intelligent assistance, perhaps through knowledge discovery and data mining techniques are required in some studies (Blaschke et al., 2011; Bielecka and Burek, 2019). The availability of increasing amounts of data and growing knowledge is, within the confines of a traditional scientific paradigm, a desirable situation as knowledge and information are believed to increase precision and reduce uncertainty. It is, therefore, unfortunate that uncertainty is ubiquitous and is inherent in geographical data (Blaschke et al., 2011). Moreover, the inherent uncertainty may be compounded in data processing and analysis. Actions such as rescaling or rezoning data sets may introduce uncertainty. Changes in administrative boundaries, for example, greatly handicap studies as it results in the use of incompatible zoning systems over time (Bielecka and Burek, 2019). This adds to problems in analyzing data sets in which locational uncertainty may be large (Mayer-Schönberger and Cukier, 2013; Bielecka and Burek, 2019). This is common in remote sensing and GIS. For example, the inability to perfectly co-register remotely sensed data sets may substantially limit the ability to accurately monitor major land cover changes (e.g., deforestation) and so limit scientific understanding of environmental processes and change.

3.2 Compatibility with existing climate models

3.2.1 Strategies for overcoming integration challenges

3.2.1.1 Calibration and validation techniques

The Soil and Water Assessment Tool (SWAT) is a continuous-time, semi-distributed and processed-based model developed and supported by the USDA Agricultural Research Service (Neitsch et al., 2011; Arnold et al., 2012; Ismail et al., 2020). The model was originally developed to evaluate the impact of land management practices on water resources, sediment and agricultural chemical yields in large complex catchments with varying soils, land use and management conditions (Neitsch et al., 2011; Daniel et al., 2011). Water balance is the major driving force behind any process in SWAT which can be combined with climate models (Randall et al., 2007; Ismail et al., 2020). Hence, besides the different components of water balance, SWAT is being used to model plant growth and the movement of sediments, nutrients, pesticides and pathogens in a catchment (Neitsch et al., 2011; Arnold et al., 2012). The model requires several input data to simulate catchment hydrologic processes, and these include a digital elevation model (DEM), land use-land cover data, soil types, and different daily weather data, including details of precipitation, maximum and minimum air temperatures, solar radiation, wind speed, and relative humidity. SWAT has received international acceptance as a robust interdisciplinary catchment-scale modeling tool which contribute to climate prediction steps. Major input data for SWAT include DEM, LULC, soil properties, and daily weather data (precipitation, maximum and minimum air temperature, relative humidity, wind speed and solar radiation). Although hydrologic models play a critical role in the management of natural resources in arid areas, their application is challenged by the scarcity of data for calibration and validation. Mengistu et al. (2019) attempted to configure, calibrate and validate SWAT model in a data-scarce catchment by using the regionalization with physical similarity approach. This approach adopted the dual calibration and validation procedure, i.e., one in the donor catchment (by using SWAT-CUP (SWAT Calibration and Uncertainty Programs). Results of the study showed that the P-factor was 0.73 and 0.65 for calibration and validation, respectively; whereas the R-factor was 0.93 for calibration and 0.66 for validation periods. Generally, good model uncertainty is expressed by a higher value of the P factor (towards 100%) and a lower value of R-factor (towards 0).

Ismail et al. (2020) in their study compared two model calibration/validation approaches and their influence on future hydrological projections under climate change by employing two climate scenarios (RCP2.6 and 8.5) projected by four global climate models. Two HMs, snowmelt runoff model + glaciers and variable infiltration capacity model coupled with a glacier model, were used to simulate streamflow in the highly snow and glacier melt-driven Upper Indus Basin. Using the conventional and enhanced calibration approaches, the monthly Nash-Sutcliffe Efficiency (NSE) for both HMs ranged from 0.71 to 0.93 and 0.79 to 0.90 in the calibration, while 0.57–0.92 and 0.54–0.83 in the validation periods, respectively. For the future impact assessment, comparison of differences based on the two calibration/validation methods at the annual scale (i.e., 2011–2099) shows small to moderate differences of up to 10%, whereas differences at the monthly scale reached up to 19% in the cold months (i.e., October-March) for the far future period (Ismail et al., 2020).

3.2.1.2 Ensemble modeling approaches

Ensemble modeling is a process where multiple diverse models are created to predict an outcome, either by using many different modeling algorithms or using different training data sets (Wang, 2023). The ensemble model then aggregates the prediction of each base model and results in once final prediction for the unseen data. The three main classes of ensemble learning methods that can be applied to overcome integration challenges are bagging, stacking, and boosting (Al Ali et al., 2013; Zhang and Yu, 2005). Bagging or bootstrap aggregating is an ensemble modeling algorithm trained with data subsets randomly selected from the training dataset to boost model variance. The moderate approach is to take a weighted vote or a weighted average in the case of a classification or a numerical prediction respectively (Zhang and Yu, 2005). It allows you to differentiate test incidents for a specific machine learning method in order to create correct predictions. Instead of gathering independent sample datasets from the domain, bagging also deals with the original training data. Bagging is unique when resampling the original training data rather than using independent domain datasets (Al Ali et al., 2013). Different models are built in parallel from multiple samples to vote for the final model with equivalent weights. The classifier is always successful from the original training data, with considerably better predictions than the actual classifier, and never performs significantly poor. This seeks to neutralize the variance by adjusting the original training data through canceling certain instances and duplicating others. By avoiding overfitting, the classifier improves consistency and reduces bias and variance errors (Al Ali et al., 2013).

Bagging is an extremely successful ensemble learner that produces diverse models on distinct random samples of the original dataset. These samples are held uniformly with replacements that are called bootstrap samples (Zhang and Yu, 2005). Since samples are held with replacement, the bootstrap sample will in general include duplicates, and therefore some of the original data points will be lost even if the bootstrap sample is of the same size as the original dataset. This is precisely what is needed, since differences between the bootstrap samples will produce diversity among the models in the ensemble. They can choose to aggregate the predictions of the various models by voting the class predicted by most models wins or by averaging, which is more fitting of the performance scores or probabilities of the base classifiers.

4 Data open access and sharing

By having access to data more easily, scientists have more resources to support and enrich their work. In its strict sense, the more data you have access to, the easier it is to identify trends and the more evidence there can be to support claims. Open data also enables collaboration which is an integral element to support scientific research. A series of advanced applications and innovative methodical insights have been exposed by the emergence of free and open access data sets (Piwowar and Chapman, 2007; Sepuru and Dube, 2018). As of the present moment, a cumulative count of eight satellites belonging to the Landsat lineage (Landsat 1, 2, 3, 4, 5, 7, 8, and 9) have been successfully launched.

The United States Geological Survey's open data policy declaration for instance with regards to Landsat images in 2008 led to a dramatic increase in remotely sensed data access from 25,000 images in 2001 to 2.5 million in 2010, including in data poor and economically challenged areas (Forkuor et al., 2018; Wulder et al., 2012). Free and open access data has transformed the means of RS data applications (Scambos et al., 2018). This shows that the extensive adoption of RS data, has served diverse functions including biomass monitoring, climate variability assessment and soil erosion mapping. The appeal of RS data e.g., Landsat data stems from its open accessibility, setting it apart from high-resolution alternatives like World-View. This growing adoption contributes to improved environmental monitoring, particularly in resource-constrained communities, thus supporting informed decision-making and environmental conservation. Much earlier, in 2002, the Open Access (OA) movement commenced with the Budapest Open Access Initiative led by a group of international scholars who were keen to allow unrestricted access to journal literature for the public good. Since then, OA has grown exponentially with the Directory of Open Access Journals (DOAJ) having reached over 3.1 million articles in nearly 12,000 peer-reviewed journals (Mauthner and Parry, 2013). Improvements in the sophistication of data-mining techniques mean that new research can combine multiple datasets from diverse but related disciplines. Open access to research data is increasingly being institutionalized across funding and research-related organizations through data sharing policies. Compliance with these policies has been hampered by reluctance, among many social and natural scientists, to release data on an open access basis. The availability of so many complex datasets is also changing the way that scientists approach their subject. Analysis of information clearly shows that the OA has opened up a series of avenues for research using open archival data which is full to meet the intended objectives.

Creating large, internationally-renowned institutional repositories will ensure that data are not lost to future researchers and will promote appropriate citation and acknowledgement of their use (Mauthner and Parry, 2013). It is also assumed that such repositories can 'enhance and accelerate' scientific advancement by enabling multiple uses of datasets and preventing duplication of efforts. There is evidence from specific fields, such as climate science, that the provision of open access data is increasing the impact of research by making it available, not only to researchers, but also to practitioners such as resource managers and policy-makers who seek to use data to inform their decisions and develop strategies. Climate change scientists need to integrate these large and varied datasets from completely different disciplines in order to understand the current situation and provide help in creating and predicting the impact of government and institutional policies. More recently, cases of scientific misrepresentation such as "climate-gate" have emphasized the importance of reliability of data and experiment reproducibility as a built-in peer review system use (Mauthner and Parry, 2013).

5 Remote sensing in climate change decision-making

5.1 Role of remote sensing in informing climate change policies and strategies

5.1.1 Case studies demonstrating the use of remote sensing data in decision-making

Remote sensing technology has brought a paradigm in climate change prediction and monitoring which informs decision makers across the world (Abera et al., 2019; Govender et al., 2022). RS can be used to measure changes in land cover, such as deforestation and the effects of urbanization and this data can help scientists better understand how these changes may affect climate patterns (Pinnington et al., 2018; Berihun et al., 2019). It also offers an unprecedented level of accuracy in climate modeling which is an imperative for decision making process. The impact of land use land cover (LULC) change and climate variability on water resources poses as a major threat in semiarid environments, especially in the sub-Saharan Africa (Govender et al., 2022). Countries in sub-Saharan Africa for example, South Africa are vulnerable to water scarcity and element of climate variability. Therefore, there is an urgent need for understanding the various methods for LULC change and climate variability assessment, to aid in water resources management at various scales (Govender et al., 2022). Various studies have modeled and assessed the effect of LULC change and climate variability on hydrological responses, using different approaches.

Earth observation has proved to be an indispensable tool providing information at large geographic scales. Gokool et al. (2016) maintains that, it allows data to be captured in data-scarce regions and can provide time-series data easily, due to the periodic updating of information. In their study they were able to estimate total evaporation at the field scale using the SEBS model and data infilling procedures (Berihun et al., 2019; Govender et al., 2022). Remote sensing images can effectively record land use situations and provide an excellent source of data, from which updated LULC information and changes can be extracted, analyzed and simulated efficiently and better inform decision makers. Additionally, the use of remote sensing-based methods for estimating ET has gained momentum in the past decade and holds great potential for monitoring water use and ET across various spatial and temporal scales, and for being integrated into operational water resource management systems in sub-Saharan Africa (Gibson et al., 2013; Govender et al., 2022). Analysis of information in these sections clearly shows that RS is a robust technology that aids decision making as it can be used to estimate parameters such as temperature and ET which are elements of the climate system.

5.1.2 Benefits and limitations of remote sensing for policymakers

Climate change is one of the most pressing challenges of our time, with severe consequences for ecosystems, economies, and societies worldwide (Govender et al., 2022). In order to address this crisis, scientists require accurate and up-to-date information about the Earth's changing environment (Abera et al., 2019). Remote sensing technology has proven to be a powerful tool in providing such data, enabling researchers to better understand and monitor the impacts of climate change (Mengistu et al., 2019). The high spatial and temporal resolution of remote sensing data allows for more accurate climate modeling and improved decision-making in response to climate change. Thermal infrared sensors on satellites measure land and sea surface temperatures, providing data on warming trends and extreme temperature events (Liping et al., 2018; Govender et al., 2022). This information helps scientists and researchers understand climate change impacts on ecosystems, agriculture, and human health, thus informing adaptation and mitigation strategies.

Remote sensing technologies like multispectral and hyperspectral imaging enable scientists to analyze vegetation health, productivity, and phenology (Liping et al., 2018). Analyzing vegetation changes helps researchers identify regions affected by climate change-related impacts such as drought, forest degradation, or shifts in growing seasons (Kumar et al., 2022). Soil moisture is critical for agriculture, water resources, and ecosystem health. Remote sensing technologies help monitor and predict floods and droughts by measuring precipitation, soil moisture, and vegetation health (Berihun et al., 2019; Mengistu et al., 2019). Data from satellites and aerial platforms can identify areas at risk, enabling disaster management agencies to take preventive measures and respond effectively to climate change-induced extremes. Pinnington et al. (2018) maintain that satellites with microwave radiometers and SAR systems can measure soil moisture, providing insights into the impacts of climate change on water availability, drought risk, and land degradation a critical element that assists decision and policy makers. RS is also a powerful tool for precipitation monitoring, for example Govender et al. (2022) observed that satellites equipped with microwave radiometers or passive microwave imagers measure precipitation rates and distribution, providing data on global and regional trends. Analyzing precipitation changes is vital for understanding climate change impacts on water resources, agriculture, and disaster risk reduction hence aiding policy formulation.

Literature review further indicate that the application of remotely sensed imagery has improved the quality of climate modeling (Pinnington et al., 2018; Govender et al., 2022). However, a significant challenge exists in handling of large volumes of data (captured since the launch in 1972 for Landsat for example) (Abera et al., 2019; Liping et al., 2018). Forkuor et al. (2018) indicate that complexity of the data formats in developing countries also act as hindrances towards the successful application of RS data in modeling climate features. This therefore implies that although remotely sensed data offer spatially extensive and repeated objective measurements problems of data handling comprises the opportunity for improving the quality of assessments. Historically, remotely sensed based climate modeling has been a daunting task due to issues of resolution which has since been improved (Angima et al., 2003; Wolka et al., 2015; Panagos et al., 2020; Mihi et al., 2020). Just like the OLI, the TM has an improved resolution of 30 m which ensures improved accuracy in climate modeling for variables such as precipitation and temperature (Govender et al., 2022). Landsat 8 OLI presents superior sensing characteristics, coupled with its distinctive spatial-temporal alignment for LULC change detection. Simultaneously, its expansive imaging footprint facilitates the implementation of wide-area applications. Guo et al. (2015) reiterates that there are also challenges related to data processing and analysis, sensor limitations, and integration with other data sources. Satellite observations have provided a wealth of data on global change research. Phinzi et al. (2021), observed that there are still some drawbacks, such as the lack of guarantees for the continuity of the data, the data uncertainty as discussed in the preceding sections. To add on the fact that the network of multisource remote sensing data has not been optimized also presents problems, that the data assimilation process lacks data validation, and that Earth observation information has not yet met the needs of model parameters (Guo et al., 2015). This underscores that the utilization of remotely sensed data comes with its own set of challenges, yet these potential drawbacks can be overshadowed by the platform's inherent strengths and the added advantage of unrestricted accessibility (Phinzi et al., 2021; Kumar et al., 2022).

5.2 Future directions and opportunities

5.2.1 Advancements in remote sensing technology and data availability

As remote sensing technology continues to advance, it will play an increasingly important role in climate change research (Guo et al., 2015; Phinzi et al., 2021). Future developments may include the deployment of new satellite missions, the integration of artificial intelligence and machine learning techniques for data analysis, and the increased use of drone-based remote sensing platforms for localized studies (Piwowar and Chapman, 2007). Through a variety of methods such as multisatellite, multisensor, and long-term time series remote sensing data, it has been possible to derive climate sensitivity factors and aid in the study of the spatial variability of terrestrial ecosystems (e.g., glaciers, lakes, and vegetation), and in the understanding how they respond to global climate change (Klump et al., 2006; Guo et al., 2015; Abera et al., 2019). Earth observation technology has provided highly useful information in global climate change research over the past few decades and greatly promoted its development, especially through providing biological, physical, and chemical parameters on a global scale (Guo et al., 2015). Earth observation data has the 4 V features (volume, variety, veracity, and velocity) of big data that are suitable for climate change research (Mauthner and Parry, 2013). It is clear that this large amount of data available from scientific satellites plays an important rolc as it aids to data availability hence its fullness.

The Earth observation big data era is also on its way. In 2012, the International Data Corporation (IDC) released the 2020 Digital Universe report (Guo et al., 2015), which stated that the total amount of data available will double every 2 years. Increasingly, there is a drive to observe the Earth in multiscale, comprehensive, and real-time perspectives, and the capability of accessing global Earth observations information has been rapidly enhanced (Piwowar and Chapman, 2007; Mauthner and Parry, 2013). Concurrently, with global climate change becoming an international issue, the study of Earth observations related to global environmental change is increasingly linked to national interests and international politics and diplomacy (Guo et al., 2015; Abera et al., 2019). Global climate change research using big data from Earth observations is urgently needed not only to understand the Earth system's evolution but also to enable the coordinated and sustainable development of humanity within the Earth system, with the added consideration of safeguarding national rights and maintaining diplomatic relations.

5.2.2 *Potential for enhanced international cooperation*

In tackling the climate crisis, international cooperation should involve holding regular discussions with other countries on issues bordering on climate change; to cocreate and share knowledge and technology, as well as develop best practices and build joint efforts. International collaboration could accelerate shared innovation and technology deployment, support the agreement of common international data sharing standards, and foster the development of joint approaches (Guo et al., 2015; Abera et al., 2019). As the technology continues to advance, international collaboration and data sharing have emerged as critical components in maximizing the benefits of remote sensing for all. As the benefits of remote sensing become increasingly apparent, international collaboration and data sharing have emerged as essential elements in maximizing its potential. By pooling resources and expertise, countries can develop more advanced remote sensing technologies and applications, leading to better decision-making and more effective solutions to global challenges (Govender et al., 2022). One notable example of international collaboration in remote sensing is the Group on Earth Observations (GEO), a global partnership of more than 100 countries and organizations working together to build a GEOSS. This ambitious initiative aims to improve the availability, access, and use of Earth observation data for a wide range of applications, from climate change adaptation to disaster risk reduction (Guo et al., 2015). It is clear that fostering collaboration and data sharing among its members, GEO is helping to ensure that the benefits of remote sensing are shared equitably and effectively across the globe.

6 Conclusion

This book chapter explored the evolving field of remote sensing and its relationship to climate variability from national to global perspectives. Remote sensing has emerged as a critical tool in the study of climate change, providing invaluable data on a wide range of environmental variables. Satellites equipped with specialized sensors collect data about the Earth's surface and atmosphere, providing valuable insights into various aspects of the climate system. Its applications span from monitoring sea level rise and glacial retreat to observing changes in vegetation and assessing shifts in precipitation patterns. Remote sensing offers comprehensive coverage, high resolution, and long-term monitoring capabilities, making it a valuable tool for studying climate change patterns and trends. As technology continues to evolve, remote sensing will undoubtedly play an even more significant role in understanding and addressing the impacts of climate change on our planet. RS activities are regulated by several laws which vary according to country with the leading being the Outer Space treaty that calls for peaceful exploitation of space activities and ensure there is no liability and ensures compensation in cases of emergencies related to space activities. Scientifically, open access data sharing is understood to promote transparency, innovation and progress which has enhanced climate variability modeling using remotely sensed data. The importance of international cooperation has been clearly described in various legal instruments including those adopted under the framework of the United Nations. However, application of RS data for several applications is frustrated by factors such as DQ, uncertainty issues, costs associated with the data as well as limited expertise hence there still remains a lot of work. The presence of big data has provided a series of avenues that can be exploited but has varying degrees of success due to problems of lack of storage and handling facilities.

6.1 Call to action for leveraging remote sensing for climate change mitigation and adaptation

Earth observation has changed the way that we understand the Earth due to its ability to provide continuous data for environmental monitoring (Guo et al., 2015). For example, it can provide the necessary information that verifies and improves our knowledge of the coupling between the El Nino Southern Oscillation and ocean currents, temperatures, atmospheric processes, the effect of snow on water circulation, the effects of global and regional factors on sea level changes, and a range of other phenomena. NASA launched the first weather satellite, TIROS 1, in 1960. The images taken by this satellite have revealed astonishing cloud features. By December 2011, a further 514 Earth observation satellites were launched worldwide and 200 more launches are planned by 2030 (Wulder et al., 2012). This huge amount of satellite data now available plays an important role in global climate change research as it can detect changes over a wide area as it can inform adaptation and mitigation strategies (Govender et al., 2022). A number of Earth observation platforms are used to collectively enable real-time observation and dynamic monitoring of the Earth's land, atmosphere, and oceans, and provide macro-scale, accurate, comprehensive, continuous, and diverse information on Earth's surface (Guo et al., 2015; Phinzi et al., 2021). This multisource Earth observation data has altered the methods by which the Earth system data and perceptions are obtained. It also plays a fundamental role in the support of scientific innovation. Effective use of multiplatform observation data with multisensors can avoid issues relating to information extraction and inversions that arise with the use of a single sensor (Wulder et al., 2012). Multisource Earth observations also provide long-term and stable spatial data for scientific research, compensating for uneven spatio-temporal observations and playing a fundamental supporting role in global change research.

References

Abera, W., Tamene, L., Abegaz, A., Solomon, D., 2019. Understanding climate and land surface changes impact on water resources using Budyko framework and remote sensing data in Ethiopia. J. Arid Environ. 167, 56–64.

Al Ali, R., Bures, T., Gerostathopoulos, I., Hnetynka, P., Keznikl, J., Kit, M., Plasil, F., 2013. DEECo Computational Model–I. Charles University in Prague (D3S).

Anderson, M.C., Norman, J.M., Kustas, W.P., Houborg, R., Starks, P.J., Agam, N., 2008. A thermal-based remote sensing technique for routine mapping of land-surface carbon, water and energy fluxes from field to regional scales. Remote Sens. Environ. 112 (12), 4227–4241.

Angima, S.D., Scott, D.E., O'Neill, M.K., Ong, C.K., Weesies, G.A., 2003. Soil erosion prediction using Rusle for central Kenyan highland conditions. Agric. Ecosyst. Environ. 97, 295–308.

Arnold, J.G., Kiniry, J.R., Srinivasan, R., Williams, J.R., Haney, E.B., Neitsch, S.L., 2012. SWAT Input/Output Documentation Version 2012. vol. 654 Texas Water Resources Institute, pp. 1–646.

Batini, C., Blaschke, T., Lang, S., Albrecht, F., Abdulmutalib, H.M., Barsi, Á., Szabó, G., Kugler, Z., 2017. Data quality in remote sensing. Int. Arch. Photogramm. Remote Sens. Spatial Inf. Sci. 42, 447–453.

Berihun, M.L., Tsunekawa, A., Haregeweyn, N., Meshesha, D.T., Adgo, E., Tsubo, M., Masunaga, T., Fenta, A.A., Sultan, D., Yibeltal, M., et al., 2019. Hydrological responses to land use/land cover change and climate variability in contrasting agro-ecological environments of the Upper Blue Nile basin, Ethiopia. Sci. Total Environ. 689, 347–365.

Bielecka, E., Burek, E., 2019. Spatial data quality and uncertainty publication patterns and trends by bibliometric analysis. Open Geosci. 11 (1), 219–235.

Blaschke, T., Hay, G.J., Weng, Q., Resch, B., 2011. Collective sensing: integrating geospatial technologies to understand urban systems—an overview. Remote Sens. 3 (8), 1743–1776.

Bradley, A.V., Thornes, J.E., Chapman, L., Unwin, D., Roy, M., 2002. Modelling spatial and temporal road thermal climatology in rural and urban areas using a GIS. Clim. Res. 22, 41–55.

Daniel, E.B., Camp, J.V., LeBoeuf, E.J., Penrod, J.R., Dobbins, J.P., Abkowitz, M.D., 2011. Catchment modeling and its applications: a state-of-the-art review. Open Hydrol. J. 5 (1), 26–50.

de Hipt, F.O., Diekkrüger, B., Steup, G., Yira, Y., Hoffmann, T., Rode, M., Näschen, K., 2019. Modeling the effect of land use and climate change on water resources and soil erosion in a tropical West African catchment (Dano, Burkina Faso) using SHETRAN. Sci. Total Environ. 653, 431–445.

Dempsey, P.S., 2016. National laws governing commercial space activities: legislation, regulation, & enforcement. Northwest. J. Int. Law Bus. 36, 1.

Elhag, M., Psilovikos, A., Manakos, I., Perakis, K., 2011. Application of the SEBS water balance model in estimating daily evapotranspiration and evaporative fraction from remote sensing data over the Nile Delta. Water Resour. Manag. 25 (11), 2731–2742.

European Space Agency, 2012. ESA Data Policy for ERS, Envisat and Earth Explorer Missions. Retrieved from The European Space Agency https://earth.esa.int/c/document_library/get_file?folderId=296006&name=DLFE3602.pdf.

Ferreira, E., Mannaerts, C.M., Dantas, A.A., Maathuis, B.H., 2016. Surface energy balance system (SEBS) and satellite data for monitoring water consumption of irrigated sugarcane. Eng. Agric. 36 (6), 1176–1185.

Forkuor, G., Dimobe, K., Serme, I., Tondoh, J.E., 2018. Landsat-8 vs. Sentinel-2: examining the added value of sentinel-2's red-edge bands to land-use and land-cover mapping in Burkina Faso. GIsci. Remote Sens. 55 (3), 331–354.

Froehlich, A., Seffinga, V., Qiu, R., 2020. Initial mandates of the committee on the peaceful uses of outer space (COPUOS) and the conference on disarmament (CD). In: The United Nations and Space Security: Conflicting Mandates between UNCOPUOS and the CD. Springer, pp. 7–28.

Gibson, L.A., 2013. The Application of the Surface Energy Balance System Model to Estimate Evapotranspiration in South Africa. University of Cape Town, Cape Town.

Gibson, L.A., Jarmain, C., Su, Z., Eckardt, F.E., 2013. Estimating evapotranspiration using remote sensing and the surface energy balance system—a South African perspective. Water SA 39 (4), 477–484.

Giorgi, F., 2006. Climate change hot-spots. Geophys. Res. Lett. 33 (8).

Gokool, S., Chetty, K.T., Jewitt, G.P., Heeralal, A., 2016. Estimating total evaporation at the field scale using the SEBS model and data infilling procedures. Water SA 42 (4), 673–683.

Govender, T., Dube, T., Shoko, C., 2022. Remote sensing of land use-land cover change and climate variability on hydrological processes in sub-Saharan Africa: key scientific strides and challenges. Geocarto Int. 37 (25), 10925–10949.

Guo, H.D., Zhang, L., Zhu, L.W., 2015. Earth observation big data for climate change research. Adv. Clim. Chang. Res. 6 (2), 108–117.

Hecker, C.A., Gieske, A.S.M., 2004. Thermal remote sensing. In: Kerle, N., Janssen, L.L.F., Huurneman, G.C. (Eds.), Principles of Remote Sensing. ITC, Netherlands.

Ismail, M.F., Naz, B.S., Wortmann, M., Disse, M., Bowling, L.C., Bogacki, W., 2020. Comparison of two model calibration approaches and their influence on future projections under climate change in the upper Indus Basin. Clim. Chang. 163, 1227–1246.

Jin, X., Guo, R., Xia, W., 2013. Distribution of actual evapotranspiration over Qaidam Basin, an arid area in China. Remote Sens. 5 (12), 6976–6996.

Klump, et al., 2006. Data publication in the open access initiative. Data Sci. J. 5, 79–83.

Kostenko, I., 2020. Current problems and challenges in international space law: legal aspects. Adv. Space Law 5 (1), 48–57.

Kumar, S., Meena, R.S., Sheoran, S., Jangir, C.K., Jhariya, M.K., Banerjee, A., Raj, A., 2022. Remote sensing for agriculture and resource management. In: Natural Resources Conservation and Advances for Sustainability. Elsevier, pp. 91–135.

Land Remote Sensing Policy Act, 1992. 102nd Congress 1992, s 3. Retrieved from https://www.congress.gov/bill/102nd-congress/house-bill/6133/text.

Li, X.Y., Ma, Y.J., Xu, H.Y., Wang, J.H., Zhang, D.S., 2008. Impact of land use and land cover change on environmental degradation in Lake Qinghai-Tibet Plateau. Land Degrad. Dev. 20, 69–83.

Lillesand, T.M., Kiefer, R.W., Chipman, J.W., 1994. Remote Sensing and Image Interpretation. Wiley & Sons, New York, p. 499.

Liping, C., Yujun, S., Saeed, S., 2018. Monitoring and predicting land use and land cover changes using remote sensing and GIS techniques—a case study of a hilly area, Jiangle, China. PLoS One 13 (7), 1–23.

Lowell, K., 2004. Why Aren't We Making Better Use of Uncertainty Information in Decision Making? Proceedings of 6th International Symposium on Spatial Accuracy Assessment in Natural Resources and Environmental Sciences. Portland, Maine, USA. https://www.spatial-accuracy.org/system/files/Lowell2004accuracy.pdf.

Lu, C., Huang, G., Wang, G., Zhang, J., Wang, X., Song, T., 2021. Long-term projection of water cycle changes over China using RegCM. Remote Sens. 13, 3832.

Marboe, I., 2019. Living in the moon village—ethical and legal questions. Acta Astronaut. 154, 177–180.

Mauthner, N.S., Parry, O., 2013. Open access digital data sharing: principles, policies and practices. Soc. Epistemol. 27 (1), 47–67.

Mayer-Schönberger, V., Cukier, K., 2013. Big Data: A Revolution that Will Transform How We Live, Work and Think. Houghton Mifflin, New York.

Meftah, M., Boutéraon, T., Dufour, C., Hauchecorne, A., Keckhut, P., Finance, A., Bekki, S., Abbaki, S., Bertran, E., Damé, L., et al., 2021. The UVSQ-SAT/INSPIRESat-5 CubeSat mission: first in-orbit measurements of the earth's outgoing radiation. Remote Sens. 13, 1449.

Mengistu, A.G., van Rensburg, L.D., Woyessa, Y.E., 2019. Techniques for calibration and validation of SWAT model in data scarce arid and semi-arid catchments in South Africa. J. Hydrol. Reg. Stud. 25, 100621.

Michon, H.K., 2008. Climate change and flooding. In: Philander, S.G. (Ed.), Encyclopaedia of Global Warming and Climate Change. Vols. 1-3. SAGE Publications Inc.

Mihi, A., Benarfa, N., Arar, A., 2020. Assessing and mapping water erosion-prone areas in northeastern Algeria using analytic hierarchy process, USLE/RUSLE equation, GIS, and remote sensing. Appl. Geomat. 12 (2), 179–191.

Murthi, K.S., Madhusudan, H.N., 2008. Strategic considerations in Indian space programme—towards maximising socio-economic benefits. Acta Astronaut. 63 (1–4), 503–508.

National Academies of Sciences, Engineering, and Medicine, 2007. Contributions of Land Remote Sensing for Decisions About Food Security and Human Health: Workshop Report. The National Academies Press, Washington, DC. https://doi.org/10.17226/11759.

Neitsch, S.L., Arnold, J.G., Kiniry, J.R., Williams, J.R., 2011. Soil and Water Assessment Tool: Theoretical Documentation, Version 2009. Texas Water Resources Institute Technical Report No. 406. Texas A&M University, Texas, USA.

Niu, J., Tang, W., Xu, F., Zhou, X., Song, Y., 2016. Global research on artificial intelligence from 1990–2014: spatially-explicit bibliometric analysis. ISPRS Int. J. Geo Inf. 5 (66). https://doi.org/10.3390/ijgi5050066.

Nouri, H., Beecham, S., Kazemi, F., Hassanli, A.M., Anderson, S., 2013. Remote sensing techniques for predicting evapotranspiration from mixed vegetated surfaces. Hydrol. Earth Syst. Sci. 10 (3), 3897–3925.

Panagos, P., Ballabio, C., Poesen, J., Lugato, E., Scarpa, S., Montanarella, L., Borrelli, P., 2020. A soil erosion indicator for supporting agricultural, environmental and climate policies in the European Union. Remote Sens. 12 (9), 1365.

Phinzi, K., Ngetar, N.S., Ebhuoma, O., 2021. Soil erosion risk assessment in the Umzintlava catchment (T32E), Eastern Cape, South Africa, using RUSLE and random forest algorithm. S. Afr. Geogr. J. 103 (2), 139–162.

Pinnington, E., Quaife, T., Black, E., 2018. Impact of remotely sensed soil moisture and precipitation on soil moisture prediction in a data assimilation system with the JULES land surface model. Hydrol. Earth Syst. Sci. 22 (4), 2575–2588.

Piwowar, Chapman, 2007. Public sharing of research data sets: a pilot study of associations. J. Informetr. 4, 148–156.

Randall, D.A., Wood, R.A., Bony, S., Colman, R., Fichefet, T., Fyfe, J., Kattsov, V., Pitman, A., Shukla, J., Srinivasan, J., Stouffer, R.J., Sumi, A., Taylor, K.E., 2007. Climate models and their evaluation. In: Solomon, S., Qin, D., Manning, M., Chen, Z., Marquis, M., Averyt, K.B., Tignor, M., Miller, H.L. (Eds.), Climate Change 2007: The Physical Science Basis. Contribution of Working Group I to the Fourth Assessment Report of the Intergovernmental Panel on Climate Change. Cambridge University Press, Cambridge, United Kingdom/New York, NY, USA.

Remote Sensing Space Systems Act, 2005. S. C. 2005, c 45. Retrieved from https://lawslois.justice.gc.ca/eng/acts/R-5.4/page-1.html#h-426988.

Rosenzweig, C., Karoly, D., Vicarelli, M., Neofotis, P., Wu, Q., Casassa, G., Menzel, A., Root, T.L., Estrella, N., Seguin, B., Tryjanowski, P., 2008. Attributing physical and biological impacts to anthropogenic climate change. Nature 453 (7193), 353–357.

Satellite Data Security, 2007. Satellitendatensicherheitsgesetz—SatDSiG. Retrieved from https://www.bmwi.de/Redaktion/DE/Downloads/S-T/satdsig-hintergrunden.pdf?__blob=publicationFile&v=1.

Scambos, T.A., Campbell, G.G., Pope, A., Haran, T., Muto, A., Lazzara, M., Reijmer, C.H., van den Broeke, M.R., 2018. Ultralow surface temperatures in East Antarctica from satellite thermal infrared mapping: the coldest places on earth. Geophys. Res. Lett. 45 (12), 6124–6133.

Schott, J.R., 2007. Remote Sensing: The Image Chain Approach. Oxford University Press.

Schowengerdt, F., Fox, R., Duke, M., Marzwell, N., McKnight, B., 2007. PISCES: developing technologies for sustained human presence on the moon and Mars. In: AIAA SPACE 2007 Conference & Exposition, p. 6293.

See, E., 2017. Commercialization of space activities the laws and implications. J. Air Law Comm. 82 (1), 145–168.

Sepuru, T.K., Dube, T., 2018. Understanding the spatial distribution of eroded areas in the former rural homelands of South Africa: comparative evidence from two new non-59 commercial multispectral sensors. Int. J. Appl. Earth Obs. Geoinf. 69, 119–132.

Solomon, S. (Ed.), 2007. Climate Change 2007—The Physical Science Basis: Working Group I Contribution to the Fourth Assessment Report of the IPCC. vol. 4. Cambridge University Press.

Sterling, J.A.L., 2007. Space copyright law: the new dimension: preliminary survey and proposals. J. Copyright Soc. USA 54 (2–3), 354–418.

Sun, Y., Joachimski, M.M., Wignall, P.B., Yan, C., Chen, Y., Jiang, H., Wang, L., Lai, X., 2012. Lethally hot temperatures during the early Triassic greenhouse. Science 338 (6105), 366–370.

Sundara, K.K., Udaya, B.P., Padmakumari, K., 2012. Estimation of land surface temperature to study urban heat island effect using Landsat ETM+ image. Int. J. Eng. Sci. Technol. 4, 771–778.

Toth, C., Grzegorz, J., 2016. Remote sensing platforms and sensors: a survey. ISPRS J. Photogramm. Remote Sens. 115, 22–36.

United Nations, 2008. Principles Relating to Remote Sensing of the Earth from Space. http://www.un.org/documents/ga/res/41/a41r065.htm. (Visited 25 August 2023).

Wang, X., 2023. Remote sensing applications to climate change. Remote Sens. 15 (3), 747.

Wolka, K., Tadesse, H., Garedew, E., Yimer, F., 2015. Soil erosion risk assessment in the Chaleleka wetland watershed, Central Rift Valley of Ethiopia. Environ. Syst. Res. 4 (1), 1–12.

Wulder, M.A., Masek, J.G., Cohen, W.B., Loveland, T.R., Woodcock, C.E., 2012. Opening the archive: how free data has enabled the science and monitoring promise of Landsat. Remote Sens. Environ. 122, 2–10.

Yang, J., Gong, P., Fu, R., Zhang, M., Chen, J., Liang, S., Xu, B., Shi, J., Dickinson, R., 2013. The role of satellite remote sensing in climate change studies. Nat. Clim. Chang. 3 (10), 875–883.

Zhang, H., Yu, B., 2005. Boosting with early stopping: convergence and consistency. Ann. Stat. 33, 1538–1579.

Zhang, K., Kimball, J.S., Running, S.W., 2016. A review of remote sensing based actual evapotranspiration estimation. Water 3 (6), 834–853.

Zhou, Y., Chen, J., Cheng, X., 2021. Glacier velocity changes in the Himalayas in relation to ice mass balance. Remote Sens. 13, 3825.

Chapter 19

Opportunities, progress, and prospects in remote sensing of climate variability

Munyaradzi Davis Shekede[a,b], Cletah Shoko[c], and Timothy Dube[d]
[a]*Department of Geography Geospatial Sciences and Earth Observation, University of Zimbabwe, Harare, Zimbabwe,* [b]*Department of Geospatial Sciences and Earth Observation, Zimbabwe National Geospatial and Space Agency, Harare, Zimbabwe,* [c]*Division of Geography, School of Geography, Archaeology and Environmental Studies, University of Witwatersrand, Johannesburg, South Africa,* [d]*Department of Earth Science, Institute for Water Studies, University of the Western Cape, Cape Town, South Africa*

Chapter outline

1 Introduction

This book offers a comprehensive overview of climate variability and change, focusing on their drivers and implications for various biophysical and socioeconomic systems. It begins by grounding the discourse in climate change and variability, discussing internal factors like El Niño and La Niña, as well as anthropogenic activities that drive climate variability. Additionally, it reviews and advocates for the use of remotely sensed datasets and geospatial techniques to understand large-scale climate patterns and their impacts on subsystems such as agriculture, water quantity and quality, natural ecosystems, and human health. Overall, the book emphasizes the role of remote sensing in characterizing and monitoring the impacts of climate variability across various spatial scales, drawing from studies conducted globally. Furthermore, it explores global initiatives aimed at mitigating and adapting to climate variability and

Remote Sensing of Climate. https://doi.org/10.1016/B978-0-443-21731-9.00020-X
Copyright © 2024 Elsevier Inc. All rights are reserved, including those for text and data mining, AI training, and similar technologies.

change across sectors, presenting progress, opportunities, challenges, and prospects in remote sensing applications for characterizing the effects of climate change and variability across natural and anthropogenic systems. Finally, it discusses the limitations associated with remotely sensed datasets and approaches, offering suggestions for future developments in the field (Wang et al., 2020).

The transformative impact of remote sensing technology on our understanding of climate variability is paramount. This book highlights the pivotal role of remote sensing in unraveling the complexities of climate variability across diverse environmental domains. Through technological advancements, analytical algorithms, and big data analytics, remote sensing enables researchers to explore the intricate dynamics of climate variability and its repercussions on ecosystems worldwide. Beginning with an exploration of remarkable technological strides that propelled remote sensing to the forefront of climate research, the chapter discusses advancements from the inception of satellite missions like Landsat and Sentinel to the deployment of cutting-edge sensors and platforms (Dube et al., 2023). These advancements provide scientists with tools to scrutinize changes in land cover, rangeland conditions, and agriculture with unprecedented accuracy.

Building upon this technological foundation, the book further delves into the myriad applications of remote sensing in dissecting climate variability's impacts on ecosystems. From agricultural landscapes to natural ecosystems, water quality, groundwater-dependent ecosystems, and urban environments, remote sensing provides invaluable insights into the intricate interplay between climate dynamics and environmental processes (Kowe et al., 2022; Rampheri et al., 2023). Moreover, the chapter scrutinizes the ramifications of climate variability on water resources, including flooding, drought, water quality, and aquatic weed proliferation. By harnessing remote sensing capabilities, researchers can monitor changes in coastal ecosystems, wetlands, and aquatic macrophytes, thereby facilitating effective management and mitigation strategies (Cohen et al., 2004; Song et al., 2013; Mulder et al., 2020). Furthermore, the chapter underscores the indispensable role of remote sensing in urban environments, where climate variability poses unique challenges. Remote sensing facilitates the monitoring of urban heat islands, land use changes, and climate-related risks, aiding urban planners and policymakers in fostering climate resilience and sustainability (Oke, 1982; Seto et al., 2020). In addition, the chapter underscores the imperative of integrating remote sensing data into climate management and policy development frameworks. By offering national to global perspectives on remote sensing and climate variability, policymakers can formulate evidence-based strategies to address the multifaceted challenges posed by a changing climate (IPCC, 2018).

In light of the above, the objective of this *Remote Sensing of Climate Variability* book chapter is to explore the opportunities, progress, and prospects within the above-highlighted research-specific domain or field of study, providing a comprehensive overview of current advancements, future directions, and potential areas of growth. Through critical analysis and examination, the chapter aims to elucidate the existing opportunities for innovation and development, highlight the progress made thus far in the chosen area, and outline the prospective avenues for further exploration and advancement. By offering insights into the landscape of opportunities, tracking the trajectory of progress, and envisioning prospects, the chapter endeavors to contribute to a deeper understanding of the subject matter and inspire continued scholarly inquiry and practical application.

2 Evolution of remote sensing techniques

The evolution of remote sensing techniques has been pivotal in shaping our understanding of climate dynamics over the years. Historically, remote sensing has played a crucial role in climate research, dating back to the early exploration of Earth's atmosphere and surface. From the pioneering days of aerial photography to the advent of satellite-based observations, remote sensing has undergone remarkable transformations, enabling scientists to gather valuable data on climate variables such as temperature, precipitation, and land cover changes (Cochrane, 2009). Advances in satellite technology have been instrumental in enhancing our capacity to monitor climate variability on a global scale. With the launch of satellites equipped with advanced sensors and platforms, researchers can now capture high-resolution imagery and multispectral data, allowing for more precise and comprehensive analysis of climate patterns and trends. Satellite missions such as Landsat, MODIS, and Sentinel series have revolutionized remote sensing capabilities, providing researchers with a wealth of data to study changes in land surface properties, atmospheric conditions, and ocean dynamics (Wulder et al., 2012; Copernicus., 2016).

Moreover, the integration of remote sensing with other data sources has further enriched our understanding of climate variability. By combining satellite observations with ground-based measurements, meteorological data, and climate models, researchers can conduct more robust analyses and validate remote sensing-derived results. This synergy between different data sources allowed for a more comprehensive and nuanced understanding of complex climate processes, facilitating informed decision-making and policy formulation (Liu et al., 2015; IPCC, 2018). In summary, the evolution of remote sensing techniques has significantly contributed to our ability to study climate variability. From its historical roots to the present-day advancements in satellite technology and data integration, remote sensing continues to play a pivotal role in advancing our understanding of climate dynamics and informing strategies for climate resilience and adaptation.

3 Remote sensing applications in climate variability studies

Remote sensing technology has undergone remarkable advancements, playing a pivotal role in studying climate variability across diverse fields. In agriculture, remote sensing techniques such as multispectral imagery and thermal infrared sensors enable the monitoring of crop health and stress factors, aiding farmers in making informed decisions amidst climate fluctuations (Dhau et al., 2021; Mpakairi et al., 2024). Similarly, in natural ecosystems, remote sensing facilitates the assessment of biodiversity changes and ecosystem dynamics in response to climate variability. By integrating satellite data with ground-based surveys, researchers gain insights into the impacts of climate variability on ecosystem services (Mushore et al., 2022). Furthermore, remote sensing contributes significantly to understanding climate variability's effects on water resources. Satellite-based measurements provide essential data for monitoring hydrological cycles, water availability, and groundwater recharge rates. This information aids water managers in anticipating fluctuations in water supply and implementing effective management strategies to address water scarcity risks. Additionally, in urban environments, remote sensing is instrumental in assessing climate impacts on infrastructure and socioeconomic systems. By mapping urban heat islands and vulnerable areas, policymakers can develop climate-resilient cities and enhance disaster preparedness (Mushore et al., 2022).

Moreover, remote sensing technology supports climate management and policy development by providing policymakers with crucial information on climate trends and vulnerabilities. Satellite observations of land use changes greenhouse gas emissions, and deforestation aid in tracking progress toward climate mitigation goals and evaluating policy effectiveness. Incorporating remote sensing data into adaptation plans strengthens global cooperation and resilience to climate variability's impacts. In summary, remote sensing serves as an indispensable tool for studying climate variability, offering valuable insights and informing evidence-based strategies for climate adaptation and mitigation.

4 Progress in remote sensing data analysis

In recent years, significant progress has been made in the development of algorithms and models for processing remote sensing data. These advancements have enabled more accurate and efficient extraction of information from satellite imagery, facilitating a deeper understanding of Earth's surface dynamics. For example, algorithms for image classification, feature extraction, and change detection have become increasingly sophisticated, allowing researchers to identify land cover changes, monitor deforestation, and assess urban expansion with greater precision (Liu et al., 2018).

Integration of machine learning (ML) and artificial intelligence (AI) techniques has further enhanced remote sensing data analysis. ML algorithms, such as deep learning neural networks, have demonstrated remarkable capabilities in image classification, object detection, and semantic segmentation tasks. By training models on large datasets, researchers can automate the process of feature identification and classification, enabling more efficient analysis of remote sensing imagery. The utilization of high-performance computing (HPC) resources has become essential for processing large-scale remote sensing datasets. HPC systems offer the computational power and storage capacity needed to handle massive volumes of satellite imagery, enabling researchers to analyze terabytes of data in a fraction of the time. This capability is particularly valuable for applications such as land cover mapping, climate modeling, and environmental monitoring, where timely analysis of big data is critical for decision-making (Han et al., 2019; Li et al., 2020).

For instance, the integration of platforms like Google Earth Engine (GEE) has revolutionized remote sensing data analysis by providing researchers with access to an extensive archive of satellite imagery and powerful computational tools. GEE's cloud-based platform allows users to efficiently process and analyze large-scale remote sensing datasets without the need for extensive computational resources. By leveraging GEE's capabilities, researchers can perform complex analyses, such as land cover classification, change detection, and time-series analysis, with ease and scalability (Gorelick et al., 2017). Additionally, GEE offers a range of prebuilt algorithms and functions, making it accessible to users with varying levels of programming expertise. This democratization of remote sensing analysis through GEE has facilitated interdisciplinary research and enabled advancements in fields such as ecohydrology, agriculture, and climate science. Overall, GEE has emerged as a valuable tool for unlocking insights into Earth's surface dynamics and addressing global environmental challenges.

Despite the advancements in remote sensing data processing, challenges remain in data fusion and integration for multisensor analysis. Integrating data from multiple sensors with varying spatial and spectral resolutions poses significant technical and methodological challenges. Researchers must develop robust fusion techniques to reconcile differences in data quality, calibration, and sensor characteristics, ensuring the accuracy and reliability of integrated datasets. However, overcoming these challenges presents opportunities for innovation and advancement in remote sensing data fusion, paving the way for more comprehensive and integrated analyses of Earth's surface processes (Chen et al., 2020).

Overall, the development of algorithms and models, integration of ML and AI, utilization of HPC, and challenges in data fusion and integration are key areas driving advancements in remote sensing data processing and analysis. By addressing these challenges and leveraging emerging technologies, researchers can unlock new insights into Earth's surface dynamics and improve our ability to monitor and understand environmental changes.

5 Contributions to climate change research

In the realm of climate change research, remote sensing technology has emerged as a crucial tool for understanding the drivers of climate variability and their impacts on both regional and global scales. By analyzing satellite imagery and derived datasets, researchers can monitor changes in key climate variables such as temperature, precipitation, and land cover dynamics. Remote sensing enables the identification of trends and anomalies, shedding light on the complex interactions between natural processes and human activities that drive climate change (Bhandari and Lee, 2019). Moreover, remote sensing plays a vital role in climate modeling and forecasting by providing valuable input data for model calibration and validation. High-resolution satellite observations contribute to improved spatial representation and parameterization of climate models, enhancing their accuracy and predictive capabilities.

One significant contribution of remote sensing to climate and variability research is the identification of climate change hotspots and vulnerable regions. Satellite-based assessments help identify areas experiencing rapid environmental changes. By pinpointing these hotspots, policymakers can prioritize resources and implement targeted interventions to mitigate the impacts of climate change and build resilience in vulnerable communities (Gao et al., 2017). Furthermore, remote sensing data support the development of climate change mitigation and adaptation strategies. By monitoring land use changes, agriculture and ecosystem dynamics, remote sensing enables policymakers to assess the effectiveness of mitigation measures and design adaptive strategies that promote sustainable resource management and resilience-building initiatives (DeFries et al., 2020).

In summary, remote sensing contributes significantly to climate change research by providing insights into the drivers of climate variability, supporting climate modeling and forecasting efforts, identifying hotspots and vulnerable regions, and informing climate change mitigation and adaptation strategies. By leveraging the capabilities of remote sensing technology, researchers and policymakers can advance our understanding of climate change dynamics and develop evidence-based strategies to address its impacts on both human and natural systems.

6 Future directions and prospects

The future of remote sensing for climate variability studies is poised for significant advancements with the integration of emerging technologies and platforms. One notable trend is the adoption of high-resolution multispectral datasets, which offer a wealth of detailed information for analyzing environmental dynamics. These datasets capture imagery across multiple spectral bands with enhanced spatial resolution, enabling more precise monitoring of land cover changes, vegetation health, and atmospheric conditions (Wang et al., 2020). Additionally, the integration of hyperspectral and LiDAR technologies promises unparalleled insights into terrain morphology, vegetation structure, and atmospheric composition, further enriching our understanding of climate variability.

Furthermore, there are opportunities to enhance the spatial and temporal resolution of remote sensing data, facilitating more nuanced analysis of climate phenomena. Advances in sensor technology, coupled with cloud computing infrastructure, enable the processing of large volumes of high-resolution imagery with unprecedented efficiency (Liu et al., 2018). The integration of artificial intelligence and machine learning algorithms further enhances data analysis capabilities, allowing for the automated extraction of meaningful information from remote sensing datasets. These techniques enable researchers to identify patterns, trends, and anomalies in climate data, empowering more accurate predictions and informed decision-making.

Moreover, the introduction of big data analytics techniques revolutionizes the analysis of remote sensing data, unlocking insights from vast datasets with unprecedented speed and accuracy. By leveraging cloud-based platforms and distributed computing resources, researchers can process and analyze massive volumes of remote sensing data in near real-time, enabling timely responses to climate events and trends (Amitrano et al., 2019). Additionally, the integration of Internet of Things (IoT) devices with remote sensing networks enhances data collection and monitoring capabilities, providing continuous, real-time observations of environmental parameters. These advancements hold great promise for advancing our understanding of climate variability and informing climate change mitigation and adaptation strategies in the years to come.

7 Conclusion

In conclusion, this book chapter provides a comprehensive exploration of the advancements and applications of remote sensing in understanding climate variability. By delving into various fields such as agriculture, natural ecosystems, water resources, and urban environments, we have witnessed the transformative impact of remote sensing technology in unraveling the complexities of climate dynamics. From monitoring land surface changes to assessing water availability and tracking vegetation patterns, remote sensing offers invaluable insights into the intricate interplay between climate variability and environmental processes. The integration of high-resolution multispectral datasets, hyperspectral, and LiDAR technologies has significantly enhanced our capacity to observe and analyze Earth's systems with unprecedented detail and precision (Wang et al., 2020).

Furthermore, the utilization of big data analytics, cloud computing, artificial intelligence, and machine learning algorithms has revolutionized remote sensing data analysis, enabling researchers to extract meaningful information from vast datasets with remarkable speed and accuracy (Amitrano et al., 2019; Liu et al., 2018). Platforms like GEE have democratized remote sensing analysis, providing researchers with access to extensive archives of satellite imagery and powerful computational tools for conducting complex analyses (Gorelick et al., 2017). These advancements, coupled with the integration of IoT devices and emerging technologies, hold great promise for advancing our understanding of climate variability and informing evidence-based strategies for climate resilience and adaptation.

Moving forward, it is imperative to continue fostering interdisciplinary collaborations and innovation in remote sensing research. By addressing challenges in data fusion, integrating multisensor analysis, and enhancing spatial and temporal resolution, we can further unlock insights into Earth's surface dynamics and address pressing environmental challenges. Moreover, policymakers must prioritize the integration of remote sensing data into climate management and policy development frameworks, leveraging national to global perspectives to inform informed decision-making (IPCC, 2018). By embracing emerging technologies and harnessing the power of remote sensing, we can pave the way for a more sustainable and resilient future in the face of climate variability and change.

8 Future recommendations

The exploration of climate variability and change is a pressing global concern, with profound implications for ecosystems, societies, and economies worldwide. Remote sensing technology has emerged as a powerful tool for understanding the complex dynamics of climate variability and informing evidence-based strategies for resilience and adaptation. This book presents a comprehensive overview of the advancements and applications of remote sensing in studying climate variability, encompassing diverse environmental domains such as agriculture, natural ecosystems, water resources, and urban environments. By delving into the transformative impact of remote sensing technology, the book chapter elucidates how technological advancements, analytical algorithms, and big data analytics enable researchers to unravel the intricacies of climate variability and its repercussions on Earth's systems. Through an exploration of historical roots, current capabilities, and future prospects of remote sensing, this chapter aims to further provide insights into the pivotal role of remote sensing in addressing the challenges posed by climate variability and fostering a more sustainable and resilient future. The book chapter thus recommends the following:

1. *Enhanced Data Integration:* Addressing the challenges associated with data fusion and integration for multisensor analysis is crucial. Developing robust fusion techniques to reconcile differences in data quality, calibration, and sensor characteristics will ensure the accuracy and reliability of integrated datasets. Additionally, exploring innovative approaches for seamless integration of remote sensing data with ground-based measurements and climate models will enhance our understanding of climate variability.
2. *Advancements in Sensor Technology*: Continued advancements in sensor technology, particularly in high-resolution multispectral, hyperspectral, and LiDAR technologies, will provide unparalleled insights into Earth's surface dynamics. Investing in research and development to improve sensor capabilities, such as enhanced spatial and spectral resolution, will enable more precise monitoring and analysis of climate variables and environmental processes.

3. *Integration of Emerging Technologies*: Embracing emerging technologies like artificial intelligence, machine learning, big data analytics, and IoT devices holds promise for advancing remote sensing capabilities. Leveraging AI and ML algorithms for automated data analysis and feature extraction from remote sensing datasets will streamline data processing and enhance analytical capabilities. Integrating IoT devices with remote sensing networks will enable real-time monitoring of environmental parameters, facilitating timely responses to climate events and trends.
4. *Spatial and Temporal Resolution Enhancement*: Further enhancing the spatial and temporal resolution of remote sensing data will facilitate more nuanced analysis of climate variability. Investing in research and development to improve sensor resolution and data processing techniques will enable researchers to capture finer details and temporal dynamics of environmental processes, enhancing our ability to monitor and understand climate variability at local and regional scales.
5. *Interdisciplinary Collaboration*: Fostering interdisciplinary collaborations between remote sensing scientists, climate researchers, policymakers, and stakeholders is essential for advancing our understanding of climate variability and informing evidence-based strategies for resilience and adaptation. Encouraging knowledge exchange and collaboration across disciplines will facilitate holistic approaches to addressing climate challenges and developing effective mitigation and adaptation strategies.
6. *Policy Integration*: Prioritizing the integration of remote sensing data into climate management and policy development frameworks is imperative. Policymakers should leverage national to global perspectives provided by remote sensing data to inform evidence-based decision-making and develop proactive strategies for climate resilience and adaptation. Establishing mechanisms for regular assessment and update of climate policies based on remote sensing insights will ensure adaptive and responsive governance in the face of a changing climate.

By focusing on these future recommendations, we can harness the full potential of remote sensing technology to advance our understanding of climate variability and inform sustainable strategies for building resilience and adapting to a changing climate.

References

Amitrano, D., Lantéri, N., Manach, C., Blanchet, J., Dupuy, J.L., 2019. Big Earth data analytics: a survey. Big Earth Data 3 (2), 138–166.

Bhandari, A., Lee, H., 2019. Utilizing remote sensing data for climate change research: a review. Remote Sens. 11 (8), 909.

Chen, C., Wang, C., Liu, X., 2020. A review of multi-sensor data fusion techniques for urban remote sensing. Remote Sens. 12 (11), 1865.

Cochrane, J., 2009. Air and spaceborne remote sensing. In: Encyclopedia of Earth Sciences Series. Springer, Berlin, Heidelberg.

Cohen, A.S., Kaufman, D.S., WoldeGabriel, G., 2004. Late Pleistocene and Holocene environmental and climatic change from Lake Abiyata, Main Ethiopian Rift valley. Palaeogeogr. Palaeoclimatol. Palaeoecol. 213 (1–2), 103–123.

Copernicus., 2016. Copernicus Program. European Commission. https://www.copernicus.eu/en.

DeFries, R.S., Karanth, K.K., Turner, B.L., Spector, S., Matthews, T., 2020. The contribution of remote sensing to global land-change science. Remote Sens. Environ. 244, 111816.

Dhau, I., Dube, T., Mushore, T.D., 2021. Examining the prospects of sentinel-2 multispectral data in detecting and mapping maize streak virus severity in smallholder Ofcolaco farms, South Africa. Geocarto Int. 36 (16), 1873–1883.

Dube, T., Seaton, D., Shoko, C., Mbow, C., 2023. Advancements in earth observation for water resources monitoring and management in Africa: a comprehensive review. J. Hydrol., 129738.

Gao, J., Hu, T., Li, Y., 2017. Remote Sensing and GIS-Based Assessment of Climate Change Vulnerability in the Future Directions and Prospects.

Gorelick, N., Hancher, M., Dixon, M., Ilyushchenko, S., Thau, D., Moore, R., 2017. Google Earth Engine: planetary-scale geospatial analysis for everyone. Remote Sens. Environ. 202, 18–27.

Han, L., Yang, C., Xu, X., 2019. A high-performance computing approach to big earth data processing in remote sensing. Remote Sens. 11 (7), 871.

IPCC, 2018. IPCC Special Report on the Impacts of Global Warming of 1.5°C Above Pre-Industrial Levels and Related Global Greenhouse Gas Emission Pathways, in the Context of Strengthening the Global Response to the Threat of Climate Change, Sustainable Development, and Efforts to Eradicate Poverty. IPCC.

Kowe, P., Dube, T., Mushore, T.D., Ncube, A., Nyenda, T., Mutowo, G., Chinembiri, T.S., Traore, M., Kizilirmak, G., 2022. Impacts of the spatial configuration of built-up areas and urban vegetation on land surface temperature using spectral and local spatial autocorrelation indices. Remote Sens. Lett. 13 (12), 1222–1235.

Li, X., Han, Y., Zeng, Z., 2020. High-performance computing in land use and land cover change detection: a review. Remote Sens. 12 (9), 1451.

Liu, J., Linderman, M., Ouyang, Z., An, L., Yang, J., Zhang, H., 2015. Ecological degradation in protected areas: the case of Wolong Nature Reserve for giant pandas. Science 292 (5514), 98–101.

Liu, Y., Zhang, L., Hu, J., Liu, R., 2018. A review of recent advances in change detection. Int. J. Remote Sens. 39 (2), 361–387.

Mpakairi, K.S., Dube, T., Sibanda, M., Mutanga, O., 2024. Remote sensing crop water productivity and water use for sustainable agriculture during extreme weather events in South Africa. Int. J. Appl. Earth Obs. Geoinf. 129, 103833.

Mulder, V.L., de Bruin, S., Schaepman, M.E., 2020. The use of remote sensing in soil and terrain mapping—a review. Geoderma 362 (4), 114168.

Mushore, T.D., Mutanga, O., Odindi, J., Sadza, V., Dube, T., 2022. Pansharpened landsat 8 thermal-infrared data for improved Land Surface Temperature characterization in a heterogeneous urban landscape. Remote Sens. Appl. Soc. Environ. 26, 100728.

Oke, T.R., 1982. The energetic basis of the urban heat island. Q. J. R. Meteorol. Soc. 108 (455), 1–24.

Rampheri, M.B., Dube, T., Dondofema, F., Dalu, T., 2023. Progress in the remote sensing of groundwater-dependent ecosystems in semi-arid environments. Phys. Chem. Earth, A/B/C 130, 103359.

Seto, K.C., Dhakal, S., Bigio, A., Blanco, H., Delgado, G.C., Dewar, D., et al., 2020. Human settlements, infrastructure and spatial planning. In: Climate Change and Land: An IPCC Special

Report on Climate Change, Desertification, Land Degradation, Sustainable Land Management, Food Security, and Greenhouse Gas Fluxes in Terrestrial Ecosystems.

Song, C., Woodcock, C.E., Seto, K.C., Lenney, M.P., Macomber, S.A., 2013. Classification and change detection using Landsat TM data: when and how to correct atmospheric effects? Remote Sens. Environ. 105.

Wang, Z., Zhao, J., Song, W., Cheng, S., 2020. High-resolution multispectral remote sensing imagery classification using a convolutional neural network. Remote Sens. 12 (1), 17.

Wulder, M.A., White, J.C., Loveland, T.R., Woodcock, C.E., Belward, A.S., Cohen, W.B., et al., 2012. The global Landsat archive: status, consolidation, and direction. Remote Sens. Environ. 122, 22–28.

Index

Note: Page numbers followed by *f* indicate figures, *t* indicate tables and *b* indicate boxes.

A

B

C

D

G

H

M

N

S